Soil Survey Laboratory Information Manual

Soil Survey Investigations Report No. 45
Version 2.0
February 2011

Compiled and Edited by Rebecca Burt

United States Department of Agriculture
Natural Resources Conservation Service
National Soil Survey Center
Lincoln, Nebraska

Cover Photos: The soil survey landscape (upper left) and pedon (upper right) are from Lyon County, Nevada. The landscape is a typical area of Cleaver soils, and the pedon is representative of the Cleaver series. Cleaver soils are classified as loamy, mixed, superactive, mesic, shallow Typic Argidurids. These well drained soils are shallow to an indurated duripan. They formed in alluvium derived from igneous rocks and are on fan remnants in the Great Basin section of the Basin and Range physiographic province. The present vegetation in the rangeland ecological site is mainly Bailey's greasewood (*Sarcobatus baileyi*), shadscale (*Atriplex confertifolia*), winterfat (*Krascheninnikovia lanata*), Nevada jointfir (*Ephedra nevadensis*), and cheatgrass (*Bromus tectorum*). *(Photos courtesy of Joseph V. Chiaretti, NRCS, National Soil Survey Center, Lincoln, Nebraska.)* Lower left: The National Soil Survey Laboratory in Lincoln, Nebraska. Lower right: A thin section.

Trade names are used in this manual solely for the purpose of providing specific information. Mention of a trade name does not constitute a guarantee of the product by the USDA nor does it imply an endorsement by the USDA.

USDA Nondiscrimination Statement

Citation: Soil Survey Staff. 2011. Soil Survey Laboratory Information Manual. Soil Survey Investigations Report No. 45, Version 2.0. R. Burt (ed.). U.S. Department of Agriculture, Natural Resources Conservation Service.

Contents

iii

Preface

For any measurement program that collects analytical data over a long period of time for comparative purposes, the quality and credibility of those data are critical (Taylor, 1988). It is equally critical that the data can be easily understood by the user. The uses of these data include, but are not limited to, routine soil characterization, special analyses, soil classification, interpretations, and soil genesis and geomorphology studies. Because of the diverse uses of these data, it follows that pedon characterization data, or any soil survey data, are more appropriately used when the operations for collection, analysis, and reporting of these data are well understood. Results differ when different methods are used, even though these methods may carry the same name or concept. Comparison of one bit of data with another is difficult without knowing how both bits were gathered. As a result, operational definitions have been developed and are linked to specific methods. *Soil Taxonomy* (Soil Survey Staff, 1999) is based almost entirely on criteria that are defined operationally, e.g., standard particle-size analysis. When *Soil Taxonomy* (Soil Survey Staff, 1975) was written, the authors knew that no conceptual definition of clay could be approximated in all soils by any feasible combination of laboratory analyses. Hence, instead of defining clay, the authors defined the operations to test the validity of a clay measurement and a default type of operation for those situations in which the clay measurement was not valid. The operational definition helps to describe a soil property in terms of operations used to measure it. This document, the *Soil Survey Laboratory Information Manual,* Soil Survey Investigations Report (SSIR) No. 45, discusses operational and conceptual definitions of Soil Survey Laboratory (SSL) procedures.

The *purpose* of this manual is to serve as a standard reference in the use and application of SSL characterization data. The manual is *intended* to help maximize user understanding of these data. Even though it presents descriptive terms or interpretive classes commonly associated with ranges of some data elements, this document is *not intended* to be an *interpretive guide*.

This manual serves as a companion manual to the *Soil Survey Laboratory Methods Manual,* Soil Survey Investigations Report No. 42 (Soil Survey Staff, 2004), and the *Soil Survey Field and Laboratory Methods Manual,* Soil Survey Investigations Report No. 51 (Soil Survey Staff, 2009). SSIR No. 42 documents the methodology and serves as a reference for the laboratory analyst, whereas SSIR No. 51 serves as a reference for the scientist in a field or field-office setting. The documentation of standard operating procedures (SOPs) ensures continuity in the analytical process. Both SSIR No. 42 and SSIR No. 51 are "how to" manuals; their respective described methods follow the same format and cover many of the same kinds of analyses. The *Soil Survey Laboratory Information Manual* (SSIR No. 45) follows the same topical outline as the *Soil Survey Laboratory Methods Manual* (Soil Survey Staff, 2004). SSIR No. 45 provides brief summaries of the SSL methods as well as detailed discussion of the use and application of the resulting data.

This manual serves to *document* the historical background of the development of many SSL methods. It is important to document this background, as methods development in soil characterization has been instrumental in developing principles and understanding of the nature and behavior of a wide range of soils. It is expected that this manual will evolve over time as new methods based on new knowledge or technologies are developed and applied. It is also expected that the scope of this manual will change over time. Currently, the *scope* of this document includes such diverse uses as soil survey, salinity, fertility, and soil quality. With the continued development of and modification to the database derived from these diverse data, it is expected that more discipline-dedicated manuals will be developed and enhanced.

This manual is divided into four major parts: *Introduction, Primary Characterization Data, Supplementary Characterization Data*, and the *Appendices*. The introduction describes general pedon information that appears on both the Primary Characterization Data Sheets and the Supplementary Characterization Data Sheets. This general information is important nonanalytical metadata. Also described in the introduction are the "Pedon Calculations" that appear on the Primary Characterization Data Sheet.

Primary data are those data that appear on the SSL data reports entitled Primary Characterization and are based primarily on analytical data. Rather than following the SSL data sheet format, the discussion of the primary data follows the discussion format presented in SSIR No. 42 (Soil Survey Staff, 2004); that is, it presents broad categories of characterization data. Method codes are not embedded in the descriptions of the primary data but are cross-referenced by method code in the table of contents in this manual. The discussion is logically and sequentially presented as follows: (1) field procedures for site and pedon description and sampling and (2) laboratory procedures used to characterize the physical, chemical, biological, and mineralogical properties of a soil and to characterize water and plant samples. The field component of this manual provides information on the rationale of the SSL field procedures. Key considerations and procedures related to site selection, geomorphology, and pedon, water, and biological sampling are discussed. Within the aforementioned categories (physical, chemical, biological, and mineralogical) of the laboratory component of this manual is discussion of specific soil properties (e.g., structure, pH, biomass, and clay mineralogy) that are commonly measured for soil survey and are indicative of soil processes. Important references related to these topics include, but are not limited to, the *Soil Survey Manual* (Soil Survey Division Staff, 1993), the *Field Book for Describing and Sampling Soils* (Schoeneberger et al., 2002), *Soil Taxonomy* (Soil Survey Staff, 1999), and peer-recognized literature (e.g., Soil Science Society of America monographs).

Supplementary data are those data that appear on the SSL data reports entitled Supplementary Characterization. These data are considered the interpretive physical data for pedons analyzed at the SSL. They are primarily derived or calculated data, using the analytical data as a basis for calculation. Unlike the primary data, the supplementary data are not discussed in SSIR No. 42 (Soil Survey Staff, 2004) and thus do not carry method codes.

The Appendices consist of example pedon data sheets, including the primary, supplementary, and taxonomy sheets as well as grain-size distribution curves and water retention curves for selected pedons. These data sheets are used in a number of example pedon calculations presented throughout this manual, such as weight to volume conversions, weighted averages, and other estimates. These examples are *intended* to improve the ability of users of SSL data to understand and apply these data.

Rebecca Burt, Editor
Research Soil Scientist
National Soil Survey Center
United States Department of Agriculture
Natural Resources Conservation Service
Lincoln, Nebraska

I. Introduction

Since 1977, the Soil Survey Laboratory (SSL) has maintained a computerized analytical laboratory database and pedon description database for soils sampled by the previous three regional laboratories (Beltsville, Riverside, and Lincoln) and by the SSL. These databases are used to generate various other special databases; reports, including Soil Survey Investigations Reports (SSIRs); and data evaluation studies. The SSL provides data in reports, such as Primary and Supplementary Characterization Data Sheets. Data have also been provided in electronic forms, including tapes, disks, and CD-ROMs and more recently through the National Cooperative Soil Survey (NCSS) Characterization Database, which is available online at http://ssldata.nrcs.usda.gov/ and is stored and maintained by the National Soil Survey Center (NSSC). This application allows users to generate, print, and download reports containing pedon data from analyses for soil characterization and research within the National Cooperative Soil Survey (NCSS). These pedon data are primarily generated by the SSL, but data from cooperators' laboratories are included.

The SSL reports are in a standard format that provides uniformity in reporting and enhances communication. This standard format has changed with time as a result of changes in established methods and the adoption of new methods. The protocols for the recording of important nonanalytical metadata differ among cooperator laboratories. For this reason, the following section describes pedon metadata as *examples*—specifically, the SSL-generated information that appears on both the Primary and Supplementary Characterization Data Sheets. These metadata include site and pedon identification numbers; SSL project numbers and names; "sampled as" and "revised to" soil names; sample layer number; depth (cm); genetic horizon; and laboratory preparation code. These metadata provide informative labels for pedons analyzed at the SSL. Also described in this section are the "Pedon Calculations" that appear on the Primary Data Sheet. These calculations follow the general pedon information but precede the reporting of SSL primary analytical data. In the *Appendices* of this manual, example SSL data sheets describe the metadata and the various pedon calculations. Each pedon example includes the respective primary, supplementary, and taxonomy data sheets, grain-size distribution curves, and water retention curves for selected pedons. Pedon calculations are as follows:

- CEC Activity, CEC-7/Clay, Weighted Average
- Weighted Particles, 0.1-75 mm, 75-mm Basis
- Volume, >2 mm, Weighted Average
- Clay, Total, Weighted Average
- Clay, Carbonate Free, Weighted Average
- LE, Whole Soil, Summed to 1 m

1.1 General Pedon Information

Refer to Appendix 1 (Chowchow Pedon, S2004WA027009)

Soil Sample Origin: County, State; or Country (if other than USA)

Example: Grays Harbor, Washington

1.1.1 Laboratory Name and Location

Example:

United States Department of Agriculture
Natural Resources Conservation Service
National Soil Survey Center
Soil Survey Laboratory
Lincoln, Nebraska 68508-3866

1.1.2 Print Date: Date when SSL Characterization Data Sheets are printed (does not reference dates of sampling or completion of analytical results)

Example: June 7 2010 11:42 AM

1.1.3 Pedon Identification (ID): Soil Survey Number, client assigned

Example: S2004WA027009

S = Special sample (used if soil is sampled)
2004 = Calendar year described/sampled
WA = Two-character (alphabetic) Federal Information Processing Standards (FIPS)
     code for state where described/sampled
027 = Three-digit (numeric) FIPS code for county where described/sampled
009 = Consecutive pedon number for calendar year for county

1.1.5 Sampled as: Pedon name and classification at time of sampling

Example: Chowchow—Loamy, isotic, dysic, isomesic Terric Haplosaprist

1.1.6 Revised to: Pedon name and classification at correlation

Example: Chowchow—Loamy, isotic, dysic, isomesic Terric Haplosaprist

1.1.7 SSL—Project: SSL project number and name

Example: C2005USWA011   Grays Harbor

The alphanumeric project code is referenced in all project data correspondence. Notations in this project code identify whether the project is considered a characterization (C), investigations (I), reference (R), or other (O) project; fiscal year (2005); alphabetic FIPS code for country (US = United States of America); alphabetic FIPS code for state (WA = Washington); and a sequential project number assigned (011) in order of project receipt. Project code is followed by project name (Grays Harbor).

1.1.8 Site ID: Site ID number is the same as the Pedon ID number (client-assigned).

Example: S2004WA027009

1.1.9 Pedon No.: SSL-assigned pedon number and layer (sample) numbers

Example: 05N0175

Immediately upon receipt, soil samples are logged into the SSL system. The assignment of unique laboratory numbers is an important step in the "chain of custody" sequence as they help to ensure the integrity of results; i.e., there has been no "mix-up" of samples. The pedon number (05N0175) and layer numbers (05N00981–05N00986) are unique laboratory-assigned numbers for the specified fiscal year (2005).

1.1.10 General Methods: General laboratory methods

Example: 1B1a, 2A1, 2B

Some SSL methods are general or are applicable to all of the samples listed on a particular data sheet. These procedures are referenced by SSL method codes, e.g., 1B1a (laboratory preparation method) and 2A1 and 2B (conventions for reporting laboratory data).

1.1.11–1.1.12 Horizon and Original Horizon: Soil horizon or layer designation, including lithological designation

Example: Oi

The horizon designation is made at the time of sampling by the sampling party. This consensus record is deemed important and is rarely changed in the database; therefore, the original horizon designation also is maintained in the database. Over time, the horizon nomenclature and other descriptive morphological features may become archaic, but the record as to what was determined at the time of sampling is deemed more important than the achievement of complete editorial uniformity. Refer to *Keys to*

Soil Taxonomy (Soil Survey Staff, 2010) for a more detailed discussion of designations for soil horizons and layers.

1.1.13 Depth (cm): Depth limits in centimeters (cm) are reported for each soil horizon or layer**.**

Example: 0-9 cm

1.1.14–1.1.16 Field Label(s) 1–3: Field labels for layers are client assigned and may or may not be derived from the Pedon ID as follows:

Example: Field Label 1 S04WA0270091

Example: Field Label 2 CHOWCHOW

Example: Field Label 3 (not designated)

1.1.17 Field Texture: Field-determined texture

Example: SIL

The field-determined texture is reported. Soil texture class names are reported as codes (abbreviations). Texture class names are based first on the distribution of sand, silt, and clay and then, for some classes, on the distribution of several size fractions of sand. In the past the field texture was reported on the Supplementary Characterization Data Sheet, but currently the field texture is reported as metadata on the Primary Characterization Data Sheet. Refer to the Supplementary Characterization Data Sheet for descriptions of texture class codes.

1.1.18 Lab Texture: Laboratory-determined texture

Example: SIL

The SSL-determined soil texture is reported. The laboratory-determined texture may or may not agree with the field-determined texture. Names are based on Particle-Size Distribution Analysis (PSDA) data to the nearest 1 percent applied to definitions of the texture classes (Soil Survey Staff, 1951). The SSL PSDA soil texture is also reported on the Supplementary Characterization Data Sheet.

1.1.19 Sample Preparation Codes: SSL sample preparation code

Example: "S" Standard air-dry preparation

Laboratory preparation codes depend on the properties of the sample and on the requested analyses. These codes carry generalized information about the characteristics of the analyzed fraction—i.e., the water content (e.g., air-dry, field-moist) and the original and final particle-size fraction (e.g., sieved <2-mm fraction processed to 75 μm)—and, by inference, the type of analyses performed. Identification numbers and preparation codes are reported on the SSL data sheets. In recent years these codes have been significantly revised; therefore, they are not described in detail in the *Soil Survey Laboratory Methods Manual* (Soil Survey Staff, 2004). Detailed information on the current preparation codes as they appear on the SSL data sheets may be obtained from the SSL upon request.

1.2 Pedon Calculations

Weighted averages are based on the control section. Refer to *Keys to Soil Taxonomy* (Soil Survey Staff, 2010) for appropriate taxonomic control section criteria.

1.2.1 CEC Activity, CEC-7/Clay, Weighted Average (based on control section)

Example: Refer to Appendix 2 (Wildmesa Pedon, S89CA027004)

Horizon	Depth (cm)	Hcm (cm)	CEC-7/Clay (%)	Product A
2Bt	15-46	31	0.82	25.42
2Btk	46-74	19	0.85	16.15
SUM		50		41.57

Control Section = 15-65 cm

Equation 1.2.1.1:

Sum of Products A = Sum of (CEC-7/Clay$_T$ x Hcm) for all soil horizons

where
CEC-7 = CEC by NH_4OAc, pH 7 (cmol(+)/kg)
Clay$_T$ = Weight percentage of total clay on <2-mm basis
Hcm = Horizon thickness, cm

Weighted Average = (Sum of Products A)/(Sum of Hcm for all horizons)

CEC Activity, CEC-7/Clay, Weighted Average = 41.57/50 = 0.83

1.2.2   Weighted Particles, 0.1-75 mm, 75-mm Basis (based on control section)

Example: Refer to Appendix 2 (Wildmesa Pedon, S89CA027004)

Horizon	Depth (cm)	Hcm (cm)	$Wt_{0.1-75mm}$ (%)<75mm	Product A
2Bt	15-46	31	24	744
2Btk	46-74	19	17	323
SUM		50		1067

Control Section = 15-65 cm

Equation 1.2.2.1:

Sum of Products A = Sum of ($Wt_{0.1-75mm}$ x Hcm) for all soil horizons

where
$Wt_{0.1-75mm}$ = Weight percentage of 0.1-75 mm on 75-mm basis
Hcm = Horizon thickness, cm

Weighted Average = (Sum of Products A)/(Sum of Hcm for all horizons)

Weighted Particles, 0.1-75 mm, 75-mm Basis = 1067/50 = 21.34 (21%)

1.2.3   Volume, >2 mm, Weighted Average (based on control section)

Example: Refer to Appendix 3 (Nuvalde Pedon, S09TX307003)

Horizon	Depth (cm)	Hcm (cm)	$Vol_{>2mm}$ (%)	Product A
Ap2	15-34	9	1	9
Bw	34-59	25	2	50
Bk1	59-90	31	20	620
Bk2	90-120	10	15	150
SUM		75		829

Control Section = 25-100 cm

Equation 1.2.3.1:

Sum of Products A = Sum of (Wt$_{>2mm}$ x Hcm) for all soil horizons

where
Vol$_{>2mm}$ = Volume percentage of >2 mm on whole-soil basis
Hcm = Horizon thickness, cm

Weighted Average = (Sum of Products A)/(Sum of Hcm for all horizons)

Volume, >2 mm, Weighted Average = 829/75 = 11.05 (11%)

1.2.4   Clay, Total, Weighted Average (based on control section)

Example: Refer to Appendix 2 (Wildmesa Pedon, S89CA027004)

Horizon	Depth (cm)	Hcm (cm)	Clay (%)	Product A
2Bt	15-46	31	34.9	1081.9
2Btk	46-74	19	38.1	723.9
SUM		50		1805.8

Control Section = 15-65 cm

Equation 1.2.4.1:

Sum of Products A = Sum of (Vol$_{>2mm}$ x Hcm) for all soil horizons

where
Vol$_{>2mm}$ = Volume percentage of >2 mm on whole-soil basis
Hcm = Horizon thickness, cm

Weighted Average = (Sum of Products A)/(Sum of Hcm for all horizons)

Clay, Total, Weighted Average = 1805.8/50 = 36.12 (36%)

Refer to the discussion under *Primary Characterization Data Sheet*, Section 3.1.2, Particles <2 mm, for an example calculation of clay percentage on a volumetric whole-soil basis. Unlike the preceding example for total clay, weighted average, this calculation uses bulk density and a coarse fragment conversion factor.

1.2.5 Clay, Carbonate Free, Weighted Average (based on control section)

Example: Refer to Appendix 3 (Nuvalde Pedon, S09TX307003)

Horizon	Depth (cm)	Hcm (cm)	$Clay_{CF}$ (%)	Product A
Ap2	15-34	9	34.6	311.4
Bw	34-59	25	37.3	932.5
Bk1	59-90	31	20.1	623.1
Bk2	90-120	10	19.6	196.0
SUM		75		2063

Control Section = 25-100 cm

Equation 1.2.5.1:

Sum of Products A = Sum of ($Clay_{CF}$ x Hcm) for all soil horizons

where
$Clay_{CF}$ = Weight percentage of carbonate-free clay, <2-mm basis (subtract carbonate clay from total clay)
Hcm = Horizon thickness, cm

Weighted Average = (Sum of Products A)/(Sum of Hcm for all horizons)

Clay, Carbonate Free, Weighted Average = 27.51 (28%)

1.2.6 LE, Whole Soil, Summed to 1 m

Example: Refer to Appendix 3 (Nuvalde Pedon, S09TX307003)

Horizon	Depth (cm)	Hcm (cm)	$COLE_{whole}$	Product A
Ap1	0-15	15	0.075	1.125
Ap2	15-34	19	0.064	1.216
Bw	34-59	25	0.062	1.55
Bk1	59-90	31	0.034	1.054
Bk2	90-120	10	0.033	0.33
SUM		100		5.275

Equation 1.2.6.1:

Product A = COLE$_{whole}$ x Hcm

Equation 1.2.6.2:

LE = Sum of Products A for all horizons up to 100 cm

where
COLE$_{whole}$ = Coefficient of Linear Extensibility
Hcm = Horizon thickness, cm
LE = Linear extensibility

LE, whole soil, summed to 1 m = 5.275 (5)

II. Primary Characterization Data

Primary Characterization Data are herein defined as those data that appear on the Primary Characterization Data Sheet and are based primarily on analytical data. While the Primary Characterization Data Sheet is composed mainly of analytical data, some calculated values also are presented. Historically, the Soil Survey Laboratory (SSL) has described and assigned method codes to only those data reported on the Primary Characterization Data Sheet, including the traditionally reported ratios, estimates, and calculations; e.g., coefficient of linear extensibility (COLE) and water retention difference (WRD). This tradition is followed in the *Soil Survey Laboratory Methods Manual*, Soil Survey Investigations Report (SSIR) No. 42 (Soil Survey Staff, 2004). The more recently reported calculated values that appear on the Primary Characterization Data Sheet (e.g., estimated organic carbon, estimated organic matter) as well as those values reported under "Pedon Calculations" are not described or assigned method codes in SSIR No. 42 (Soil Survey Staff, 2004) but are described in the introduction to this manual. Refer to *Keys to Soil Taxonomy* (Soil Survey Staff, 2010) for more information on the use of "Pedon Calculations" (e.g., weighted clay average) and other derived data.

Important metadata are shown on the Primary Characterization Data Sheet as well as on the Supplementary Characterization Data Sheet. Examples include site and pedon identification numbers; SSL project numbers and names; "sampled as" and "revised to" soil names; sample layer number; depth (cm); genetic horizon; and laboratory preparation code. Refer to the introduction to this manual for a more detailed discussion of the significance of these metadata.

Rather than following the SSL data sheet format, the discussion of the Primary Characterization Data follows the discussion format presented in SSIR No. 42 (Soil Survey Staff, 2004); that is, it presents broad categories of characterization data. Method codes are not embedded in the following descriptions of the Primary Characterization Data but are cross-referenced by method code in the table of contents in this manual. The discussion is logically and sequentially presented as follows:

- Sample collection and preparation
- Conventions
- Soil physical and fabric-related analyses
- Soil and water chemical extractions and analyses
- Soil biological and plant analyses
- Mineralogy

Soil properties are characteristics described by measurements. Within the aforementioned categories (physical, chemical, biological, and mineralogical) are specific soil properties (e.g., structure, pH, biomass, and clay mineralogy) that are commonly measured for soil survey and are indicative of soil processes. Within a soil process, a particular outcome is a quantitative/qualitative point on a continuum unique to each soil, reflective of the relations between a soil property and soil processes or some aspect of soil function, such as plant growth. Included in the discussion herein of selected SSL methods and applications are references to case studies and datasets that

11

serve as evidentiary examples of actions/practices that have promoted or diminished certain soil processes. The transition between the actions/practices that promote soil processes and those that diminish soil processes is the ongoing development and understanding of the cause-and-effect relationships of these processes and the appropriate methods of constraint/stress alleviation, restoration, and quality enhancement. These actions/practices are not intended to be an exhaustive historical list, but they illustrate some significant examples. Many of the soil characteristics discussed herein are described in the United States Department of Agriculture (USDA) soil survey program, using data resulting from decades of collection. Sources of these data include the USDA National Soil Information System (NASIS) and the USDA National Cooperative Soil Survey (NCSS) Soil Characterization Database, which contains laboratory data and pedon descriptions for nearly 50,000 pedons in the United States as well as internationally. For more detailed discussions of these soil characteristics and their applications, refer to the *Soil Survey Manual* (Soil Survey Division Staff, 1993) and *Keys to Soil Taxonomy* (Soil Survey Staff, 2010). For detailed descriptions of the SSL methods which are cross-referenced by method code in the table of contents in this manual, refer to the *Soil Survey Laboratory Methods Manual*, SSIR No. 42 (Soil Survey Staff, 2004), which is available online at http://soils.usda.gov/technical/lmm/. For descriptions of field methods as used in Natural Resources Conservation Service (NRCS) soil survey offices, refer to the *Soil Survey Field and Laboratory Methods Manual*, SSIR No. 51 (Soil Survey Staff, 2009), which is available online at http://www.soils.usda.gov/technical/.

1 Sample Collection and Preparation

This section describes the various SSL procedures for field and laboratory sample collection and preparation. Information is provided on the rationale for these procedures. The field component describes key considerations and procedures related to site selection, geomorphology (Schoeneberger and Wysocki, 2004), and pedon, water, and biological sampling (Soil Survey Staff, 2004). Other important references related to these topics include, but are not limited to, the *Soil Survey Manual* (Soil Survey Division Staff, 1993); *Field Book for Describing and Sampling Soils* (Schoeneberger et al., 2002); *Soil Survey Field and Laboratory Methods Manual*, SSIR No. 51 (Soil Survey Staff, 2009); *Soil Taxonomy* (Soil Survey Staff, 1999); and *Keys to Soil Taxonomy* (Soil Survey Staff, 2010). The intent of the laboratory component is not to detail all possibilities of the universe but to provide some information on the master preparation procedures that are typically requested for analysis at the SSL. Method selection depends on the properties of the sample and on the requested analyses. Soil procedures include, but are not limited to, the preparation of the <2-mm and >2-mm particle-size fractions as well as air-dry and field-moist preparations. For detailed descriptions of the SSL methods which are cross-referenced by method code in the table of contents in this manual, refer to SSIR No. 42 (Soil Survey Staff, 2004), which is available online at http://soils.usda.gov/technical/lmm/. Finally, this section briefly discusses the historical background of the development of classification systems and soil survey in the United States.

1.1 Field Sample Collection and Preparation
1.1.1 Site Selection
1.1.1.1 Geomorphology
1.1.1.2–1.1.1.4 Pedon, Water, and Biological Sampling
1.1.2 Classification Systems and Soil Survey

General: The NCSS program has prepared soil maps for much of the United States. Both field and laboratory data are used to design map units and provide supporting information for scientific documentation and predictions of soil behavior. A soil map delineates areas occupied by different kinds of soil, each of which has a unique set of interrelated properties characteristic of the material in which it formed, its environment, and its history (Soil Survey Division Staff, 1993). The soils mapped by the NCSS are identified by names that serve as references to a national system of soil taxonomy (Soil Survey Staff, 2010). Coordination of mapping, sampling site selection, and sample collection in this program contributes to the quality assurance process for laboratory characterization (Burt, 1996). Requisites to successful laboratory analysis of soils occur long before the sample is analyzed (USDA/SCS, 1984; Soil Survey Staff, 1996). In the field, these requisites include site selection, descriptions of site and soil pedon, and careful sample collection. A complete description of the sampling site not only provides a context for the various soil properties determined but also is a useful tool in the evaluation and interpretation of the soil analytical results (Patterson, 1993). Landscape, landform, and pedon documentation of the sampling site serves as a link in a continuum of analytical data—sampled horizon, pedon, landscape, and overall soil survey area.

The objectives of a project or study form the basis for the design of the sampling strategy. A carefully designed sampling plan is required to provide reliable samples for the purpose of the sampling. The plan should address site selection, depth of sampling, type and number of samples, details of collection, and the sampling and subsampling procedures to be followed. The SSL primarily serves the NCSS, which is conducted jointly by the NRCS, the Bureau of Land Management (BLM), the Forest Service, and representatives of American universities and Agricultural Experiment Stations. In this context, the primary objectives of SSL sampling programs have been to select sites and pedons that are representative of a soil series or landscape segment and to collect samples that are representative of horizons within the pedon to support the objectives of soil survey.

There are various kinds of sampling plans, e.g., intuitive and statistical, and many types of samples, e.g., representative, systematic, random, and composite. In the field, the SSL has more routinely used intuitive sampling plans to obtain representative samples. The intuitive sampling plan is one based on the judgment of the sampler, wherein general knowledge of similar materials, past experience, and present information about the universe of concern, ranging from knowledge to guesses, are used (Taylor, 1988). A representative sample is one that is considered to be typical of the universe of concern; its composition can be used to characterize the universe with respect to the parameter measured (Taylor, 1988).

In the laboratory, the primary objectives of sample collection and preparation are to homogenize and obtain a representative soil sample to be used in chemical, physical, and mineralogical analyses. The analyst and the reviewer of data assume that the sample is representative of the soil horizon being characterized. Concerted effort is made to keep analytical variability small. Precise laboratory work means that the principal variability in characterization data resides in sample variability; i.e., sampling is the precision-limiting variable. As a result, site selection and sample collection and preparation are critical to successful soil analysis.

Geomorphic considerations: Soils form a vital, complex continuum across the landscape. The primary goal of the soil survey program is to segregate the soil continuum into individual areas that have similar properties and, therefore, similar use and management. Soils cannot be fully understood or studied using a single observation scale. Instead, soil scientists use multiple scales to study and segregate soils and to transfer knowledge to soil users. To accomplish the task of the soil survey at reasonable cost and within a reasonable timeframe, soil scientists extend knowledge from point observations and descriptions to larger land areas.

Soil map unit delineations are the individual landscape areas defined during and depicted in a soil survey. Soil observation, description, and classification occur at the pedon scale (1 to ≈ 7 m) and represent a small portion of any map unit (tens to thousands of hectares). Further, pedons selected, described, and sampled for laboratory analysis represent only a small subset of the observation points. Pedon descriptions and classifications, along with measured lab data, accurately apply to a named soil map unit or to landscape areas (soil component) within the map unit. Soil scientists can reliably project ("scale up") pedon information to soil map units based on experience and the strong linkages among soils, landforms, sediment bodies, and geomorphic processes. Thus, soil geomorphology serves several key functions in soil survey. These functions can be summarized as:

- Providing a scientific basis for quantitatively understanding soil-landscape relationships, stratigraphy, parent materials, and site history.
- Providing a geologic and geographic context or framework that explains regional soil patterns.
- Providing a conceptual basis for understanding and reliably predicting soil occurrence at the landscape scale.
- Effectively and succinctly communicating the location of a soil within a landscape.

During a soil survey, soil scientists achieve these functions both tacitly and by deliberate effort. Geomorphic functions are best explained by citing examples. The first function listed above involves planned, detailed studies of soil landscapes (e.g., Ruhe et al., 1967; Daniels et al., 1970; Gamble et al., 1970; Parsons et al., 1970; Gile et al., 1981; Lee et al., 2001, 2003a, 2003b), which are an important component of the soil survey. Such studies quantify and explain the links between soil patterns and stratigraphy, parent materials, landforms, surface age, landscape position, and hydrology. Studies of this nature provide the most rigorous, quantitative, and complete information about soil patterns and landscapes. The time and effort required for these studies are significant but are justified by the quantitative information and scientific understanding acquired as a result. Soil survey updates by major land resource area (MLRA) can and should involve similar studies.

The three remaining geomorphic functions are tacit and to a degree inherent in a soil survey. A number of earth science sources (Fenneman, 1931, 1938, 1946; Hunt, 1967; Wahrhaftig, 1965) identify and name geomorphic regions, which are grouped by geologic and landform similarity. The value of relating soil patterns to these regions is self-evident. Such terms as Basin and Range, Piedmont, Columbia Plateau, and Atlantic Coastal Plain provide both a geologic and geographic context for communicating regional soil and landform knowledge.

Soil occurrence can be accurately predicted and mapped using observable landscape features (e.g., landforms, vegetation, slope inflections, parent material, bedrock outcrops, stratigraphy, drainage, and photo tonal patterns). During a soil survey, soil scientists develop a tacit knowledge of soil occurrence generally based on landscape relationships. Soil occurrence is consistently linked to a number of geomorphic attributes. Among these are landform type, landscape position, parent material distribution, slope shape and gradient, and drainage pattern. This tacit soil-landscape knowledge model is partially encapsulated in block diagrams and map unit and pedon descriptions. In turn, a clear, concise geomorphic description effectively conveys to other soil scientists and land users an understanding of the location of a soil within a landscape. Recent publications (Soil Survey Staff, 1998; Schoeneberger et al., 2002; Wysocki et al., 2000) provide a comprehensive and consistent system for describing geomorphic and landscape attributes for soil survey. Geomorphic Description Systems (GDS) are not discussed here. For more detailed information, refer to Soil Survey Staff, 1998; Wysocki et al., 2000; and Schoeneberger et al., 2002.

Geomorphology is an integral part of all processes and stages of soil survey. Preliminary or initial knowledge of soil patterns is commonly based on landscape or geomorphic relationships. Observations during a soil survey refine existing landscape models and can compel and create new models. Map unit design includes landform recognition and naming and observations on landscape position, parent materials, and landscape and soil hydrology. Soil scientists capture this observational and expert knowledge through soil map unit and pedon descriptions, which should convey information about soil properties, soil horizons, landscape and geomorphic relationships, and parent material.

Any study plan, site selection, or pedon sampling must take geomorphology into consideration. Study or sampling objectives can vary. Descriptions of every sampled pedon should include complete descriptions of both the soil and the geomorphology. In a characterization project, the sample pedons should be representative of the landscape unit (e.g., stream terrace and backslope) upon which they occur. Note that the landscape unit that is sampled can be multiscale. The unit could be a landform (e.g., stream terrace, dune, or drumlin), a geomorphic component (e.g., nose slope), a hillslope position (e.g., footslope), or all of these.

The sampled pedon represents both a taxonomic unit and a landscape unit. Both the landscape unit and the taxonomic unit should be considered in site selection. Note that a single landscape unit (e.g., backslope) may contain one or more taxonomic units. A landscape unit is more easily recognizable and mappable in the field than a soil taxonomic unit. For a characterization project, select the dominant taxonomic unit within a given landscape unit. The existence of other soils or taxa can and should be included in the soil description and in the map unit description.

Soil patterns on landscapes follow catenary relationships. It is important to characterize both individual pedon properties and the relationships to soils in the higher and lower areas on the landscape. This goal requires that soils be sampled as a catenary sequence (i.e., multiple samples across the same hillslope). This sampling scheme appears intensive, but it serves multiple purposes. A sample pedon or set of pedons provides vital characterization data and also can quantify the catenary pattern and processes; this approach is thus an efficient use of sampling time and effort and of laboratory resources. Moreover, this approach provides an understanding of the entire soil landscape.

Lastly, and perhaps most importantly, soil geomorphic relationships deserve and sometimes demand specific study during a soil survey. Crucial problems can be addressed by appropriately designed geomorphic, stratigraphic, or parent material study. For example, a silty or sandy mantle over adjacent soils and/or landforms may be of eolian origin. A well-designed geomorphic study can test this hypothesis. In another geomorphic setting, soil distribution and hydrology may be controlled by stratigraphic relationships rather than by elevation or landscape patterns. A drill core or backhoe pit sequence can address this hypothesis. These studies need not be elaborate, but they require forethought and planning. Such studies are applicable and necessary to the MLRA approach to soil survey.

Pedon sampling: The *pedon* is defined as a unit of sampling within a soil, i.e., the smallest body of one kind of soil large enough to represent the nature and arrangement of horizons and variability in the other properties that are preserved in samples (Soil Survey Division Staff, 1993). In the NCSS program, laboratory pedon data combined with field data (e.g., transects and pedon descriptions) are used to define map unit components, establish ranges of component properties, establish or modify property ranges for soil series, and answer taxonomic and interpretive questions (Wilson et al., 1994).

In the early 1950s, field and laboratory soil scientists of the Soil Conservation Service began sampling "paired pedons." Instructions specified that these pedons be selected from the middle of the range of a single phase of a series (Mausbach et al., 1980). Paired pedons were morphologically matched as closely as possible through field observations within practical restrictions of time, size of area, access to site, and inherent variability of the parent material; the variability within these pairs represents variability within a narrow conceptual range (Mausbach et al., 1980). Evaluation of vertical distribution of properties of important horizons has been performed in soil survey by sampling one complete pedon plus *satellite samples* of these horizons. According to Mausbach et al. (1980), in order to assess a single horizon efficiently, one should sample only that horizon in several pedons. Sampling of paired pedons is a good first-approach technique for studying soils in an area. Important early literature on soil variability includes Robinson and Lloyd (1915), Davis (1936), and Harradine (1949). As series concepts narrowed, variability studies of properties and composition of mapping units were made, including those by Powell and Springer (1965), Wilding et al. (1965), McCormack and Wilding (1969), Beckett and Webster (1971), Nielsen et al. (1972), Crosson and Protz (1974), Amos and Whiteside (1975), and Bascomb and

Jarvis (1976). Studies of variability of properties within a series include those by Nelson and McCracken (1962), Andrew and Stearns (1963), Wilding et al. (1964), Ike and Cotter (1968), and Lee et al. (1975).

A site that meets the objectives of the laboratory sampling is selected. The site and the soil pedon are described and georeferenced. Included in these descriptions are complete soil and geomorphic descriptions. The soil descriptions include observations of specific soil properties, such as texture, color, slope, and depth. Descriptions may also include inferences of soil quality (soil erodibility and productivity) as well as soil-forming factors (climate, topography, vegetation, and geologic material). The sampled pedons should be representative of the landscape unit upon which they occur and can be multiscale (Fig. 1.1.1).

A soil pit is commonly excavated with a backhoe (Fig. 1.1.2). Depth and breadth of the pit depend on the soil material and on the objectives of sampling. Soil horizons or zones of uniform morphological characteristics are identified for sampling (Fig. 1.1.3). Photographs are typically taken of the landform or landform segment and the soil profile. Photographs of the soil profile, with photo tapes showing vertical scale (metric and/or feet), are taken after the layers have been identified (Fig. 1.1.4) but before the extraction of the vertical section by the sampling process (Fig. 1.1.5).

The variable nature or special problems inherent in the soil (as may be the case with Vertisols, Histosols, or soils affected by permafrost) may require the use of specific excavation and sampling techniques. For example, the shear failure that forms slickensides in Vertisols also disrupts the soil to the point that conventional soil horizons do not adequately describe the morphology.

Representative samples are collected and mixed for chemical, physical, and mineralogical analyses. A representative sample is collected using the boundaries of the horizon to define the vertical limits and the observed short-range variability to define the lateral limits. The tag on the sample bag is labeled to identify the site, pedon, and soil horizon for the sample.

In the field, the 20- to 75-mm fraction is generally sieved, weighed, and discarded. In the laboratory, the <20-mm fraction is sieved and weighed. The SSL estimates weight percentages of the >2-mm fractions from volume estimates of the >20-mm fractions and weight determinations of the <20-mm fractions.

Undisturbed clods are collected for bulk density and micromorphological analysis. Clods are obtained in the same part of the pit as the mixed, representative sample. Bulk density clods are used for water retention data, to convert from a weight to volume basis, to determine the coefficient of linear extensibility (COLE), to estimate saturated hydraulic conductivity, and to identify compacted horizons. Microscope slides of soil prepared for micromorphology are used to identify fabric types, skeleton grains, intensity of weathering, and illuviation of argillans and to investigate the genesis of soil or pedological features.

Figure 1.1.1.—Landscape of selected site for sampling.

Figure 1.1.2.—Excavated pit for pedon sampling.

Figure 1.1.3.—Soil horizons or zones of uniform morphological characteristics are identified for sampling.

Figure 1.1.4.—Photographs are typically taken of the soil profile after the layers have been identified but before the vertical section by the sampling process. Note scale in metric units.

Figure 1.1.5.—Pedon sampling activities.

Water sampling: Water samples are analyzed by the SSL on a limited basis in the support of specific research projects. These projects are typically undertaken in conjunction with soil investigations and have involved monitoring seasonal nutrient flux to evaluate movement of N and P via subsurface and overland flow from agricultural lands into waterways and wetlands. Choice of water sampling sites depends not only on the purpose of the investigation but also on local conditions, depth, and the frequency of sampling (Velthorst, 1996). Specific recommendations are not applicable, as the details of collection can vary with local conditions. Nevertheless, the primary objective of water sampling is the same as that of soil and biological sampling, i.e., to obtain a representative sample in laboratory analyses. Water samples require expedited transport under ice or gel packs and are refrigerated (4 °C) immediately upon arrival at the laboratory.

Biological sampling: Biological samples also are collected for analysis at the SSL, either in conjunction with pedon sampling or for specific research projects. Measurable biological indices have been considered as a component in the assessment of soil quality (Gregorich et al., 1997; Pankhurst et al., 1997). A large number of soil biological properties have been evaluated for their potential use as indicators of soil quality/health (Doran and Parkin, 1994; Pankhurst et al., 1995). The NRCS has utilized soil biology and carbon data in macronutrient cycling, soil quality determinations, resource assessments, global climate change predictions, long-term soil fertility assessments, impact analysis for erosion effects, conservation management practices, and carbon sequestration (Franks et al., 2001). Soil quality was identified as an emphasis area of the NRCS in 1993. Soil quality publications and technical notes are available online at http://soils.usda.gov/sqi/.

As with pedon sampling, sampling for root biomass includes selecting a representative site, sampling by horizon, and designating and sampling a subhorizon if root mass and morphology change. The same bulk sample collected for soil mineralogical, physical, and chemical analyses during pedon sampling can be used for some soil biological analyses. Alternatively, a separate biobulk sample can be collected in the field. Surface litter and O horizons are sampled separately, as with pedon sampling. If certain biological analyses (e.g., microbial biomass) are requested, these samples require expedited transport under ice or gel packs and are refrigerated (4 °C) immediately upon arrival at the laboratory to avoid changes in the microbial communities.

Classification systems and soil survey: It has long been recognized that an inventory of natural resources and the management of those resources required a land classification system, a process of arranging or ordering information about land units that improves our understanding of their similarities and relationships (Bailey, 1996). Such terms as "Corn Belt" and "Cotton Belt" were coined by early farmers and ranchers in the United States, who realized that the different soils and climates they encountered required them to grow certain types of crops in order to survive economically (USDA/NRCS, 2006). These land delineations were the early versions of land resource areas. As the USDA soil survey program mapped soils across the country, soil scientists and natural resource managers subdivided the land into resource units based on similar soils, climate, and vegetation or crop types. Scientists and managers were then able to provide many landowners soil interpretations and soil conservation recommendations that were based on regionalized information (USDA/NRCS, 2006). These early efforts resulted in the publication of Agricultural Handbook 296 in 1965 (USDA/SCS, 1965), in which the U.S. was subdivided into a number of land resource regions consisting of major land resource areas. The USDA classification system helps natural resource planners target efforts in education and financial and technical assistance (USDA/NRCS, 2006) and is used to make decisions about regional and national agricultural issues. It also serves as the basis for organizing and operating natural resource conservation programs. Today, the organization of the NRCS soil survey program is designed to serve these groups of major land resource areas.

One of the best known classification systems is the USDA Land Capability Classification System (Klingebiel and Montgomery, 1961). This interpretive system uses the USDA soil survey map as a basis for classifying individual soil map units in groups that have similar management requirements. The system shows the suitability of soils for agricultural uses and classifies soils for mechanized production of the more commonly cultivated field crops, e.g., corn, small grains, cotton, hay, potatoes, and field-grown vegetables.

The establishment of the soil survey program in the United States was an important development in evaluating and predicting the effects of land use on the environment. Soil surveys were first authorized in the United States in 1899. Since then, many surveys have been completed and published cooperatively by the USDA and State and Federal agencies through the National Cooperative Soil Survey (NCSS) (Soil Survey

Division Staff, 1993). Soil survey describes the characteristics of the soils in a given area, classifies the soils according to a standard system of classification, plots the boundaries of the soils on a map, and makes predictions about the behavior of the soils (Soil Survey Division Staff, 1993).

In the 1920s and 1930s, the work in soil classification was primarily qualitative, but gradually a system with more quantitative class limits was used. Work began on this system in 1945. The system was adopted in 1965, and the work culminated in the publication of *Soil Taxonomy* in 1975 (Soil Survey Staff, 1975). This publication was revised in 1999 (Soil Survey Staff, 1999). In the United States, soil surveys vary in scale and in intensity of observation. According to the *Soil Survey Manual* (Soil Survey Division Staff, 1993), the components of map units are designated by the taxa identified in *Soil Taxonomy*. These taxa are further subdivided by specific surface characteristics of the geographic unit of land being mapped. Developments in the Canadian soil classification system somewhat paralleled those in the United States. The National Soil Survey Committee first met in Ontario in 1945 to formulate ways and means of utilizing soil survey information and to propose a new soil classification system.

Over the years, the soil survey program in the United States has broadened in application, precision, and discipline. Before 1950, the primary applications of soil survey were for farming, ranching, and forestry. In the 1950s and 1960s, the applications of soil survey increased along with increases in nonagricultural uses of the soil, e.g., urban development, highways, and other engineering projects (Bartelli et al., 1966). In the 1970s, the authorities for soil survey were expanded to include urban lands. More recently, soil survey information has been used in environmental studies. Beginning in the 1930s and early 1940s, the use of aerial photographs has greatly increased the precision of soil maps, and even greater detail has more recently been provided as a result of advances in satellite imagery. The modern soil survey utilizes many disciplines, including soil chemistry, mineralogy, physics, hydrology, geochemistry, genesis, pedology, geomorphology, and environmental science (Jenny, 1941; Baver, 1956; Jackson, 1958; Alexiades and Jackson, 1966; Ruhe, 1975; Small, 1975).

Another important development in the assessment of soils in the United States was the establishment of USDA soil laboratories to provide analytical data in support of such activities as soil survey and to address specific soil problems, such as salinity. In 1976, the U.S. soil survey laboratories were combined to form the National Soil Survey Laboratory (SSL) in Lincoln, Nebraska. The SSL primarily serves the NCSS. In recognition that saline and alkali soil conditions reduced the value and productivity of considerable areas of land in the U.S., the United States Salinity Laboratory was created in 1947. In 1954, the first handbook on the diagnosis and improvement of saline and alkali soils was published (U.S. Salinity Laboratory Staff, 1954). The Salinity Laboratory was instrumental in developing analytical methods and concepts (e.g., saturated paste and its relationship to field water content) and in providing soil indexes as indicators of and criteria for alkalinity, sodicity, and salinity as related to plant growth and yield (e.g., exchangeable sodium percentage, sodium adsorption ratio, electrical conductivity, and soluble salts). These concepts and laboratory data were later used in *Soil Taxonomy* (Soil Survey Staff, 1975 and 1999) for the identification and classification of these soils. The establishment of these and other Federal soil laboratories, along with soil mapping (delineation on the landscape) and soil classification, was instrumental in the production of data and the development of a long-term, comprehensive assessment of agricultural soils in the United States.

Over the last three decades, there has been an evolution toward the assemblage and development of long-term soil resource assessment technologies that are land based or are based on ecological considerations and away from the management of individual resources (e.g., soils). This trend is especially noticeable in forestry management in both the U.S. and Canada (Hills, 1952; Wertz and Arnold, 1972; Bailey, 1976, 1996; Jordan, 1982; Rowe, 1980, 1984; Jones, 1983; Driscoll, 1984; Pregitzer and Barnes, 1984; Spies and Barnes, 1985; Cleland et al., 1985; O'Neill et al., 1986; McNab, 1987; Smalley, 1986).

In general, as management strategies of natural resources in the United States have moved toward systems that are land based or are based on ecological considerations, there has also been a growing recognition that soils play an important role in these strategies. Soil is one of the basic natural resources, and thus its inventory and assessment are critical. Other examples of this recognition are the establishment in 1980 by the National Science Foundation of Long Term Ecological Research (LTER) sites in the United States. The LTER network supports research on long-term ecological phenomena over large temporal and broad spatial scales; the soils component is an important part of this research (Robertson et al., 1999). There are currently 26 LTER sites in the United States. Another example is the changing philosophy (definition and scope) of rangeland management in the United States (Orr, 2006). Over time, this philosophy has ranged from focusing on ecological principles (e.g., succession and grazing systems) and considering rangeland use primarily for domesticated livestock (Sampson, 1923) to incorporating soil science, geomorphology, climate, ecology, and animal science and establishing multiple-use relationships (Stoddart et al., 1975) and to an "overriding goal (not just the effects and management of domestic animals) of rangeland resource rehabilitation, protection, and management for multiple objectives including biological diversity, preservation, and sustainable development for people" (Heady and Child, 1994). Soil and site stability, hydrologic function, and biotic integrity are considered the attributes or indicators of rangeland health by the NRCS (USDA/NRCS, 2005). Refer to the "National Range and Pasture Handbook" (USDA/NRCS, 2003) for information on range and pasture management.

1.2   Laboratory Sample Collection and Preparation
1.2.1   Soil Samples

Soil samples, purpose and interferences: The purpose of any soil sample is to obtain information about a particular soil and its characteristics. Sampling provides a means to estimate the parameters of these soil characteristics with an acceptable accuracy at the lowest possible cost (Petersen and Calvin, 1986). Subsampling also may be used, as it permits the estimation of some characteristics of the larger sampling unit without the necessity of measurement of the entire unit. Subsampling reduces the cost of the investigation but typically reduces the precision with which the soil characteristics are estimated. Efficient use of subsampling depends on a balance between cost and precision (Petersen and Calvin, 1986).

Soil variability and sample size are interferences to sample collection and preparation. The objective of laboratory preparation is to homogenize the soil samples used in chemical, physical, and mineralogical analyses. At each stage of sampling, an additional component of variability, the variability within the larger units, is added to the sampling error (Petersen and Calvin, 1986). Soil material must be adequate in amount and thoroughly mixed if a representative sample is to be obtained.

Soil samples, identification numbers and preparation codes: The SSL receives bulk soil samples from across the United States and internationally for a wide variety of chemical, physical, and mineralogical analyses. The SSL also typically receives natural fabrics, clods, and cores. Undisturbed clods are used to investigate micromorphology and to determine some physical properties, e.g., bulk density. All soils from quarantined areas are strictly controlled under Animal and Plant Health Inspection Service (APHIS) quarantine regulations 7 CFR 330.

Laboratory identification numbers and preparation codes are assigned to bulk soil samples. These identification numbers are unique client- and laboratory-assigned numbers that carry important information about the soil sample (e.g., pedon, soil horizon, location, and year sampled). Laboratory identification number and preparation codes are also assigned to natural fabrics, clods, and cores. These identification numbers typically relate to a corresponding bulk sample. Laboratory preparation codes depend on the properties of the sample and on the requested analyses. These codes carry generalized information about the characteristics of the analyzed fraction, i.e., the water content (e.g., air-dry, field-moist) and the original and final particle-size fraction (e.g., sieved <2-mm fraction processed to 75 μm) and, by inference, the types of analyses performed. Identification numbers and preparation codes are reported on the SSL Primary Characterization Data Sheets. Since the publication of SSIR No. 42, Version 3.0 (Soil Survey Staff, 1996), these preparation codes have been significantly revised. The revised preparation codes are not described in detail in SSIR No. 42, Version 4.0 (Soil Survey Staff, 2004). Detailed information on the current preparation codes as they appear on the SSL Primary Characterization Data Sheets may be obtained from the SSL upon request.

In SSIR No. 42, Version 3.0 (Soil Survey Staff, 1996), laboratory preparation procedures were described as stand-alone methods based on various procedures summarized by specific preparation codes that are reported on the SSL Primary Characterization Data Sheets. In SSIR No. 42, Version 4.0 (Soil Survey Staff, 2004), however, a different approach is used. A process approach is appropriate in that any one sample received from the field may result in a number of laboratory subsamples being collected and prepared based on analytical requests and type of materials. This approach is the logic base whereby laboratory procedures are described in SSIR No. 42, Version 4.0 (Soil Survey Staff, 2004). The intent of these descriptions is not to detail all possibilities of the universe but to describe some of the master preparation procedures that are typically requested for analyses at the SSL. Examples of SSL master collection and preparation procedures include, but are not limited to, air-dry, <2-mm; field-moist, <2-mm particles; and air-dry, >2-mm fractions.

Soil samples, preparation: Soil is mixed by moving it from the corners to the middle of the processing area and then by redistributing the material. This process is repeated four times. Enough soil material must be sieved and weighed to obtain a statistically accurate rock fragment content. In order to accurately measure rock fragments with a maximum particle diameter of 20 mm, the minimum specimen size ("dry" weight) that must be sieved and weighed is 1.0 kg. Refer to American Society for Testing and Materials (ASTM) Standard Practice D 2488-06 (ASTM, 2008c). A homogenized soil sample is more readily obtained from air-dry material than from field-moist material. Whenever possible, "moist" samples or materials should have weights two to four times as large as those for "dry" specimens (ASTM, 2008c). The minimum specimen sizes ("dry" weights) for particle-size analysis are as follows:

Table 1.2.1 Minimum dry weights for particle-size analysis[1]

Maximum particle size Sieve opening	Minimum specimen size Dry weight
4.75 mm (No. 4)	100 g (0.25 lb)
9.5 mm (⅜ in)	200 g (0.5 lb)
19.0 mm (¾ in)	1.0 kg (2.2 lb)
38.1 mm (1½ in)	8.0 kg (18 lb)
75.0 mm (3 in)	60.0 kg (132 lb)

[1] ASTM, 2008c.

Soil samples, air-dry preparation: Any one soil sample received from the field may result in a number of laboratory subsamples being collected and prepared based on the properties of the sample and on the requested analyses. For most standard chemical, physical, and mineralogical analysis, the field sample is air-dried, crushed, and sieved to <2 mm. Air-dry is generally the optimum water content to handle and to process soil. In addition, the weight of air-dry soil remains relatively constant, and biological activity is low during storage. For routine soil analyses, most U.S. and Canadian laboratories homogenize and process samples to pass a 2-mm sieve (Bates, 1993). For some standard air-dry analyses, the <2-mm fraction is further processed so as to be in accordance with a standard method, e.g., Atterberg limits; to meet the sample preparation requirements of the analytical instrument, e.g., total C, N, and S; or to achieve greater homogeneity of sample material, e.g., total elemental analysis and carbonates and/or gypsum. Additionally, some standard air-dry analyses by definition may require nonsieved material, e.g., whole-soil samples, for aggregate stability.

Soil samples, field-moist preparation: Field-moist, fine-earth fraction samples are processed by forcing the material through a 2-mm screen by hand or with a large, rubber stopper and are placed in a refrigerator for future analysis. A field-moist, <2-mm sample is prepared when the physical properties of a soil are irreversibly altered by air-drying, e.g., water retention, particle-size analysis, and plasticity index for Andisols and Spodosols, and/or when moist chemical analyses are appropriate. Some biological analyses require field-moist samples, as air-drying may cause significant changes in the microbial community. The decomposition state of organic materials is used in *Keys to Soil Taxonomy* (Soil Survey Staff, 2010) to define sapric, hemic, and fibric organic materials; therefore, the evaluation of these materials (Histosol analysis) requires a field-moist, whole-soil sample.

Rock fragments: Knowing the amount of rock fragments is necessary for several applications, e.g., available water capacity and linear extensibility. Generally, the >2-mm fractions are sieved, weighed, and discarded and are excluded from most chemical, physical, and mineralogical analyses. Exceptions include, but are not limited to, samples containing coarse fragments with carbonate- or gypsum-indurated material or material from Cr and R soil horizons. In these cases, the coarse fragments may be crushed to <2 mm and analytical results are reported on that fraction, e.g., 2 to 20 mm, or the coarse fragments and fine-earth material are homogenized and crushed to <2 mm and laboratory analyses are made on the whole-soil. Additionally, depending on the type of soil material, samples can be tested for the proportion and particle size of air-dry rock fragments that resist abrupt immersion in tapwater.

1.2 Laboratory Sample Collection and Preparation
1.2.2 Water Samples

As with soil samples, laboratory identification numbers and preparation codes are assigned to water samples. Long periods between collection and laboratory analysis of water samples should be avoided. To prevent significant changes (e.g., degradation, volatilization), water samples should be transported rapidly under ice or gel packs and refrigerated (4 °C) immediately upon arrival at the laboratory. The freezing of water samples should be avoided because it can influence pH and the separation of dissolved organic matter from the water phase.

Some water analyses, e.g., electrical conductivity, total C, and inorganic C, need to be performed promptly, as optimal preservation is not possible (Velthorst, 1996). Upon completion of these analyses, sample filtration (0.45-μm membrane) is used to separate dissolved material from suspended material. The sample is then split into two subsamples, with one acidified to pH 2 for cation analyses (e.g., Al, Fe, Mn) and the other for anion analyses. These other water analyses also need to be performed as promptly as possible.

1.2 Laboratory Sample Collection and Preparation
1.2.3 Biological Materials

As with soil samples, laboratory identification numbers and preparation codes are assigned to biological materials. Some biology samples arrive at the laboratory as part of the soil bulk sample. If this is the case, biological subsamples are collected and prepared. In other cases, biology bulk samples may be split in the field and are separate sampling units from the soil bulk sample. Additionally, some biological samples, e.g., microbial biomass, are separate units from the soil bulk or other biology samples; require expedited transport under ice or gel packs; and should be refrigerated (4 °C) immediately upon arrival at the laboratory.

Long periods between collection and laboratory analysis of biological samples should be avoided so as to prevent significant changes (e.g., microbial community). Refer to the section on soil biological and plant analyses for additional information on the further processing and preparation of these biological samples for laboratory analysis.

2 Conventions

This section discusses the importance of using standard operating procedures (SOPs) documented through method codes and linked with analytical results stored in the NCSS Characterization Database, which is available online at http://ssldata.nrcs.usda.gov/. In addition, this section covers the types of data as well as the significant figures and rounding procedures; data sheet symbols; and sample weight and particle-size fraction basis for reporting data on the SSL Soil Characterization Data Sheets. For detailed descriptions of the SSL methods which are cross-referenced by method code in the table of contents in this manual, refer to the *Soil Survey Laboratory Methods Manual*, SSIR No. 42 (Soil Survey Staff, 2004), which is available online at http://soils.usda.gov/technical/lmm/.

2.1 Methods and Codes

Standard operating procedures: The SSL ensures continuity in its analytical measurement process by using standard operating procedures (SOPs). A standard method is defined herein as a method or procedure developed by an organization, based on consensus opinion or other criteria and often evaluated for its reliability by a collaborative testing procedure (Taylor, 1988). An SOP is a procedure written in a standard format and adopted for repetitive use in the performance of a specific measurement or sampling operation; i.e., an SOP may be a standard one or one developed by a user (Taylor, 1988).

The use of SOPs provides consistency and reproducibility in soil preparations and analyses and helps to ensure that these preparations and analyses provide results of known quality. SSIR No. 42, Version 4.0, replaces as a methods reference all earlier versions (Soil Survey Staff, 1989, 1992, 1996). It also replaces *Procedures for Collecting Soil Samples and Methods of Analysis for Soil Survey*, SSIR No. 1 (Soil Conservation Service, 1984). All SSL methods are performed with methodologies appropriate for the specific purpose. The SSL SOPs are standard methods, peer-recognized methods, SSL-developed methods, and/or methods specified in *Keys to Soil Taxonomy* (Soil Survey Staff, 2010). SSIR No. 42 also serves as the primary document from which its companion manual, the *Soil Survey Laboratory Information Manual* (SSIR No. 45), was developed. The current manual, SSIR No. 45, is the second version; the original version was published in 1995 (Soil Survey Staff, 1995). SSIR No. 45 describes the application of SSL data in more detail than SSIR No. 42.

Method codes: Included in SSIR No. 42 (Soil Survey Staff, 2004) are descriptions of current as well as obsolete methods, all of which are documented by method codes and linked with analytical results that are stored in the SSL database. This linkage between laboratory method codes and the respective analytical results is reported on the SSL data sheets. Reporting the method by which the analytical result is determined helps to ensure user understanding of SSL data. In addition, this linkage provides a means of technical criticism and traceability if data are questioned in the future.

The methods in current use at the SSL are described in SSIR No. 42 in enough detail that they can be performed in many laboratories without reference to other sources. Descriptions of the obsolete methods are located at the back of the methods manual. Because information is not available, the descriptions of some obsolete procedures are not as detailed as those of current laboratory methods.

Since the publication of SSIR No. 42, Version 3.0 (Soil Survey Staff, 1996), there has been a significant increase in the number and kind of methods performed at the SSL. As a result, the method codes have been restructured. As in past versions of SSIR No. 42, the current method codes are hierarchical and alphanumerical. The older method code structure had a maximum of only four characters, e.g., 6A1b, whereas the new structure allows more characters, which provide more information about the method, e.g., particle-size and sample weight bases for reporting data. SSIR No. 42, Version 4.0 (Soil Survey Staff, 2004), carries not only the new method codes but also the older ones. These older codes are cross-referenced in a table preceding the descriptions of the obsolete SSL methods. It is important to maintain this linkage between the two method code systems as many older SSL data sheets and scientific publications report the older codes.

Data sheets: The SSL provides data in reports, e.g., Primary and Supplementary Characterization Data Sheets and, more recently, Taxonomy and Dynamic Soil Properties Characterization Data Sheets. Data are also provided in electronic forms, including tapes, disks, and CD-ROMs, and are available online at http://ssldata.nrcs.usda.gov/. While the Primary Characterization Data Sheet is mainly composed of analytical data, some calculated values also are presented. Historically, the SSL has described and assigned method codes to only those data reported on the Primary Characterization Data Sheet (as opposed to the Supplementary Characterization Data Sheet). This tradition was followed in SSIR No. 42 (Soil Survey Staff, 2004). Some of the more recently developed calculated values on the Primary Characterization Data Sheet are not described or assigned method codes in SSIR No. 42 (Soil Survey Staff, 2004). For more detailed information about the calculation and application of these derived values, refer to other sections of this manual (SSIR No. 45) and to *Keys to Soil Taxonomy* (Soil Survey Staff, 2010).

2.2 Data Types

The methods described in this section identify the specific type of analytical or calculated data. Most of these methods are analytical in nature, i.e., quantitative or semiquantitative measurements, and include physical, chemical, mineralogical, and biological analyses. Sample collection and preparation in the field and in the laboratory also are described. Historically, SSIR No. 42 has described some derived values, e.g., coefficient of linear extensibility (COLE) and water retention difference (WRD), and reported these values along with the analytical data on the SSL Primary

Characterization Data Sheets. SSIR No. 42 (Soil Survey Staff, 2004) follows this tradition. The more recently developed calculated values that appear on the Primary Characterization Data Sheet (e.g., estimated organic carbon, estimated organic matter, and "Pedon Calculations") are not described or assigned method codes in SSIR No. 42 (Soil Survey Staff, 2004) but are described in the introduction to this manual. Also refer to *Keys to Soil Taxonomy* (Soil Survey Staff, 2010) for more information on the use of "Pedon Calculations," e.g., weighted clay average, and other derived data. The SSL Taxonomy Data Sheet is a mixture of distinct analytical data as well as data repeated from the Primary Characterization Data Sheet for user convenience. The Supplementary Characterization Data Sheet is considered to contain the interpretive physical data for pedons analyzed at the SSL. These data are primarily calculated data; the analytical data are used as the basis for calculation.

2.3   Particle-Size Size-Fraction Basis for Reporting Data
2.3.1   Particles <2 mm
2.3.2   Particles <Specified Size >2 mm

Unless otherwise specified, all SSL data are reported on the basis of the <2-mm material. Other size fractions reported on the Primary Characterization Data Sheets include, but are not limited to, the <0.4-mm, <20-mm, <75-mm, and whole-soil basis. The maximum coarse-fragment size for the >2-mm basis varies. The basis usually includes those fragments as large as 75 mm (3 in), if they occur in the soil. The maximum size for fragments >75 mm, commonly termed whole soil, includes boulders with maximum horizontal dimensions less than those of the pedon. The maximum particle-size set is recorded in parentheses in the column heading. The basis with which to calculate the reported >2-mm percentages includes all material in the sample smaller than the particle size recorded in the column heading.

2.4   Sample Weight Basis for Reporting Data
2.4.1   Air-Dry/Oven-Dry
2.4.2   Field-Moist/Oven-Dry
2.4.3   Correction for Crystal Water

Unless otherwise specified, all SSL data are reported on an oven-dry weight or volume basis for the designated particle-size fraction. The calculation of the air-dry/oven-dry (AD/OD) ratio is used to adjust AD results to an OD weight basis and, if required in a procedure, to calculate the sample weight that is equivalent to the required OD soil weight. The AD/OD ratio is converted to a crystal water basis for soils with gypsum (Nelson et al., 1978). The calculation of the field-moist/oven-dry (FM/OD) ratio is used to adjust FM results to an OD weight basis and, if required in a procedure, to calculate the sample weight that is equivalent to the required OD soil weight.

AD and OD weights are defined herein as constant sample weights obtained after drying at 30±5 °C (≈ 2 to 7 days) and at 110±5 °C (≈ 12 to 16 h), respectively. As a general rule, air-dry soils contain about 1 to 2 percent water and are drier than soils at 1500-kPa water content. FM weight is defined herein as the sample weight obtained without drying prior to laboratory analysis. In general, these weights are reflective of the water content at the time of sample collection.

2.5 Significant Figures and Rounding

Unless otherwise specified, the SSL uses the procedure of significant figures to report analytical data. Historically, significant figures are said to be all digits that are certain plus 1, which contains some uncertainty. If a value is reported as 19.4 units, the 0.4 is not certain; i.e., repeated analyses of the same sample would vary more than one-tenth of a whole unit but generally less than a whole unit.

2.6 Data Sheet Symbols

The analytical result of "zero" is not reported by the SSL. The following symbols are used or have been used for trace or zero quantities and for samples not tested.

tr, Tr, TR Trace; either is not measurable by quantitative procedure used or is less than reported amount.

tr(s) Trace; detected only by qualitative procedure more sensitive than quantitative procedure used.

- Analysis run but none detected.

-- Analysis run but none detected.

-(s) None detected by sensitive qualitative test.

blank Analysis not run.

nd Not determined; analysis not run.

< Either none is present or amount is less than reported amount; e.g., <0.1 is in fact <0.05 since 0.05 to 0.1 is reported as 0.1.

3 Soil Physical and Fabric-Related Analyses

This section describes the SSL methods for soil physical and fabric-related analyses and their specific method applications and interferences, as follows:

- Particle-size distribution analysis
- Bulk density
- Water retention

- Ratios, estimates, and calculations associated with particle-size distribution analysis, bulk density, and water retention
- Micromorphology
- Aggregate stability
- Particle density
- Atterberg limits

3.1 Particle-Size Distribution Analysis

This section on particle-size distribution analysis (PSDA) provides general information about the various soil classification systems as well as definitions of particle-size limits and the historical background for the development and/or modifications to these limits. Applications of PSDA data and the calculations derived from these data also are discussed. Particle-size distribution (soil texture) is a major soil property affecting a soil's susceptibility to erosion and as such is a key parameter in any soil erosion prediction model. For these reasons, the process of soil loss is described and references to case studies/datasets are presented as evidentiary examples of the actions/practices that have promoted or diminished this soil process. Major developments in the knowledge, science, and technology of soil and water conservation also are discussed.

Procedures for particles <2 mm in diameter using the pipet and hydrometer methods are described. The sieve and pipet method is the standard SSL method, whereas the hydrometer method is used by the USDA Soil Mechanics Laboratory (SML) as well as by NRCS soil survey offices. Included in the discussions about these PSDA methods is information about the use of air-dry versus field-moist samples as well as routine versus nonroutine pretreatment and dispersion techniques. Also described in this section are the SSL procedures for >2-mm particles using weight estimates by field and laboratory weighing; weight estimates from volume and weight estimates; and volume estimates. For detailed descriptions of the SSL methods which are cross-referenced by method code in the table of contents in this manual, refer to SSIR No. 42 (Soil Survey Staff, 2004), which is available online at http://soils.usda.gov/technical/lmm/. Also refer to the *Soil Survey Field and Laboratory Methods Manual*, SSIR No. 51 (Soil Survey Staff, 2009; available online at http://www.soils.usda.gov/technical/), for detailed descriptions of field methods as used by NRCS soil survey offices.

3.1 Particle-Size Distribution Analysis
3.1.1 Classification Systems

Particle-size distribution analysis: Perhaps the single most important physical property of a soil and one of the most requested SSL characterization analyses is particle-size distribution analysis. The behavior of most physical soil properties and many chemical soil properties is sharply influenced by particle-size distribution classes and their relative abundance. Precise meaning is given to the term "soil texture" only through the concept of particle-size distribution (Skopp, 1992). Particle-size distribution analysis is a measurement of the size distribution of individual particles in a soil sample. These data may be presented on a cumulative PSDA curve. These distribution curves are used in many kinds of investigations and evaluations, e.g., geologic, hydrologic, geomorphic, engineering, and soil science (Gee and Bauder, 1986). Cumulative curves have the advantage of allowing comparison of particle-size analyses that use different particle-size classes. Most commonly, the cumulative percentage of particles finer than a given particle size is plotted against the logarithm of "effective" particle diameter (Gee and Bauder, 1986). In soil science, particle size is used as a tool to explain soil genesis, quantify soil classification, and define soil texture. Refer to Appendix 2 (Wildmesa Pedon) and Appendix 5 (Caribou Pedon) for example particle-size distribution curves.

USDA classification system: In the *USDA soil classification system*, soil texture refers to the relative proportions of clay, silt, and sand on a <2-mm basis. The system also recognizes proportions of five subclasses of sand (Soil Survey Division Staff, 1993). The USDA classification scheme uses a textural triangle to show the percentages of clay, silt, and sand. Refer to Appendix 6 (Guide for Textural Classification). The USDA soil classification system classifies soil particles (soil separates) according to size, as follows: Very coarse sand, 2.0 to 1.0 mm; coarse sand, 1.0 to 0.5 mm; medium sand, 0.5 to 0.25 mm; fine sand, 0.25 to 0.10 mm; very fine sand, 0.10 to 0.05 mm; silt, 0.05 to 0.002 mm; and clay, <0.002 mm. In soil science, the terms *clay, silt, very fine sand, fine sand,* and *coarse sand* are used to define not only soil separates but also specific soil classes. In addition, the term *clay* is used to define a class of soil minerals (Sumner, 1992). The PSDA data by the SSL are soil separates reported as weight percentages on a specified basis.

Other classification systems: In addition to the USDA soil classification scheme, other classification systems include the particle-size classes for differentiation of families in *Soil Taxonomy* (Soil Survey Staff, 1975); the International Union of Soil Science (IUSS); the Canadian Soil Survey Committee (CSSC); and the American Society for Testing and Materials (ASTM). In reporting and interpreting data, it is important to recognize that these other classification systems are frequently cited in the literature, especially engineering systems, e.g., the American Association of State Highway and Transportation Officials (AASHTO) and the ASTM Unified Soil Classification System (USCS) (Gee and Bauder, 1986). The AASHTO system, developed in 1929 by the Bureau of Public Roads, currently uses seven major groups of soils (A1 to A7) and provides a general rating of the soil as a subgrade for road construction. Developed by Casagrande in 1942, the USCS is widely used by geotechnical engineers. The AASHTO and USCS engineering classification systems as applied in soil survey are discussed in more detail in the "National Soil Survey Handbook," available online at http://www.soils.usda.gov/technical/handbook/. The National Soil Information System (NASIS), available online at http://www.soils.usda.gov/technical/nasis/, serves as the depository of all soil survey information, thereby integrating information on soil properties and qualities as well as groupings for engineering properties and AASHTO and USCS classes.

Particle-size classes: In general, the term *particle size* is used to characterize the grain-size composition of the mineral portion of a whole soil, while the term *texture* is used in describing its fine-earth fraction (Soil Survey Staff, 2010). As used herein, the

32

fine-earth fraction refers to particles <2 mm in diameter and the whole soil is all particle-size fractions, including boulders with maximum horizontal dimensions less than those of the pedon. The term *rock fragments* means particles of the whole soil that are ≥2 mm in diameter and includes all particles with horizontal dimensions smaller than the size of the pedon (Soil Survey Division Staff, 1993; Soil Survey Staff, 2010). At one time, the term *rock fragments* was differentiated from the term *coarse fragments*, which excluded stones and boulders with diameters ≥250 mm (Soil Survey Staff, 1951, 1975). The rationale for this distinction was that particles <250 mm were generally regarded as part of the "soil mass;" i.e., they affect moisture storage, infiltration, runoff, root growth, and tillage (Soil Survey Staff, 1951). In the descriptions of soil horizons, particles ≥250 mm were excluded from the soil textural class name but phase names for stoniness and rockiness, although not a part of the textural class names, were used to modify the soil-class part of the soil-type name, e.g., Gloucester *very stony loam* (Soil Survey Staff, 1951). Refer to Soil Survey Staff (1951) for additional discussion of the rationale for this particle-size distinction. Refer to Soil Survey Division Staff (1993) for additional discussion on rock fragments. Refer to Soil Survey Staff (2010) for additional discussion on particle-size classes.

3.1 Particle-Size Distribution Analysis
3.1.2 Particles <2 mm

Clay, historical concepts and class limits: The definition of *clay* has been debated for many years. Early concepts of clay attempted to characterize clay on the basis of its chemical nature and its effects upon the soil (Baver, 1956). Osborne (1887), who developed the beaker method of soil mechanical analysis in 1886, defined clay as follows: "*True clay* is here meant that material derived from the decomposition of feldspars and similar silicates, which is capable of uniting with a considerable amount of water, and thus assuming a gelatinous condition in which it exerts a powerful binding action upon the particles of sand in the soil. To some extent, probably, this action is also exerted by iron and alumina hydroxides, as well as by colloid organic bodies."

The purely chemical definition of clay by Osborne (1887) was eventually replaced by one that was colloidal in meaning (Baver, 1956). The colloidal concept of clay was developed when the ideas of disperse systems were applied to the study of soils by Oden (1921-1922) and other investigators. Oden (1921-1922) defined clay as "disperse formations of mineral fragments in which particles of smaller dimensions than 2 μm (0.002 mm) predominate;" i.e., clay consists of primary mineral fragments together with the secondary products of weathering as long as the individual particle sizes are small enough (Baver, 1956). The definition of clay with an upper size limit of 2 μm was first introduced by Atterberg (1912). Refer to the discussion of clay versus colloidal clay under the data element *fine clay*.

Atterberg classification system, scientific rationale: The *Atterberg definition of clay* and the classification of other soil particles according to size were accepted by the International Society of Soil Science in 1913. This classification of soil particles according to size is as follows: Gravel, 20 to 2.0 mm; coarse sand, 2.0 to 0.2 mm; fine sand, 0.2 to 0.02 mm; silt, 0.02 to 0.002 mm; and clay, <0.002 mm. Atterberg's scientific rationale for setting up the various size limits and for characterizing clay as <2 μm is described by Baver (1956) as follows:

> The 20- to 2-mm limit is between the points where no water is held in pore spaces between particles and where water is weakly held in the pores. The lower limit of the 2- to 0.2-mm is the point where water is held in the pores by the forces of capillary attraction. The lower limit of the 0.2- to 0.02-mm fraction is given the theoretical significance that smaller particles cannot be seen by the naked eye; do not have the usual properties of sand; and can be coagulated to form the *crumbs* that are so significant in the mechanical handling of soils, i.e., there are the limits between dry sand which gives poor soils, and adequately moist sand, which forms productive sandy soils. The lower limit of the 0.02- to 0.002-mm fraction is established on the basis that particles smaller than 2 μm (clay) exhibited Brownian movement in aqueous suspension. Capillary movement of water is very slow for <2-μm particles, and the properties of *stiff* clays are strongly manifested. Thus, silt is visualized as a range of particle-sizes from the point where sand begins to assume some clay-like properties to the upper limit of clay.

Atterberg definition of clay, scientific justification: The Atterberg definition of clay as a soil separate with an upper size limit of 2 μm has scientific justification in mineralogical studies of soils (Marshall, 1935; Robinson, 1936; Truog et al., 1936). Robinson (1936) determined that the <2-μm fraction is primarily composed of colloidal products of weathering and is truly the chemically active portion of the soil. Marshall (1935) and Truog et al. (1936) found that very few unweathered primary minerals exist in the <2-μm fraction. Baver (1956) later modified the definition of clay by Oden (1921-1922) as follows: "Clays are disperse systems of the colloidal products of weathering in which secondary particles of smaller dimensions than 2 μm predominate."

USDA classification system, historical: In 1896, investigators in the USDA Bureau of Soils modified the beaker method developed by Osborne. They extended the separation of the smallest particles from 0.1 to 0.005 mm (5 μm) and gave the latter limit the designation of clay. The choices of the different limits were arbitrarily made, based apparently on the convenience of calibration with the particular eyepiece micrometer that was used (Baver, 1956), as illustrated by the following statements: "With the microscope used in this Division the 1-in eyepiece and 3/4-in objective, three of the 0.1 mm spaces of the eyepiece micrometer measure 0.05 mm on the stage. With the same eyepiece and 1/5-in objective, two spaces of the micrometer are equal to 0.01 mm, and one space to 0.005 mm. These three values are sufficient for the beaker separation" (Whitney, 1896). This classification of soil separates was used in the United States until 1937.

USDA classification system, revisions, clay: In 1937, the USDA Bureau of Chemistry and Soils changed the size limits for clay from <5 to <2 μm. It was hoped that this change to 2 μm as the upper limit for clay would make the data from mechanical analysis more useful by effecting a better correlation between field textural classification and classification from the data of mechanical analysis (Soil Science Society of America, 1937). The reduction in size limits tended to reduce the percentage of clay, thus offsetting, in part, the higher percentage obtained by modern dispersion methods (Soil Science Society of America, 1937). Additionally, this change made the definition for the clay separate the same for the USDA and International classification systems.

USDA classification system, revisions, silt: In 1937, the Bureau of Chemistry and Soils also changed the size limits for silt to that fraction between 0.002 and 0.05 mm (2 and 50 μm). In addition, an extra pipetting at 0.02 mm (20 μm) was added, making it possible to compare data with those reported under either the former American system or the International system (Soil Science Society of America, 1937). The split at 20 μm is a class limit between the sand and silt fractions in the International system proposed by Atterberg (1912). The split at 20 μm is the class limit between fine silt and coarse silt in the USDA classification system.

Particle-size distribution analysis, objectives: Particle-size analysis (mechanical analysis) consists of isolating various particle sizes or size increments and then measuring the amount of each size-fraction. The major features of PSDA include the destruction or dispersion of soil aggregates <2 mm in diameter into discrete units by chemical, mechanical, or ultrasonic means followed by the separation or fractionation of particles according to the size limits by sieving and sedimentation (Gee and Bauder, 1986). The primary objectives of dispersion are the removal of cementing agents, rehydration of clays, and the physical separation of individual soil particles (Skopp, 1992). Chemical dispersion usually involves the use of hydrogen peroxide and sodium hexametaphosphate. The hydrogen peroxide oxidizes the organic matter. The sodium hexametaphosphate complexes any calcium in solution and replaces it with sodium on the ion exchange complex, which results in the repulsion of individual particles (Skopp, 1992). Upon completion of the chemical treatments, mechanical agitation is used to enhance separation of particles and facilitate fractionation. Fractionation data provide the size or range of sizes that a measurement represents and the frequency or cumulative frequency with which the size occurs. The most common methods of fractionation are sieving and sedimentation by the hydrometer or pipet method. The Kilmer and Alexander (1949) pipet method was chosen by the USDA Soil Conservation Service (now the USDA/NRCS) because it is reproducible in a wide range of soils.

Particle-size distribution analysis, interferences: The sedimentation equation is derived from Stokes' Law and relates the time of settling to the particle size sampled. The sedimentation equation follows.

Equation 3.1.2.1:

$$v = 2r^2g(\rho_s-\rho_l)/(9\eta)$$

where
v = Velocity of fall
r = Particle radius
g = Acceleration due to gravity
ρ_s = Particle density
ρ_l = Liquid density
η = Fluid viscosity

Assumptions used in applying Stokes' Law to soil sedimentation measurements are as follows:

1. Terminal velocity is attained as soon as settling begins.

2. Settling and resistance are entirely due to the viscosity of the fluid.

3. Particles are smooth and spherical.

4. There is no interaction between individual particles in the solution (Gee and Bauder, 1986; Gee and Or, 2002).

Since soil particles are not smooth and spherical, the radius of the particle is considered an equivalent rather than an actual radius. Effective or equivalent diameters are used to represent either an average value or the replacement of the actual value by a value representative of simplified geometry (Skopp, 1992). The use of effective diameters also emphasizes that determinations of particle sizes are biased by the measurement technique (Skopp, 1992). Identical particles measured by different techniques commonly appear to have different diameters.

Gypsum interferes with PSDA by causing flocculation of particles. The SSL removes gypsum by stirring and washing the soil with reverse osmosis water. This procedure is effective if the soil contains <25 percent gypsum. Currently, the SSL and New Mexico State University (NMSU), an NCSS cooperator, are developing other PSDA methods appropriate for soils with >25 percent gypsum. The SSL is developing a method utilizing 70 percent ethanol and sonication. It has been theorized that the smallest gypsum crystal size that can form in nature is approximately 5 μm. Since 2 μm is the upper limit of clay, gypsum particles in the clay-size fraction would be fractured crystals. This SSL method assumes that clay-size gypsum particles are not a significant fraction and are ignored. New Mexico State University is investigating the use of $CaSO_4$-saturated solutions. For other PSDA laboratory methods developed for gypsic soils, refer to Coutinet (1965), Loveday (1974), Hesse (1974), Matar and Douleimy (1978), and Vieillefon (1979). In general, these other methods call for the pretreatment of gypsic soils with $BaCl_2$ to coat gypsum with $BaSO_4$ prior to PSDA.

Partial flocculation may occur in some soils if excess H_2O_2 is not removed from the soil after its use in organic matter oxidation.

Treatment of micaceous soils with H_2O_2 causes exfoliation of the mica plates and a matting of particles when dried in the oven. Since exfoliation occurs in these soils, a true measurement of fractions is uncertain (Drosdoff and Miles, 1938).

Air-dry versus field-moist samples: The standard SSL procedure for particles <2 mm in diameter is the air-dry method. While a homogenized sample is more easily obtained from air-dry material than from moist material, some soils irreversibly harden when dried; therefore, moist PSDA may be used upon the request of the project coordinator. The phenomenon of aggregation through oven drying or air drying is an important example of irreversibility of colloidal behavior in the soil-water system (Kubota, 1972; Espinoza et al., 1975). Drying such soils decreases the measured clay content. This effect can be attributed to the cementation upon drying (Maeda et al., 1977). The magnitude of the effect varies with the particular soil (Maeda et al., 1977).

Pretreatments: The results of particle-size distribution analysis are dependent on the pretreatments used to disperse the soil. In the standard SSL PSDA method, a 10-g sample of <2-mm air-dry soil is pretreated to remove organic matter and soluble salts. Complete dispersion is often prevented in the presence of cementing agents, such as carbonates, Fe, and Si. In these cases, special pretreatment procedures may be performed upon request on either an air-dry or field-moist sample. However, these special techniques in themselves may interfere with PSDA. These five nonstandard SSL procedures are as follows:

(1) Carbonate removal, pretreatment: Soils high in carbonate content do not readily disperse. Pretreatment of these soils with acid removes the carbonates (Grossman and Millet, 1961; Jackson, 1969; Gee and Bauder, 1986; Gee and Or, 2002). The determination of particle-size distribution after the removal of carbonates is used primarily for studies of soil genesis and parent material. The removal of carbonates with 1 N NaOAc (pH 5) results in sample acidification. This pretreatment can destroy the primary mineral structure of clay (Gee and Bauder, 1986).

(2) Iron removal, pretreatment: Iron and other oxides coat and bind particles of sand, silt, and clay and form aggregates. Soils with iron cementation do not readily disperse. The iron oxides are removed using bicarbonate-buffered sodium dithionite-citrate solution (Mehra and Jackson, 1960; Gee and Bauder, 1986; Gee and Or, 2002). If in the removal of iron the temperature of the water bath exceeds 80 °C, elemental S can precipitate (Mehra and Jackson, 1960). This pretreatment can destroy primary mineral grains in the clay fraction (El-Swaify, 1980).

(3) Silica removal, pretreatment: Soils that are cemented by Si do not completely disperse with hydrogen peroxide pretreatment and sodium hexametaphosphate. A pretreatment with a weak base dissolves the Si bridges and coats and increases the soil dispersion. The determination is used for soil parent material and genesis studies. The effects of Si removal with 0.1 N NaOH on the clay fraction and particle-size distribution are unknown.

(4) Ultrasonic dispersion, pretreatment: Soils that do not completely disperse with standard PSDA can be dispersed using ultrasonic dispersion (Gee and Bauder, 1986; Gee and Or, 2002). Pretreatments coupled with ultrasonic dispersion yield maximum clay concentrations (Mikhail and Briner, 1978). This is a developmental procedure, as no standard method has been adopted using ultrasonic dispersion. Ultrasonic dispersion has been reported to destroy primary soil particles. Watson (1971) summarized studies that reported the destruction of biotite and breakdown of microaggregates by ultrasonic dispersion. Saly (1967), however, reported that ultrasonic vibration did not cause the destruction of the clay crystalline lattice or the breakdown of primary grains. The samples ranged from sandy to clayey soils. The cementing agents represented humus, carbonates, and hydroxides of Fe and Al. No standard procedures have been adopted using ultrasonic dispersion.

(5) Water dispersible, pretreatment: The phenomena of flocculation and dispersion (deflocculation) are very important in determining the physical behavior of the colloidal fraction of soils and thus, indirectly, have a major bearing on the physical properties which soils exhibit (Sumner, 1992). In the standard SSL PSDA method, soils are

pretreated to remove organic matter and soluble salts. Samples are chemically treated with hydrogen peroxide and sodium hexametaphosphate to effect dispersion. Water dispersible particle-size distribution analysis may be determined from a soil suspension without the removal of organic matter or soluble salts or without the use of a chemical dispersant. Upon omitting the procedural steps of removing organic matter or soluble salts, or without the use of a chemical dispersant, the remainder of the standard SSL PSDA method is performed. This method provides a means of evaluating the susceptibility of a soil to water erosion. The degree to which a soil disperses without the oxidation of organic matter, the removal of soluble salts, or the addition of a chemical dispersant may be compared with results from chemical dispersion (Bouyoucos, 1929). The standard SSL water dispersible PSDA for particles <2 mm in diameter is by pipet analysis on air-dry samples. Water dispersible PSDA may also be determined on field-moist samples for those soils that irreversibly harden when dried.

Dispersion and fractionation: Upon completion of the chemical pretreatments (removal of organic matter and soluble salts) in the standard SSL PSDA method, the sample is then dried in the oven to obtain the initial weight, dispersed with sodium hexametaphosphate solution, and mechanically shaken. The sand fraction is separated from the suspension by wet sieving and then fractionated by dry sieving. The clay and fine silt fractions are determined using the suspension remaining from the wet sieving process. This suspension is diluted to 1 L in a sedimentation cylinder and is stirred, and 25-mL aliquots are removed with a pipet at calculated predetermined intervals based on Stokes' Law (Kilmer and Alexander, 1949). Particle density is assumed to be 2.65 g cc^{-1}.

3.1 Particle-Size Distribution Analysis
3.1.2 Particles <2 mm
3.1.2.1 Pipet Analysis

Pipet and sieve analysis: The standard SSL PSDA method is by pipet and sieve analysis. The pipet method was chosen by the USDA/NRCS because it is reproducible in a wide range of soils (Kilmer and Alexander, 1949). The SSL routinely uses this method to determine the soil separates of total sand (0.05 to 2.0 mm), silt (0.002 to 0.05 mm), and clay (<2 μm), with five subclasses of sand (very coarse, coarse, medium, fine, and very fine) and two subclasses of silt (coarse and fine). The coarse silt fraction is a separate with 0.02- to 0.05-mm particle diameter. The fine silt fraction is a soil separate with 0.002- to 0.02-mm particle diameter. In addition to the routine soil separates of sand, silt, and clay, the SSL determines the fine-clay and/or carbonate-clay fractions, depending on analytical requests and properties of the sample. The fine-clay fraction consists of mineral soil particles with an effective diameter of <0.0002 mm (<0.2 μm). Carbonate clay is a soil separate with <0.002 mm (<2 μm) particle diameter. The SSL reports these various soil separates as weight percentages on a <2-mm basis.

PSDA, process: In SSIR No. 42, Version 3.0 (Soil Survey Staff, 1996), the analysis of <2-mm size fractions that were not routinely reported (e.g., fine-clay and/or carbonate-clay) as well as nonroutine pretreatment and dispersion techniques were described as stand-alone methods. In SSIR No. 42, Version 4.0 (Soil Survey Staff, 2004), these procedures are described more as a procedural process. This approach is appropriate in that certain procedural steps may be modified, omitted, or enhanced by the investigator, depending on the properties of the sample and on the requested analyses. The process by which specific procedural steps are selected for sample analysis is based upon knowledge or intuition of certain soil properties or related to specific questions, e.g., special studies of soil genesis and parent material. In the following section, the soil separates analyzed by the SSL are further defined and discussed.

PSDA, measurements: In the following section, the SSL PSDA method for particles <2 mm in diameter by sieve and pipet analysis is described. The hydrometer method as used by the USDA Soil Mechanics Laboratory as well as by NRCS soil survey offices also is discussed.

3.1   Particle-Size Distribution Analysis
3.1.2   Particles <2 mm
3.1.2.1   Pipet Analysis
3.1.2.1.1   Total Clay, <0.002 mm (<2 μm)

Total clay, definition: *Clay* is a soil separate with a particle diameter of <0.002 mm (<2 μm). The SSL determines total clay by pipet analysis. The total clay value determined by the SSL includes the carbonate-clay and fine-clay fractions. *Clay* is also used to define a class of soil minerals. Refer to Table 3.1.2.1.1.1 (Sumner, 1992) for particle dimensions, thickness, and surface area of some clay minerals.

Table 3.1.2.1.1.1   Comparison of clay particle diameter dimensions, thickness, and surface area[1]

Mineral	Particle dimensions	Particle thickness	Surface area
	μm	μm	$m^2\ g^{-1}$
Montmorillonite	0.03*	0.001	600–800
Micas	0.3–1	0.02–0.07	60–200
Vermiculite	0.03	0.001	400–800
Hydroxy-interlayered vermiculite	0.2–1*	0.02–0.07	80–150
Kaolinite	0.3–2*	1–4	5–40
Halloysite, tubular	0.07*	0.04–1**	21–43
Halloysite, spheroidal	0.02–1		
Goethite	0.02*	0.05–0.1**	30–200
Hematite	0.02–0.05	0.01–0.02	50–120
Gibbsite	0.1	0.005	10–30
Allophane hollow spheres	0.003–0.005*		1000
Imogolite hollow filiform	0.002–0.003	1–3	1000

[1] Sumner, M.E. 1992. "The Electrical Double Layer and Clay Dispersion," pp. 1-32 in *Soil Crusting: Chemical and Physical Processes*. M.E. Sumner and B.A. Stewart, eds. Taylor & Francis Group LLC–Books. Reproduced with permission of Taylor & Francis Group LLC–Books.
* Diameter
** Length

Clay percentage, volumetric, whole-soil basis, calculation: Clay percentages or any data may be calculated volumetrically on a whole-soil basis according to horizon thickness. Refer to Appendix 2 (Wildmesa Pedon) for laboratory data used in the following clay percentage calculation based on control section 15-65 cm.

Equation 3.1.2.1.1.1:

Product A = (Hcm x ρ_{B33} x Cm)

Equation 3.1.2.1.1.2:

Product B = (Product A x Clay)

where
Hcm = Horizon thickness, cm
ρ_{B33} = Bulk density at 33-kPa water content on <2-mm soil basis (g cm^{-3})
Clay = Weight percentage of clay on <2-mm soil basis
Cm = Coarse fragment conversion factor. If no coarse fragments, Cm = 1. If coarse
    fragments are present, calculate Cm as follows:

Equation 3.1.2.1.1.3:

Cm = [Vol $_{moist<2\text{-mm fabric}}$ (cm^3)]/[Vol$_{moist\ whole\ soil}$ (cm^3)]

OR (alternatively)

Equation 3.1.2.1.1.4:

Cm = (100–Vol$_{>2mm}$)/100

where
Vol$_{>2mm}$ = Volume percentage of the >2-mm fraction

Equation 3.1.2.1.1.5:

Weighted Average = (Sum of Products B)/(Sum of Products A)

where
Sum of Products A = Sum of (Hcm x ρ_{B33} x Cm) for all soil horizons
Sum of Products B = Sum of (Product A x Clay) for all soil horizons

Example: Refer to Appendix 2 (Wildmesa Pedon, S89CA027004)

Horizon	Depth (cm)	Hcm (cm)	ρ_{B33} (g cm^{-3})	Cm	Product A	Clay (%)	Product B
2Bt	15-46	31	1.45	0.99	44.50	34.9	1553.07
2Btk	46-74	19	1.38	1.00	26.22	38.1	998.98
SUM		50			70.72		2552.05

Weighted Average = 2552.05/70.72 = 36 percent clay

Refer to the discussion in the introduction to this manual, Section 1.2.4, for an example calculation of clay percentage, weighted average, as it appears under "Pedon Calculations" on the SSL Primary Characterization Data Sheet. In the pedon calculations of weighted averages, bulk density and Cm values are not used in the equations.

3.1   Particle-Size Distribution Analysis
3.1.2   Particles <2 mm
3.1.2.1   Pipet Analysis
3.1.2.1.2   Total Silt, 0.002 to 0.05 mm

Total silt, definition: *Total silt* is a soil separate with 0.002- to 0.05-mm particle diameter. Total silt is the sum of the fine silt and coarse silt fractions. The SSL determines the fine silt separate by pipet analysis and the coarse silt separate by difference. Total silt is reported as a weight percentage on a <2-mm basis.

3.1   Particle-Size Distribution Analysis
3.1.2   Particles <2 mm
3.1.2.1   Pipet Analysis
3.1.2.1.3   Total Sand, 0.05 to 2.0 mm

Total sand, definition: *Total sand i*s a soil separate with 0.05- to 2.0-mm particle diameter. The SSL determines the total sand fraction by sieve analysis. Total sand is the sum of the very fine sand (VFS), fine sand (FS), medium sand (MS), coarse sand (CS), and very coarse sand (VCS) fractions. The rationale for five subclasses of sand and the expansion of the texture classes of sand, e.g., sandy loam and loamy sand, is that the sand separates are the most visible to the naked eye and the most detectable by "feel" by the field soil scientist. Total sand is reported as a weight percentage on a <2-mm basis.

Total sand, weight to volume conversion: Particle-size analysis data by the standard SSL procedure are reported as a weight percentage on a <2-mm *mineral* soil basis, i.e., free of organic matter and salts. Using total sand as an example, PSDA data can be converted from a weight to volume basis as follows:

Equation 3.1.2.1.3.1:

$$V_{sand} = [Wt_{sand} \times \rho_{B33} \times (1 - (V_{om}/100))]/2.65 \text{ g cm}^{-3}$$

where
V_{sand} = Volume percentage of sand (0.05- to 2.0-mm diameter) on <2-mm soil basis
Wt_{sand} = Weight percentage of sand (0.05- to 2.0-mm diameter) on <2-mm soil basis
ρ_{B33} = Bulk density at 33-kPa water content on <2-mm soil basis (g cm^{-3})
2.65 = Assumed particle density for sand (g cm^{-3})
V_{om} = Volume percentage of organic matter on <2-mm basis. Calculate V_{om} as follows:

Equation 3.1.2.1.3.2:

$$V_{om} = (Wt_{oc} \times 1.724 \times \rho_{B33})/1.1 \text{ g cm}^{-3}$$

where
Wt_{oc} = Weight percentage of organic C on <2-mm soil basis
1.724 = "Van Bemmelen factor"
ρ_{B33} = Bulk density at 33-kPa water content on <2-mm soil basis (g cm^{-3})
1.1 = Assumed particle density of organic matter (g cm^{-3})

3.1   Particle-Size Distribution Analysis
3.1.2   Particles <2 mm
3.1.2.1   Pipet Analysis
3.1.2.1.4   Fine Clay, <0.0002 mm (<0.2 μm)

Fine clay, definition: The *fine clay* fraction consists of mineral soil particles with an effective diameter of <0.0002 mm (<0.2 μm). Fine clay amounts are never greater than total clay. The SSL determines the fine clay fraction by centrifuging, followed by pipet analysis using the soil suspension from the standard PSDA method. The time of centrifugation is determined from the following equation modified from Stokes' Law (Jackson, 1969).

Equation 3.1.2.1.4.1:

$$t_m = (63.0 \times 10^8 \eta \log (rs^{-1})) (N_m^2 \, D\mu^2 \, \Delta\rho)^{-1}$$

where
t_m = Time in minutes
η = Viscosity in poises
r = Radius in cm from center of rotation to sampling depth (3 cm + s)
s = Radius in cm from center of rotation to surface of suspension
N_m = rpm (1500)
$D\mu$ = Particle diameter in microns (0.2 μm)
$\Delta\rho$ = Difference in specific gravity between solvated particles and suspension liquid
63.0×10^8 = Combination of conversion factors for convenient units of time in minutes, t_m, N_m as rpm, and particle diameter in microns, $D\mu$

Colloidal clay, definition: *Colloids* are small particles which, due to their size, tend to remain suspended in solution and exhibit unique physical and chemical properties compared to other soil particle-size classes (Bohn et al., 1979). They have a large surface area per unit of mass and are chemically active with an electrical field that extends into the soil solution. Many of the properties that a soil exhibits are related to the types (both inorganic and organic) and amounts of colloidal materials that are present in the soil.

Colloidal clay versus clay: The distinction between *clay* and *colloidal clay* has been debated for many years. Some early separations set the upper limit of the colloidal range at 0.5 µm (Freundlich, 1926), at 1 µm (Brown and Byers, 1932; Bray, 1934), or at 0.2 µm (according to many colloidal chemists at the time). Prior to 1937, the U.S. Bureau of Soils and Chemistry termed particles <0.002 mm (<2 µm) as colloids (Soil Survey Staff, 1951). Other investigators (DeYoung, 1925; Joseph, 1925) stated that clay and colloidal contents were identical if the sample was completely dispersed. Baver (1956) considered 0.1 to 0.2 µm a more reliable estimate of the upper limit of the colloidal range. Such colloidal material not only conformed more closely to the accepted standards of colloidal chemistry but also possessed a much greater chemical and physical activity per unit weight than coarser fractions (Baver, 1956). The 0.0002-mm (<0.2-µm) separate reported as fine clay most closely corresponds to those estimates of the upper colloidal range proposed by Baver (1956) and others. More recently, the 0.001-µm (1 nm) to 1-µm range has been used to define colloidal particles (van Olphen, 1977; Singer and Munns, 1987). It is difficult to establish exact size limits for colloidal soil particles since activity of a colloid is determined not only by the composition, size, and shape of the colloid but also by the concentration and composition of the soil solution.

Fine clay, taxonomic significance: The percentage of fine clay is determined for soils that are suspected of having illuviated clay or argillic horizons or as a tool to help explain soil genesis. As soil genesis occurs, an argillic horizon may form through clay translocation or the neoformation of minerals. The fine clay to total clay ratio is used as an index of argillic development; i.e., this ratio is normally one-third higher than in the overlying eluvial horizon or in the underlying horizon (Soil Survey Staff, 2010).

3.1   Particle-Size Distribution Analysis
3.1.2   Particles <2 mm
3.1.2.1   Pipet Analysis
3.1.2.1.5   Carbonate Clay, <0.002 mm (<2 µm)

Carbonate clay, definition: *Carbonate clay* is a soil separate with <0.002-mm (<2-µm) particle diameter. Using the soil suspension from the standard PSDA method, the SSL determines the carbonate-clay fraction by pipet analysis followed by acid treatment in a closed system. The pressure is measured with a monometer and related linearly to the CO_2 content in the carbonates. This determination is semiquantitative as it is assumed that all of the carbonates in a soil sample are converted to CO_2; i.e., not only the carbonates of Ca but also the carbonates of Mg, Na, and K react with the acid.

Carbonate clay, soil-related factors: The carbonate-clay fraction is considered important in PSDA because clay-size (<2 µm) carbonate particles have properties that are different from those of noncarbonate clay. The cation-exchange capacity of carbonate clay is very low compared to that of noncarbonate clay. Saturation percentage, Atterberg limits, and 15-bar water retention for carbonate clay are ≈ 2/3 the corresponding values for the noncarbonated clays (Nettleton et al., 1991). Since carbonate clay is a diluent, it is often subtracted from the total clay in order to make inferences about soil genesis and clay activities. Total clay is routinely estimated and carbonates measured by soil scientists in the field. Generally, the amount of carbonate clay, as estimated by hand texture, is underestimated by ≈ 1/2 (Nettleton et al., 1991). In *Keys to Soil Taxonomy*, carbonates of clay size are not considered to be clay for soil texture but are treated as silt in all particle-size classes (Soil Survey Staff, 2010).

3.1 Particle-Size Distribution Analysis
3.1.2 Particles <2 mm
3.1.2.1 Pipet Analysis
3.1.2.1.6 Fine Silt, 0.002 to 0.02 mm

Fine silt, definition: *Fine silt* is a soil separate with 0.002- to 0.02-mm particle diameter. The SSL determines the fine-silt fraction by pipet analysis. The fine silt is reported as a weight percentage on a <2-mm basis.

3.1 Particle-Size Distribution Analysis
3.1.2 Particles <2 mm
3.1.2.1 Pipet Analysis
3.1.2.1.7 Coarse Silt, 0.02 to 0.05 mm

Coarse silt, definition: *Coarse silt* is a soil separate with 0.02- to 0.05-mm particle diameter. The SSL determines the coarse-silt fraction by difference. Coarse silt = (100 − (% total clay + % fine silt + % total sand). The 0.02 mm (20 μm) is the break between sand and silt in the International classification system. The particle-size separation at 20 μm also has significance in optical microscopy, as this class limit represents the optical limits of the polarizing light microscope. The coarse silt is reported as a weight percentage on a <2-mm basis.

3.1 Particle-Size Distribution Analysis
3.1.2 Particles <2 mm
3.1.2.1 Pipet Analysis
3.1.2.1.8 Very Fine Sand, 0.05 to 0.10 mm

Very fine sand, definition: *Very fine sand* is a soil separate with 0.05- to 0.10-mm particle diameter. The SSL determines the very fine sand fraction by sieve analysis. The SSL reports the very fine sand as a weight percentage on a <2-mm basis.

Very fine sand, taxonomic significance: Particle-size classes are a compromise between engineering and pedologic classes (Soil Survey Staff, 2010). In engineering classifications, the limit between sand and silt is a 0.074-mm diameter. The break between sand and silt is 0.05 and 0.02 mm in the USDA and International classification systems, respectively. In engineering classes, the very fine sand (VFS) separate is split. In particle-size classes of the U.S. soil taxonomic system, the VFS is allowed to *float*; i.e., the VFS is treated as sand if the texture is fine sand, loamy fine sand, or a coarser class and is treated as silt if the texture is very fine sand, loamy very fine sand, sandy loam, silt loam, or a finer class (Soil Survey Staff, 2010).

3.1 Particle-Size Distribution Analysis
3.1.2 Particles <2 mm
3.1.2.1 Pipet Analysis
3.1.2.1.9 Fine Sand, 0.10 to 0.25 mm

Fine sand, definition: *Fine sand* is a soil separate with 0.10- to 0.25-mm particle diameter. The SSL determines the fine sand fraction by sieve analysis. The SSL reports the fine sand as a weight percentage on a <2-mm basis.

3.1 Particle-Size Distribution Analysis
3.1.2 Particles <2 mm
3.1.2.1 Pipet Analysis
3.1.2.1.10 Medium Sand, 0.25 to 0.50 mm

Medium sand, definition: *Medium sand* is a soil separate with 0.25- to 0.50-mm particle diameter. The SSL determines the medium sand fraction by sieve analysis. The SSL reports the medium sand as a weight percentage on a <2-mm basis.

3.1 Particle-Size Distribution Analysis
3.1.2 Particles <2 mm
3.1.2.1 Pipet Analysis
3.1.2.1.11 Coarse Sand, 0.5 to 1.0 mm

Coarse sand, definition: *Coarse sand* is a soil separate with 0.5- to 1.0-mm particle diameter. The SSL determines the coarse sand fraction by sieve analysis. The SSL reports the coarse sand as a weight percentage on a <2-mm basis.

3.1 Particle-Size Distribution Analysis
3.1.2 Particles <2 mm
3.1.2.1 Pipet Analysis
3.1.2.1.12 Very Coarse Sand, 1.0 to 2.0 mm

Very coarse sand, definition: *Very coarse sand* is a soil separate with 1.0- to 2.0-mm particle diameter. In 1947, the class name for the 1.0- to 2.0-mm fraction in the USDA classification system was changed from fine gravel to very coarse sand. The SSL determines the very coarse sand fraction by sieve analysis. The SSL reports the very coarse sand as a weight percentage on a <2-mm basis.

3.1 Particle-Size Distribution Analysis
3.1.2 Particles <2 mm
3.1.2.2 Hydrometer Analysis

Hydrometer method, Bouyoucos: The *hydrometer method,* like the pipet method, depends fundamentally on Stokes' Law (Gee and Bauder, 1986). The hydrometer method is based on the decrease in density that occurs at a given depth as a dispersed suspension settles. The rate of decrease in density at any given depth is related to the settling velocities of the particles, which in turn are related to their sizes (Kilmer and Alexander, 1949). Since the introduction of the hydrometer by Bouyoucos (1927), this method has been widely adopted for particle-size analysis of soils and other materials (Kilmer and Alexander, 1949). The primary reasons for the popularity of this method have been the rapidity with which a mechanical analysis can be made and the simplicity of the equipment required (Kilmer and Alexander, 1949). Hydrometer readings at 40 s and 2 h have been used to estimate sand and clay percentages, respectively.

The correlations between sand and clay and the 40-s and 2-h readings are empirical (Gee and Bauder, 1986). Basic sedimentation theory indicates that the 2-h reading is a better estimate of the 5-μm limit than it is of the 2-μm limit (Gee and Bauder, 1986). Errors in clay content using the 2-h reading often exceed 10 percent by weight for clay soils, and differences between sieve and the 40-s hydrometer measurement often exceed 5 percent by weight (Gee and Bauder, 1979, 1986). These errors are primarily attributable to the fact that in 1937 the Bureau of Chemistry and Soils changed the size limits for silt and clay. With the change to 2 μm as the upper limit for clay and the lower limit for silt, a better correlation was determined between field textural classification and laboratory classification by mechanical analysis (Soil Science Society of America, 1937). The reduction in size limits to <2 μm tended to decrease the clay percentage, thus offsetting in part the higher percentage obtained by modern dispersion methods (Soil Science Society of America, 1937).

Hydrometer, ASTM method: Over time, modifications to the Bouyoucos hydrometer specifications and procedure have been suggested and adopted (Casagrande, 1934; Bouyoucos, 1951, 1962; ASTM, 1958, 1963; Day, 1956, 1965; Gee and Bauder, 1986). The NRCS Soil Mechanics Laboratories (SML) in Lincoln, Nebraska, and Fort Worth, Texas, use ASTM-designated methods for particle-size analysis by hydrometer. Refer to ASTM D 422-63 for the standard test method for particle-size analysis of soils (ASTM, 2008i). This test method covers the quantitative determination of the distribution of particle sizes in soils. The distribution of particle sizes >0.075 mm (retained on No. 200 sieve) is determined by sieving. The distribution of particle sizes <0.075 mm is determined with a hydrometer by a sedimentation process. Separation may be made on the No. 4 (4.75 mm), No. 40 (0.425 mm), or No. 200 (0.075 mm) sieve instead of the No. 10 (2 mm) sieve. The procedure specifies an ASTM hydrometer, graduated to read in specific gravity of the suspension or g L^{-1} suspension and conforming to the requirements for hydrometers 151H and 152H in Specifications E 100. Dimensions of both hydrometers are the same; the scale is the only item of difference.

3.1 Particle-Size Distribution Analysis
3.1.2 Particles <2 mm
3.1.2.3 Water Dispersible

Water dispersible PSDA: The phenomena of flocculation and dispersion (deflocculation) are very important in determining the physical behavior of the colloidal fraction of soils and thus, indirectly, have a major bearing on the physical properties which soils exhibit (Sumner, 1992). Particle-size distribution analysis by mechanical means in distilled water without the removal of organic matter and soluble salts and without the use of a chemical dispersant is referred to as water dispersible PSDA.

Water dispersible PSDA, pipet and hydrometer methods: On some SSL data sheets, water dispersible PSDA by hydrometer is not designated by a specific SSL method but rather by "SML" (Soil Mechanics Laboratory). There is no method documented in SSIR No. 42 for water dispersible PSDA by hydrometer, as the standard water dispersible PSDA is determined by the pipet method at the SSL. Refer to Soil Survey Staff (2004) for the standard water dispersible PSDA method by pipet. Also refer to ASTM D 4221-99 for the standard test method for dispersive characteristics of clay soil by double hydrometer (ASTM, 2008h). This test method, when used in conjunction with a test performed by ASTM D 422-63 on a duplicate soil sample, provides an indication of the natural dispersive characteristics of clay soils. This test method is applicable only to soils with a plasticity index (PI) >4, as determined in accordance with ASTM D 4318-05 (ASTM, 2008k), and to soils in which >12 percent of the soil fraction is finer than 0.005 mm, as determined in accordance with Method 422-63. Test method ASTM D 4221-99 (ASTM, 2008h) is similar to ASTM D 422-63, except that the former determines the percent soil particles <0.005 mm in diameter in a soil-water suspension without mechanical agitation and without the addition of a dispersing agent. The amount of <0.005-mm particles by this method compared with the total amount of <0.005-mm particles as determined by ASTM D 422-63 is a measure of the dispersive characteristics of the soil. Test method ASTM D 4221-99 does not identify all dispersive clay soils.

3.1   Particle-Size Distribution Analysis
3.1.2   Particles <2 mm
3.1.2.3   Water Dispersible
3.1.2.3.1   Total Clay, <0.002 mm (<2 μm)

Water dispersible clay, definition: *Water dispersible clay (WDC)* is a soil separate with <0.002-mm (<2-μm) particle diameter. The clay percentage determined by mechanical means without the removal of organic matter and soluble salts and without the use of a chemical dispersant is referred to as WDC. The SSL determines the WDC by the pipet method. The SSL reports the WDC as a weight percentage on a <2-mm basis.

Water dispersible clay, application: Middleton (1930) suggested a relationship between the easily dispersed silt and clay (dispersion ratio) and soil erodibility. The WDC measurement was evaluated as a predictor in the USDA/NRCS Water Erosion Prediction Program (WEPP). This measurement has also been suggested as a parameter for evaluating positive charge in tropical soils (Gillman, 1973). The WDC is a significant factor in the physical condition of a soil in that many of the soil properties that affect soil erodibility, aggregate stability, and crust formation are those properties that affect the propensity of the clay fraction to disperse in water (Brubaker et al., 1992). These properties include organic matter content; relative amounts of various cations on exchange sites; presence of soluble salts, such as gypsum; clay mineralogy; nature of charge on colloids; and antecedent soil moisture content (Brubaker et al., 1992). Water dispersible clay values can be useful when relationships can be developed between these data and total clay amounts, as determined by standard PSDA.

Water dispersible clay, data assessments: In a study of 54 sites across the United States, representing approximately 60 soil series, the soil property most strongly correlated with WDC was total clay; less significantly correlated properties included 1500-kPa water content, dithionite-citrate extractable Fe and Al, coefficient of linear extensibility, Wischmeier's M, the content of very fine sand, the ratio of cation-exchange capacity (CEC) to total clay, Bouyoucos's clay ratio, and the CEC (Brubaker et al., 1992). The best model for estimating WDC when all of the data were used in the regression analysis included the total clay content and the ratio of CEC corrected for organic (CCEC) to total clay ($R^2 = 0.723$); however, sorting the data by the ratio of CCEC to total clay instead of including it in the model significantly improved the overall fit of the model ($R^2 = 0.879$) (Brubaker et al., 1992). Results seem to indicate that low-activity clays are about twice as dispersible as high-activity clays.

3.1 Particle-Size Distribution Analysis
3.1.2 Particles <2 mm
3.1.2.3 Water Dispersible
3.1.2.3.2 Total Silt, 0.002 to 0.05 mm

Water dispersible silt, definition: *Water dispersible silt* is a soil separate with 0.002- to 0.05-mm particle diameter. The silt percentage determined by mechanical means without the removal of organic matter and soluble salts and without the use of a chemical dispersant is referred to as water dispersible silt. The SSL determines the water dispersible silt by the pipet method. The SSL reports the water dispersible silt as a weight percentage on a <2-mm basis.

Water dispersible silt, application: Middleton (1930) cited the ratio of water dispersible silt plus clay to total silt plus clay (dispersion ratio) as "probably the most valuable single criterion in distinguishing between erosive and non-erosive soils."

3.1 Particle-Size Distribution Analysis
3.1.2 Particles <2 mm
3.1.2.3 Water Dispersible
3.1.2.3.3 Total Sand, 0.05 to 2.0 mm

Water dispersible sand, definition: *Water dispersible sand* is a soil separate with 0.05- to 2.0-mm particle diameter. The sand percentage determined by mechanical means without the removal of organic matter and soluble salts and without the use of a chemical dispersant is referred to as water dispersible sand. The SSL determines the water dispersible sand by sieve analysis. The SSL reports the water dispersible sand as a weight percentage on a <2-mm basis.

3.1 Particle-Size Distribution Analysis
3.1.2 Particles <2 mm
3.1.2.4 Soil Loss Through Water Erosion and Wind Erosion, Processes, Case Studies, and Major Developments

Soil erosion, processes: *Particle-size distribution (soil texture)* is a major soil property that affects a soil's susceptibility to erosion and as such is a key parameter in any soil erosion prediction model. For these reasons, the process of soil loss is described and references to case studies/datasets are presented as evidentiary examples of the actions/practices that have promoted or diminished this soil process. Major developments in the knowledge, science, and technology of soil and water conservation also are discussed.

Soil erosion has been defined as the detachment or breaking away of soil particles from land surface by some erosive agent (e.g., water and wind) and the subsequent transportation of the detached particles to another location (Flanagan, 2002). Soil erosion, a major cause of the degradation of water quality throughout the United States, is the result of several factors, including rainfall intensity, steepness of slope, length of slope, vegetative cover, and management practices (O'Geen et al., 2006). The inherent properties of a soil also play a major role in erosion. This intrinsic property is the soil's *erodibility* (O'Geen et al., 2006). Four major soil properties govern erodibility: texture (particle-size distribution), structure, organic matter content, and permeability. These properties have been identified through nationwide studies performed by the USDA Agricultural Research Service (ARS) using rainfall simulation tests (USDA/NRCS, 2009b), and the soil survey staff measures these properties and uses them to predict the potential of the soil for erosion by water (O'Geen et al., 2006). This potential is called the K factor, or soil erodibility.

Soil erosion is both a human-induced process and a natural process, the latter of which is a critical factor in the formation of soil from rock parent material. Human-induced erosion is caused by removal or reduction of plant and residue cover related to such activities as crop removal, tillage, and livestock grazing. Water erosion can occur on rainfed and irrigated lands and can result from snowmelt. Wind erosion has been driven by cycles of climatic change over geologic time, resulting in transport and accumulation of eolian sediments. It is a common phenomenon today in regions of arid and semiarid climates and sparse vegetation (Busacca and Chandler, 2002). Erosion has a range of impacts, both onsite and offsite. It removes fertile topsoil, organic matter, and nutrients, thereby decreasing the tilth, water-holding capacity, and general productivity of a soil for onsite agricultural production (Flanagan, 2002). In addition, erosion impacts pollution of natural waters and environments through the transport of agricultural inputs.

Soil erosion, case studies: The Dust Bowl of the 1930s is perhaps the most famous area in the U.S. for historians studying erosion (Bonnifield, 1979; Worster, 1979; Hurt, 1981). This area encompasses western Kansas, southeastern Colorado, northeastern New Mexico, and the panhandles of Oklahoma and Texas. It was during this period that the USDA Soil Conservation Service (SCS) was created by public law (1935), which declared soil erosion as a menace to the natural welfare. In 1936, the USDA/SCS began cost sharing for soil conservation practices. The 3- to 10-year term contracts called for a number of conservation practices in the Great Plains, e.g., field and wind stripcropping, windbreaks, waterways, terraces, diversions, erosion-control dams and grade-stabilization structures, water-spreading systems, the reorganization of irrigation systems, well and water storage facilities, the use of fencing to distribute grazing, and control of shrubs. Perhaps the most far-reaching USDA/SCS recommendation in the Great Plains was the conversion of cropland on highly erodible soils back to grassland, thereby improving rangeland and pasture and further diversifying farming-ranching (Helms, 1990). This recommendation was based on surveys in the 1930s showing that failure in the Great Plains was related to two groups: strict dryland farmers who had no cattle, and cattlemen who grew no feed. While technology has changed through the years, these essential elements still guide the Great Plains conservation program (Helms, 1990). These new technologies impacting the Great Plains included conservation tillage, which was introduced in the 1970s. In 1988, the acreage planted using conservation tillage in the southern and northern plains was estimated at 23 percent and 32 percent, respectively (National Association of Conservation Districts, 1988).

The Dust Bowl era was not the last of the episodes of wind erosion in the Great Plains. The drought that struck the Great Plains in the 1950s led not only to emergency drought measures but also eventually to new long-term conservation programs and policies (Helms, 1990). The USDA/SCS recommended that farmers be assisted in converting cropland back to grassland by paying 50 percent of the cost, provided that these lands remained in grass for at least 5 years (USDA/SCS, 1955). While dust storms are not common generally, several years of drought can set the stage for these storms. Such a situation occurred in Kansas on March 14, 1989 (Helms, 1990). The 1988-1989 wind erosion season was the worst since 1954-1955, when SCS started keeping records (USDA, 1989).

Another area in the United States severely impacted by erosion is the Southeast. While soil loss in semiarid areas of cropland is primarily the result of wind erosion, soil loss in the Southeast is most directly linked to water erosion. The history of the Southeastern United States is largely a story of depletion, erosion, runoff, and farm abandonment that can still be witnessed today in the behavior of the soils of this region (Miller and Radcliffe, 1992). The intensive cultivation history in this area over the last two centuries has transformed an area of deep soils and clear waters into a marginal agricultural region of exposed subsoils and turbid, sediment-clogged streams (Miller and Radcliffe, 1992). The study by Trimble (1974) is one of the classic investigations of the long-term effects of erosion (1700-1970) on some highly erodible soils in this area under continuous intensive cultivation. Trimble reported that soils of the Southern Piedmont were stripped of their topsoil and dissected by gullies, with the entire region (about 150,000 km^2) having lost an average of 0.17 m of topsoil. Trimble attributed this erosion to the advent of clean-cultivated cash crops, e.g., tobacco and cotton, and the exploitative nature of land clearing. The decline in erosive land use in this area from 1920-70 was largely due to a decline in agriculture, mainly resulting from the unsuitability of small, sloping, and irregular fields for modern machinery. Also, crops formerly grown on the Piedmont could be grown more economically elsewhere; nonfarm employment was available both within and outside the region; and in some cases farms were so damaged by erosion that continued cultivation was no longer

51

profitable. Poor physical land condition was the reported reason for 31 percent of the land abandonment in the Southern Piedmont between 1930 and 1940 (U.S. Bureau of the Census, 1943). The decline in agriculture from 1919 to 1967 is evidenced by row crop acreages, e.g., 159,000 acres versus 9,000 acres, respectively, in Chambers County, Alabama; 91,000 acres versus 1,000 acres, respectively, in Jasper County, Georgia; 137,000 acres versus 9,000 acres, respectively, in Gwinnett County, Georgia; and 33,000 acres versus 4,000 acres, respectively, in Stokes County, North Carolina (Trimble, 1974). After 1935, soil conservation management (e.g., contour plowing, terracing, crop rotation) improved dramatically and became widespread across the Southern Piedmont, primarily due to the efforts of the USDA/SCS (Trimble, 1974). Despite this apparent success in this area, there were few reports released from 1935 to 1970 relating quantitative measures to the effectiveness of the applied conservation practices.

Over the years, estimates of soil loss in North America as a result of erosion have differed or are variable over time, reflective of the types of agricultural practices and applications of conservation techniques (Geiger, 1957; De Bivort, 1975; Pimentel et al., 1976; Howard, 1981; Larson, 1981; Crosson, 1985). Estimates of soil loss in North America have been derived from national, regional, or site-specific studies, and the site-specific ones are often related to areas of increased concern, sometimes termed "hot spots." In the Cordillera region of Canada, Stichling (1973) estimated the soil loss rate to be greater than 3 t ha^{-1} yr^{-1}. As a result of logging and clearcutting and mining, the Oldman River basin in Alberta was shown to have stream instability; undercut, massive landslides; gullying; and bank erosion (Northwest Hydraulic Consultants, Ltd., 1980). Brown (1984) reported that the Mississippi River carries 331 million tons of soil into the Gulf of Mexico each year. Other areas of concern in the United States are areas with sandy soils caused by the lack of binding agents for aggregation; peats in Florida, which may erode when drained, cleared, and cultivated; and light loess soils that occur in some parts of Washington State and parts of the Great Plains (Warren, 2002). Frazier et al. (1983) reported that soils of the Palouse region, which formed in loess, are some of the most vulnerable soils in the United States.

At the national level, for over five decades the USDA/NRCS (formerly the USDA/SCS) has conducted periodic inventories of natural resources. The 1945 Soil and Water Conservation Needs Inventory (CNI), a reconnaissance study, was the foundation for the 1958 and 1967 CNIs, the agency's first efforts to collect data nationally for scientifically selected field sites. The 1975 Potential Cropland Study focused on identifying lands best suited for cultivation. One of the more recent efforts by the NRCS is the National Resources Inventory (NRI), a statistical survey of natural resource conditions and trends on nonfederal land in the U.S. The NRI was conducted every 5 years from 1977 to 1997; the inventory is currently transitioning to an annual process. In 1982, soil loss was estimated at 3.1 billion tons annually on U.S. cropland, with 29 percent of this land eroding at excessive rates; but the estimates for 1997 and 2001 were 1.9 billion tons and 1.8 billion tons, respectively (USDA/NRCS, 2001). From 1982 to 2001, the rate of sheet and rill erosion dropped by almost 41 percent and the rate of wind erosion dropped by 43 percent. The acreage of highly erodible cropland declined from 123.9 million acres in 1982 to 101.1 million acres in 2001. Reductions in soil loss during this 20-year period may be due in part to the enactment of the Food Security Act of 1985, which linked farmers' eligibility for USDA programs (e.g., price support payments and crop insurance) to conservation preservation, especially on highly erodible lands. Title XII of this act was the Conservation Reserve Program (CRP), a program with roots in the Soil Bank Act of 1956. The CRP was a voluntary long-term cropland retirement program which provided participants (farm operators or tenants) with annual per-acre rent plus half the cost of establishing a permanent land cover (usually grass or trees). In exchange, the participant retired highly erodible or environmentally sensitive cropland from production for 10 to 15 years.

The Food Security Act of 1985 was in effect a reversal of the "plant fence row to fence row" philosophy of the 1970s, which in retrospect was detrimental to the gains that conservation programs had made in the previous 40 years (Cain and Lovejoy, 2004). A 1977 Congressional study found that 26 percent of farmers in the Great Plains Conservation Reserve Program had plowed up their newly established grasslands for wheat production after their contracts expired. This study emphased the difficulty of maintaining long-term conservation practices, especially in land retirement programs (Doering, 1997).

In general, soil erosion is more severe in North America than in some countries in Europe, partly as a result of differences in climate, e.g., higher intensity rains and climatic extremes (hot summers, cold winters), which increase the soil's susceptibility to water erosion (Lal, 1990). Other reasons for this difference are related to intensive land use, monocropping without frequent use of soil-conserving cover crops, continuous cropping, and the excessive and often unnecessary use of heavy machinery (Lal, 1990).

In the last 50 years, great strides have been made in the development and application of soil and water conservation techniques in North America. In a somewhat parallel manner, data collection and erosion assessment technologies have also improved during this period. More recently, advances have been made in estimating on-farm economic costs of erosion, e.g., the Productivity Index (PI) model developed by Pierce et al. (1983) and the Erosion Productivity Impact Calculation (EPIC) model (USDA/SCS, 1989). All of these are important ingredients in the ongoing development and understanding of the cause-and-effect relationships of the soil erosion process and the appropriate methods that provide constraint/stress alleviation, restoration, and quality enhancement.

Major developments in knowledge, science, and technology in soil and water conservation: Since 1945, soil and water conservation technologies in the U.S., individually and in combination, have been developed and refined. These technologies are most commonly presented in conservation plans. Earlier efforts to reduce soil erosion were promoted by the Department of the Interior's soil erosion program, which evolved to the USDA/SCS in 1935 and to the USDA/NRCS in 1994. These vegetative or mechanical technologies, which address erosion by water and by wind, include but are not limited to: terraces, interseeding, crop rotations, stripcropping, vegetative waterways, buffer strips, filter strips, cover crops, conservation tillage (e.g., no-till, ridge-tillage, mulch-till), residue management, contour cropping, management-intensive grazing (MIG) systems, tree windbreaks, herbaceous windbreaks, artificial barriers, land reshaping to reduce erosion on knolls, and maintaining clods or stable aggregates at the soil surface (USDA, 1957; Troeh et al., 1980; Weesies et al., 2002; Tibke, 2002).

In the 1970s, conservation tillage became a major part of the conservation program in the U.S., due in large part to advances in herbicide developments that took place in the 1960s. Prior to this time, lack of weed control had defeated previous practical attempts to utilize crop residue for its known erosion-control potential (Doren, 1986). Since this time, the application of conservation tillage has been modified and adapted to encompass a wide range of tillage practices, climates, and soils. Additionally, conservation tillage has become integrated as one component of an overall soil management system. *Conservation tillage* has been described as any tillage and planting system in which 30 percent or more of the soil surface is covered with crop residue after planting, thus reducing the hazard of soil erosion by water (Weesies et al., 2002). Where soil erosion by wind is the primary concern, any system that maintains the equivalent of at least 1120 kg/ha (1000 lb/acre) of flat, small grain residue on the soil surface throughout the critical wind erosion period qualifies as *conservation tillage*.

Assessment methods for soil and water erosion have changed dramatically over the years, i.e., from experimental research plots (e.g., Columbia, Missouri, in 1917) to prediction models, e.g., empirical, physical or process-based, and hybrid models. Empirical models have used mathematical equations or sets of equations developed and used since the 1950s, such as the Wind Erosion Equation (Woodruff and Siddoway, 1965), the Universal Soil Loss Equation (water) (Wischmeier and Smith, 1978), and the Revised Universal Soil Loss Equation (RUSLE) (Renard et al., 1997). The development of T factors (soil loss tolerance) in the 1950s and 1960s was based on cropland and was not applicable to permanent pasture or rangeland. The development of process-based models to predict soil erosion by wind or water is largely a result of advances in computer technology. These process-based models include the USDA/ARS Water Erosion Prediction Program (WEPP) (Flanagan and Nearing, 1995) and the USDA/ARS Wind Erosion Prediction System (WEPS) (Hagan, 1991). The WEPP model is a steady-state model that uses numerous U.S. databases of climate, soils, tillage, and crops information and incorporates information on disturbed sites (e.g., forest roads and burned areas). Runoff in the WEPP model is generated from rainfall input using an infiltration equation (Lane and Nearing, 1989). The WEPS model is a daily time-step model that predicts soil erosion via simulation of the physical processes that control wind erosion. It is intended for use in soil conservation and environmental planning. Hybrid models have also been developed, such as EPIC (Erosion Productivity Impact Calculator, Environmental Policy Integrated Climate), which has evolved over time (Williams et al., 1984; Sharpley and Williams, 1990). Other developments in the assessment of erosion and deposition include cesium-157, derived from weapons testing after 1945 and used as a tracer; remote sensing; and monitoring of fields and sediment yield of rivers (Boardman, 2002). The advantages and limitations of some of these methods/models for the assessment of erosion are reviewed by Rose, 1998, 2002; Arnold et al., 2002; and Laflen, 2002.

The publication of scientific papers in the Soil Science Society of America (SSSA) Division S-6, Soil and Water Management and Conservation, closely parallels the temporal and spatial shifts of emphasis in this area. There was an overall decline in scientific publications in soil and water conservation from 1962 to 1965 (9.7 percent) and from 1966 to 1973 (7.4 percent) but an increase from 1982 to 1985 (12.9 percent) (Doren, 1986), somewhat coinciding with the renewed interest in soil erosion and the enactment of the USDA Farm Bill of 1985. Over 49 percent of the papers in this division were from the North Central Region (1962 to 1965). From 1970 to 1977, the emphasis shifted to the West with 59 percent. The average for each of these two regions was 34 percent from 1978 to 1986 (Doren, 1986).

3.1 Particle-Size Distribution Analysis
3.1.3 Particles >2 mm

Particles >2 mm, definitions: *Rock and pararock fragments* are defined as particles *>2 mm* in diameter and include all particles with horizontal dimensions less than the size of a pedon (Soil Survey Division Staff, 1993). Rock fragments are further defined as strongly cemented or more resistant to rupture, whereas pararock fragments are less cemented than the strongly cemented class; most of these fragments are broken into particles 2 mm or less in diameter during the preparation of samples for particle-size analysis in the laboratory. Rock fragments are generally sieved and excluded from most chemical, physical, and mineralogical analyses. Exceptions are described in SSIR No. 42 (Soil Survey Staff, 2004). It is necessary to know the amount of particles >2 mm in diameter for several applications, e.g., available water capacity and linear extensibility (Grossman and Reinsch, 2002). Refer to the "National Soil Survey Handbook" (USDA/NRCS, 2009b) for a detailed description of rock fragments, their significance, classes, size, shape, hardness, etc. *Nonflat rock fragment classes* are defined as follows:

- The *2- and 5-mm* fraction corresponds to the size openings in the No. 10 and No. 4 screen (4.76 mm), respectively, used in engineering. Coarse fractions with 2- to 5-mm particle diameter correspond to the rock fragment division *fine gravel*.

- The *5- and 20-mm* fraction corresponds to the size of openings in the No. 4 screen (4.76 mm) and the 3/4-in screen (19.05 mm), respectively, used in engineering. Coarse fractions with 5- to 20-mm particle diameter correspond to the rock fragment division *medium gravel*.

- The *20- and 75-mm* fraction corresponds to the size of openings in the 3/4-in screen (19.05 mm) and the 3-in screen (76.1 mm), respectively, used in engineering. Coarse fractions with 20- to 75-mm particle diameter correspond to the rock fragment division *coarse gravel*.

- The *75-mm* fraction corresponds to the size of opening in the 3-in screen (76.1 mm) used in engineering. The 0.1 and 75 mm division is for taxonomic placement of particle-size class, i.e., to distinguish loamy and silty family particle-size classes.

Particles >2 mm, measurements: In this section, the SSL PSDA methods for >2-mm diameter particles are described. These include weight estimates by field and laboratory weighing; weight estimates from volume and weight estimates; and volume estimates. For detailed descriptions of the SSL methods which are cross-referenced by method code in the table of contents in this manual, refer to SSIR No. 42 (Soil Survey Staff, 2004), which is available online at http://soils.usda.gov/technical/lmm/.

3.1   Particle-Size Distribution Analysis
3.1.3   Particles >2 mm
3.1.3.1   Weight Estimates
3.1.3.1.1   By Field and Laboratory Weighing
3.1.3.1.2   From Volume and Weight Estimates
3.1.3.2   Volume Estimates

Weight estimates by field and laboratory weighing: The SSL determines weight percentages of the >2-mm fractions by field and laboratory weighing. In the field or in the laboratory, the sieving and weighing of the >2-mm fraction are limited to the <75-mm fractions. In the field, fraction weights are usually recorded in pounds, whereas in the laboratory, fraction weights are recorded in grams. The 20- to 75-mm fraction is generally sieved, weighed, and discarded in the field. This is the preferred and usually the most accurate method. Less accurately, the 20- to 75-mm fraction is estimated in the field as a volume percentage of the whole soil. If this fraction is sieved and weighed in the laboratory, the results are usually not reliable because of small sample size.

Weight estimates from volume and weight estimates; volume estimates: The SSL estimates weight percentages of the >2-mm fractions from volume estimates of the >20-mm fractions and weight determinations of the <20-mm fractions. The volume estimates are visual field estimates. Weight percentages of the >20-mm fractions are calculated from field volume estimates of the 20- to 75-mm, 75- to 250-mm, and >250-mm fractions. The >250-mm fraction includes stones and boulders that have horizontal dimensions that are smaller than the size of the pedon. Weight measurements for the 2- to 20-mm fraction are laboratory measurements. Weight measurements of the 20- to 75-mm fractions in the field are more accurate than visual volume estimates. Weight measurements of this fraction in the laboratory are not reliable. The volume estimates that are determined in the field are converted to dry weight percentages. For any >2-mm fractions estimated by volume in the field, the SSL calculates weight percentages. The visual volume estimates of the >20-mm fraction are subjective. The conversion of a volume estimate to a weight estimate assumes a particle density of 2.65 g cc^{-1} and a bulk density for the fine-earth fraction of 1.45 g cc^{-1}. Measured values can be substituted in this volume to weight conversion, if required. Unless otherwise specified, the SSL reports the particle-size fractions 2 to 5, 5 to 20, 20 to 75, and 0.1 to 75 mm on a <75-mm oven-dry weight percentage basis. The total >2-mm fraction is reported on a whole soil oven-dry weight percentage basis.

Weight and volume estimates, interferences: Soil variability and sample size are interferences to weight determinations of the >2-mm particles. Enough soil material needs to be sieved and weighed to obtain statistically accurate rock fragment content. In order to accurately measure rock fragments with maximum particle diameters of 20 and 75 mm, the minimum dry specimen sizes that need to be sieved and weighed are 1.0 and

60.0 kg, respectively. Refer to ASTM method D 2488-06 (ASTM, 2008c). Whenever possible, the field samples or "moist" material should have weights two to four times as large (ASTM, 2008c). Therefore, sieving and weighing the 20- to 75-mm fraction should be done in the field. The <20-mm fractions are sieved and weighed in the laboratory. Refer to Table 1.2.1 for minimum dry weights for particle-size analysis.

The visual volume estimates of the >75-mm fractions are subjective. The conversion of a volume estimate to a weight estimate assumes a particle density of 2.65 g cc^{-1} and a bulk density for the fine-earth fraction of 1.45 g cc^{-1}. If particle density and bulk density measurements are available, they are used in the calculations.

3.2 Bulk Density

This section describes the SSL field and laboratory methods for bulk density and information on key definitions and applications of resulting data. There are two broad groupings of SSL bulk density methods: (1) those for soil materials coherent enough that a field-sample can be removed and (2) those for soils too fragile for removal of a sample, in which case an excavation operation must be performed. The SSL uses bulk density notations to designate the water state of the sample when the volume was measured as follows: ρ_f, ρ_{B33}, ρ_{Bod}, and ρ_{Br} for field-state, 33-kPa, oven-dry, and rewet, respectively. This section also describes the soil process of compaction, and references to case studies/datasets are presented as evidentiary examples of the actions/practices that have promoted or diminished this soil process. For detailed descriptions of the SSL methods which are cross-referenced by method code in the table of contents in this manual, refer to SSIR No. 42 (Soil Survey Staff, 2004), which is available online at http://soils.usda.gov/technical/lmm/. Refer to SSIR No. 51 (Soil Survey Staff, 2009; available online at http://www.soils.usda.gov/technical/) for descriptions of field methods as used in NRCS soil survey offices.

3.2 Bulk Density
3.2.1 Assessments and Predictions

Bulk density, definition: *Density* is defined as mass per unit volume. *Soil bulk density* of a sample is the ratio of the mass of solids to the total or bulk volume. This total volume includes the volume of both solids and pore space. Bulk density is distinguished from *particle density*, which is mass per unit volume of only the solid phase (Blake and Hartge, 1986b). Particle density excludes pore spaces between particles.

Bulk density, general applications: In the USDA soil survey program, bulk density has been studied and related to soil genesis, classification, and interpretations as follows: to convert data from a weight to a volume basis, to determine the coefficient of linear extensibility (COLE), to estimate saturated hydraulic conductivity, to detect the presence of significant amounts of volcanic ash and pumice in soil material, to estimate the degree of weathering of rocks and soils, to follow volume changes with soil genesis, and to study gains and losses of soil materials (Buol et al., 1980). A bulk density of

<0.90 g cc^{-1} (g cm^{-3}) at 33-kPa water retention is a diagnostic criterion for andic soil properties (Soil Survey Staff, 2010).

Bulk density, data assessments: Bulk densities of Histosols range from 0.05 to 0.15 g cm^{-3} for fibric and most of the hemic materials (Lynn et al., 1974). For sapric materials, the range is wider, but densities >0.25 g cm^{-3} are limited to organic soils with <7 percent rubbed fiber, of which most are from cultivated surface soil. Bulk density measurements have also been commonly used to assess soil compaction. Relationships have been established between high bulk density and lack of root penetration (Veihmeyer and Hendrickson, 1948; Grossman et al., 1994). Vrindts et al. (2005) found that dry bulk density of >1.6 g cm^{-3} limited winter wheat yields, but otherwise no relation was observed between the yield and dry soil bulk density. Bulk densities ≥1.8 g cc^{-1} have been related to root growth impedance, and densities of 1.6 to 1.8 g cc^{-1} may indicate that aeration and water movement are too low for optimum growth (National Soil Survey Laboratory Staff, 1975). Some plow layers approach densities of 1.8 g cc^{-1}, and some natural formations, e.g., duripans, fragipans, and petrocalcic horizons (Soil Survey Staff, 2010), have densities this high or higher (National Soil Survey Laboratory Staff, 1975). Bulk density has also been used as a key parameter in the development of a numerical index to quantify soil productivity and assess long-term changes due to erosion (25, 50, 100 yr) in Major Land Resource Area (MLRA) 105 in Minnesota (Pierce et al., 1983). This model was based on the assumption that soil is a major factor of crop yield (other factors being climate, management, and plant genetic potential) because of its effects on root growth, i.e., resistance to root growth as expressed by bulk density. Refer to additional discussion on physical root limitations (Grossman et al., 1994) and on nonlimiting, restriction-initiation, and root-limiting bulk densities for <2-mm family particle-size classes (Pierce et al., 1983).

Bulk density, predictions: In a study of pedotransfer functions to estimate bulk density (ρ_B) using existing Brazilian soil survey data, Benites et al. (2007) found that ρ_B could be predicted from other properties, i.e., total N*, clay*, sum of basic cations (SB)*, C:N, water dispersible clay, Al_2O_3, and Ca + Mg. These variables explained 70 percent of ρ_B variance, with * variables as the strongest contributors. A simplified regression model using only soil organic carbon (SOC), clay, and SB described 66 percent of the ρ_B variation in all soils at all depths, and partitioning the dataset (n = 1002) into groups by soil depth and soil order did not lead to remarkable improvements in ρ_B prediction. On the other hand, using the 1997 SSL characterization data (N = 47,000, subsoil + surface samples), Heuscher et al. (2005) found that partitioning the database by soil suborder improved the regression relationships (R^2 = 0.62, p <0.001). In a stepwise regression procedure, SOC was the strongest contributor to ρ_B prediction; other significant variables included clay and water contents and, to a lesser extent, silt content and depth. In general, the accuracy of regression equations was greater for suborders containing more SOC (most Inceptisols, Spodosols, Ultisols, and Mollisols) and more poorly predicted for suborders of Aridisols and Vertisols that contain little or no SOC. Heuscher et al. (2005) concluded that regression equations are a feasible alternative for bulk density estimation.

3.2 Bulk Density
3.2.2 Soil Compaction, Process and Case Studies

Soil compaction, process: The term *soil compaction* refers to the compression of a soil (reduced spaces between soil particles) resulting in reduced pore space, decreased movement of water and air into and within the soil, decreased water storage, and increased surface runoff and erosion.

Soil compaction, case studies: Human-induced compaction has increased dramatically over recent decades, largely resulting from mechanical stress caused by off-road wheel traffic and machinery traffic (Hakansson and Voorhees, 1998). This trend is related to increasing mechanization and use of larger, more efficient farm vehicles, which lead to increased soil densification and a corresponding reduction of productivity in some regions. Reductions in content of organic matter related to overtillage and extensive use of inorganic fertilizers have also been related to increased susceptibility to compaction.

Soil compaction (traffic pans, hardpans, and plowpans resulting from constant depth plowing) and associated yield reductions have been observed since the 1930s, and the literature on this topic has increased over the last 30 years (Hakansson and Voorhees, 1998). Because of complex and interrelated soil, management, and climatic factors, however, the challenge has been to directly and quantitatively relate compaction to yield reductions and the resulting economic impact.

Many experiments have evaluated the effects of soil pans on crop yields. In some cases yields did not increase upon pan disruption, whereas in other cases yields increased substantially upon tillage. Those cases in which yields did not respond to disruption of a discernible pan may have been due in part to other more yield-limiting factors. It is difficult to isolate and evaluate the direct effects of physical resistance because of its interaction with other environmental factors; i.e., sufficient physical and chemical factors need to be measured and interactions understood in order to assign a probable cause-effect relationship to excessive soil strength (Taylor, 1971).

Those cases in which yields did respond to tillage have been shown for many crops and across geographical areas representing a wide range of soil types, as follows: cotton in the Central Valley of California (Carter and Tavernetti, 1968) and Big Spring, Texas (Taylor and Burnett, 1964); grass species (Barton et al., 1966); corn in Iowa (Phillips and Kirkham, 1962); grain sorghum in the southern Great Plains (Taylor and Burnett, 1964); and sugar beets (Taylor and Bruce, 1968). Lal and Ahamdi (2000) directly related tillage effects and axle load to higher soil bulk density and yield reductions on some soils in Ohio. Similarly, Raghavan et al. (1978) related delays in development and reductions in yield to magnitude of vehicle contact pressure and number of passes, with yield reductions over 50 percent, suggesting that careful traffic planning was essential to better production in agricultural fields in Quebec. Rogers and Thurlow (1973) directly related the effects of soil compaction to reductions in soybean yields during the critical pod-filling period, when rainfall is usually poorly distributed in Alabama and much of the Southeast. These compaction effects were exacerbated in dry years compared to normal rainfall years, with 10 percent and 60 percent relative yields, respectively. This study recommended ameliorative practices of deeper plowing; operating the tractor with its wheels on unplowed land rather than in a furrow when the soil is turned; limiting wheel traffic to certain rows after planting; and making fewer trips over the field. This study and resulting recommendations predated the development and wide application of conservation tillage (no-till or reduced tillage) and its associated benefits, e.g., reduced machinery traffic and increased residue.

Going a step further in establishing these cause-effect relationships, Mehuys (1984) attributed 85 percent of the economic impact of soil degradation in Quebec to compaction, estimating its impact to be 15 percent of potential yields, representing a $100 million farm revenue loss. Gill (1971) estimated U.S. on-farm losses through land compaction at $1.2 billion per year. Eswaran et al. (2001) estimated that soil compaction causes yield reductions of 25 to 50 percent in North America. In Ohio, reductions in crop yields were estimated over a 7-year period at 25 percent for maize, 20 percent for soybeans, and 30 percent for oats (Lal, 1993).

Soil compaction induces or accelerates other soil degradation processes, such as runoff and erosion. Chancellor (1976) found higher operational costs of irrigation due to poor infiltration and presumably higher evaporative losses on some compacted soils in California. Lindstrom et al. (1981) attributed higher runoff and erosion rates to topsoil compaction under different long-term continuous corn (*Zea mays L.*) tillage systems (conventional, conservation, no-till) in south-central Minnesota. This study was conducted at a time when conservation tillage systems (no-till or reduced tillage) were receiving considerable attention as measures for controlling water runoff and soil erosion. An interesting finding of this study was that the no-till system, while effectively absorbing the energy of falling raindrops, was not capable of retaining water from heavy rainstorms, due in part to a consolidated soil surface condition prior to the establishment of the no-tillage system. This prior surface compaction could not be corrected through 10 years of normal amelioration processes (e.g., freezing and thawing, wetting and drying, and soil fauna activity).

While the various causes and effects of soil compaction are interrelated and are often difficult to assess, it is generally considered that this process and its amelioration are understood well enough that systems for its management on the farm can be reasonably formulated. This generalization is of course tempered with the recognition that practices to reverse subsoil compaction (e.g., subsoiling) compared to surface compaction are more costly and in some cases counterproductive if not irreversible, resulting in denser recompaction due to destabilization from the mechanical energy input from the subsoiling operation (Zoebisch and Dexter, 2002).

Soil compaction has been managed through the use of controlled traffic, in which heavy traffic is confined to specific lanes through the crop and from year to year. In recent years the controlled traffic approach has been facilitated by the development and use of GPS-based guidance systems (Reeder, 2002). Shallow compaction resulting from random wheel traffic has been shown to reduce cotton yields in areas planted using a no-till system (Burmester et al., 1995). Restricting compaction to trafficked lanes removes some of the problems with no-till systems, in which the potential for compaction by driving on wet soil can be a concern. Additionally, controlled traffic helps to retain the long-term benefits of subsoiling for alleviating compaction (Reeder, 2002).

The assessment of surface sealing and crusting and compaction has received less attention, both globally and within the United States, than assessments of other degradation processes, e.g., erosion and salinity. Dryland salinity maps were developed for the northern Great Plains of Canada and the United States (Vander Pluym, 1978). Erosion maps were prepared for Canada on the basis of suspended sediment load (Stichling, 1973). Maps and statistical surveys of erosion in the United States are available (USDA/NRCS, 2001). Even though soil compaction and crusting are common in areas of cropland and rangeland, maps at any scale of their distribution are currently not known to exist, perhaps because these phenomena are so common and occur haphazardly (Dregne, 1998). Additionally, the effects of these processes have historically been considered "temporal" properties rather than inherent soil properties and are not captured in taxonomic classification (Soil Survey Staff, 2010). In the 1990s, the USDA/NRCS developed a soil quality initiative to address and incorporate these types of assessments into soil survey. Research findings and practical technologies of this USDA initiative can be found at http://soils.usda.gov/sqi/.

Technologies in the production and application of maps in the United States have improved dramatically over the last decade, as evidenced by the USDA soil survey program. National and global assessments are important in providing generalized information on the extent, severity, and location of land degradation, and maps are typically the most useful way to present this information (Dregne, 1998). These maps, however, do not provide the specificity that is needed to understand and address these problems in any meaningful way, due in part to scale. In general, maps of any soil degradation process are scarce at any scale (Dregne, 1998). These kinds of maps are critical to any long-term assessment, monitoring, or restoration of soil quality and its interrelated components. The production of maps is likely less cost prohibitive than the interpretations of these maps, generated from onsite field investigations and laboratory analyses.

3.2 Bulk Density
3.2.3 Saran-Coated Clods
3.2.3.1 Field-State (ρ_{Bf})
3.2.3.2 33-kPa Equilibration (ρ_{B33})
3.2.3.3 Oven-Dry (ρ_{Bod})
3.2.3.4 Rewet (ρ_{Br})
3.2.4 Reconstituted
3.2.4.1 33-kPa Equilibration
3.2.4.2 Oven-Dry
3.2.5 Compliant Cavity
3.2.6 Ring Excavation
3.2.7 Frame Excavation
3.2.7.1 Field-State
3.2.8 Soil Cores
3.2.8.1 Field-State

Bulk density, soil water content: Bulk density may be highly dependent on soil conditions at the time of sampling. Changes in soil volume due to changes in water content will alter bulk density. Soil mass remains fixed, but the volume of soil may change as water content changes (Blake and Hartge, 1986a). Bulk density, as a soil characteristic, is actually a function rather than a single value. Therefore, subscripts are added to the bulk density notation, ρ_B, to designate the water state of the sample when the volume was measured. The SSL uses the bulk density notations of ρ_{Bf}, ρ_{B33}, ρ_{Bod}, and ρ_{Br} for field-state, 33-kPa equilibration, oven-dry, and rewet, respectively.

Bulk density methods, groupings: In general, there are two broad groupings of bulk density methods: one for soil materials coherent enough that a field sample can be removed and the other for soils too fragile for the removal of a sample, in which case an excavation operation must be performed. Under the former condition, there are clod methods in which the sample has an undefined volume, the sample is coated, and the volume is determined by submergence. Also, there are various methods in which a cylinder of known volume is obtained of soil sufficiently coherent that it remains in the cylinder. The complete cylinder may be inserted, or only part of the cylinder is inserted, and the empty volume is subtracted from the total volume of the core (e.g., variable height method, Grossman and Reinsch, 2002). Three SSL excavation procedures have been used to determine ρ_{Bf} as follows: (1) compliant cavity, (2) ring excavation, and (3) frame excavation (Grossman and Reinsch, 2002; Soil Survey Staff, 2004).

Bulk density, measurements: The following section describes the SSL field and laboratory procedures for bulk density as well as their specific method applications and interferences. Using the broad groupings previously discussed and their associated water contents, the SSL method categories for bulk density (g cm^{-3}) are as follows:

- For soil materials coherent enough that a field sample can be removed, use the following:
  - saran-coated natural clods at water contents—
    - field-state (ρ_{Bf})
    - 33-kPa equilibration (ρ_{B33})
    - oven-dry (ρ_{Bod})
    - rewet (ρ_{Br})
  - soil cores at field-state (ρ_{Bf})
- For soils subject to tillage or other mechanical disturbances followed by an extreme water-state cycle, use the following:
  - reconstituted bulk density
    - 33-kPa equilibration (ρ_{B33})
    - oven-dry (ρ_{Bod})
- For soils too fragile for the removal of a sample and for which an excavation operation must be performed, use the following:
  - compliant cavity
    - field-state (ρ_{Bf})
  - ring excavation
    - field-state (ρ_{Bf})
  - frame excavation
    - field-state (ρ_{Bf})

3.2 Bulk Density
3.2.3 Saran-Coated Clods
3.2.3.1 Field-State (ρ_{Bf})
3.2.3.2 33-kPa Equilibration (ρ_{B33})
3.2.3.3 Oven-Dry (ρ_{Bod})
3.2.3.4 Rewet (ρ_{Br})

Saran-coated natural clods, field-state (ρ_{Bf}), definition and measurement: ρ_{Bf} is the bulk density of a soil sample at field-soil water content at the time of sampling. ρ_{Bf} is particularly useful if the soil layers are at or above field capacity and/or the soils have low extensibility and do not exhibit desiccation cracks even if below field capacity. ρ_{Bf} using saran-coated clods is determined by collecting field-occurring fabric (clods) from the face of an excavation. One coat of plastic lacquer is applied in the field. Additional coats of plastic lacquer are applied in the laboratory. In its field-water state or after equilibration, the clod is weighed in air to measure its mass and in water to measure its volume. After the clod is dried in an oven at 110 °C, its mass and volume are determined again.

Saran-coated natural clods, 33-kPa equilibration (ρ_{B33}), definition and measurement: ρ_{B33} is the bulk density of a soil sample that has been desorbed to 33 kPa. Field-occurring fabric (clods) is collected from the face of an excavation. One coat of plastic lacquer is applied in the field. Additional coats of plastic lacquer are applied in the laboratory. The clod is desorbed to 33 kPa. After equilibration, the clod is weighed in air to measure its mass and in water to measure its volume. After the clod is dried in an oven at 110 °C, its mass and volume are determined again.

Saran-coated natural clods, oven-dry (ρ_{Bod}), definition and measurement: ρ_{Bod} is the bulk density of a soil sample that has been dried in an oven at 110 °C. Field-occurring fabric (clods) is collected from the face of an excavation. One coat of plastic lacquer is applied in the field. Additional coats of plastic lacquer are applied in the laboratory. The clod is dried in an oven at 110 °C and then weighed in air to measure its mass and in water to measure its volume.

Saran-coated natural clods, rewet (ρ_{Br}), definition and measurement: ρ_{Br} is the bulk density of a soil sample that has been equilibrated, air dried, and reequilibrated. The ρ_{Br} is used to determine the irreversible shrinkage of soils and subsidence of organic soils. Field-occurring fabric (clods) is collected from the face of an excavation. One coat of plastic lacquer is applied in the field. Additional coats of plastic lacquer are applied in the laboratory. After equilibration, the clod is weighed in air to measure its mass and in water to measure its volume. The clod is air dried and reequilibrated, and its mass and volume are remeasured. After the clod is dried in an oven at 110 °C, its mass and volume are determined again.

Saran-coated natural clods, interferences: The complication concerning the difference between bulk density of the soil and that of the sample is particularly important for the clod method as presented herein, which permits determination of the volume at different water contents and, hence, volumes. If the water content is at or near field capacity, desiccation cracks are closed and the bulk density (ρ_{B33} or ρ_{Bf} if field-water is near field capacity) of the soil and of the sample are considered the same; however, if the sample is at water content below field capacity through drying after sampling or because the sample was taken below field capacity, then desiccation cracks that occur in place are excluded from the soil and the sample bulk density exceeds that of the soil. If the sample is large and inclusive of the desiccation cracks, as in some excavation procedures, then again the sample and soil bulk density are the same. The difference between sample bulk density and soil bulk density is particularly large for oven-dry clods (ρ_{Bod}) of soils with high extensibility and may also be large for soils subject to a large increase if taken through a rewet cycle. Grossman and Reinsch (2002) discuss the manipulation of clod bulk densities (the sample) at water contents below field capacity to obtain an estimate of the soil bulk density at such water contents. Similarly, estimates of soil bulk density at intermediate field-water contents between field capacity and oven dryness inclusive of desiccation crack space are discussed by Grossman et al. (1990).

Errors are caused by nonrepresentative samples. Only field-occurring fabric (clods) should be sampled. The whole bulk density may be overestimated because sampled clods frequently exclude the crack space between clods (Grossman and Reinsch, 2002).

The penetration of plastic lacquer into the voids of sandy or organic soil interferes with the corrections of mass and volume of the plastic coat and with the accuracy of water content determinations. Penetration can be reduced by spraying water on the clod and then immediately dipping the clod in the plastic lacquer.

Loss of soil during the procedure will void the analyses because all calculations are based on the oven-dry soil mass. Holes in the plastic coating, which are detected by escaping air bubbles from the submerged clod, introduce errors in volume measurement. An inadequate evaporation of the plastic solvent results in overestimation of the soil mass. A drying time of 1 h is usually sufficient time for evaporation of solvent; however, clods with high organic matter content may need to dry longer.

As bulk density (ρ_B) is usually reported for the <2-mm soil fabric, the mass and volume of rock fragments are subtracted from the total mass and volume (Brasher et al., 1966; Blake and Hartge, 1986a). This correction for rock fragments with >2-mm diameter requires either knowledge or an assumption of the rock fragment density. Estimate or measurement errors of rock fragment density affect the accuracy of the soil bulk density value. The porosity of the rock fragments also is a factor that must be considered when the values for soil bulk density and water-holding capacity are corrected. In SSL bulk density calculations, corrections are made for the mass and volume of rock fragments and, if applicable, for plastic coatings (Brasher et al., 1966; Blake and Hartge, 1986a; Grossman and Reinsch, 2002).

3.2 Bulk Density
3.2.4 Reconstituted
3.2.4.1 33-kPa Equilibration
3.2.4.2 Oven-Dry

Reconstituted, 33-kPa equilibration (ρ_{B33}), oven-dry (ρ_{Bod}), definition and measurement: Some models and programs require one bulk density to represent a given horizon. *Reconstituted bulk density* provides a single, reproducible value for horizons that are subject to tillage or other mechanical disturbances followed by an extreme water-state cycle (Reinsch and Grossman, 1995). In this procedure, a <2-mm sample is formed into a clod by wetting and desiccation cycles that simulate reconsolidating by water in a field setting. Plastic lacquer is applied in the laboratory to form an impermeable coat on the clod. The clod is desorbed to 33 kPa. After equilibration, the clod is weighed in air to measure the mass and in water to measure the volume. After the clod is oven dried at 110 °C, its mass and volume are determined again (Brasher et al., 1966; Blake and Hartge, 1986a; Grossman and Reinsch, 2002). Bulk density by 33-kPa equilibration and oven-dry are reported for this bulk density method.

3.2 Bulk Density
3.2.5 Compliant Cavity
3.2.6 Ring Excavation
3.2.7 Frame Excavation
3.2.7.1 Field-State

Compliant cavity, field-state (ρ_{Bf}), definition, measurement, and interferences: The compliant cavity (ρ_{Bf}) is designed for fragile cultivated near-surface layers and O horizons of forestland soils. This method has the important advantage that it is not necessary to flatten the ground surface on steep slopes or to remove irregularities; i.e., the surficial zone is usually not altered (Grossman and Reinsch, 2002). The cavity volume on the zone surface is lined with thin plastic, and water is added to a datum level. Soil is quantitatively excavated in a cylindrical form to the required depth. The difference between the initial volume and the volume after excavation is the sample volume. The excavated soil is dried in an oven and then weighed. Bulk density by compliant cavity can be made on soils with rock fragments but is more complex (Grossman and Reinsch, 2002).

Ring excavation, field-state (ρ_{Bf}), definition, measurement, and interferences: Ring excavation (ρ_{Bf}) is a robust, simple, and rapid method. This method is good for O horizons in the woods where local variability is large and rock fragments are common. The diameter can range down to 15 cm and up to 30 cm or more. It is not necessary to excavate from the whole area within the ring. A limit of 2 cm on the minimum thickness of the sample should be considered. The size of the 0.1 m^2 is sufficient to encompass considerable local variability. A 20-cm ring is inserted into the ground. A piece of shelf standard is placed across the ring near to a diameter. The distance to the ground surface is measured at eight points equally spaced along the diameter using the depth-measurement tool to measure the distance. The piece of shelf is rotated 90 degrees, and eight more measurements are made. The 16 measurements are then averaged. The soil is excavated to the desired depth, and the distance measurements are repeated. The change in distance is calculated on the removal of the soil. This change in distance is then multiplied by the inside cross-sectional area of the ring to obtain the volume of soil. The excavated soil is oven dried and weighed. Rock fragments may make it impossible to insert the ring into the ground.

Frame excavation, field-state (ρ_{Bf}), definition and measurement: Frame excavation (ρ_{Bf}) is appropriate for O horizons in the woods where local variability is large and rock fragments are common. The size of the 0.1 m^2 is sufficient to encompass considerable local variability. The assembled frame is placed on the ground surface. The four threaded rods are pushed through the holes in the corners of the frame deep enough to hold. The frame is then secured onto the soil surface by screwing down wing nuts, and plastic is placed over the frame and secured. The depth-measurement tool is placed on top of a slot to measure the distance to the soil surface. The slots are traversed, and measurements of the distance to the ground surface are made at about 40 regularly spaced intervals. The plate is then removed, and soil is excavated and retained. Measurements of the distance to the ground surface are repeated. The volume of soil is determined by taking the difference in height and multiplying by 1000 cm^2. The rock fragments up to 20 mm are included in the sample. Excavated soil is oven dried and weighed.

3.2 Bulk Density
3.2.8 Soil Cores
3.2.8.1 Field-State

Soil cores, field-state (ρ_{Bf}), measurement: Soil cores (ρ_{Bf}) also are determined by the SSL. A metal cylinder is pressed or driven into the soil. The cylinder is removed, and a sample of known volume is extracted. The moist sample weight is recorded. The sample is then dried in an oven and weighed.

3.3 Water Retention

This section describes the standard SSL procedures and their specific method applications. These procedures include pressure-plate (6, 10, 33, 100, 200 kPa) and pressure-membrane (1500 kPa) extractions as well as water retention at field-state. Sample materials for these various procedures include, but are not limited to, <2-mm particles, natural clods, and soil cores. This section also provides information on key definitions, historical development of these terms, and expressions and calculations related to water content. For detailed descriptions of the SSL methods which are cross-referenced by method code in the table of contents in this manual, refer to SSIR No. 42 (Soil Survey Staff, 2004), which is available online at http://soils.usda.gov/technical/lmm/. Also refer to SSIR No. 51 (Soil Survey Staff, 2009; available online at http://www.soils.usda.gov/technical/) for detailed descriptions of field methods as used by NRCS soil survey offices.

3.3 Water Retention
3.3.1 Definitions and Data Assessments

Water content, definition: In soil science, *water content* has traditionally been expressed as either a dimensionless ratio of two masses or two volumes or as a mass per unit volume (Gardner, 1986). When either of these dimensionless ratios is multiplied by 100, the values become percentages and the basis (mass or volume) is stated. Conversions from gravimetric to volumetric basis or vice versa require a measure or an estimate of bulk density. In either case (mass or volume basis), the amount of water in the sample must be determined by either the removal or measurement of the water or by determination of the sample mass before and after water removal, i.e., dried to a constant weight (Gardner, 1986). In addition, when precision is critical, there must be criteria for determining the point at which the sample is considered "dry." The SSL defines air-dry and oven-dry weights as constant sample weights obtained after drying at 30±5 °C (≈ 2 to 7 days) and 110±5 °C (≈ 12 to 16 h), respectively.

Water content, data assessments: Direct or indirect (index) determinations of soil water content are generally required in many soil studies. In the field, measurements or estimates of soil water content are required to determine plant-available water. In the laboratory, soil water data are necessary for determining and reporting many physical and chemical properties (Gardner, 1986). In addition, soil water content may be used to help determine the water retention function, the water-holding capacity, the pore-size distribution, and the porosity of a soil sample at a specific water content and to calculate unsaturated hydraulic conductivity.

Water retention, definitions: *Water retention* is defined as the soil water content at a given soil water suction (Gardner, 1986). By varying the soil suction and recording the changes in soil water content, a water retention function or curve is determined. This relationship is dependent on particle-size distribution, clay mineralogy, organic matter, and structure or physical arrangement of the particles as well as hysteresis, i.e., whether the water is absorbing into or desorbing from the soil. The data collected in these procedures are from water desorption (Gardner, 1986). Water retention or desorption curves are useful directly and indirectly as indicators of other soil behavior traits, such as drainage, aeration, infiltration, plant-available water, and rooting patterns (Topp et al., 1993).

The relation between the soil water content and the soil water suction is a fundamental part of the characterization of the hydraulic properties of a soil (Klute, 1986). For many purposes, water retention properties of individual soil horizons are more usefully combined to form a complete profile, and the importance of a large or small value for available water or air capacity varies in relation to properties of neighboring horizons (Hall et al., 1977). Agricultural, pedological, and hydrological interpretations depend mainly on the assemblage of properties of the whole profile (Hall et al., 1977).

Water retention, 33-kPa, definition: *Water retention, 33-kPa*, has become identified with field capacity in some soils (Richards and Weaver, 1944) and as such the upper limit of plant-available water. *Water retention at 10 kPa* may be used as the upper limit of plant-available water for coarse materials. *Coarse materials* are defined (Soil Survey Division Staff, 1993) as follows: if strongly influenced by volcanic ejecta, soil material must be nonmedial and weakly or nonvesicular; if not strongly influenced by volcanic ejecta, soil material must meet the sandy or sandy-skeletal family particle-size criteria and also be coarser than loamy fine sand with <2 percent organic C and <5 percent water at 1500-kPa suction; and computed total porosity of <2-mm fraction must be >35 percent. Refer to Soil Survey Division Staff (1993) and Grossman et al. (1994) for additional discussion of coarse materials and the significance of soil water content at lower suctions, e.g., 5 and 10 kPa, as well as suggestions for the selection of these lower suctions for the determination of water retention difference (WRD).

Water retention, 1500-kPa, definition: *Water retention, 1500-kPa*, has become identified with the permanent wilting point (PWP) and is frequently used as an index of PWP (Richards and Weaver, 1943; Kramer, 1969). The maximum size pore filled with water at 1500 kPa is 0.2-μm diameter. This diameter is in the clay-size range. For this reason, a high correlation usually exists between this water content and clay percentage (National Soil Survey Laboratory Staff, 1983). Clay percentages may be estimated by subtracting the percent organic C from the 1500-kPa water content and then multiplying by 2.5 or 3 (Soil Survey Staff, 2010). Refer to Soil Survey Staff (2010) for the appropriate use of these estimates, e.g., criteria for oxic and kandic horizons. The percent water retained at 1500-kPa suction (dried and undried samples) is also used as a criterion for modifiers that replace particle-size classes, e.g., ashy and medial classes, and for strongly contrasting particle-size classes, e.g., ashy over medial-skeletal (Soil Survey Staff, 2010). Refer to Soil Survey Staff (2010) for a more detailed discussion of these criteria.

Water retention, 5-kPa, definition: Some investigators (Hall et al., 1977) have defined the upper limit of plant-available water as the percentage of water retained in a core sample when equilibrated at *5-kPa suction (retained water capacity)*. This application of 5-kPa water is supported by some investigations of field moisture regimes under British conditions (Thomasson, 1967; Webster and Beckett, 1972). Even in very permeable, well drained soils, the suction in surface horizons commonly is in the range of 3 to 7 kPa (0.03 to 0.07 bar) during winter and spring, when the soil moisture deficit is effectively zero (Webster and Beckett, 1972). Interest in water retention at small suctions, e.g., 5 kPa, is a result of the need to identify a moisture content near to the field-capacity state and to measure differences in pore-size distribution in the >10-μm diameters, the fraction of pores considered by some as critical for water movement, aeration, root growth, and soil fauna (Hall et al., 1977). Pores in this size range are sensitive to soil structural condition. Subsequently, the water content between 5 and 1500 kPa has been viewed by some as the more appropriate approximation of AWC. As the fine pores are mainly associated with the clay fraction of a soil, the correlation between clay content and water retention increases with increasing suction, e.g., 1500 kPa (decreasing pore size). Silt and organic matter appear to have more effect on coarse-pore distribution; therefore, it is at the lower suctions, e.g., 5 kPa, that they are most significant in accounting for variation in retained water. Both silt content and organic matter content are positively correlated with water retention, and in topsoils, bulk density has been related as the major single factor explaining variance in water retention after clay, silt, and organic matter have been considered (Hall et al., 1977).

Field capacity, definition: The term *field capacity* was first introduced by Veihmeyer and Hendrickson (1931) and has been used widely to refer to the relatively stable soil water content after which drainage of gravitational water has become very slow—generally within 1 to 3 days after the soil has been thoroughly wetted by rain or irrigation. The intent of this concept was twofold: (1) to define the upper limit of plant-available water retained by the soil, and (2) to provide a concept that would encourage farmers in irrigated regions not to irrigate excessively (Cassel and Nielsen, 1986). This water that is slowly draining is assumed to be subject to interception by most plant roots and therefore plant available (Salter and Williams, 1965). There are several unstated assumptions to the field capacity concept; i.e., the soil is deep and permeable, no evaporation occurs from the soil surface, and no water table or slowly permeable barriers occur at shallow depths in the profile (Cassel and Nielsen, 1986).

The term *in situ field water capacity* is defined by the Soil Science Society of America (SSSA) (2010) as the content of water, on a mass or volume basis, remaining in a soil 2 or 3 days after having been wetted and after free drainage is negligible. A problem with this definition is the difficulty in defining when the drainage rate is negligible. Many factors affect the field capacity measurement, including the conditions under which it is measured, e.g., initial saturation or presence of wetting front, as well as the characteristics of the soil itself, e.g., degree of nonuniformity.

Field capacity, measurements: Laboratory determinations of the field capacity of a soil are useful data but are not necessarily reliable indicators of this value in the field because of the effects of soil profile and structure. Laboratory determinations are usually made by simulating the tension that develops during drainage in the field by use of pressure membranes or tension tables. There has been considerable debate as to the appropriate tension to apply. In a study by Richards and Weaver (1944), the average soil moisture content at 33-kPa pressure for 71 different soils (<2 mm) approximated the moisture equivalent or field capacity of the soils. Water content at field capacity may be overestimated from sieved-sample data (Young and Dixon, 1966). Some studies have indicated that the upper limit of plant-available water may be more appropriately represented in some soils by the moisture contents at 10- or 5-kPa water retention. As field capacity has no fixed relationship to soil water potential, it cannot be considered as a soil moisture constant (Kramer, 1969). The amount of water retained at field capacity decreases as the soil temperature increases (Richards and Weaver, 1944). Field capacity is not a true equilibrium measurement but rather a soil condition of slow water movement with no appreciable changes in moisture content between measurements (Kramer, 1969).

Some investigators have attempted to remove the term *field capacity* from technical usage (Richards, 1960; Sykes and Loomis, 1967). The usage of this term persists, however, in both technical and practical applications; to date, no alternative concept or term has been advanced to identify the upper limit of plant-available water (Cassel and Nielsen, 1986). It has been argued from a practical standpoint that the concept of field capacity should be clarified and maintained until a viable alternative is advanced (Cassel and Nielsen, 1986).

Permanent wilting point, definition: The term *permanent wilting percentage or point* (PWP) has been widely used to refer to the lower limit of soil water storage for plant growth. The establishment of this lower limit of available water retained by the soil reservoir is of considerable practical significance (Cassel and Nielsen, 1986). Briggs and Shantz (1912) defined this lower limit, first termed *wilting coefficient*, as the water content at which plants remain permanently wilted (assuming that leaves exhibit visible wilting), unless water is added to the soil. Briggs and Shantz (1911, 1912) conducted a large number of measurements on a wide variety of plants and found little variation in the soil water content at which wilting occurred (Kramer, 1969). Other investigators (Richards and Wadleigh, 1952; Gardner and Nieman, 1964) determined that the soil water potential at wilting for indicator plants, e.g., dwarf sunflower, approximated -1000 to -2000 kPa with a mean value of -1500 kPa. The percentage of water at 1500-kPa retention has become identified with PWP and is frequently used as an index of PWP (Richards and Weaver, 1943; Kramer, 1969).

The PWP criteria (Briggs and Shantz, 1912) were later modified by Furr and Reeve (1945) to include the incipient wilting point, the water content at which the first (usually lower) leaves wilted, and the permanent wilting point, a much lower soil water potential at which all the leaves wilted. The incipient wilting percentage is related to the lower limit at which soil water is available for plant growth; i.e., water extraction may occur at lower contents. In addition, there is no physical reason why continued water extraction may not occur after growth ceases or even after plant death (although much reduced because of stomatal closure) (Kramer, 1969). The PWP is defined by the SSSA (2010) as the water content of a soil when indicator plants growing in the soil wilt and fail to recover when placed in a humid chamber (usually estimated by the water content at -1500-kPa soil matric potential). In general, there is a considerable range in water content between the incipient and the permanent wilting percentage (Gardner and Nieman, 1964).

Permanent wilting point, soil-related factors: There are many factors that may affect the onset of wilting and the visible wilting of plants in the field. These factors include the soil water conductivity as well as the transient inability of the water supply system in the plant to meet evaporative demand (as opposed to conditions associated with permanent wilting) (Kramer, 1969). Slatyer (1957) criticized the concept of PWP as a soil constant and defined wilting as the loss of turgor (zero point of turgor), which is primarily associated with osmotic characteristics of the leaf tissue sap; i.e., wilting occurs when there is a dynamic balance between the plant and soil water potentials. Soil water potential at wilting can vary as widely as the variation in osmotic potential in plants, which can range from -500 to -20000 kPa (Kramer, 1969). Furthermore, in the equilibrium measurement (Briggs and Shantz, 1912), the PWP is merely a function of the index plant for any given soil. Because of the shape of the water potential/water content curve of soils, however, marked changes in water potential often accompany small changes in water content, so that for practical purposes, the PWP or the percentage at 1500-kPa retention can still be viewed as an important soil value (Kramer, 1969). This approximation is particularly appropriate for most crop plants, as the osmotic potentials of many species range from -1000 to -2000 kPa (Kramer, 1969).

Available water capacity, definition: The term *available water capacity (AWC)* refers to the availability of soil water for plant growth and is usually considered the amount of water retained in a soil between an upper limit termed *field capacity* and a lower limit termed *permanent wilting percentage* (PWP). The SSSA (2010) defines available water as the portion of water in a soil that can be absorbed by plant roots and is the amount of water released between *in situ field water capacity* and the PWP, usually estimated by water content at soil matric potential of -1500 kPa (-15 bar). These upper and lower limits represent a range which has been used in determining the agricultural value of soils. The importance of AWC relates to the water balance in the soil during the growing season, i.e., the difference between evapotranspiration and precipitation.

The range of water available for plant survival is substantially greater than that available for good growth. In addition, within the range of available water, the degree of availability usually tends to decline as soil water content and potential decline (Richards and Wadleigh, 1952; Kramer, 1969). There is no sharp limit between available and unavailable water. The PWP is only a convenient point on a curve of decreasing water potential and decreasing availability (Kramer, 1969). The range of soil water between field capacity and PWP, however, constitutes an important field characteristic of soils when interpreted properly (Kramer, 1969). Refer to the "National Soil Survey Handbook" (USDA/NRCS, 2009b) for additional discussion related to estimates, significance, and classes.

Available water capacity, soil-related factors: Available water capacity varies widely in different soils. In general, finer textured soils have a wider range of water between field capacity and permanent wilting percentage than do coarser textured soils. In addition, in finer textured soils, the slope of the curve for water potential over water content indicates a more gradual water release with decreasing water potential, whereas coarser soil materials, with their large proportion of noncapillary pore space and predominance of larger pores, usually release most of their water within a narrow range of potential (Kramer, 1969). Available water capacity only approximates the soil's ability to retain or store water and does not provide an estimate of the supplying capacity of a soil or even the amount that plants extract. The supplying capacity is affected by many factors, e.g., hydraulic conductivity, stratification, runoff, run-on, irrigation, rainfall, osmotic potential, and the plants themselves. Caution is required when readily available water data are used because the availability of water depends on many factors. For example, deep rooting in the whole soil profile can compensate for a narrow range of available water in one or more soil horizons as opposed to restricted root distribution combined with a narrow range of available water.

Available water capacity estimate, water retention difference, between 33, 10, or 5 kPa and 1500 kPa: In interpretations, the interest is usually not the water retention differences (WRD) but the AWC (Grossman et al., 1994). The calculation of the WRD can be used in the approximation of the AWC. The first step in the estimation of the AWC is the selection of the suction to approximate the water retention at field capacity. Usually, this is 33 kPa (10 or 5 kPa for coarse soil materials). The second step in the AWC estimate is the selection of the WRD for the fine-earth fraction (WRD_f). The third step is to adjust the WRD_f for salts (WRD_{fs}) (Baumer, 1992). The fourth step is to adjust downward for the volume percentage of the >2-mm fraction. The fifth and final step in the AWC estimate is to adjust for root restriction (Grossman et al., 1994). Refer to Baumer (1992) and Grossman et al. (1994) for additional discussion of AWC estimates that are either calculated using measured water retention data or estimated using other soil properties, e.g., family particle-size classes (>2-mm fraction excluded), bulk density, and clay mineralogy.

Available water capacity estimate, ratio of 33 minus 1500-kPa water to silt content: The ratio of 33 minus 1500-kPa water to silt content has been used in estimating AWC (National Soil Survey Laboratory Staff, 1983). The water retained between these two suctions has been correlated with 0.2- to 10-µm diameter pores (National Soil Survey Laboratory Staff, 1983). Hence, the amount of silt is important to the concept of plant-available water. Ratios of 33 minus 1500-kPa water to silt content range from 0.12 to 0.25 in many soils with silicate clays, quartz and feldspar silts and sands, and modest amounts of organic matter. Higher ratios in soils may be associated with amorphous material and significant amounts of organic matter (National Soil Survey Laboratory Staff, 1983).

Air-filled porosity, 5 kPa: The air-filled porosity at 5 kPa has been used as a measure of tilth (McKeague et al., 1982). This measurement is related to an approximation of that fraction of coarse pores (>60 µm) in the soil. These pores are normally air filled, except during short periods following heavy rainfall. Air capacity is that percent of sample volume occupied by air at a specified suction. Total pore space is that volume of sample not occupied by solid soil material and is available to water and/or air. Clay contents and bulk density have been associated with the most variation in air capacity, both of which are negatively correlated with air capacity (Hall et al., 1977). Air capacity has received less attention than available water, but as a measurement of coarse porosity, it is a useful indicator of saturated hydraulic conductivity and aeration. Divergence between water content at 5 kPa and field capacity 48 h after saturation is likely to be greatest in coarse-textured soils with shallow ground water and in peaty soils. Elsewhere, many believe that the difference would be small and often within the random error inherent in physical measurements of this nature. It has been suggested that an important advantage of using laboratory-measured 5-kPa water content, instead of a field measurement, e.g., tensiometers, is that horizons of a similar nature can be compared on a common basis, i.e., standardized laboratory procedure. Similarly, air capacity at 5 kPa has been considered a much more reproducible value than air voids at field capacity, even though a field measurement would be useful in an intensive investigation of an experimental site (Hall et al., 1977). A large difference in air capacity between two soil horizons would be expressed as a large difference in conductivity.

Air-filled porosity, 5 kPa, calculation: The air-filled porosity at 5 kPa (Soil Survey Division Staff, 1993; Grossman et al., 1994) may be calculated as follows:

Equation 3.3.1.1:

$$\text{AFP} = \{[100 - (100 \times \rho_{B33})/\rho_p] - (W_5 \times \rho_{B33})\} \times [(1 - V_{>2mm})/100]$$

where
AFP = Air-filled porosity at 5-kPa water content. Total porosity minus volume fraction of water at 5 kPa.
ρ_{B33} = Bulk density at 33-kPa water content on a <2-mm soil basis (g cm^{-3})
W_5 = Weight percentage of water at 5 kPa. Data obtained from soil water retention curve. Refer to Appendix 2 (Wildmesa Pedon) and Appendix 5 (Caribou Pedon) for example water retention curves.
$V_{>2mm}$ = Volume percentage of >2-mm fraction
ρ_p = Particle density (g cm^{-3}). Calculate ρ_p as follows:

Equation 3.3.1.2:

$$\rho_p = 100/\{[(\text{SOC} \times 1.7)/\rho_{p1}] + [(\text{Fe} \times 1.6)/\rho_{p2}] + [100 - (\text{Fe} \times 1.6) + (\text{SOC} \times 1.7)/\rho_{p3}]\}$$

where

ρ_p = Particle density (g cc^{-1})

SOC = Weight percentage of soil organic C on a <2-mm soil basis

Fe = Weight percentage of dithionite-citrate extractable Fe on a <2-mm soil basis

ρ_{p1} = 1.4 g cm^{-3}, assumed particle density of organic matter

ρ_{p2} = 4.2 g cm^{-3}, assumed particle density of the minerals from which dithionite-citrate extractable Fe originates

ρ_{p3} = 2.65 g cm^{-3}, assumed particle density of material exclusive of organic matter and minerals contributing to the dithionite-citrate extractable Fe

Equation 3.3.1.1 may be used to calculate air-filled porosity at any suction by substitution of the weight percentage of water at the specified suction for W_5. The water content for any suction can be computed from the calculated soil water characteristic.

Available water, water retention, and porosity, general trends: Some general trends in water retention, available water, and air capacity have been cited (Hall et al., 1977) for both topsoil and subsoil samples, as follows: unavailable water increases with rising clay content; air capacity increases with increasing amounts of sand; available water is at a maximum for silty classes and at a minimum for sandy classes; and bulk density (inversely related to total pore space) is generally higher for sandy particle-size classes. The main difference between topsoils and subsoils is that the amount of available water is higher in topsoils than in subsoils because of the higher content of organic matter and the inherently lower density of the topsoils. Other trends (Hall et al., 1977) are that clayey soils release small amounts of water at low tensions, retaining about one-half of the available water at suctions >200 kPa (2 bar); clay loams reflect the same general release characteristics, although they hold slightly more water at suctions <200 kPa; sandy soils hold small quantities of water at high suctions and hence the majority is easily available, approximately one-half at suctions <400 kPa (4 bar); and silt loams and sandy silt loams have a more even distribution of water throughout the available range. In practice, the proportion of easily available water (<200 kPa, 2 bar) in a soil may be agronomically significant as total available water in assessing droughtiness. Total porosity may not necessarily be the best indicator of aeration or water movement. More recently, emphasis has been on pore-size distribution. The >10-μm diameter range has been considered by some as critical for water movement, aeration, root growth, and soil fauna, and the >60-μm diameter range has also been used as a good indicator of saturated hydraulic conductivity (Hall et al., 1977).

Expressions related to water content: Calculating the amount of pore space and the amount of water in the pore space is often a complex soil physics problem. Some general definitions and relationships, e.g., bulk density and porosity, are required so that comparisons between soils are appropriate. Some of these definitions and relationships as well as techniques to calculate soil water content (Skopp, 1992) are as follows:

Equation 3.3.1.3:

$$\rho_B = M_s/V_{s+v}$$

73

where
ρ_B = Bulk density of soil
M_s = Mass of solids
V_{s+v} = Volume of solids + volume of voids = volume of soil

Bulk density is highly dependent on soil conditions at the time of sampling. Changes in soil swelling due to changes in water content alter the bulk density. Once the bulk density is specified, then the relative amount of pore space also is fixed. The amount of pore space is usually described in terms of volumes (ratio of volumes), as follows:

Equation 3.3.1.4:

$$\varepsilon = V_v/V_{s+v}$$

where
ε = Total porosity
V_v = Volume of voids
V_{s+v} = Volume of solids + volume of voids = volume of soil

Using the definitions for bulk density and particle density, the derivation of a formula for porosity based on these properties is as follows:

Equation 3.3.1.5:

$$\varepsilon = 1 - (\rho_B/\rho_p)$$

where
ε = Total porosity
ρ_B = Bulk density of soil
ρ_p = Particle density of soil

This relationship is not empirical but is the result of definitions that confirm that for every value of bulk density for a specified soil there is one possible value of porosity. However, a soil does not have one possible value for bulk density.

Porosity is usually defined as a ratio of volumes which is dimensionless and, thus, can just as easily be defined as a ratio of equivalent depths. In order to make this relationship, a comparison is required based on equal cross-sectional areas (A) which comprise the volumes as follows:

Equation 3.3.1.6:

$$\varepsilon = V_v/V_s = Ad_v/Ad_s = d_v/d_s$$

where
ε = Total porosity

74

V_v = Volume of voids
V_s = Volume of soil
A = Cross-sectional area
d_v = Depths of voids
d_s = Depths of soil

Unlike voids, which are usually related in terms of volume, the amount of soil water can be expressed on either a mass (gravimetric) or volumetric basis as follows:

Equation 3.3.1.7:

$$\theta_m = M_w/M_s$$

where
θ_m = Gravimetric water content
M_w = Mass of water
M_s = Mass of solids

Equation 3.3.1.8:

$$\theta_v = V_w/V_s$$

where
θ_v = Volumetric water content
V_w = Volume of water
V_s = Volume of soil

The gravimetric water is based on dry solids, whereas the volumetric water is based on the volume of the soil (solids, water, and gas) at the moisture content at the time of measurement. These water content values can be related as follows:

Equation 3.3.1.9:

$$\theta_v = (\theta_m \ x \ \rho_B)/\rho_w$$

where
θ_v = Volumetric water
θ_m = Gravimetric water
ρ_B = Bulk density
ρ_w = Particle density of water

The depth of water can be related to the volumetric water as follows:

Equation 3.3.1.10:

$$d_w = (\theta_v \ x \ d_s)$$

where
d_w = Depth of water
θ_v = Volumetric water
d_s = Depth of soil

The maximum soil water content (saturation) is the point at which all the voids are filled with water. Saturation may be defined as follows:

Equation 3.3.1.11:

$$\theta_v = E$$

where
θ_v = Volumetric water
E = Total porosity

In reality, saturated soils are uncommon since a small amount of gas is typically present even after prolonged wetting; i.e., the soil is satiated (Skopp, 1992). The water content of a satiated soil has no fixed value and will change with time (as gas diffuses out of soil) and is strongly dependent on the soil water content prior to wetting as well as the manner of wetting (Skopp, 1992).

A number of other expressions are used to characterize the amount of water or air in the soil. These expressions (Skopp, 1992) are as follows:

Equation 3.3.1.12:

$$\theta_A = V_a/V_{s+v}$$

where
θ_A = Air-filled porosity
V_a = Volume of air
V_{s+v} = Volume of solids + volume of voids = volume of soil

Equation 3.3.1.13:

$$\theta_R = (\theta_v/E)$$

where
θ_R = Relative saturation
θ_v = Volumetric water
E = Total porosity

Equations 3.3.1.14 and 3.3.1.15:

$$\lambda = V_v/V_s = \lambda = \varepsilon/(1 - \varepsilon)$$

where
λ = Void ratio
V_v = Volume of voids
V_s = Volume of solids
ε = Total porosity

Water retention methods and general applications: Two desorption procedures—suction and pressure methods—are commonly used to measure water retention. The SSL uses the pressure method (U.S. Salinity Laboratory Staff, 1954) with either a pressure-plate or pressure-membrane extractor (Soil Survey Staff, 2004). Except for the 1500-kPa measurement, the pressure-plate extractor is used for all SSL water retention procedures. These procedures are used for the water retention function, water-holding capacity, pore-size distribution, porosity, and saturated conductivity of a soil sample at specific water contents.

Water retention measurements: In this section, the SSL water retention procedures are described. The SSL reports water retention as percent gravimetric water. Major method categories and their associated water retention and material types are as follows:

- Pressure-plate extraction
    - 6, 10, 33, 100, 200 kPa, <2-mm, air-dry sieved samples
    - 6, 10, 33, or 100 kPa, natural clods
    - 6, 10, 33, or 100 kPa, soil cores
    - 33 kPa, rewet, natural clods
- Pressure-membrane extraction
    - 1500-kPa, <2-mm (sieved), air-dry or field-moist soil sample
- Field-state
    - cores, clods, or bulk

3.3 Water Retention
3.3.2 Pressure-Plate Extraction
3.3.2.1–5   **6, 10, 33, 100, or 200 kPa**
3.3.2.1–5.1.1   **<2-mm (Sieved), Air-Dry**
3.3.2.1–4.2   **Natural Clods**
3.3.2.1–4.3   **Soil Cores**
3.3.2.1.3.4   **Rewet**
3.3.2.1.3.5   **Reconstituted**

Pressure-plate extraction, 6, 10, 33, 100, or 200 kPa, <2-mm (sieved), air-dry samples, measurements: A <2-mm (sieved), air-dry soil sample of nonswelling loamy sand or coarser soil and of some sandy loams is placed in a retainer ring sitting on a porous ceramic plate in a pressure-plate extractor. The plate is covered with water to wet the sample by capillarity. The sample is equilibrated at the specified pressure (6, 10,

33, 100, or 200 kPa; 0.06, 0.1, 1/3, 1, or 2 bar, respectively). The pressure is kept constant until equilibrium is obtained (Klute, 1986).

Pressure-plate extraction, 6, 10, 33, or 100 kPa, natural clods, measurements: Natural clods are placed on a tension table and equilibrated at a 5-cm tension at the base of the sample. The clods are then transferred to a porous ceramic plate, which is placed in a pressure-plate extractor. The sample is equilibrated at the specified pressure (6, 10, 33, or 100 kPa). The pressure is kept constant. Equilibrated samples are weighed, oven dried at 110 °C overnight, and then weighed again. This procedure is usually used in conjunction with the SSL method for bulk density at 33 kPa (ρ_{B33}).

Pressure-plate extraction, 6, 10, 33, or 100 kPa, soil cores, measurements: A metal cylinder is pressed or driven into the soil. Upon removal from the soil, the cylinder extracts a sample of known volume. The sample weight is recorded. The sample is dried in the oven and then weighed. The soil core is placed on a tension table and equilibrated at a 5-cm tension at the base of the sample. The core is then transferred to a porous ceramic plate, which is placed in a pressure-plate extractor. The sample is equilibrated at the specified pressure (6, 10, 33, or 100 kPa). The pressure is kept constant until equilibrium is obtained. This procedure is usually used in conjunction with the SSL method for bulk density at 33 kPa (ρ_{B33}). The equilibrated sample is oven dried at 110 °C overnight and then weighed.

Pressure-plate extraction, 33 kPa, rewet, natural clods, measurements: Natural clods are equilibrated at 33 kPa, air dried, and reequilibrated. The resulting data are called rewet water retention and are usually used in conjunction with the rewet bulk density to estimate changes in physical properties of a soil as it undergoes wetting and drying cycles. Natural clods are placed on a tension table and equilibrated at a 5-cm tension at the base of the sample. The clods are then transferred to a porous ceramic plate, which is placed in a pressure-plate extractor. The samples are equilibrated at 33 kPa. The pressure is kept constant until equilibrium is obtained. The clods are air dried and then placed on a tension table and desorbed again. The equilibrated samples are oven dried at 110 °C overnight and then weighed.

Pressure-plate extraction, 33 kPa, reconstituted, measurement: Natural clods are placed on a tension table and equilibrated at a 5-cm tension at the base of the sample. The clods are then transferred to a porous ceramic plate, which is placed in a pressure-plate extractor. The samples are equilibrated at 33 kPa. The pressure is kept constant until equilibrium is obtained. The equilibrated samples are oven dried at 110 °C overnight and then weighed.

3.3   Water Retention
3.3.3   Pressure-Membrane Extraction
3.3.3.1   1500 kPa
3.3.3.1.1   <2-mm (Sieved)
3.3.3.1.1.1–2   Air-Dry or Field-Moist

Pressure-membrane extraction, 1500 kPa, <2-mm (sieved), air-dry or field-moist samples, measurements: A <2-mm (sieved), air-dry soil sample is placed in a retainer ring sitting on a cellulose membrane in a pressure-membrane extractor. The membrane is covered with water to wet the sample by capillarity. The sample is equilibrated at 1500 kPa. The pressure is kept constant until equilibrium is obtained. The equilibrated sample is oven dried at 110 °C overnight and then weighed. Soils with gypsum are a special case because gypsum ($CaSO_4 \cdot 2H_2O$) loses most of its two water molecules at 105 °C (Nelson et al., 1978). Properties of soils with gypsum, such as 1500-kPa water content, that are reported on an oven-dry weight basis are converted to include the weight of crystal water in gypsum.

Pressure-plate and pressure-membrane methods, interferences: Laboratory-determined water retention data are usually higher than field-determined water retention data because the confining soil pressure is not present in the laboratory (Bruce and Luxmoore, 1986). Water retention data for soils with expansive clay are overestimated when sieved samples are used in place of natural soil fabric for tensions of 6, 10, and 33 kPa (Young and Dixon, 1966).

Aerated 0.005 M $CaSO_4$ has also been recommended for use in determining water retention (Dane and Hopmans, 2002), especially for fine-textured soils that contain significant amounts of swelling clays. Distilled or deionized water can possibly promote dispersion of clays in samples, and freshly drawn tapwater is often supersaturated with air, affecting the water content at a given pressure head (Dane and Hopmans, 2002).

3.3   Water Retention
3.3.4   Field-State
3.3.4.1   Cores
3.3.4.2   Clods
3.3.4.3   Bulk

Field-state, measurements and interferences: Soil samples (cores, clods, or bulk) are collected in the field. The samples are stored in plastic or metal containers to prevent drying and are then transported to the laboratory. Field water content is determined by weighing, drying, and reweighing a soil sample. The resulting data are used to estimate the water content at the time of sampling. Field-moist water content is calculated as follows:

Equation 3.3.3:

$$H_2O \% = 100 \text{ x } (M_{s+w} - M_s)/(M_s - M_c)$$

where
$H_2O \%$ = Percent gravimetric water content
M_{s+w} = Weight of solids and H_2O (g) + container (g)
M_s = Oven-dry weight of solids (g) + container (g)
M_c = Weight of container (g)

Leaks in the plastic or metal storage containers can cause the samples to dry, resulting in an underestimation of the field water content.

3.4 Ratios, Estimates, and Calculations Related to Particle-Size Analysis, Bulk Density, and Water Retention

This section describes the ratios, estimates, and calculations provided by the SSL related to particle-size analysis, bulk density, and water retention. These methods include the ratios for air-dry/oven-dry (AD/OD), field-moist/oven-dry (FM/OD), and correction for crystal water. This section also includes the calculations of the coefficient of linear extensibility (COLE), the water retention difference (WRD) at various suctions, the 1500-kPa water content/total clay ratio, and various particle-size fractions (e.g., total silt, total sand, and >2-mm fractions). In addition, this section provides definitions of terms as well as applications of these ratios, estimates, and calculations. For detailed descriptions of the SSL methods which are cross-referenced by method code in the table of contents in this manual, refer to SSIR No. 42 (Soil Survey Staff, 2004), which is available online at http://soils.usda.gov/technical/lmm/.

3.4 Ratios, Estimates, and Calculations Related to Particle-Size Analysis, Bulk Density, and Water Retention
3.4.1 Air-Dry/Oven-Dry Ratio (AD/OD)
3.4.2 Field-Moist/Oven-Dry Ratio (FM/OD)
3.4.3 Correction for Crystal Water

Air-dry/oven-dry ratio and field-moist/oven-dry ratio: Soil properties generally are expressed on an oven-dry weight basis. The calculation of the air-dry/oven-dry (AD/OD) ratio or field-moist/oven-dry (FM/OD) ratio is used to adjust all results to an oven-dry basis and, if required in a procedure, to calculate the sample weight that is equivalent to the required oven-dry soil weight.

The *air-dry (AD) and oven-dry (OD) weights* are defined herein as constant sample weights obtained after drying at 30 ± 5 °C ($\approx$ 2 to 7 days) and at 110 ± 5 °C ($\approx$ 12 to 16 h), respectively. As a general rule, air-dry soils contain about 1 to 2 percent water and are drier than soils at 1500-kPa water content. FM weight is defined herein as the sample weight obtained without drying prior to laboratory analysis. In general, these weights are reflective of the water content at the time of sample collection.

A sample is weighed, dried to a constant weight in an oven, and reweighed. The moisture content is expressed as a ratio of the air-dry to the oven-dry weight (AD/OD) or as a ratio of field-moist to the oven-dry weight (FM/OD).

Correction for crystal water: Soils with gypsum are a special case because gypsum ($CaSO_4 \cdot 2H_2O$) loses most of its two water molecules at 105 °C. Properties of soils with gypsum that are reported on an oven-dry weight basis should be converted to include the weight of crystal water in gypsum. The AD/OD ratio is calculated. This ratio is used to convert soil properties to an oven-dry basis. The *AD/OD ratio is converted to a crystal water basis* (Nelson et al., 1978). The inclusion of weight of crystal water in gypsum allows the properties of soils with gypsum to be compared with those properties of soils with no gypsum. This conversion also avoids the possible calculation error of obtaining >100 percent gypsum when the data are expressed on an oven-dry basis (Nelson, 1982).

Properties of soils with gypsum that are reported on an oven-dry weight basis are converted to include the weight of the crystal water. When the water content of soils with gypsum is reported, the crystal water content must be subtracted from the total oven-dry water content. The AD/OD ratio is corrected to a crystal water basis when the gypsum content of the soil is ≥1 percent.

3.4 Ratios, Estimates, and Calculations Related to Particle-Size Analysis, Bulk Density, and Water Retention

3.4.4 Coefficient of Linear Extensibility (COLE)

Coefficient of linear extensibility, definition: *Coefficient of linear extensibility (COLE)* is a derived value that denotes the fractional change in the clod dimension from a moist to a dry state (Franzmeier and Ross, 1968; Grossman et al., 1968; Holmgren, 1968). COLE may be used to make inferences about shrink-swell capacity and clay mineralogy. The COLE concept does not include irreversible shrinkage, such as that occurring in organic soils and some andic soils. Certain soils with relatively high contents of smectite clay have the capacity to swell significantly when moist and to shrink and crack when dry. This shrink-swell potential is important for soil physical qualities (large, deep cracks in dry seasons) as well as for genetic processes and soil classification (Buol et al., 1980). Greene-Kelly (1974) found that soils with equal amounts of kaolinite and smectite are similar to those with smectite alone. In a study of Vertisols in El Salvador, Yerima et al. (1985, 1987) found that kaolinite-rich, fine clay soils have physical behavior (shrink-swell) similar to that of smectitic soils because of their large surface area.

COLE can also be expressed as percent, i.e., *linear extensibility percent (LEP)*. LEP = COLE x 100. The LEP is not the same as LE. In *Keys to Soil Taxonomy* (Soil Survey Staff, 2010), *linear extensibility (LE)* of a soil layer is the product of the thickness, in centimeters, multiplied by the COLE of the layer in question. The LE of a soil is defined as the sum of these products for all soil horizons (Soil Survey Staff, 2010). Refer to Soil Survey Staff (2010) for additional discussion of LE.

Coefficient of linear extensibility, air-dry or oven-dry to 33-kPa tension, calculation: The SSL calculates the COLE for the whole soil (air-dry or oven-dry to 33-kPa suction). The COLE value is reported in cm cm^{-1}. Calculate COLE when coarse fragments are present as follows:

Equation 3.4.4.1:

$$COLE_{ws} = \{1/[Cm \times (\rho_{B33<2mm}/\rho_{Bod<2mm}) + (1 - Cm)]\}^{1/3} - 1$$

where
$COLE_{ws}$ = Coefficient of linear extensibility on whole-soil basis
$\rho_{B/33<2mm}$ = Bulk density at 33-kPa water content on <2-mm soil basis (g cm^{-3})
$\rho_{Bod<2mm}$ = Bulk density, oven-dry or air-dry, on <2-mm soil basis (g cm^{-3})
Cm = Coarse fragment (moist) conversion factor

Equation 3.4.4.2:

If no coarse fragments, Cm = 1. If coarse fragments are present, calculate Cm as follows:

$$Cm = Vol_{<2mm}/Vol_{whole}$$

where
$Vol_{<2mm}$ = Volume moist <2-mm fabric (cm^3)
Vol_{whole} = Volume moist whole soil (cm^3)

OR (alternatively)

Equation 3.4.4.3:

$$Cm = (100 - Vol_{>2mm})/100$$

where
$Vol_{>2mm}$ = Volume percentage of the >2-mm fraction

Equation 3.4.4.4:

If no coarse fragments, Cm = 1, the previous equation reduces as follows:

$$COLE_{ws} = (\rho_{Bod<2mm}/\rho_{B33<2mm})^{1/3} - 1$$

where
$COLE_{ws}$ = Coefficient of linear extensibility on whole-soil basis
$\rho_{Bod<2mm}$ = Bulk density, oven-dry or air-dry, on <2-mm soil basis (g cm^{-3})
$\rho_{B33<2mm}$ = Bulk density at 33-kPa water content on <2-mm soil basis (g cm^{-3})

3.4   Ratios, Estimates, and Calculations Related to Particle-Size Analysis, Bulk Density, and Water Retention
3.4.5   Water Retention Difference (WRD), Whole Soil

Water retention difference, definition: The calculation of the *water retention difference (WRD)* is considered the initial step in the approximation of the *available water capacity (AWC)*. Refer to the section on water retention for more information on AWC estimates. The WRD does not allow for restriction of roots from the soil layer or osmotic pressure. The volume of rock fragments is usually considered a diluent containing no water between the suctions that define WRD. The WRD, as defined by the SSL, is a calculated value that denotes the volume fraction for water in the whole soil that is retained between 1500-kPa suction and an upper limit of usually 33- or 10-kPa suction. The upper limit (lower suction) is selected so that the volume of water retained approximates the volume of water held at field capacity. The 10-, 33-, and

1500-kPa gravimetric water contents are then converted to a whole soil volume basis by multiplying by the bulk density (ρ_{B33}) and adjusting downward for the volume fraction of rock fragments, if present in the soil. The lower suctions, e.g., 10- or 5-kPa, are used for coarse materials. Refer to Soil Survey Division Staff (1993) and Grossman et al. (1994) for additional discussion on coarse materials and the significance of soil water content at lower suctions, e.g., 5 kPa and 10 kPa, as well as suggestions for the selection of these lower suctions for the determination of water retention difference (WRD).

Water retention difference, between 33 and 1500 kPa, calculation: The SSL calculates the WRD between 33- and 1500-kPa suctions in the whole soil. The WRD is reported as centimeters of water per centimeter of depth of soil (cm cm^{-1}), but the numbers do not change when other units, e.g., in in^{-1} or ft ft^{-1}, are needed. The WRD with W_{33} as the upper limit is reported as cm cm^{-1}. This WRD is calculated on a whole-soil basis as follows:

Equation 3.4.5.1:

$$WRD_{ws} = [(W_{33<2mm} - W_{1500<2mm}) \times (\rho_{B33<2mm}) \times Cm]/(P_w \times 100)$$

where
WRD_{ws} = Volume fraction (cm^3 cm^{-3}) of water retained in the whole soil between 33-kPa and 1500-kPa suction reported in cm cm^{-1}
$W_{33<2mm}$ = Weight percentage of water retained at 33-kPa suction on <2-mm soil basis
$W_{1500<2mm}$ = Weight percentage of water retained at 1500-kPa suction on <2-mm soil basis. If available, moist 1500 kPa is the first option in the WRD calculation; otherwise, dry 1500 kPa is used.
$\rho_{B33<2mm}$ = Bulk density at 33-kPa water content on a <2-mm soil basis (g cm^{-3})
P_w = Density of water (1 g cm^{-3})
Cm = Coarse fragment material conversion factor. If no coarse fragments, $Cm = 1$. If coarse fragments are present, calculate Cm as follows:

Equation 3.4.5.2:

$$Cm = Vol_{<2mm}/Vol_{whole}$$

where
$Vol_{<2mm}$ = Volume moist <2mm fabric (cm^3)
Vol_{whole} = Volume moist whole soil (cm^3)

OR (alternatively)

Equation 3.4.5.3:

$$Cm = (100 - Vol_{>2mm})/100$$

where
$Vol_{>2mm}$ = Volume percentage of the >2-mm fraction

Water retention difference (between 10 and 1500 kPa), calculation: The SSL also calculates the WRD between 10-kPa (W_{10}) and 1500-kPa suctions (W_{1500}). This WRD value can be calculated by substituting the W_{10} in place of W_{33} in equation 3.4.5.1. The W_{10} may be used as the upper limit of plant-available water for coarse soil materials.

Water retention difference (between 33 kPa rewet and 1500 kPa), calculation: The SSL also calculates the WRD between 33 kPa rewet (W_r) and W_{1500}. This WRD value can be calculated by substituting the W_r in place of W_{33} in equation 3.4.5.1. The W_r is used for organic materials.

3.4 Ratios, Estimates, and Calculations Related to Particle-Size Analysis, Bulk Density, and Water Retention

3.4.6 1500-kPa Water Content/Total Clay

1500-kPa water content/total clay, calculation: Divide the percent 1500-kPa water retention by the total clay percentage. This ratio is reported as a dimensionless value. In the past, the ratios of 1500-kPa water:clay have been reported as g g^{-1}. For more detailed information on the application of this ratio, refer to Soil Survey Staff (2010).

1500-kPa water content/total clay, data assessments: Water retention at 1500 kPa is considered the wilting point for many agricultural plants and has been equated with water retained in pores $\leq$0.2-μm diameter and on particle surfaces. Therefore, due to the much greater amount of surface area of clay-sized materials on a per weight basis relative to silt and sand, a high correlation exists between 1500-kPa water and clay content. Thus, this ratio is a good tool for data assessment for dispersion of clays during particle-size analysis (National Soil Survey Laboratory Staff, 1983). A reference point for soils dominated by silicates that disperse well in the standard PSDA is as follows:

Equation 3.4.6.1:

$$W_{1500}/Clay_T \approx 0.4$$

OR (alternatively)

Equation 3.4.6.2:

$$Clay_T \approx 2.5 \times W_{1500}$$

where
W_{1500} = Weight percentage of water retained at 1500-kPa suction on a <2-mm soil basis
$Clay_T$ = Weight percentage of total clay on a <2-mm soil basis

A number of soil-related factors can cause deviation from this 0.4 reference point. Low-activity clays, e.g., kaolinites, chlorites, and some micas, tend to lower the ratio to

≤0.35. High-activity clays, e.g., smectites and some vermiculites, tend to increase this ratio. The relationship between 1500-kPa water and the amount of clay has been characterized for groups of soils dominated by different kinds of clay minerals (National Soil Survey Laboratory Staff, 1990), and some average ratios are as follows: 0.45 for smectite ($r^2 = 0.88$, n = 547); 0.42 for clay mica ($r^2 = 0.90$, n = 493); and 0.32 for Bt horizons of Paleudults ($r^2 = 0.98$, n = 18). Whether these differences in the ratios are caused by differences in clay mineralogy or by differences in other properties associated with the different clay mineralogies is not known. Likewise, Wilson et al. (2002) studied a group of soils in eastern Oregon and found that horizons with a subsoil clay mineralogy dominated by smectite and vermiculite had a better correlation of measured and calculated clay using a factor of 1.7 (1500 kPa/clay = 0.59).

Keys to Soil Taxonomy (Soil Survey Staff, 2010) uses this ratio to determine the adequacy of laboratory-measured clay for family particle-size classes; i.e., the 1500 kPa/clay ratio should be between 0.4 and 0.6 to document adequate dispersion. Failure of this relationship in the majority of the particle-size control section results in the use of calculated clay as follows:

Equation 3.4.6.3:

$$\text{Clay (\%)} = (1500\text{-kPa water} - \% \text{ organic C}) \times 2.5$$

Due to the low water retention capacity of minerals (e.g., kaolinite, Fe oxides) in highly weathered soils, the calculation when used for oxic horizon criteria (Soil Survey Staff, 2010) is as follows:

Equation 3.4.6.4:

$$\text{Clay (\%)} = (1500\text{-kPa water} - \% \text{ organic C}) \times 3.0$$

Keys to Soil Taxonomy (Soil Survey Staff, 2010) recognizes that soil organic matter (SOM) increases the 1500-kPa water retention and subsequently increases the 1500 kPa/clay ratio. An increased 1500 kPa to clay ratio can be expected if the soil organic carbon (SOC) percent is >0.1 of the percent clay (National Soil Survey Laboratory Staff, 1983). In a study of 34 Borolls (National Soil Survey Laboratory Staff, 1990), each percentage increase in SOC increased the 1500-kPa percentage approximately 1.5 percent ($r^2 = 0.67$). In a study of 53 Xerolls (National Soil Survey Laboratory Staff, 1990), each percentage increase in SOC increased the 1500-kPa percentage approximately 1.3 percent ($r^2 = 0.72$).

Poorly crystalline materials also tend to increase this ratio. If this ratio is >0.6 and SOC does not adequately explain the increased value, incomplete dispersion in PSDA may be a factor (National Soil Survey Laboratory Staff, 1983). Soil components which act as cements and cause poor dispersion include gypsum, Fe oxides, and poorly crystalline Si. Soils from volcanic materials or with andic soil properties (Soil Survey Staff, 2010) can be dominated by poorly crystalline minerals, such as allophane or imogolite. These minerals lack a discrete and well defined particle shape as typical of most phyllosilicate minerals and thus do not react in the same manner during particle-size analysis. These minerals may be more porous or gel-like and may form coatings on other particle grains. Thus, the standard theory used for particle-size measurement is not functional and calculations of clay content based on 1500-kPa water retention are inadequate. Therefore, Andisols and soils that have andic soil properties or are composed of a large amount of pyroclastic materials (volcanic glass, pumice, or cinders) have been assigned a substitute particle-size family class in *Keys to Soil Taxonomy* (Soil Survey Staff, 2010).

Clay-sized carbonate tends to decrease the ratio in most cases. The 1500-kPa water retention for carbonate clays is ≈ 2/3 the corresponding value for the noncarbonate clays (Nettleton et al., 1991). Clay used for determination of family particle-size classes is silicate clay, and thus carbonate clay measured in the laboratory must be subtracted from the total (measured or calculated) clay for family particle-size class (Soil Survey Staff, 2010).

In sandy textured soils, the low amount of clays results in SOC and surface area of other nonclay constituents having a greater influence on the 1500-kPa water retention. Any small increase or decrease in measured clay can result in a large change of the 1500 kPa/clay ratio. For these reasons, ratios above 0.5 for some samples with less than 5 to 10 percent clay may erroneously indicate poor PSDA dispersion (National Soil Survey Laboratory Staff, 1990) and application of this ratio must used judiciously.

Poor dispersion by the SSL standard particle-size distribution analysis (PSDA) is typical in Andisols. In some soils, however, poor dispersion emphasizes one of the fundamental guidelines of the laboratory, i.e., standard methods. Documenting the response of a particular soil to a standard operating procedure is necessary in order to determine differences between soils. These comparisons have been critical factors in developing many relationships used in understanding soils and in the development of the U.S. soil taxonomic system (Soil Survey Staff, 2010). Not all soils are composed of well defined particles that could be dispersed into their appropriate fraction if only the "correct method" were used. In addition, there is no exacting measure of poor dispersion. The 1500 kPa/clay ratio is one measure, and the comparison of laboratory-versus field-determined soil textures and clay contents is another (Nettleton et al., 1999). Neither measure nor indicator is perfect. Alternative and additional pretreatments may extract additional clay from a soil sample, but are these pretreatments freeing or creating clay particles? The fact that a particular soil sample reacts differently to a standard method in itself provides information concerning the soil's properties.

3.4   Ratios, Estimates, and Calculations Related to Particle-Size Analysis, Bulk Density, and Water Retention
3.4.7   Total Silt Fraction

Total silt, definition and application: *Total silt* is a soil separate with 0.002- to 0.05-mm particle diameter. The SSL determines the fine silt separate by pipet analysis and the coarse silt separate by difference. The silt to clay ratio is an important criterion for classification of soils in the Tropics (Van Wambeke, 1962) and for evaluating such phenomena as clay migration, stage of weathering, and age of parent material (Ashaye, 1969). In general, soils with high silt contents are associated with unweathered soils that are fertile. Sandy clay and sandy clay loam textures (low silt contents) are common in highly weathered materials. Total silt is reported as a weight percentage on a <2-mm soil basis.

3.4   Ratios, Estimates, and Calculations Related to Particle-Size Analysis, Bulk Density, and Water Retention
3.4.8   Total Sand Fraction

Total sand, definition and application: *Total sand* is a soil separate with 0.05- to 2.0-mm particle diameter. The SSL determines the sand fractions by sieve analysis. Total sand is the sum of the very fine sand (VFS), fine sand (FS), medium sand (MS), coarse sand (CS), and very coarse sand (VCS) fractions. The rationale for five subclasses of sand and the expansion of the texture classes of sand, e.g., sandy loam and loamy sand, is that the sand separates are the most visible to the naked eye and the most detectable by "feel" by the field soil scientist. Total sand is reported as a weight percentage on a <2-mm soil basis.

3.4   Ratios, Estimates, and Calculations Related to Particle-Size Analysis, Bulk Density, and Water Retention
3.4.9   2- to 5-mm Fraction

2- to 5-mm fraction, definition: The *2- to 5-mm fraction* corresponds to the size of opening of the No. 10 screen and the No. 4 screen (4.76 mm), respectively, used in engineering. Coarse fractions with 2- to 5- mm particle diameter correspond to the nonflat rock fragment class *fine gravel* (USDA/NRCS, 2009b). Coarse fractions with 2- to 5-mm particle diameter are reported as a weight percentage on a <75-mm basis.

3.4   Ratios, Estimates, and Calculations Related to Particle-Size Analysis, Bulk Density, and Water Retention
3.4.10   5- to 20-mm Fraction

5- to 20-mm fraction, definition: The *5- to 20-mm fraction* corresponds to the size of opening of the No. 4 screen (4.76 mm) and the 3/4-in screen (19.05 mm), respectively, used in engineering. Coarse fractions with 5- to 20-mm particle diameter correspond to the nonflat rock fragment class *medium gravel* (USDA/NRCS, 2009b). Coarse fractions with 5- to 20-mm particle diameter are reported as a weight percentage on a <75-mm basis.

3.4   Ratios, Estimates, and Calculations Related to Particle-Size Analysis, Bulk Density, and Water Retention
3.4.11   20- to 75-mm Fraction

20- to 75-mm fraction, definition: The *20- to 75-mm fraction* corresponds to the size of opening of the 3/4-in screen (19.05 mm) and the 3-in screen (76.1 mm), respectively, used in engineering. Coarse fractions with 20- to 75-mm particle diameter correspond to the nonflat rock fragment class *coarse gravel* (USDA/NRCS, 2009b). Coarse fractions with 20- to 75-mm particle diameter are reported as a weight percentage on a <75-mm basis.

3.4 Ratios, Estimates, and Calculations Related to Particle-Size Analysis, Bulk Density, and Water Retention
3.4.12   0.1- to 75-mm Fraction

0.1- to 75-mm fraction, definition: The *75-mm fraction* corresponds to the size of opening in the 3-in screen (76.1 mm) used in engineering. These data are listed for taxonomic placement for particle-size class, i.e., to distinguish loamy and silty family particle-size classes. Refer to Soil Survey Staff (2010) for additional discussion on particle-size classes. Coarse fractions with 0.1- to 75-mm particle diameter are reported as a weight percentage on a <75-mm basis.

3.4 Ratios, Estimates, and Calculations Related to Particle-Size Analysis, Bulk Density, and Water Retention
3.4.13   >2-mm Fraction

>2-mm fraction, definition: Coarse fractions with >2-mm particle diameter are reported as a weight percent on a whole-soil basis. For more information on these data, refer to Soil Survey Division Staff (1993) and Soil Survey Staff (2010).

3.5 Micromorphology
3.5.1   Thin Sections

This section summarizes the method for preparation of a thin section. In addition, it provides background information related to micromorphology, description and terminology of microfabrics, and interpretations of these fabrics (Nettleton, 2004).

3.5 Micromorphology
3.5.1   Thin Sections
3.5.1.1   Preparation

Micromorphology is used to identify fabric types, skeleton grains, weathering intensity, and illuviation of argillans and to investigate genesis of soil or pedological features. In this method, a soil clod is impregnated with a polymer resin (Innes and Pluth, 1970). A flat surface of the soil sample is glued to a glass slide. The soil sample is cut and ground to a thickness of ≈ 30 μm. The thin section is examined with a petrographic microscope (Anonymous, 1987; Cady, et al., 1986).

3.5 Micromorphology
3.5.1 Thin Sections
3.5.1.2 Interpretations

Background

Micromorphology may be defined as the study of soils or regolith samples in their natural undisturbed arrangement using microscopic techniques (Cady et al., 1986; Stoops, 2003). This technique, also termed microfabric analysis, entails descriptive terminology that has been developed over the past 50 years. The science and terminology of microfabric analysis were initially documented by Kubiena (1938). Since then, important publications documenting terminology have included Brewer (1964); FitzPatrick (1984, 1993); Bullock et al. (1985); and Stoops (2003). Methodological descriptions for producing thin sections can be found in Cady et al. (1986); FitzPatrick (1984); Murphy (1986); Fox et al. (1993); and Fox and Parent (1993). An excellent book on examination of mineral weathering in thin sections is Delvigne (1998), and the Soil Science Society of America (1993) has a CD collection of images that illustrate many features of microfabrics.

Examination of thin sections with a polarizing light microscope can be considered an extension of field morphological studies. The level of resolution increases from field examination to optical microscopic examination and finally to submicroscopic techniques (electron microscopy), but this sequence of techniques increasingly sacrifices field of view (Cady et al., 1986). Thus, the results of micromorphological studies are most useful when they are combined with other field (landscape description, pedon morphological description) and laboratory data (Cady, 1965). Micromorphology is used to identify types and sequences of active processes occurring in soils via identification of argillans, fabric types, skeleton grains, and weathering intensity. It is an ideal tool for investigating genesis of soil or pedological features.

Initially, the investigator should scan the overall features of a thin section and determine those features that require emphasis. This initial scanning may include all the thin sections from a soil profile or all those related to a particular problem. Different kinds of illumination should be used with each magnification. Strong convergent light with crossed polarizers elucidates structures in dense or weakly birefringent material that may appear opaque or isotropic. Structures in translucent specimens become more clearly visible if plain light is used and the condensers are stopped down. Everything should be viewed in several positions of the stage or during slow rotation with cross-polarized light.

A thin section is a two-dimensional slice through a three-dimensional body. The shapes of mineral grains and structural features are viewed in one plane, and the true shapes must be inferred. A grain that appears needle-shaped may be a needle or the edge of a flat plate. An elliptical pore may be an angular slice through a tube. A circular unit is probably part of a sphere. With a three-dimensional perspective in mind as well as an awareness of section thickness, repeated viewing of similar features that appear to be cut at different angles is the best way to accustom oneself to a volume rather than a planar interpretation of shape. A well prepared section is 20- to 30-μm thick. Grains smaller in thickness are stacked and cannot be viewed as individual grains. Similarly, pores smaller than 20 to 30 μm cannot be seen clearly. A pore size of 20-μm diameter equates to a soil moisture tension of 15 kPa (0.15 bar) (Rode, 1969) so that visible pores in thin section are mostly drained at water contents below field capacity.

Sand and silt grains in thin sections are identified by standard methods presented in petrography texts. The general analytical approach is the same for grain studies as it is for thin sections; however, in grain studies the refractive index is used only as a relative indicator, and other optical and morphological properties are more important. Furthermore, in thin sections, a concern with minerals that occur in small quantities or an attempt to quantify mineralogical analysis is seldom necessary. The separate particle-size fractions should be used for the identification and mineralogical analyses that are important to a study, whereas the thin sections should be used mainly for information about component arrangement. Recognition of aggregates, concretions, secondary pseudomorphs, and weathered grains is more important in thin section studies than in sand and silt petrography. Recognition of these components in thin section is easier because interior structures are exposed. Although grain studies are important in soil genesis studies, the arrangement of components is destroyed or eliminated by sample preparation procedures that separate the sand, silt, and clay.

In the United States, the emphasis in micromorphology has been on clay arrangement. Clay occurs not only in the form of aggregates but also in massive interstitial fillings, coatings, bridges, and general groundmass. Even though the clay particles are submicroscopic, they can be described, characterized, and sometimes identified; e.g., the 1:1 and 2:1 lattice clays can be distinguished. Completely dispersed, randomly arranged clay of less than 1 μm exhibits no birefringence and appears isotropic in cross-polarized light. Clay in a soil is seldom all random and isotropic. Clay develops in oriented bodies, either during formation or as a result of pressure or translocation. If enough plate-shaped particles are oriented together in a body that is large enough to see, birefringence can be observed.

With the exception of halloysite, the silicate clay minerals in soils are platy. The a and b crystallographic axes are within the plane of the plate, and the c axis is almost perpendicular to this plane. Even though the crystals are monoclinic, the minerals are pseudohexagonal, as the distribution of stems along the a and b axes is so nearly the same and the c axis is so nearly perpendicular to the other axes. The optical properties, crystal structure, and general habit of clay are analogous to those of the micas, which can be used as models to analyze and describe clay properties.

The speed of light that travels in the direction of the c axis and vibrates parallel to the a axis is almost the same as that light that vibrates parallel to the b axis. Therefore, the refractive indices are very close, and the interference effects in cross-polarized light are small when observed along the c axis. Light that vibrates parallel to the c axis travels faster than in other directions. Hence, the refractive index is lower. If the edge of the crystal or aggregate of crystals is viewed along the a-b plane between crossed polarizers, two straight extinction positions are viewed and interference colors are manifested in other positions. If a clay concentration is organized so that most of the plates are parallel, the optical effects can be observed. The degree and quality of optical effects depend on the purity, continuity, and orientation process of the clay body.

Kaolinite has low birefringence and has refractive indices slightly higher than those of quartz. In the average thin section, interference colors for kaolinite are gray to pale yellow. In residual soils that are derived from coarse-grained igneous rocks, kaolinite occurs as booklike and accordionlike aggregates of silt and sand size.

Even though halloysite can form oriented aggregates, it should not show birefringence because of its tubular habit (Churchman et al., 1984). Halloysite may show very faint, patternless birefringence, which is caused by impurities or by refraction of light at the interfaces between particles.

The 2:1 lattice minerals (Fig. 3.5.1.2.1) have high birefringence and show bright, intermediate-order interference colors if the edges of aggregates are viewed. In the clay-sized range, distinctions among smectite, mica, vermiculite, and chlorite in thin section are seldom possible. These clay minerals are usually mixed in the soil and seldom occur pure. In many soils, these clay minerals are stained and mixed with iron oxide and organic matter.

Residual clay has been in place since its formation by weathering. Although it may have been transported within fragments of weathered material, it remains in place relative to the fabric of these fragments. This clay may be random, have no orientation, and thus be isotropic; however, more often, it shows some birefringence. In transported materials, silt-sized flakes and other small aggregates are common. In many residual materials, clay is arranged either in forms that are pseudomorphs of rock minerals or in definite bodies of crystal aggregates, e.g., vermicular or accordionlike kaolin books. The regular, intact arrangement of these materials is usually diagnostic of residual material.

Clay rearrangement may result from differentially applied stress that produces shear (Fig. 3.5.1.2.2). Platy particles become oriented by slippage along a plane, e.g., slickenside faces in a Vertisol or in clayey layers. Platy particles also are oriented inside the blocks. Root pressure, mass movement, slump, and creep can produce stress orientation. If the faces on structural units are smooth and do not have separate coatings, stress orientation can be inferred. Otherwise, in plain light, stress orientation cannot be observed. In plain light, clay in the thin section may be homogeneous and featureless. In cross-polarized light, the orientation pattern is reticulate, consisting of bright lines showing aggregate birefringence, often intersecting at regular angles. The effect is that of a network in a plaid pattern. There may be numerous sets of these slippage planes, which appear in different positions as the stage is turned. Stress-oriented clay may be near rigid bodies, e.g., quartz grains, or along root channels. Stress-oriented clay is often strongly developed on ped faces. Stress can also orient mica flakes and any other small platy grains.

Location features that distinguish translocated clay from residual clay are its occurrence in separate bodies, usually with distinct boundaries, and its location on present or former pore walls, channel linings, or ped faces. Translocated clay may have a different composition than matrix clay, especially if its origin is another horizon. This clay is more homogeneous and is usually finer than the matrix clay. Translocated clay displays lamination, indicating deposition in successive increments, and manifests birefringence and extinction, indicating that these translocated clay bodies are oriented aggregates. If these bodies are straight, they have parallel extinction. If these bodies are curved, a dark band is present wherever the composite c axis and the composite a and b axes are parallel to the vibration planes of the polarizers. When the stage is rotated, these dark bands sweep through the clay aggregate.

Other substances, such as goethite, gibbsite, carbonate minerals (Fig. 3.5.1.2.3), and gypsum, may form pore linings and ped coatings. These substances can be identified by their mineralogical properties.

Amorphous coatings of organic matter, with or without admixed Fe and Al, are common, especially in spodic horizons. This material is dark brown to black, is isotropic or faintly birefringent, and is often flecked with minute opaque grains. Amorphous coatings of organic matter occur as the bridging and coating material in B horizons of sandy Spodosols and as thin coatings or stains on pore and ped faces in other soils.

Description of Microfabrics

Terms have been defined for distribution patterns of the components of soil thin sections (Brewer, 1964 and 1976; Stoops and Jongerius, 1975; Brewer et al., 1983; Bullock et al., 1985; Stoops, 2003). As these terms have become more widely adopted in the literature, the SSL increasingly uses them in Soil Survey Investigations Reports (SSIRs) and in soil project correspondence. Micromorphological descriptions often contain terminology from different sources to describe properties of the fabric.

Related distribution patterns: The five "coarse-fine related distribution patterns" of Stoops and Jongerius (1975) are in common usage. The nomenclature of these distribution patterns, as described by Stoops and Jongerius (1975), is intended to be broadly defined. There are no restrictions on material type, absolute size, orientation, granulation, or origin. The system may be used to describe the distribution of primary particles, e.g., quartz grains, as well as compound units, e.g., humic microaggregates. The coarser particles may be silt, sand, or gravel, whereas the finer material may be clay, silt, or sand. Figure 3.5.1.2.4 shows the average textures, linear extensibilities (LE), and drained pore to filled pore (DP/FP) ratios of some related distribution patterns of a number of U.S. soils.

The *monic type* (granic type of Brewer et al., 1983) consists of fabric units of only one size group, e.g., pebbles, sand, lithic fragments (coarse monic), or clays (fine monic). In the *gefuric type*, the coarser units are linked by bridges of finer material but are not surrounded by this material. In the *chitonic type* (chlamydic type of Brewer et al., 1983), the coarser units are surrounded by coatings of finer material. In the *enaulic type*, the larger units support one another and the interstitial spaces are partially filled with finer material. The enaulic fabric consists of material finer than that found in either the gefuric or chitonic type but not so fine as that found in the porphyric type. In the end member of the sequence, the *porphyric type*, the large fabric units occur in a dense groundmass of smaller units and there is an absence of interstitial pores. This type is equivalent to the earlier porphyroskelic class of Brewer (1964) or to the current porphyric class (Brewer et al., 1983). The class may be divided into types based on the spacing of the coarser units.

Plasma fabrics: Brewer (1976) divided soil materials into three groups for descriptive purposes: peds, pedological features, and s-matrices. *Peds* are the basic units in soils that contain organized structural units and are composed of skeleton grains, plasma, and pedological features. The *s-matrix* is the material within which pedological features occur, having no definite boundary, size, shape, or orientation (Brewer, 1976).

Skeleton grains of a soil material are individual grains larger than colloidal size. The *soil plasma* includes all the colloidal size material as well as relatively soluble material not bound in skeleton grains.

The description of plasmic fabrics is based on the interpretations of optical properties under cross-polarized light, especially extinction phenomena. Plasma concentrated or crystallized into pedological features is not included in the description of plasmic fabrics. In general, the descriptive terms for the s-matrix are those defined by Brewer (1976). The s-matrix plasma fabrics are divided into two groups: the asepic and sepic types. *Asepic fabrics* are those with anisotropic plasma in which the *domains*, i.e., the plasma separations, are not oriented relative to each other. *Sepic fabrics* are those with anisotropic domains with various orientation patterns visible under cross-polarized light. Figure 3.5.1.2.5 shows some plasma fabrics and their clay, silt + clay, and linear extensibility averages for a number of U.S. soils.

Eswaran (1983) characterized the <25-μm^2 size domains of monomineralic soils using a scanning electron microscope (SEM). These features are smaller than some domains described by Brewer (1976); however, these small features provide the detail expected of the interparticle relationships present in the larger separations. The domains in allophanic soils are composed of globular aggregates. The halloysitic soils differ in that the halloysite tubes generally may be seen as protrusions from globular forms. The domains in micaceous soils retain the face-to-face packing that is common in micas and may retain some of the booklike forms as well. The domains in montmorillonitic soils are bent to conform to the shape of skeleton grains. The packing is essentially face to face, however, and, upon drying, the fabric is very dense and compact. In kaolinitic soils, the domains frequently are present as booklets that are packed face to face, unless iron hydrous oxide has disrupted the platelets, in which case the platelets may still be packed face to face in subparallel stacks.

Asepic plasmic fabrics are subdivided into two groups: argillasepic and silasepic types. *Argillasepic fabrics* are dominated by anisotropic clay minerals and have a random orientation pattern of clay-sized domains. Overall, asepic fabrics have flecked extension patterns. *Silasepic fabrics* have a wider range of particle sizes than the argillasepic types; however, a careful observer may view silt-sized domains or plasma bodies that give the matrix an overall flecked extinction pattern (Fig. 3.5.1.2.6).

The *sepic plasmic fabrics* have recognizable domains with various patterns of orientation. Internally, the *domains*, i.e., plasma separations, have striated extinction patterns. Brewer (1964) recognizes seven kinds, most of which are widely adopted. *Insepic fabrics* consist of isolated, striated plasma domains within a flecked plasma matrix (Fig. 3.5.1.2.7). *Mosepic fabrics* consist of plasma domains with striated orientation that may adjoin each other or may be separated by small plasma areas with flecked orientation that are not oriented relative to each other (Fig. 3.5.1.2.8). The fabric is *vosepic* when the plasma separations with striated orientation are associated with channel or pore (void) walls. The fabric is *skelsepic* when the plasma separations occur at the skeleton grain-matrix contact (Fig. 3.5.1.2.9).

The remaining three sepic plasmic fabrics are most common in fine-textured soils. In *masepic fabrics*, the plasma separations occur as elongated zones within the s-matrix and apparently are not associated with void walls or skeleton grains (Fig. 3.5.1.2.10). The striations have orientations parallel to zone length. *Lattisepic fabrics* are similar to masepic fabrics, except that the acicular and prolate domains occur in latticelike patterns. In *omnisepic fabrics*, all of the plasma has a complex striated orientation pattern.

Three other kinds of plasmic fabrics are characteristic of particular minerals or kinds of soils. *Undulic plasmic fabrics* have practically isotropic extinction patterns at low magnification, and the domains are indistinct even at high magnification. *Isotic plasmic fabrics* have isotropic plasma, even at highest magnifications with high light intensity. *Crystic plasmic fabrics* have anisotropic plasma with recognizable crystals, typically of soluble materials.

Pedological features, cutans: The term *cutan* and definitions of its respective types (Brewer, 1964) have been widely adopted by soil scientists. *Cutan* is defined by Brewer as a modification of the texture, structure, or fabric at natural surfaces in soil materials due to the concentration of particular soil constituents or as in-place modification of the plasma (Fig. 3.5.1.2.11). Generally, the cutans are subdivided on the basis of their location, composition, and internal fabric. Cutan locations are surfaces of grains, peds, channels, or voids. The mineralogical nature of cutans is characterized, e.g., argillans, ferri-argillans, or organo-argillans. *Argillans* are composed dominantly of clay minerals, *ferri-argillans* have iron oxides as a significant part of their composition, and *organo-argillans* have significant color addition by inclusion of organic matter.

Sesquan is a general term used for a cutan of sesquioxides or hydroxides. Sesquans that are specific for goethite, hematite, and gibbsite are called *goethans*, *hematans*, and *gibbsans*, respectively. Similarly, cutans of gypsum, carbonate, calcite, halite, quartz, silica, and chalcedony are called *gypsans, calcans, calcitans, halans, quartzans, silans,* and *chalcedans*, respectively. Skeleton grains that adhere to the cutanic surface are called *skeletans*.

Pedological features, glaebules: *Glaebules* (Brewer, 1964) are three-dimensional pedological units (e.g., Fe oxide or carbonate nodules) within the s-matrix; their morphology is incompatible with the composition of the present matrix material. (The name is derived from the Latin term *glaebula*, meaning a small lump or aggregate of earth.) Glaebules are typically prolate to equant. A glaebule is recognized as a unit either because of a greater concentration of a constituent, or difference from the s-matrix fabric, or because of the presence of distinct boundaries of a constituent within the enclosing s-matrix. Glaebules include papules, nodules, concretions, and pedodes. *Papules* are pedogenic features composed of clay minerals with continuous and/or laminar fabric, sharp external boundaries, and commonly prolate to equant, somewhat rounded shapes. *Nodules* (Fig. 3.5.1.2.12) are pedological features with undifferentiated internal fabric. *Concretions* are pedological features with concentrically laminated structures about a center. *Pedodes* are pedological features with hollow interiors, commonly lined with crystals.

Pedological features, voids: *Voids* are the empty spaces within the s-fabric. Those voids with diameters of 20 μm to >2 mm can be studied and measured in thin section. Brewer (1976) classifies these voids as follows: (1) *simple packing voids* (empty spaces due to random packing of single skeleton grains), (2) *compound packing voids* (Fig. 3.5.1.2.13) (empty spaces between peds or other compound individuals), (3) *vughs* (Fig. 3.5.1.2.14) (relatively large spaces that are not formed by packing of skeleton grains,

(4) *vesicles* (Fig. 3.5.1.2.15) (relatively large empty spaces with smooth, regular outlines), (5) *chambers* (empty spaces with smooth, regular outlines that connect to other voids), (6) *joint planes* (plane, empty spaces that traverse the s-matrix in a regular pattern), (7) *skew planes* (plane, empty spaces that traverse the s-matrix in an irregular pattern), (8) *craze planes* (plane, empty spaces that traverse the s-matrix in a highly irregular pattern of short, flat or curved planes), and (9) *channels* (mostly cylindrical, empty spaces that are larger than packing voids).

Interpretations

Related distribution patterns: Usually, the basic descriptive terms for soil fabrics do not imply any specific genesis of the feature. Modifiers commonly are added, however, when fabric descriptions are complete enough to explain the means of formation; i.e., stress cutan, or in-place plasma modification, is the result of differential forces, e.g., shearing, whereas an illuviation cutan is formed by movement of material in solution or suspension and later deposited (Brewer, 1964).

The average properties of some related distributions we have described are given in Figure 3.5.1.2.4. In an experimental study of soil microfabrics by anisotropic stresses of confined swelling and shrinking, Jim (1986) showed that with an increase in the activity and proportion of the clay fraction, the related distribution patterns alter from dominantly *matrigranic* (*monic*, with the units being aggregates) to *matrigranodic* (*enaulic*) to *porphyric*. Similarly, our data for some U.S. soils show that the relative pore volumes at 30 kPa for some soil coarse-fine distributions increase from *enaulic* through *open porphyric* (Fig. 3.5.1.2.16).

Some *monic* fabrics are inherited, including soil fabrics that formed in sand dunes, sandy sediments deposited by streams and rivers, beach deposits, and gruss. Fauna can produce monic fabrics that are mostly fecal pellets. Monic fabrics can also form by fracturing and flaking of organic coatings in the upper B horizons of the Spodosols (Flach, 1960) and by freezing and thawing (Brewer and Pawluk, 1975).

Several kinds of finer material (plasma) can bridge the coarser particles (skeleton grains) to form *gefuric* related distribution patterns. Gefuric patterns are common in weakly developed argillic and spodic horizons and in duripans. Silicate clays can bridge skeleton grains in some argillic horizons; the organic matter, iron, and aluminum complexes in some kinds of spodic horizons; and the amorphous silica in some kinds of duripans.

In soils that are slightly more developed than those with gefuric patterns, *chitonic* related distribution patterns form. These are common in argillic and spodic horizons and in duripans. Bridges and complete coatings of skeleton grains are present. Typically, the cement or plasma is material that adheres to skeleton grains. These cements have covalent bonds and commonly include silica (Fig. 3.5.1.2.17), iron, aluminum, and organic matter (Chadwick and Nettleton, 1990).

The *enaulic* related distribution patterns are more common in soil material in which the cement bonds to itself more strongly than to skeleton grains. In sandy soils, ionic-bonded calcite and gypsum tend to bond to themselves more strongly than to skeleton grains (Fig. 3.5.1.2.3), thereby producing *open porphyric* related distribution patterns (Chadwick and Nettleton, 1990). Even though organic matter has covalent bonds and typically surrounds grains, organic material forms pellets in void spaces between skeleton grains in some spodic horizons.

Porphyric related distribution patterns form as a result of the normal packing of grains in materials with a high proportion of fine material. These patterns can be the end member of several kinds of sequences (Brewer et al., 1983). In porphyric related distribution patterns, there may or may not be skeleton grains of primary minerals, pedorelicts, organics, lithic fragments of shale, sandstone, or other rocks. In the porphyric related distribution patterns, the material consists of silt and clay and the interstices tend to be filled with coatings exhibiting minimal formation. In precursors of the porphyric related distribution patterns, the silt to clay ratio is used to identify the kind of sequences by which the porphyric pattern forms (Brewer et al., 1983). The porphyric patterns are common in loessial soils, especially in argillic and petrocalcic horizons, duripans, and ortstein.

Plasmic fabrics: The *asepic plasmic fabrics* differ in composition mainly in silt to clay ratios. *Argillasepic fabrics* have the higher clay contents, typically <30 percent but in some cases as much as 70 percent (Brewer et al., 1983). Organic matter or iron stains, resulting in a flecked distribution pattern, mask the birefringence of the plasma. Argillasepic fabrics are important fabrics in many fine-textured B horizons. *Silasepic plasmic fabrics* have low clay contents and have more silt than clay. The silasepic fabrics are common in porphyric related distribution patterns in A and B horizons of Solonetz, Solodized Solonetz, and Solodic Soils, Soloths, Red Podzolic Soils, and Lateritic Podzolic Soils and are also associated with some sedimentary deposits (Brewer et al., 1983). Silasepic plasma fabrics are common in A and B horizons of loessial soils in association with other kinds of plasma separations. Even if there is high clay content, the horizons with asepic plasmic fabrics have low effective linear extensibilities (LE) either because the clays are low-swelling types or because the soils do not dry enough to undergo the full range of laboratory-measured LE.

In soils that form in the same climate, the kinds of *sepic plasmic fabrics* form a sequence relative to increasing linear extensibility (Nettleton et al., 1969; Holzhey et al., 1974). In increasing order of shrink-swell stress, the plasmic fabric sequence is insepic, mosepic, lattisepic, omnisepic, and masepic. Using x-ray diffraction (Clark, 1970) and scanning electron microscopy (Edil and Krizek, 1976), observations of deformation experiments indicate that the degree of clay orientation increases with an increase in applied stress. In an experimental study of soil microfabrics by anisotropic stresses of confined swelling and shrinking, Jim (1986) shows that with an increase in the activity and content of the clay fraction there is an increase in the long and narrow plasma separations, i.e., a progression from *insepic* to *mosepic* to *masepic* plasmic fabrics.

Insepic plasmic fabrics are very common in finer grained porphyric B horizons of a wide range of soil groups (Brewer et al., 1983). Soil horizons with insepic fabrics generally have an LE of <4 percent. In some insepic plasmic fabrics, the plasma islands or papules are pseudomorphs of some weatherable mineral, whereas in other insepic fabrics the papules are clay skin fragments or are eolian sand-sized clay aggregates (Butler, 1974). In some samples, the pseudomorphs do not disperse well in particle-size distribution analysis (PSDA).

Mosepic plasmic fabrics commonly have more clay than insepic fabrics do because they contain more islands of plasma. In mosepic plasmic fabrics, however, LE also remains low. Shrink-swell forces have not been sufficient or have not operated long enough to have homogenized the islands of plasma into the soil matrix.

Vosepic plasmic fabrics occur in soil horizons that have undergone stress as a result of either shrink-swell forces or tillage. Even though root growth is adequate to increase the percentage of oriented clay near the root-soil interface (Blevins et al., 1970), root growth does not appear adequate to form vosepic or other highly stressed plasmic fabrics. Typically, vosepic fabrics are present in soil horizons in which the main fabric type is masepic or skelsepic. The vosepic plasmic fabric rarely occurs as the only fabric in a soil horizon.

There are at least two types of origins for orientation of plasma on sands. One is a result of clay illuviation. By definition, this type would not be included with skelsepic fabric. The distribution patterns associated with this fabric commonly are *monic*, *gefuric*, or *enaulic*. The other origin is commonly the porphyric related distribution patterns with LEs that are >4 percent for dryland soils, i.e., soils in aridic, xeric, or ustic soil moisture regimes. These are the true *skelsepic* fabrics. Shrink-swell forces have been involved in their formation as shown by relatively few papules or clay skins remaining, and there are vosepic areas.

Masepic, lattisepic, and *omnisepic* plasmic fabrics are evidence of stress of >4 percent in dryland soils. Clay contents are typically >35 percent, but the threshold amount is dependent on clay mineral type and on degree of dryness common to the environment. In masepic, lattisepic, and omnisepic plasmic fabrics, papules and clay skins are rare but areas of *skelsepic* and *vosepic* areas commonly occur.

Undulic plasmic fabrics seem to be associated with basic parent materials, especially basalt, and with moderate to strong weathering (Brewer et al., 1983). The fabric commonly is stained deeply by iron minerals, and kaolinite and halloysite are the important clay minerals. Clays in these horizons do not disperse well in PSDA, but high 1500-kPa (15-bar) water contents suggest that the horizons belong in clayey families. Some papules and clay skins commonly are present, but these plasma separations also are stained deeply by iron.

Isotic plasmic fabrics are common in spodic horizons and in Andisols. The clays in these horizons are amorphous and disperse poorly in PSDA. The water-holding capacities of these soil horizons are relatively high. Some unweathered volcanic ash may be present.

Crystic plasmic fabrics are common in B horizons of soils that formed in dryland areas. In soil horizons with large areas of interlocking crystals, soil permeability is restricted, unconfined compressive strength is increased, and particle dispersion is limited, depending on the degree of cementation.

Cutans and pedogenic features: Most *argillans* (Fig. 3.5.1.2.11) are formed, at least in part, by illuviation. The content of strongly oriented clay (typically argillans plus *papules*) in texture-contrast soils (soils with argillic horizons) is typically <5 percent of the soil volume (Brewer et al., 1983). In some sandy soils that are low in silt, the argillans and papules are as much as 30 percent of the soil material (Brewer et al., 1983). The measured illuviated clay rarely accounts for the difference in clay content between the A and B horizons. Some of the clay may originate from weathering in place and some from a destruction of argillans and papules.

If argillans and papules are present in argillic horizons in dryland soils, the soil LE is typically <4 percent (Nettleton et al., 1969). In some humid environments, argillans and papules may be present even where the LE is >4 percent. As soils in humid environments do not dry to the same degree as those in the desert, the clay skins may survive because only part of the linear extensibility is effective.

Papules may originate by the weathering of primary minerals, the isolation of clay skins by the channel and void migration within the soil matrix (Nettleton et al., 1968; Nettleton et al., 1990), or the introduction of eolian sands and silts that are composed of clays (Butler, 1974; Brewer and Blackmore, 1976). The comparison of size and shape of papules and minerals, as well as of parent material, may help to determine whether the papules are pseudomorphs of one of the primary minerals. Internal fabric resemblances and residual parts of the primary mineral within the papules help to determine whether a papule is a pseudomorph.

The determination of whether or not a papule is an illuvial feature is important for classification purposes. Arcuate forms and laminar internal fabrics are evidence that the feature is illuvial. If the feature partially surrounds an oval body of silt, illuvial origin of the feature is relatively certain (Nettleton et al., 1968).

The origin of the papule as eolian may be determined by studying its size and shape, its internal fabric, and the number and degree of its alterations relative to other particles. Microlaminae may suggest an origin as sediment. Unlike soil pedorelicts or rock fabrics (lithorelicts), nodules, or glaebules rich in soluble plasma, probably form by accretion (Brewer, 1976). Most concretions, as well as pedodes, are accretionary and typically form in place.

A study of soil *voids* may be useful in predicting the clay activity and shrink-swell behavior of soils. In an experimental study of soil microfabrics by anisotropic stresses of confined swelling and shrinking, Jim (1986) shows that with an increase in the activity and content of the clay fraction there is a drastic decrease in void volume, especially the >30-μm fraction. Furthermore, the void shapes change from compound packing voids to planar voids and vughs. With an increase in stress from shrink-swell forces, aggregates become flattened at contacts, resulting in more angular and eventually fused compound units.

Possible objectives of micromorphological studies are the measurement of porosity and the prediction not only of soil water content at various suctions but also of hydraulic conductivity. In thin section studies of voids in sands and sandy soils, there is a close correlation between microscopic and suction methods (Swanson and Peterson, 1942); however, in those soils whose volumes change with changes in water content, pore size distribution is undefined and no constant void size distribution exists (Brewer, 1976). Furthermore, there are several invalidated assumptions that commonly are made in relating porosity to permeability (Nielsen et al., 1972, p. 11). The assumptions that especially relate to soil fabric are that no pores are sealed off, pores are distributed at random, and pores are generally uniform in size. A more serious difficulty may be that a thin section, even if reduced to a 20-μm thickness, may make the examination of the <20-μm diameter pores impossible if these pores pass through the section at an angle of $\leq$45 degrees. Under these conditions, many voids that are involved in unsaturated waterflow in soils will not be visible in thin section (Baver, 1956, p. 271).

The size, shape, and arrangement of skeleton grains determine the nature of simple packing voids, but the origin of compound packing voids is not so straightforward. The unaccommodated peds of the compound packing voids may be formed by faunal excreta, shrink-swell action, human activities, or unknown causes.

Vughs typically occur in soil materials with a wide range in size of particles, including silicate clays. Some vughs form by the weathering and removal of carbonate, and others form by faunal activity or the normal packing of plasma and skeleton grains. The very regular outline of *vesicles* is of interest (Nettleton and Peterson, 1983). Lapham (1932) states that in Sierozems (Aridisols) the vesicles that are near the surface are the result of air entrapment by rainfall following dry, dusty periods. Laboratory studies verify this phenomenon (Springer, 1958). If soils high in silt are allowed to dry before each irrigation, the vesicle size increases with the number of irrigations (Miller, 1971). As a result of studies of infiltration rates and sediment production in rangeland in central and eastern Nevada, Blackburn and Skau (1974) and Rostagno (1989) conclude that the infiltration rates are the lowest and the sediment yields are the highest on sites that have vesicular surface horizons. The failure of most vesicles to connect to other voids and the low strength of the crust in which vesicles occur help to explain the low infiltration rates and high sediment yields that are common in these soils.

Joint planes (Fig. 3.5.1.2.18) are produced in relatively uniform fine-textured soils by a relatively regular system of cracking upon drying (Brewer, 1976). Once formed, these joint planes tend to open in the same place during successive drying cycles. *Skew planes* are produced in more heterogeneous materials or by irregular drying (Brewer, 1976). *Craze planes* commonly occur in Chernozems (Mollisols), possibly as a result of the high humic acid content (Brewer, 1976). Because of their size, cross-sectional shape, and kind of branching pattern, channels probably form by faunal activity, plant root systems, or certain geological processes (Brewer and Sleeman, 1963).

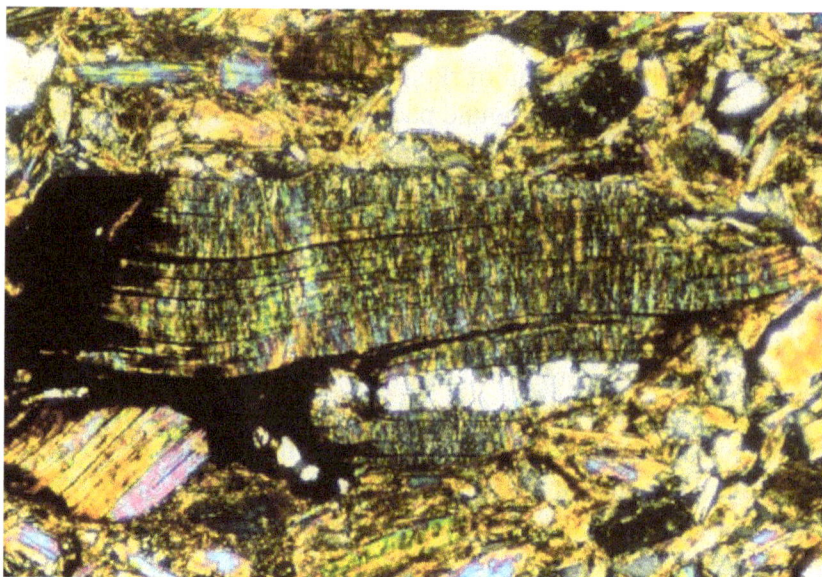

Figure 3.5.1.2.1.—Large biotite grain undergoing expansion from weathering. Note the high birefringence due to the orientation of the grain in thin section. Frame width = 1.0 mm. (Series name not designated, Fremont County, WY; Pedon 98P0456, Bt2 horizon under cross-polarized light)

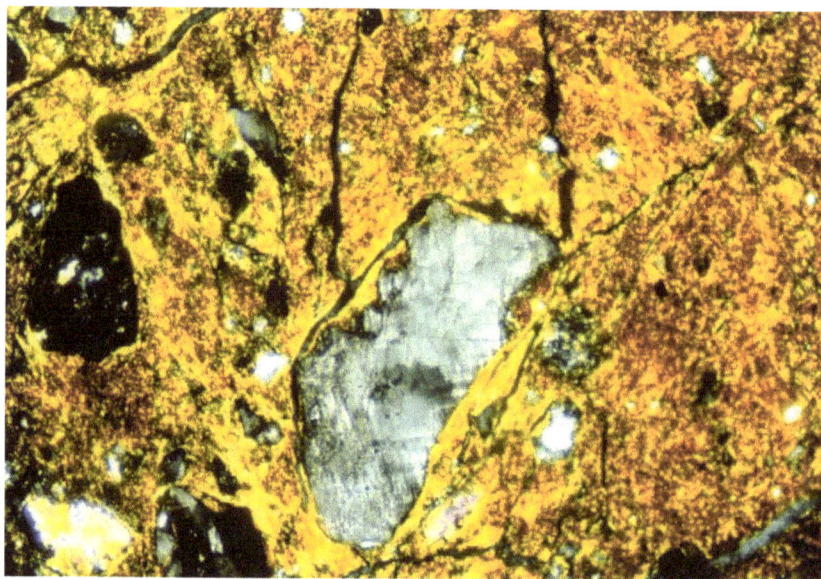

Figure 3.5.1.2.2.—Horizons with a high percentage of clay of expandable aluminosilicate clay-sized minerals become aligned through shrink-swell processes. This alignment results in preferred orientation of clay particles, making the plasma anisotropic (visible under cross-polarized light). This process results in a loss of argillans along ped faces in many soils. Frame width = 0.9 mm. (White House pedon, Cochise County, AZ; Pedon 40A001, BCtk horizon under cross-polarized light)

Figure 3.5.1.2.3.—Calcium carbonate around skeleton (sand-sized) grains in a coarse-textured matrix. These carbonate coatings are referred to as calcitans (Brewer, 1976). Their formation illustrates attraction to and deposition of carbonates on mineral surfaces (e.g., quartz or feldspar grains) accessible to percolating water. Frame width = 0.9 mm. (Cax pedon, San Bernardino County, CA; Pedon 97P0420, Bkg2 horizon under cross-polarized light)

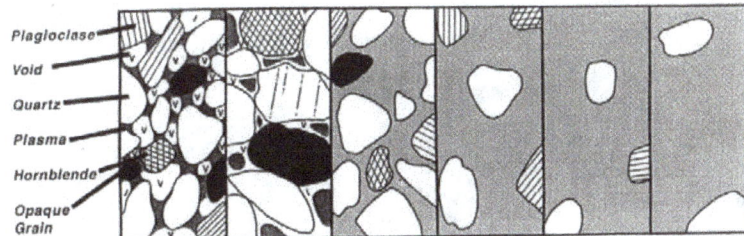

C/f Patterns	Chitonic	Enaulic	Close Porphyric	Single-Space Porphyric	Double-Space Porphyric	Open Porphyric
Clay, %	20C[‡]	23BC	20C	25BC	34AB	41A
Silt & Clay, %	37D	50BC	40CD	54B	74A	84A
LE, %	4.4AB	2.2B	2.9AB	2.7AB	4.0AB	6.6A
DP/FP	1.1BC	1.3B	0.7CD	0.6CD	0.5D	0.3D
Nos.	8	26	23	112	53	94

† C/F Patterns, are related distribution patterns of coarse and fine constituents, LE, linear extensibility; DP/FP, ratio of drained to filled pores at 33 kPa suction.
‡ Means with the same letter are not significantly different at the 95% confidence level (SAS Institute, 1988).

Figure 3.5.1.2.4.—Kinds of related distribution patterns and a listing of their physical properties. Frame width of each idealized kind of fabric is 0.5 mm. The lower size limit of coarse material in the C/F patterns was set at about 50 μm for most of the slides.

101

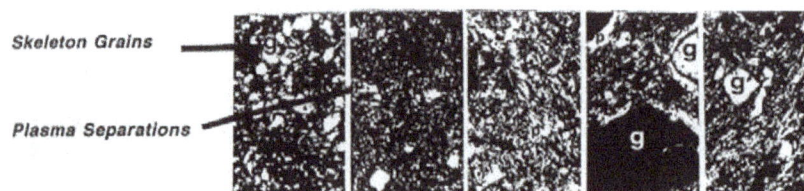

Variable[†]	Silasepic	Insepic	Mosepic	Skelsepic	Masepic
Clay, %	21C[††]	23C	32B	33B	52A
Silt + Clay, %	58B	59B	65B	61B	84A
LE, %	2.2CD	2.2CD	3.5BCD	5.8B	9.2A
DP/FP	0.8A	0.8A	0.5AB	0.7AB	0.2B
LE ÷ Clay x 100	1.0A	1.1A	1.1A	1.6A	1.8A
Number Observed	97	49	46	67	37

[†] LE, linear extensibility; DP/FP, ratio of drained to filled pores.
[††] Means with the same letter are not significantly different at the 95% confidence level (SAS Institute, 1988).

Figure 3.5.1.2.5.—Kinds of plasma fabrics and a listing of their physical properties. Frame width of each idealized kind of fabric is 0.5 mm.

Figure 3.5.1.2.6.—Silasepic plasma fabric. Frame width = 1.3 mm. (Southridge pedon, Allamakee County, IA; Pedon 87P0075, Ap horizon under cross-polarized light)

Figure 3.5.1.2.7.— Insepic plasma fabric. Frame width = 1.3 mm. (Mexico pedon, Macon County, MO; Pedon 87P0771, BE horizon under cross-polarized light)

Figure 3.5.1.2.8.—Mosepic plasma fabric. Frame width = 1.3 mm. (Leonard pedon, Macon County, MO; Pedon 87P0770, 2Btg3 horizon under cross-polarized light)

Figure 3.5.1.2.9.—Skelsepic plasma fabric. Frame width = 1.3 mm. (Redding pedon, San Diego County, CA; Pedon 40A2847, Bt horizon under cross-polarized light)

Figure 3.5.1.2.10.—Masepic plasma fabric. Frame width = 1.3 mm. (Gloria pedon, Monterey County, CA; Pedon 40A2845, Bt horizon under cross-polarized light)

104

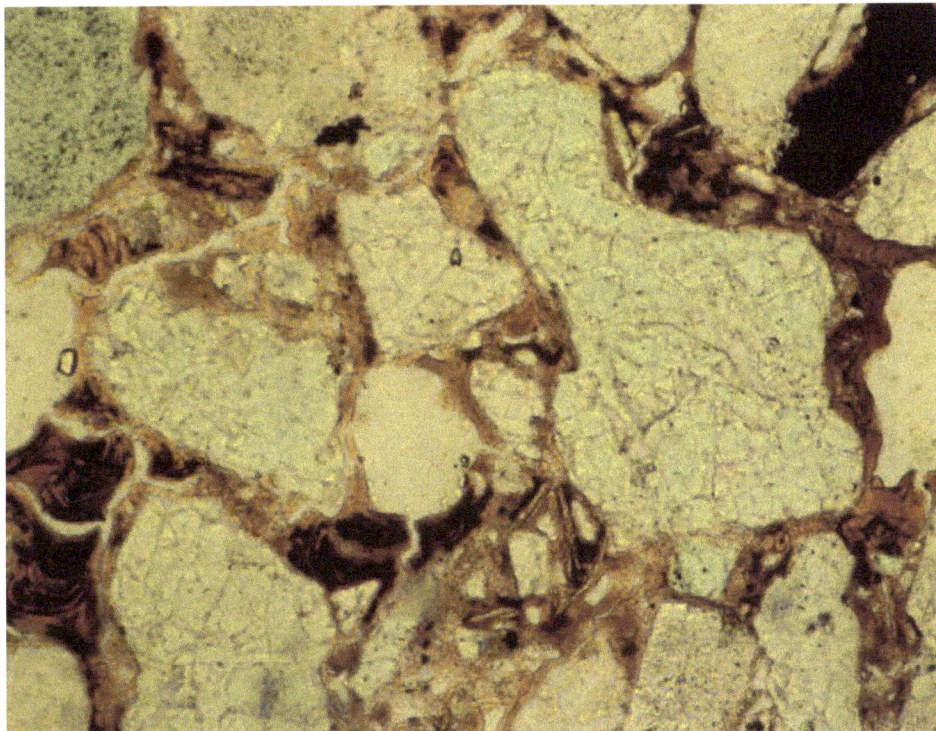

Figure 3.5.1.2.11.—Oriented illuvial clay (argillans) surrounding skeleton grains. Frame width = 1.1 mm. (Paxon pedon, New York County, NY; Pedon 00P0001, 2Cd1 horizon under plane polarized light)

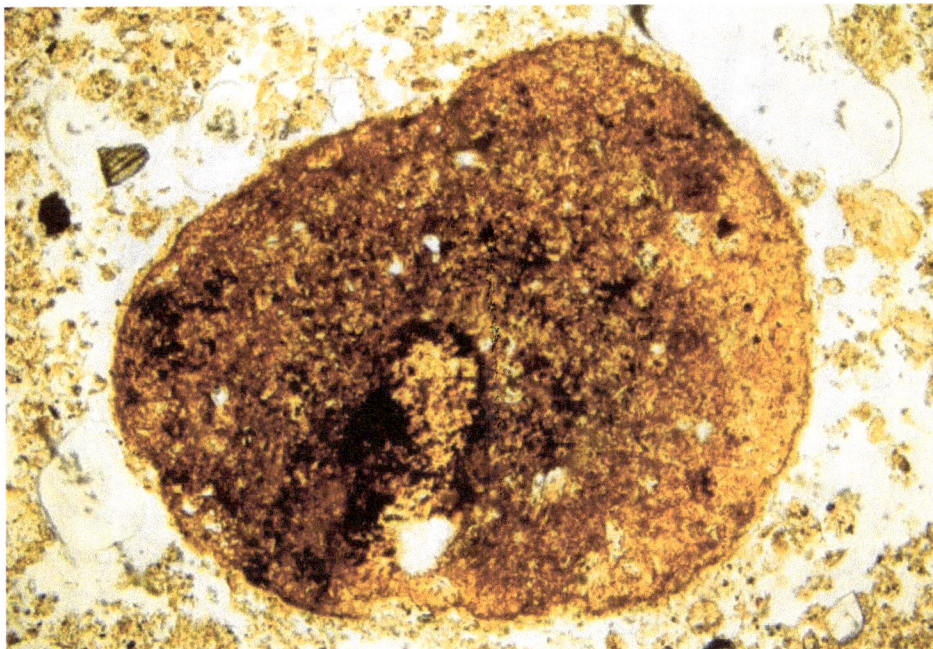

Figure 3.5.1.2.12.—Fe oxide nodule from an Andisol in Blue Mountains of eastern Oregon. Frame width = 2.5 mm. (Tower pedon, Umatilla County, OR; Pedon 97P0547, Bw1 horizon under plane polarized light)

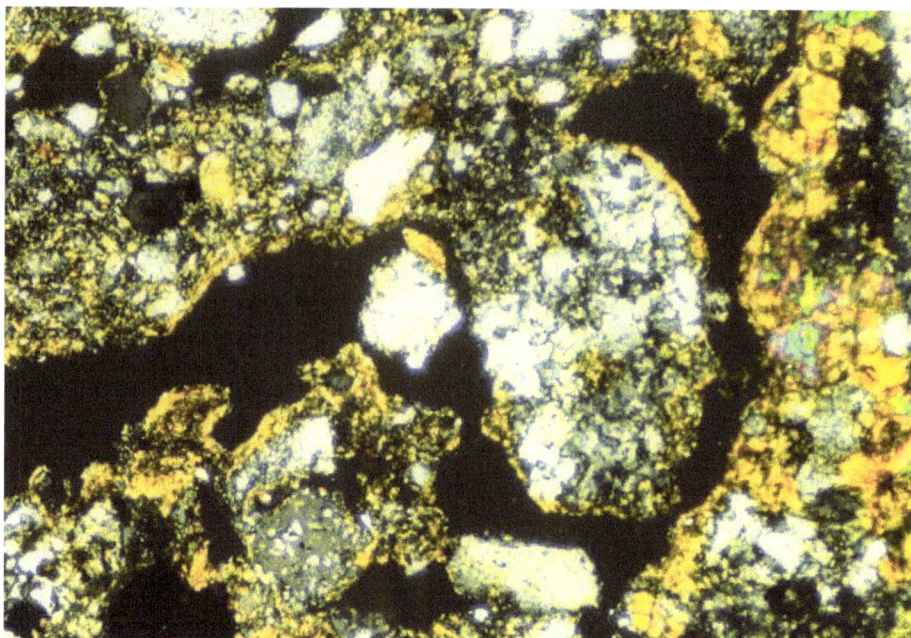

Figure 3.5.1.2.13.—Compound packing voids surrounded by illuvial clay (argillans). Clay lining channels is anisotropic due to orientation during deposition, while clay (plasma) in the s-matrix is partially anisotropic due to stress orientation (shrink-swell processes). Frame width = 1.0 mm. (Endlich pedon, Gunnison County, CO; Pedon 99P0001, Bt1 horizon under cross-polarized light)

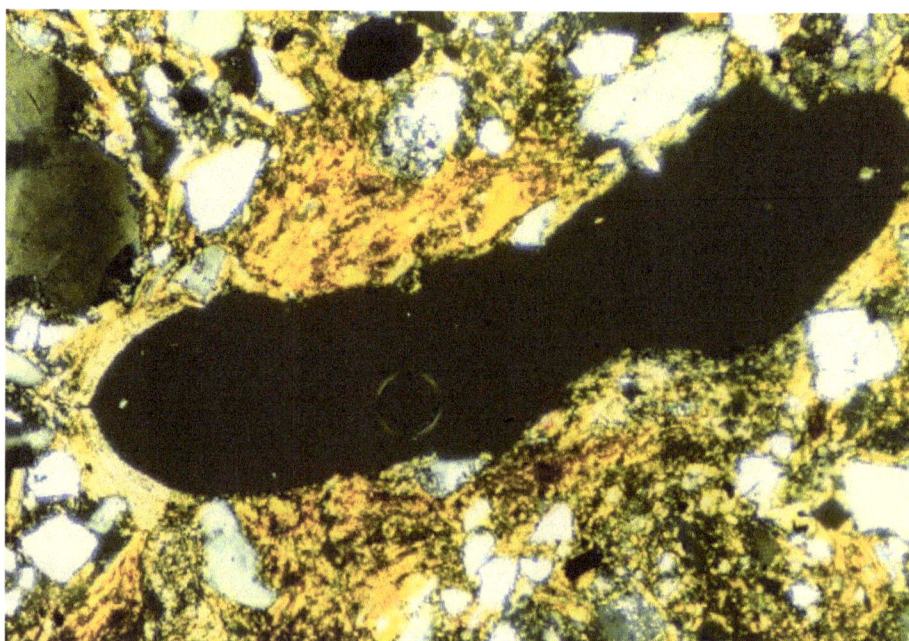

Figure 3.5.1.2.14.—Void that has smooth edges and is elongated. This void type is described by Brewer (1976) as a vugh. Frame width = 1.0 mm. (Troutville pedon, Gunnison County, CO; Pedon 99P0002, E&Bt horizon under cross-polarized light)

Figure 3.5.1.2.15.—The walls consisting of "smooth, simple curves" indicate that this void is a vesicle. These vesicles were formed in the thin surface crust of a Typic Haplargid. Frame width = 3.2 mm. (Dera pedon, Juab County, UT; Pedon 81P0610, A1 horizon under cross-polarized light)

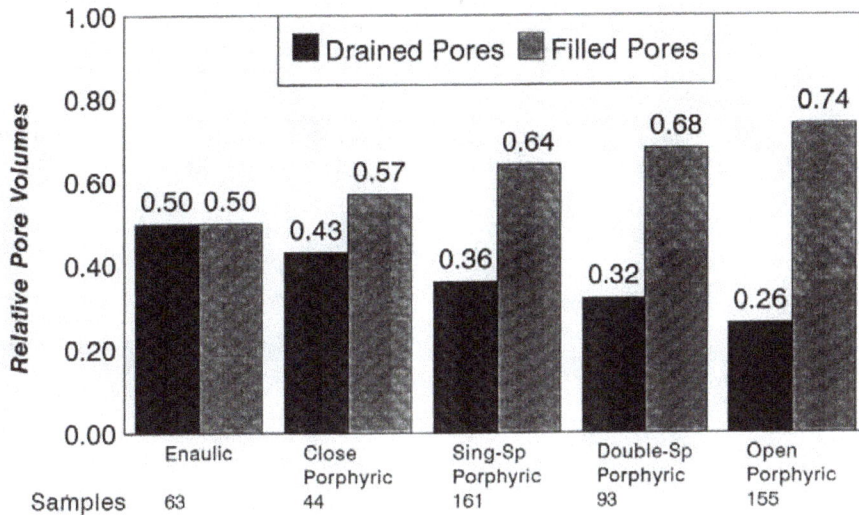

Figure 3.5.1.2.16.—Relative pore volumes at 30 kPa for soil fabric coarse-fine distributions for some U.S. soils.

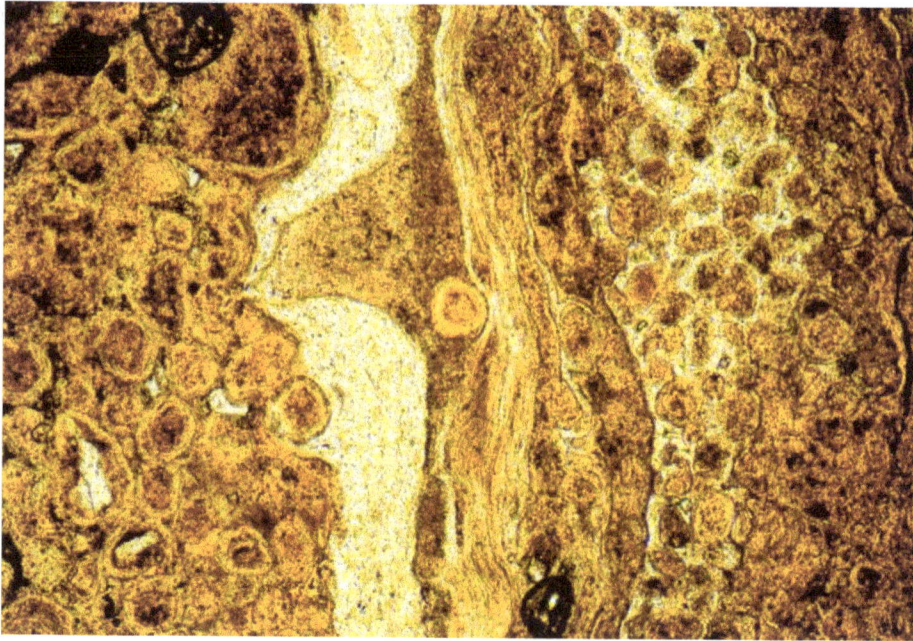

Figure 3.5.1.2.17.—Horizon with duripan exhibiting silica cementation. The fabric has an opal and chalcedony laminar cap. The matrix above and below is composed of durinodes (noncrystalline silica) surrounded by moderately oriented silicate clays. Clay can provide the initial absorption surface for silica in soil solution. The absorption of silica onto established silica phases leads to the formation of nodules. Frame width = 1.0 mm. (Series name not designated, Jefferson County, OR; Pedon 87P0513, 2Bkqm horizon under plane polarized light)

Figure 3.5.1.2.18.—Joint planes (platy structure) formed in the surface horizon of a soil. Frame width = 2.5 mm. (Frisite pedon, Fremont County, WY; Pedon 98P0453, A horizon under cross-polarized light)

3.6   Aggregate Stability

This section describes the SSL method for aggregate stability, wet sieving (2 to 1 mm, 2- to 0.5-mm aggregates retained). The SSL aggregate method is compared to the Soil Quality Institute procedure, wet sieving (<2 mm, >0.25-mm aggregates retained). Information is provided on key definitions related to soil structure and aggregation. The soil processes of sealing, crusting, and disaggregation also are described, and references to case studies/datasets are presented as evidentiary examples of the actions/practices that have promoted or diminished these processes. For a detailed description of the SSL method which is cross-referenced by method code in the table of contents in this manual, refer to SSIR No. 42 (Soil Survey Staff, 2004), which is available online at http://soils.usda.gov/technical/lmm/. Refer to the Soil Quality Test Kit Guide (Soil Quality Institute, 1999; available online at http://soils.usda.gov/sqi/assessment/files/test_kit_complete.pdf) and SSIR No. 51 (Soil Survey Staff, 2009; available online at http://www.soils.usda.gov/technical/) for descriptions of field methods as used in NRCS soil survey offices.

3.6   Aggregate Stability
3.6.1   Structure and Aggregates

Structure, definition and assessment: *Soil structure* is defined as the physical constitution of a soil material as expressed by the size, shape, and arrangement of elementary particles and voids (Brewer, 1964). Structure is one of the most difficult physical properties to determine for quantitative evaluation (Lal, 1981). As used in most soil classification systems, soil structure is more of a qualitative evaluation, e.g., description of soil peds, and does not provide a precise means to predict soil behavior in different management systems (Lal, 1981). An assessment of soil structure is typically the qualitative or visual evaluation of its physical constitution, often complemented by quantitative analyses of the stability of this particular arrangement (aggregate stability) to a disruptive force (e.g., hand manipulation, water, wind, wheel traffic) (Kladivko, 2002).

Aggregates, definition and assessment: *Soil aggregate* is defined as a group of primary articles that cohere to each other more strongly than to other surrounding soil particles (Soil Science Society of America, 2010). Disaggregation of soil mass into aggregates requires the application of a disrupting force. Aggregate stability is a function of whether the cohesive forces between particles can withstand the applied disruptive force. Aggregate stability is the result of complex interactions among biological, chemical, and physical processes in the soil (Tisdall and Oades, 1982; Diaz-Zorita et al., 2002; Marquez et al., 2004). Due to the empirical relationship between laboratory-determined aggregate-size distribution and the distribution as it exists in the field, most investigators use the stability of aggregates rather than aggregate-size distribution as an index of soil structure in the field (Kemper and Rosenau, 1986). Whereas soil texture cannot be changed, at least over a short period of time, by any economical means, successful management of some soils, e.g., tropical soils, depends on the management of soil structure (Lal, 1979).

Structure and aggregate stability, applications: Soil structure and aggregation play an important role in an array of soil processes, such as erodibility, organic matter protection, and soil fertility. Soil structure and its stability govern soil-water relationships, aeration, crusting, infiltration, permeability, runoff, interflow, root penetration, leaching losses of plant nutrients, and, therefore, the productive potential of a soil. The assessment of soil structure can be used to help evaluate the soil's ability to support plant growth; cycle C and nutrients; receive, store, and transmit water; and resist soil erosion and the dispersal of chemicals of anthropogenic origin. It can also be used to help evaluate the effects of various agricultural techniques, e.g., tillage and organic matter additions (Kay and Angers, 2002; Nimmo and Perkins, 2002). The analysis of soil aggregation can be used to evaluate or predict the effects of various agricultural techniques, e.g., tillage and organic matter additions, and erosion by wind and water (Nimmo and Perkins, 2002). Immediately after cultivation, most soils contain an abundance of large pores, which favor high infiltration rates, good tilth, and adequate aeration for plant growth. The continued existence of these large pores in the soil, however, depends on the stability of the aggregates. The measurement of aggregate stability can serve as a predictor of infiltration and soil erosion potential but not as an indicator of the soil erosion hazard (Lal, 1981). Erodibility of soils increases as aggregate stability decreases (Kemper and Rosenau, 1986). In a Zimbabwe study of paired pedons (row cropped versus pasture management) derived from granite, important indicators of soil degradation and susceptibility to erosion were determined to be aggregate stability, surface horizon thickness, and ratios of water dispersible clay to total clay and organic C to silt + clay (Burt et al., 2001b).

3.6   Aggregate Stability
3.6.2   Sealing, Crusting, and Disaggregation, Processes and Case Studies

Sealing, crusting, and disaggregation, processes: The terms *soil crusting* and *soil sealing* have been used synonymously in the literature, but some authors draw distinctions between these two soil processes and their effects, i.e., specific stages of soil compaction (Valentin and Bresson, 1998). Surface sealing is associated with the initial or wetting phase in crust formation, and crusting is associated with the hardening of the surface seal in the subsequent drying phase (Arndt, 1965; Remley and Bradford, 1989). Seals form very thin (1 to 5 mm) dense elastic layers that clog soils and seal the soil surface, whereas crusts, formed by the same processes as seals, form thicker layers (5 to 20 mm) that are platy in arrangement, crack upon drying, and can be separated from the soil surface (Gabriels et al., 1998; Zoebisch and Dexter, 2002). Impacts of crusting and sealing are reduced porosity and high penetration resistance, resulting in surface erosion and runoff and obstruction of seedling emergence (Valentin and Bresson, 1998). Soil crusting can be assessed directly through macro/micromorphological examination or indirectly through decreases in infiltration and increases in surface strength (Valentin and Bresson, 1998).

The following definitions of the various soil crusts are themselves descriptions of the actions/practices that have promoted this soil process.

(1) Sedimentary (depositional) crusts are formed by the transport and deposition (or suspension and deposition) of fine particles by surface flow, e.g., erosion and surface irrigation (Chen et al., 1980). Transport may be local (e.g., ridge to furrow, clod surface to interclod areas) or long distance (e.g., rills and sheet flow) (West et al., 1991).

(2) Structural crusts are the result of physical forces (e.g., raindrop impact, sprinkler irrigation, animal trampling, wheel traffic, and flooding). These crusts are often consequences on barren or unprotected soil surfaces (Zoebisch and Dexter, 2002).

(3) Slaking crusts are formed as a result of the breakdown of soil aggregates into smaller aggregates when they are immersed in water. These aggregates may subsequently disperse (e.g., chemical dispersion of the clay due to the presence of exchangeable Na).

Sealing, crusting, and disaggregation, case studies: Soil crusting and sealing phenomena have been studied for nearly six decades by extensive experimental investigation as well as simulation models. These phenomena are considered major contributing processes to agricultural and environmental degradation in the Western, North-Central, and Southeastern United States (Sumner and Stewart, 1992). Soil susceptibility to rainfall-induced sealing and crusting depends upon a combination of soil physical, chemical, and biological processes highly affected by climatic and soil conditions prevailing during seal formation (Bradford and Huang, 1992). In general, cultivated soils are structurally unstable, and surface seals and crusts are common phenomena of these soils (Shainberg, 1992).

Susceptibility to seal and crust formation is a property suggested to be common to many of the soils of the Western U.S. (El-Swaify et al., 1984). Extensive irrigated agriculture and dryland farming under marginal precipitation, abundance of sodium-affected soils, and inherently low organic matter are several characteristics of western agriculture that differentiate this area from the rest of the country and in part explain the susceptibility of some of these soils to sealing and crusting (Singer and Warrington, 1992). While seals and structural crusts are often the consequences of rainfall impact or sprinkler irrigation on barren or unprotected soils, these processes are enhanced by Na-induced clay dispersion.

In 1989, according to the U.S. Department of Commerce, the Western Region had 15.2 million hectares of irrigated land, or approximately 81 percent of the total irrigated land in the country. While sodic soils occur naturally in the arid and semiarid regions of the West, the problems associated with these soils are exacerbated under irrigation systems using poor-quality water (e.g., high Na) and with inadequate delivery and/or drainage systems. The accumulation of Na can result in a dispersing effect on clay and organic matter, leading to disaggregation, crust formations, and decreased permeability. Law et al. (1972) estimated that 20 percent of the total water delivered for irrigation in the U.S. was lost by seepage from conveyance and irrigation canals. These seepage waters typically percolate into the underlying strata, dissolving additional salts in the process, flow to lower elevation lands or waters, and add to the problem of salt-loading associated with on-farm irrigation (Rhoades, 2002).

The generalization of low organic matter content in many arid and semiarid western soils is supported by a study of benchmark soils (Singer and Warrington, 1992) from eight states in the West, compiled from the USDA Soil Survey Investigations Reports from those states for the years 1966 (KS, MT, ND, and WY), 1967 (CO and OK), 1970 (NV), and 1973 (CA) (USDA/SCS, 1966a, 1966b, 1966c, 1966d, 1967a, 1967b, 1970, 1973). These data showed relatively low surface organic C, ranging from <0.1 to 1.5 percent, with the highest values found in poorly drained soils and soils under native meadow or prairie vegetation. Loss of organic matter from soils inherently low in these materials, through such practices as overgrazing of rangelands and intensive cultivation, increases the soil's susceptibility to surface sealing and crusting (Smith and Elliott, 1990). The Bureau of Land Management (BLM) is the single largest manager of publicly owned grazing lands, with 270 million acres grazed in 16 Western states, or about one-eighth of the acreage in the United States. In 1990, the BLM determined by its own monitoring and evaluation data that two-thirds of its managed rangeland was in unsatisfactory condition (fair or poor), largely due to overgrazing of these lands.

A number of soil management practices and irrigation practices (Singer and Warrington, 1992; Rhoades, 2002) have been used to reduce or ameliorate crusting problems in the West, including shallow tillage to disrupt crusts; addition of crop residue and manure to increase organic matter; surface mulches used to intercept raindrop and sprinkler drop impact energy; rangeland reseeding and biota establishment; controlled grazing; streambank restoration; chemical amendments (e.g., gypsum and phosphogypsum) to increase electrical conductivity of irrigation and decrease Na content; use of high-quality water; efficient irrigation schemes; site modification using settling basins and alteration in canopy configuration; and adequate irrigation drainage systems.

Susceptibility to surface sealing has also been determined for important agricultural soils in the North-Central region. In a regional study by agricultural experiment stations (IL, IN, IA, MI, MN, NE, ND, OH, SD, and WI), 28 representative soils were found to have reductions in steady state infiltration rates and potential for surface sealing if impacted by rainfall when the surface was barren (NCRRP, 1979). A large part of the arable Corn Belt on sloping topography, represented by 58 soils in Indiana and Wisconsin, also was shown to be susceptible to extensive water erosion and surface sealing (Mannering, 1967).

The occurrence and effects of soil crusting on soils in the Southeast, though not widely acknowledged in comparison to those in the West, have played an important role in affecting seedling emergence and determining runoff and erosion behavior of cultivated soils under rainfall (Miller and Radcliffe, 1992). Over the last 100 years, agriculturalists and soil scientists have observed the tendency of sandy soils predominant in Georgia, the Carolinas, and Virginia to form hard-setting surface layers with high runoff rates; however, little research has documented specific crusting processes on these soils and the resulting effects (Miller and Radcliffe, 1992).

In response to a growing recognition of soil erosion as a problem in the Southeast in the 1930s, field plot experiments were conducted to evaluate soil erosion and runoff. One of the earliest and most informative studies, in 1935, evaluated cropping and tillage effects on the Cecil soil in Georgia under natural rainfall. The data from this study showed that untilled plots with bare surfaces had, by tenfold, the highest runoff (42.5

percent of rain) and soil loss (179 mt ha y^{-1}). From these data, Miller and Radcliffe (1992) concluded that crust formation on these soils had enhanced runoff, resulting in accelerated erosion. Peele et al. (1945) found similar results for soils under continuous cotton cultivation and natural rainfall in the South Carolina Piedmont, with the greatest runoff (38 percent of rain) and soil loss (43 mt ha y^{-1}) from those soils (Cecil sandy loam) forming compact, relatively impermeable surface layers.

Bennett et al. (1964) showed that crusting of a Greenville fine sandy loam from the Georgia Coastal Plain resulted in the emergence of only 10 percent of planted cotton seeds after a crust had formed. Edwards (1966) confirmed similar problems for cotton seedling emergence on some crusting soils in Mississippi. While it is difficult to document declines in yields resulting from the formation of crusts, the historical observations on crusting and runoff in conjunction with crop emergence data in the Piedmont and Coastal Plain are supported by more recent runoff and dispersion measurements on 25 Southeastern soils. These measurements link the colloidal phenomenon of clay dispersion with the process of crust formation (Miller and Radcliffe, 1992).

Crusting has also had an impact on irrigation efficiency and becomes more important as the use of center-pivot sprinkler irrigation increases in the Coastal Plain area (Miller and Radcliffe, 1992). The water application rates of this high energy impact irrigation system are often limited by low infiltration rates due to crust formation. Minimizing crusting would result in more efficient irrigation and would allow higher sprinkling rates and less runoff.

Dispersive soil conditions and associated soil crusting in the Southeast have been linked to the production of dispersed clay in runoff waters. This dispersed clay is readily transported and could account for 10 to 25 percent of the interrill sediment load in sandy soils (Miller and Baharuddin, 1987). This dispersed clay can also transport sorbed agricultural chemicals (e.g., P, trace metals, and pesticides) to surface waters, a primary source of potable water in many parts of the Southeast (Miller et al., 1988). More recent agricultural practices in the Southeast, including no-till (West et al., 1991), residue management systems (e.g., winter cover crops) that add organic matter and improve macroaggregation, and additions of amendments, such as gypsum, and synthetic organic polymers, such as polyacrylamide (PAM) (Azzam, 1980), have been used to reduce the amount of exposed tilled soil, reducing susceptibility to crusting and thus increasing infiltration and reducing soil loss.

In the last century, much of the evidence of sealing and crusting and their effects on U.S. agricultural soils has been anecdotal; however, there is strong indication in the literature that these phenomena have been major contributing processes to agricultural and environmental degradation in the Western, North-Central, and Southeastern United States, occurring in many geographical areas of the country and affecting a wide range of soil types. In more recent years, progress has been made in identifying and understanding this soil process, but more research is needed to identify those soil properties and their interactions with the environment that will more effectively predict soil susceptibility to crusting. One of the problems with predicting this susceptibility and evaluating its impacts is the existence of a wide variety of available methods to measure and assess the impacts of soil crusting (Valentin and Bresson, 1992). In general, this problem is common to most assessment methods of soil quality. The problem is further complicated by physical properties that are often best measured *in situ*, unlike chemical and mineralogical properties, which are more easily and accurately measured in a laboratory. Soil physical properties are typically disturbed during the process of sample extraction for laboratory analysis, and this disturbance can disrupt the very physical relationships and arrangements that are of interest. For proper comparison among datasets, standardized methods of assessing soil quality that are efficient, reproducible, and accurate are needed.

3.6 Aggregate Stability
3.6.3 Wet Aggregate Stability, Wet Sieving

Wet aggregate stability, wet sieving (2 to 1 mm, 2- to 0.5-mm aggregates retained) measurement: A simple procedure for stability analysis involves the use of one size fraction. The SSL uses the 2- to 1-mm fraction with 2- to 0.5-mm aggregates retained and the sand weight subtracted. The SSL method provides a measure of aggregate stability following a disruption of initially air-dry aggregates by abrupt submergence in water overnight followed by wet sieving. The aggregate stability is reported as a percent of aggregates (2 to 0.5 mm) retained after wet sieving. Determinations are not reported if the 2- to 0.5-mm fraction is ≥50 percent of the 2- to 1-mm sample.

Wet aggregate stability, wet sieving (<2 mm, >0.25-mm aggregates retained) measurement: The Soil Quality Test Kit (Soil Quality Institute, 1999) provides a method that measures the 0.25-mm (<250-μm) aggregates retained after wet sieving and thus is different from the standard SSL method (Soil Survey Staff, 2004). Marquez et al. (2004) define soil aggregates with diameters >250 μm as macroaggregates. Large macroaggregates have diameters >2000 μm, and small macroaggregates range between 250 and 2000 μm in diameter. Microaggregates have diameters between 53 and 250 μm. The mineral fraction is <53 μm in diameter. In essence, the method derived from the Soil Quality Institute (1999) captures a greater portion of the (water-stable) macroaggregates.

3.7 Particle Density

This section describes the SSL method for particle density by pycnometer, gas displacement. Information is provided on particle density estimates for various minerals and parent materials. For a detailed description of the SSL method which is cross-referenced by method code in the table of contents in this manual, refer to SSIR No. 42 (Soil Survey Staff, 2004), which is available online at http://soils.usda.gov/technical/lmm/. Refer to SSIR No. 51 (Soil Survey Staff, 2009; available online at http://www.soils.usda.gov/technical/) for descriptions of field methods as used in NRCS soil survey offices.

3.7 Particle Density
3.7.1 Estimates for Various Minerals and Parent Materials

Particle density, definition: *Density* is defined as mass per unit volume. *Particle density* refers to the density of the solid particles collectively (Flint and Flint, 2002). In contrast, *grain density* refers to the density of specified grains; *bulk density* includes the volume of the pores created between particles and pores that exist within individual particles; and *specific gravity* is the ratio of particle density to that of water at 3.98 °C (1.0000 g cm^{-3}) or other specified temperature and as such is unitless (Flint and Flint, 2002).

Particle density, general applications: Particle density affects many of the interrelationships of porosity, bulk density, air space, and rates of sedimentation of particles in fluids. Particle-size analyses that use sedimentation rate, as well as calculations involving particle movement by wind and water, require information on particle density (Blake and Hartge, 1986b). Particle density is also required for calculations of heat capacity and soil volume or mass and for mathematically correcting bulk soil samples containing significant amounts of rock fragments so as to determine fine-soil density, water content, or other soil properties affected by volume displacement of rock fragments (Flint and Childs, 1984; Childs and Flint, 1990).

Particle density, estimates: Even though there is a considerable range in the density of individual soil minerals, in most mineral soils that are predominantly quartz, feldspar, and the colloidal silicates, the densities fall within the narrow limits of 2.60 to 2.75 g cc^{-1} (Brady, 1974). The particle density of volcanic glass is approximately 2.55 g cc^{-1} (Van Wambeke, 1992). With unusual amounts of heavy minerals present, e.g., magnetite, garnet, epidote, zircon, tourmaline, and hornblende, the particle density may exceed 2.75 g cc^{-1} (Brady, 1974). Organic matter weighs much less than an equal volume of mineral solids. Organic matter has a particle density of 1.2 to 1.5 g cc^{-1}; i.e., the amount of organic matter in a soil markedly affects the particle density of the soil (Brady, 1974).

Knowledge of parent material is useful in estimating particle density. Mineral composition may also be used. Tables 3.7.1.1 and 3.7.1.2 provide particle densities for parent materials and various minerals, respectively (Flint and Flint, 2002). For example, if a sample has quartz (90 percent) and feldspar (10 percent), the particle density estimate is determined as follows:

$$0.90 \ (2.65 \ \text{g cm}^{-3}) + 0.10 \ (2.5 \ \text{to} \ 2.8 \ \text{g cm}^{-3}) = 2.385 + (0.25 \ \text{to} \ 0.28) = 2.63 \ \text{to} \ 2.67 \ \text{g cm}^{-3}$$

Table 3.7.1.1   Particle densities for various parent materials[1]

Material	Particle density
	g cm^{-3}
Agate	2.5–2.7
Basalt	2.4–3.1
Dolomite	2.84
Flint	2.63
Granite	2.64–2.76
Humus	1.5
Limestone	2.68–2.76
Marble	2.6–2.84
Sandstone	2.14–2.36
Serpentine	2.5–2.65
Slate	2.6–3.3

[1] After Flint and Flint (2002) and reproduced with permission by Soil Science Society of America, Madison, Wisconsin.

Table 3.7.1.2   Particle densities for various minerals[1]

Material	Particle density
	g cm^{-3}
Apatite	3.2
Calcite	2.21
Clay	1.8–3.1
Illite	2.8
Kaolinite	2.65
Montmorillonite (smectite)	2.5
Chlorite	3.0
Feldspar	2.5–2.8
Orthoclase	2.56
Glass	2.4–2.75
Gypsum	2.31–2.33
Mica	2.6–3.2
Biotite	2.7–3.1
Muscovite	2.83
Mordenite	2.13
Opal	1.9
Pyrite	5.02
Quartz	2.65

[1] After Flint and Flint (2002) and reproduced with permission by Soil Science Society of America, Madison, Wisconsin.

3.7 Particle Density
3.7.2 Pycnometer, Gas Displacement

Particle density, measurement: The SSL determines particle density by pycnometer gas displacement. This determination is accomplished by employing Archimedes' principle of fluid displacement to determine the volume. The displaced fluid is a gas that can penetrate the finest pores, thereby assuring maximum accuracy (Quantachrome Instruments, 2003). Helium gas is the most commonly recommended gas since its small atomic dimensions assure penetration into crevices and pores approaching 1 Angstrom (10^{-10} m) in dimension, and its behavior as an ideal gas also is desirable. Particle density (g cm^{-3}) is reported as g cm^{-3} on an oven-dry basis to the nearest 0.01 unit on either the <2-mm or >2-mm particle-size fraction.

3.8 Atterberg Limits
3.8.1 Liquid Limit
3.8.2 Plasticity Index

Atterberg limits, definition: Early ideas on soil consistency and procedures for its measurement were developed by Atterberg in 1910 (Carter and Bentley, 1991). Originally, Atterberg defined five limits (1911), but only three (shrinkage limit, plastic limit, and liquid limit) are used in soil mechanics; thus, *Atterberg limits* is a general term that encompasses *liquid limit* (LL), *plastic limit* (PL), and, in some references, *shrinkage limit* (SL). The methods of measurement for these limits are operationally defined and have changed little since 1910.

Liquid limit, definition: *Liquid limit* (LL) is the percent water content of a soil at the arbitrarily defined boundary between the liquid and plastic states. This water content is defined as the water content at which a pat of soil placed in a standard cup and cut by a groove of standard dimensions will flow together at the base of the groove for a distance of 13 mm (1/2 in) when subjected to 25 shocks from the cup being dropped 10 mm in a standard LL apparatus operated at a rate of 2 shocks s^{-1}. Refer to ASTM method D 4318-05 (ASTM, 2008k). The LL is reported as percent water on a <0.4-mm basis (40-mesh) for an air-dry or field-moist sample.

Plastic index, definition: *Plastic index* (PI) is the range of water content over which a soil behaves plastically. Numerically, the PI is the difference in the water content between the LL and the plastic limit (PL). The PL is the percent water content of a soil at the boundary between the plastic and brittle states. The boundary is the water content at which a soil can no longer be deformed by rolling into 3.2-mm (1/8-in) threads without crumbling. Refer to ASTM method D 4318-05 (ASTM, 2008k). The PI is reported as percent water on a <0.4-mm basis for an air-dry or field-moist sample.

Shrinkage limit, definition: *Shrinkage limit* (SL) represents the moisture content at which further drying of the soil causes no further reduction in volume (Carter and Bentley, 1991). In electrochemical terms, the clay mineral particles are far enough apart at the LL to reduce the electrochemical attraction to almost zero, and at the PL there is the minimum amount of water present to maintain the flexibility of the bonds (Carter and Bentley, 1991). The SL test is less likely determined in soil mechanics than the LL and PL. The SSL only reports the LL and PL. The SL test is difficult to carry out, and results vary according to the test method used; in some cases the results depend on the initial moisture of the test specimen (Carter and Bentley, 1991).

Engineering classification systems: The test method for Atterberg limits by the American Society for Testing and Materials (ASTM) has the designation of D 4318-05 (ASTM, 2008k). This test method is used as an integral part of several engineering classification systems, e.g., American Association of State Highway and Transportation Officials (AASHTO) and the ASTM Unified Soil Classification System (USCS), to characterize the fine-grained fractions of soils—ASTM D 2487-63 (ASTM, 2008b) and D 3282-93 (ASTM, 2008a)—and to specify the fine-grained fraction of construction materials—ASTM D 1241-00 (ASTM, 2008d). The LL and PI of soils also are used extensively, either individually or together with other soil properties, to correlate with engineering behavior, e.g., compressibility, permeability, compactability, shrink-swell, and shear strength. The LL and PI are closely related to amount and kind of clay, CEC, 1500-kPa water, and engineering properties, e.g., load-carrying capacity of the soil.

In general, the AASHTO engineering system is a classification system for soils and soil-aggregate mixtures for highway construction purposes, e.g., earthwork structures, particularly embankments, subgrades, subbases, and bases. The USCS classification is used for general soils engineering work by many organizations, including the NRCS.

Liquid limit, calculations: If the LL is not measured, it can be estimated for use in engineering classification through the use of algorithms. Many algorithms have been developed that are applicable to a particular region or area of study. Some equations developed by the National Soil Survey Laboratory Staff (1975) are as follows:

Equation 3.8.1.1:

$$LL = 0.9 \times Clay + 10$$

OR (alternatively)

Equation 3.8.1.2:

$$LL = 2 \times W_{1500} + 10$$

where
LL = Liquid limit
Clay = Weight percentage of clay on a <2-mm soil basis
W_{1500} = Weight percentage of water retained at 1500-kPa suction on a <2-mm soil basis

Plastic index, calculations: If either the LL or the PL cannot be determined, or if PL is ≥ LL, the soil is reported as nonplastic (NP). If the PI is not measured, it can be estimated for use in engineering classification through the use of algorithms. Many algorithms have been developed that are applicable to a particular region or area of study. Some equations developed by the National Soil Survey Laboratory Staff (1975) are as follows:

Equation 3.8.2.1:

When <15 percent clay
PI = Clay x 0.3

Equation 3.8.2.2:

When 15 to 35 percent clay
PI = Clay x 0.4

Equation 3.8.2.3:

When 35 to 55 percent clay
PI = Clay - 21

Equation 3.8.2.4:

When >55 percent clay
PI = Clay - 15

where
PI = Plasticity index
Clay = Weight percentage of clay on <2-mm soil basis

4   Soil and Water Chemical Extractions and Analyses

This section describes the SSL methods for soil and water chemical extractions and analysis and their specific method applications and interferences as follows:

- Ion exchange and extractable cations
- Ratios, estimates, and calculations associated with ion exchange and extractable cations
- Soil pH
- Soil test analyses
- Carbonate and gypsum
- Electrical conductivity and soluble salts
- Ratios, estimates, and calculations associated with electrical conductivity and soluble salts
- Selective dissolutions
- Total analysis
- Ground water and surface water analysis

4.1 Ion Exchange and Extractable Cations

Ion exchange (anion and cation exchange), processes and components: The SSL procedures to determine cation-exchange capacity (CEC) and effective CEC (ECEC), both analytical and calculated values, are described in this section. These procedures include:

- CEC by NH_4OAc, pH 7 (CEC-7)
- CEC by sum of cations (CEC-8.2)
- ECEC by NH_4Cl, neutral unbuffered
- ECEC by summing NH_4OAc extractable bases plus 1 N KCl extractable Al

Extractable and exchangeable bases and extractable (potential) and exchangeable (active) acidity, definitions: Information is provided on extractable cations in relation to factors affecting their deficiencies, toxicities, and relative abundance in soils as well as their role as essential plant nutrients. The SSL methods to determine these extractable cations are described in this section. These procedures include:

- NH_4OAc extractable bases (Ca, Mg, K, and Na)
- $BaCl_2$-triethanolamine, pH 8.2 extractable acidity
- 1 N KCl extractable Al and Mn

Ratios, estimates, and calculations related to ion exchange and extractable cations: These values include calculated CEC and ECEC values, base saturation, sum of bases, aluminum saturation, and CEC/clay ratio. In addition, this section provides definitions of terms as well as applications of these ratios, estimates, and calculations. For detailed descriptions of SSL methods which are cross-referenced by method code in the table of contents in this manual, refer to SSIR No. 42 (Soil Survey Staff, 2004), which is available online at http://soils.usda.gov/technical/lmm/. Refer to SSIR No. 51 (Soil Survey Staff, 2009; available online at http://www.soils.usda.gov/technical/) for detailed descriptions of field methods as used in NRCS soil survey offices.

Soil properties, pH, and capacity to provide plant-available nutrients: These soil properties are discussed together as their assessments are typically addressed simultaneously and any interpretation or ameliorative action thereof commonly requires measurements on all of these soil features. In addition, the causes for the degradative processes of these properties (nutrient depletion/deficiency and acidification, respectively) are commonly interrelated and their effects conjunctively expressed. The soil processes of nutrient depletion/deficiency and acidification are described, and references to case studies/datasets are presented as evidentiary examples of the actions/practices that have promoted or diminished these processes. In addition, major developments in the knowledge, science, and technology related to soil fertility are discussed.

4.1 Ion Exchange and Extractable Cations
4.1.1 Cation-Exchange Capacity

Ion exchange, definition: *Ion exchange* is a reversible process by which one cation or anion held on the solid phase is exchanged with another cation or anion in the liquid phase, and if two solid phases are in contact, ion exchange may also take place between two surfaces (Tisdale et al., 1985). In most agricultural soils, the *cation-exchange capacity* (CEC) is generally considered to be more important than the *anion-exchange capacity* (AEC); the anion molecular retention capacity of these soils is typically much smaller than the CEC (Tisdale et al., 1985).

Anion-exchange capacity, definition: Anion exchange sites arise from protonation of hydroxyls on surfaces of clays and by ligand exchange or the replacement of hydroxyls by other anions (Foth and Ellis, 1988). This hydroxyl replacement by other anions is a significant component of positive charge or AEC. Hydroxyl replacement is pH dependent, increasing with increasing acidity and decreasing pH; i.e., AEC is related to both the extent of ligand exchange and protonation of exposed hydroxyls, both of which are pH dependent (Foth and Ellis, 1988). The *zero point of charge* (ZPC) has been used to characterize the relative abundance of positive and negative charge on colloids. The ZPC is the pH at which negative and positive charge of a colloid are equal (Bohn et al., 1979). In some highly weathered soils in acidic environments with abundant goethite and gibbsite, e.g., oxic horizons or subsoils of Oxisols (Soil Survey Staff, 2010), the CEC and AEC may approach equality (i.e., CEC to AEC ratio approaches 1.0) as pH approaches the ZPC. In these soils, the soil organic matter may be low, contributing little to the negative charge and resulting in a net charge of zero or a small positive charge (Foth and Ellis, 1988). Refer to Soil Survey Staff (2010) for more discussion of the oxic horizon and Oxisols. The CEC increases as soil pH increases, and AEC increases as pH decreases (Foth, 1984).

Plants absorb as many anions as cations. Anions, such as sulfate, nitrate, and phosphate, are very important in soil-plant nutrition relationships involving the mineralization of organic matter; sulfate and phosphate are significant components in AEC in soils (Foth, 1984). Soils with net positively charged colloids may weakly adsorb anions, such as nitrate and chloride, which readily leach from soils. Also, they may adsorb sulfate and strongly adsorb or fix phosphate ions involved in ligand exchange. On the other hand, cations in these soils (e.g., calcium, magnesium, and potassium) may be repelled and thereby become susceptible to leaching in the soil solution (Foth and Ellis, 1988).

Cation-exchange capacity, definition: Soil mineral and organic colloidal particles have negative valence charges that hold dissociable cations and thus are "colloidal electrolytes" (Jackson, 1958). *Cation-exchange capacity* is usually defined as a measure of the quantity of readily exchangeable cations that neutralize negative charges in the soil (Rhoades, 1982a). More specifically, the CEC is a measure of the total quantity of negative charges per unit weight of the material and is commonly expressed in units of milliequivalents per 100 g of soil (meq 100 g^{-1}) or centimoles per kg of soil (cmol(+) kg^{-1}). The SSL reports cmol(+) kg^{-1} on a <2-mm basis. The CEC can range from less than 1.0 to greater than 100 cmol(+) kg^{-1} soil. These two units for expressing CEC are equivalent, as centimoles (in this conversion) are centimoles of monovalent charge. The term *equivalent* is defined as "1 gram atomic weight of hydrogen or the amount of any other ion that will combine with or displace this amount of hydrogen." The milliequivalent weight of a substance is one thousandth of its atomic weight. Since the equivalent weight of hydrogen is about 1 gram, the term *milliequivalent* may be defined as "1 milligram of hydrogen or the amount of any other ion that will combine with or displace it" (Tisdale et al., 1985).

Cation-exchange capacity, components: The CEC is a reversible reaction in soil solution, dependent upon negative charges of soil components arising from permanently charged or pH-dependent sites on organic matter and mineral colloid surfaces (Fig. 4.1.1.1). The mechanisms for these negative charges are isomorphic substitution within layered silicate minerals, broken bonds at mineral edges and external surfaces, dissociation of acidic functional groups in organic compounds, and preferential adsorption of certain ions on particle surfaces (Rhoades, 1982a). Isomorphic substitution produces permanent charge. The other charge mechanisms produce variable charge, which is dependent on the soil solution phase as affected by soil pH, electrolyte level, valence of counter-ions, dielectric constant, and nature of anions (Rhoades, 1982a). The total charge of soil particles commonly varies with the pH at which the charge is measured. The positive charge developed at low pH and the excess negative charge developed at high pH are collectively known as *pH-dependent charge* (Bohn et al., 1979). The soil's total charge is the algebraic sum of its negative and positive charges. As a result of the variable charge in soils, the CEC is a property dependent on the method and conditions of determination. The method of determination is routinely reported with CEC data. Common CEC values for some soil components (National Soil Survey Laboratory Staff, 1975) are as follows:

Soil component	cmol(+) kg^{-1}
Organic matter	200 to 400
"Amorphous" clay	160 (at pH 8.2)
Vermiculite	100 to 150
Smectite	60 to 100
Halloysite $4H_2O$	40 to 50
Illite	20 to 40
Chlorite	10 to 40
Kaolinite	2 to 16
Halloysite $2H_2O$	5 to 10
Sesquioxides	0

These very broad CEC ranges are intended only as general guidelines. More narrow groupings of CEC values are possible as data are continually collected and correlated. For example, the CEC of organic matter in Mollisols in the Western United States ranges from 100 to 300 cmol(+) kg^{-1} (average 200), and the CEC of organic matter in Histosols ranges from 125 to 185 cmol(+) kg^{-1} and increases with decomposition of the organic matter (National Soil Survey Laboratory Staff, 1975).

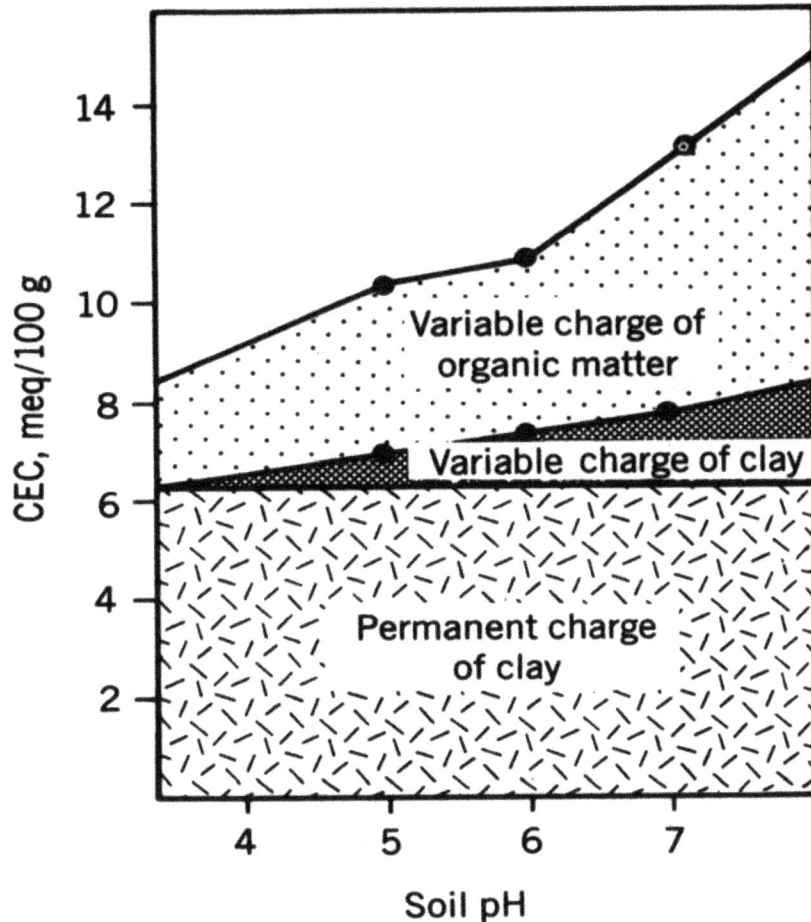

Figure 4.1.1.1.—The average source of negative charge in 60 Wisconsin soils. After Foth and Ellis (1988) and data from Helling et al. (1964).

Cation-exchange capacity, measurements: Many procedures have been developed to determine CEC. These CEC measurements vary according to the nature of the cation employed, the concentration of salt, and the equilibrium pH. The CEC measurement should not be thought of as highly exact but rather as an equilibrium measurement under the conditions selected (Jackson, 1958). Knowledge of the operational definition (procedure, pH, cation, and concentration) is necessary before the CEC measurement is evaluated (Sumner and Miller, 1996). The more widely adopted methods of CEC determination are classified (Rhoades, 1982a) as follows:

(1) cation summation
(2) direct displacement
(3) displacement after washing
(4) radioactive tracer

The SSL performs a number of CEC methods using several different reagents and pH levels. The CECs most commonly reported by the SSL are CEC by NH₄OAc, pH 7 (CEC-7), CEC by sum of cations (CEC-8.2), and effective cation-exchange capacity (ECEC). The ECEC can be determined by summing NH₄OAc extractable bases plus

KCl extractable Al or direct measurement by NH_4Cl. The ECEC by NH_4Cl is less commonly used at the SSL. The CECs most commonly reported by the SSL are CEC-7, CEC-8.2, and ECEC. As a general rule, the CEC-8.2 > CEC-7 > ECEC. The SSL reports all CEC values in cmol(+) kg^{-1}. In the past, these values were reported as meq 100 g^{-1}.

4.1 Ion Exchange and Extractable Cations
4.1.1 Cation-Exchange Capacity
4.1.1.1 NH_4OAc, pH 7 (CEC-7)

Cation-exchange capacity, NH_4OAc, pH 7 (CEC-7), application: The CEC-7 is a commonly used method and has become a standard reference to which other methods are compared (Peech et al., 1947). An advantage of using this method is that the extractant is highly buffered so that the extraction is performed at a constant and known pH (pH 7.0). In addition, the NH_4^+ on the exchange complex is easily determined. This pH represents the neutrality of the soil and is an ideal pH for the production of many important agricultural crops. CEC-7 is an analytically determined value and is usually used in calculating the CEC-7/clay ratios, although many SSL Primary Characterization Data Sheets predating 1975 show CEC-8.2/clay.

Cation-exchange capacity, NH_4OAc, pH 7 (CEC-7), measurement: Displacement after washing is the basis for this method. The CEC is determined by saturating the exchange sites with an index cation (NH_4^+) using a mechanical vacuum extractor (Holmgren et al., 1977), washing the soil free of excess saturated salt, displacing the index cation (NH_4^+) adsorbed by the soil, and measuring the amount of the index cation (NH_4^+). The extract is weighed and saved for analyses of the cations (Ca^{2+}, Mg^{2+}, K^+, and Na^+). The NH_4^+ saturated soil is rinsed with ethanol to remove the NH_4^+ that was not adsorbed on exchange sites. The soil is then rinsed with 2 M KCl. This leachate is then analyzed by steam distillation and titration to determine the NH_4^+ adsorbed on the soil exchange complex.

Cation-exchange capacity, NH_4OAc, pH 7 (CEC-7), interferences: Incomplete saturation of the soil with NH_4^+ and insufficient removal of NH_4^+ are the greatest interferences to this method. Ethanol removes some adsorbed NH_4^+ from the exchange sites of some soils. Isopropanol rinses have been used for some soils in which ethanol removes adsorbed NH_4^+. Soils that contain large amounts of vermiculite can irreversibly "fix" NH_4^+. Soils that contain large amounts of soluble carbonates can change the extractant pH and/or can contribute to erroneously high cation levels in the extract. This method overestimates the "field" CEC of soils with pH <7 (Sumner and Miller, 1996).

Cation-exchange capacity, NH_4OAc, pH 7 (CEC-7), prediction: There have been many studies using multiple regression models to predict CEC from clay and organic C. Results have shown that >50 percent of the variation in CEC can explained by the variation in clay and organic C in some New Jersey soils (Drake and Motto, 1982), for sandy soils (Yuan et al., 1967), for some Philippine soils (Sahrawat, 1983), and soils in Mexico (Bell and van Keulen, 1995). Wilding and Rutledge (1966) found that fine clay (<0.2 μm) explained a greater percentage of the CEC variation than the total clay content. In some gleyed subsoil horizons of lowland soils in Quebec, surface area was found to be a better predictor of CEC than total clay and variations in mineralogical composition were sufficient to explain nearly 50 percent of the variation in CEC (Martel et al., 1978).

Many of the aforementioned studies examined clay and organic C as single predictor variables for CEC. These studies were primarily specific to a region or area and used only a few soil types. Seybold et al. (2005), on the other hand, using data from the SSL characterization database, developed CEC (pH 7 NH_4OAc) prediction models that function comprehensively for the range of U.S. soils. Data were stratified into more homogeneous groups, and models were developed based on organic C, pH, taxonomic family mineralogy class, CEC activity class, and taxonomic order. Organic matter and noncarbonate clay served as the main predictor variables, and 1500-kPa water was used in lieu of clay content for four groups. Results indicated that between 43 and 78 percent of CEC variation could be explained for the high organic C; between 53 and 84 percent could be explained for the mineralogy groups; between 86 and 95 percent could be explained for the CEC activity class; and between 53 and 86 percent could be explained for the taxonomic orders. Using the data stratification, a decision tree was developed to guide selection of a predictive model to use for a soil layer (Seybold et al., 2005). See Tables 4.1.1.1.1 and 4.1.1.1.2 and Figure 4.1.1.1.1. Validation results indicated that models, in aggregate, provided a reasonable estimate of CEC for most U.S. soils (Seybold et al., 2005).

4.1   Ion Exchange and Extractable Cations
4.1.1   Cation-Exchange Capacity
4.1.1.2   Sum of Cations (CEC-8.2)

Cation-exchange capacity, sum of cations (CEC-8.2), application and calculation: CEC-8.2 is calculated by summing the NH_4OAc extractable bases (Ca^{2+}, Mg^{2+}, K^+, and Na^+) plus the $BaCl_2$-TEA, pH 8.2 extractable acidity. A pH of 8.2 was chosen because it is the pH that represents the equilibrium between free carbonates in the soil and CO_2 concentration in the atmosphere. Cation summation is the basis for this procedure. The CEC-8.2 minus the CEC-7 is considered the pH-dependent charge from pH 7.0 to pH 8.2. The CEC-8.2 is not reported if carbonates, gypsum, or significant quantities of soluble salts are present in the soil since the NH_4OAc extracts cations from the dissolution of these soil constituents.

CEC-8.2 is calculated as follows:

Equation 4.1.1.2.1:

CEC-8.2 = NH_4OAc extractable bases + $BaCl_2$-TEA extractable acidity

Table 4.1.1.1.1 Cation-exchange capacity (CEC) linear models and R^2, root mean square error (RMSE), and n values for the high organic carbon (OC) and mineralogy/CEC-activity stratification groups[1]

Grouping	Linear model†	R^2	RMSE	n
	OC > 8% and pH ≤7.0			
Eq. [1]; Oa horizons	2.12(totalC) + 9.992(pHCaCl$_2$) – 10.684	0.52	27.85	283
Eq. [2]; Oe horizons	2.03(totalC) + 3.396(pHCaCl$_2$) – 2.939	0.63	19.61	286
Eq. [3]; Oi horizons	1.314(totalC) + 27.047	0.43	17.01	300
Eq. [4]; OC ≤14.5%	1.823(totalC) + 0.398(nclay) + 15.54	0.42	10.41	133
	OC > 8% and pH > 7.0			
Eq. [5]; OC ≤14.5%	exp[1.316(ln totalC) + 1.063(ln nclay) – 3.211]	0.77	0.476‡	275
Eq. [6]; OC > 14.5%	4.314(totalC) – 26.492	0.78	16.62	30
	OC ≤8%			
Ferruginous	2.48(OC) + 0.128(silt) + 3.208	0.80	2.01	121
Amorphic	exp[0.182(ln OC) + 0.817(ln w15bar) + 0.736(ln pHw) – 0.608]	0.84	0.262‡	247
Glassy	exp[0.102(ln OC) + 1.219(ln w15bar) – 0.005]	0.76	0.495‡	257
Carbonatic	exp[0.253(ln OC) + 0.828(ln nclay) + 0.321]	0.78	0.348‡	406
Magnesic	2.38(OC) + 0.555(nclay) – 0.219(silt) + 10.428	0.59	6.27	80
Parasesquic	exp[0.13(ln OC) + 0.65(ln nclay) + 0.340(ln pHw) – 0.406]	0.58	0.325‡	258
Micaceous	exp[0.251(ln OC) + 0.205(ln clay) + 0.538(pHw) – 1.241]	0.64	0.464	41
Kaolinitic	exp[0.206(ln OC) + 0.618(ln nclay) + 0.303(ln silt) + 0.491(ln pHw) – 1.786]	0.56	0.431‡	1204
Smectitic	exp[0.033(ln OC) + 0.861(ln nclay) + 0.246]	0.75	0.186‡	1803
Illitic	exp[0.102(ln OC) + 0.596(ln nclay) – 1.108(ln pHw) + 2.892]	0.67	0.249‡	249
Vermiculitic	0.365(nclay) – 9.724(pHw) + 90.293	0.75	8.49	40
Isotic	exp[0.163(ln OC) + 0.683(ln w15bar) + 0.812(ln pHw) – 0.299]	0.78	0.329‡	635
Superactive	exp[0.039(ln OC) + 0.901(ln nclay) + 0.131]	0.90	0.184‡	12685
Active	exp[0.015(ln OC) + 0.987(ln nclay) – 0.576]	0.96	0.133‡	4580
Semiactive	exp[0.02(ln OC) + 0.974(ln nclay) – 0.927]	0.94	0.189‡	1648
Subactive	exp[0.009(ln OC) + 1.02(ln nclay) – 1.675]	0.91	0.289‡	256

[1] After Seybold et al. (2005) and reproduced with permission by Soil Science Society of America, Madison, Wisconsin.

† nclay, Noncarbonate clay; pHCaCl$_2$, pH in CaCl$_2$; pHw, pH in water; w15bar = -1500-kPa water.

‡ Root mean square error (RMSE) or standard deviation of the mean on the natural log transformed scale.

Table 4.1.1.1.2 Cation-exchange capacity (CEC) linear models and R^2, root mean square error (RMSE), and n values for the taxonomic order stratification groups[1]

Grouping	Linear model[†]	R^2	RMSE	n
Alfisols				
OC $\leq$ 0.3%	exp[0.911(ln nclay) − 0.308]	0.73	0.381[‡]	4129
OC > 0.3%	exp[0.158(ln OC) + 0.805(ln nclay) + 0.216]a	0.72	0.305[‡]	3206
Andisols	exp[0.088(ln OC) + 0.885(ln w15bar) + 0.867(ln pHw) − 0.985]	0.77	0.384[‡]	1181
Aridisols	exp[0.042(ln OC) + 0.828(ln nclay) + 0.236]	0.75	0.300[‡]	4114
Entisols	exp[0.078(ln OC) + 0.873(ln nclay) + 0.084]	0.85	0.350[‡]	1910
Gelisols	exp[0.359(ln OC) + 0.49(ln clay) + 1.05]b	0.72	0.509[‡]	97
Inceptisols	exp[0.134(ln OC) + 0.794(ln nclay) + 0.239]a	0.71	0.421[‡]	1921
Mollisols				
OC $\leq$ 0.3%	exp[0.932(ln nclay) − 0.174]	0.79	0.285[‡]	3284
OC > 0.3%	exp[0.113(ln OC) + 0.786(ln nclay) + 0.475]	0.74	0.203[‡]	8132
Oxisols	2.738(OC) + 0.103(nclay) + 0.123(silt) − 2.531	0.67	2.79	781
Spodosols	exp[0.045(ln OC) + 0.798(nclay) + 0.029]	0.71	0.311[‡]	243
	exp[0.999(ln w15bar) + 0 317]	0.86	0.315[‡]	636
Ultisols	exp[0.184(ln OC) + 0.57(ln nclay) + 0.365(ln silt) − 0.906]	0.76	0.350[‡]	499
Vertisols	exp[0.059(ln OC) + 0.86(ln nclay) + 0.312]	0.55	0.213[‡]	2109
Histosols	exp[0.319(ln OC) + 0.497(ln nclay) + 1.075]b	0.78	0.358[‡]	60
Gelisols and Histosols	exp[0 346(ln OC) + 0.49(ln nclay) + 1.064]	0.73	0.207[‡]	157
Alfisols (OC > 0.3%) and Inceptisols	exp[0.141(ln OC) + 0.797(ln nclay) + 0.235]	0.72	0.125[‡]	5127

[1] After Seybold et al. (2005) and reproduced with permission by Soil Science Society of America, Madison, Wisconsin.

† Equations with the same letters are not significantly different from each other. nclay, Noncarbonate clay; pHw, pH in water; w15bar, -1500-kPa water; OC, organic carbon.

‡ Root mean square error (RMSE) or standard deviation of the mean on the natural log transformed scale.

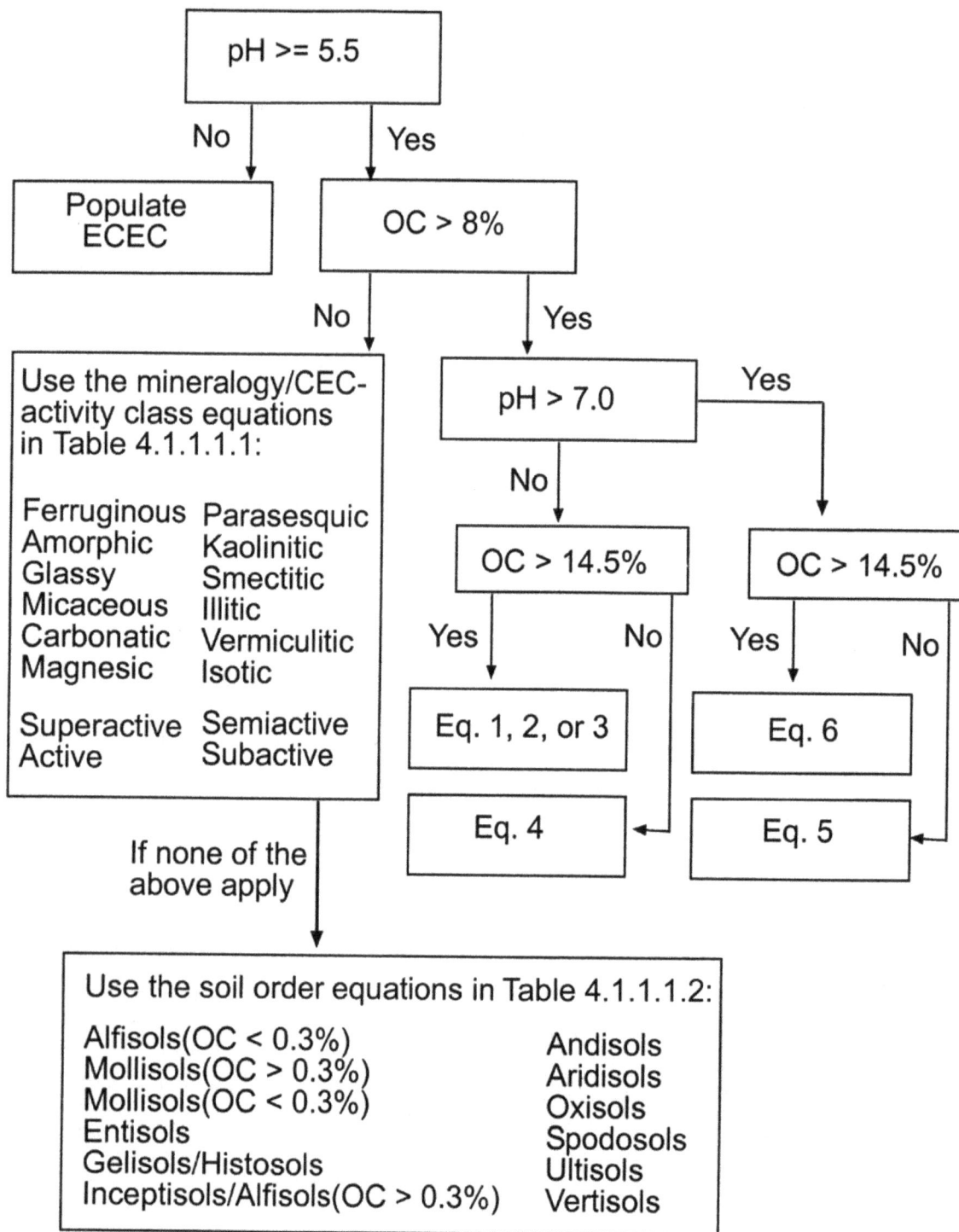

Figure 4.1.1.1.1.—A decision tree selects which cation-exchange capacity (CEC) predictive model should be used for a soil layer based on soil pH (in water), organic carbon content, and taxonomic soil classification. Equation numbers refer to those in Table 4.1.1.1.1. ECEC = effective cation-exchange capacity. After Seybold et al. (2005) and reproduced with permission by Soil Science Society of America, Madison, Wisconsin.

4.1 Ion Exchange and Extractable Cations
4.1.1 Cation-Exchange Capacity
4.1.1.3 NH₄Cl, Neutral Unbuffered

Effective cation-exchange capacity, NH₄Cl, neutral unbuffered, application:
The CEC using a neutral unbuffered salt (NH₄Cl) also is an analytically determined value. The CEC by NH₄Cl provides an estimate of the effective cation-exchange capacity (ECEC) of the soil (Peech et al., 1947). For a soil with a pH of <7.0, the ECEC value should be < CEC measured with a buffered solution at pH 7.0. The NH₄Cl CEC is ≈ equal to the NH₄OAc extractable bases plus the KCl extractable Al for noncalcareous soils. This ECEC method is less commonly used at the SSL.

Effective cation-exchange capacity, NH₄Cl, neutral unbuffered, measurement:
Displacement after washing is the basis for this method. The CEC is determined by saturating the exchange sites with an index cation (NH_4^+), washing the soil free of excess saturated salt, displacing the index cation (NH_4^+) adsorbed by the soil, and measuring the amount of the index cation (NH_4^+). A sample is leached using 1 N NH₄Cl and a mechanical vacuum extractor (Holmgren et al., 1977). The extract is weighed and saved for analyses of the cations. The NH_4^+ saturated soil is rinsed with ethanol to remove the NH_4^+ that was not adsorbed. The soil is then rinsed with 2 M KCl. This leachate is then analyzed by steam distillation and titration to determine the NH_4^+ adsorbed on the soil exchange complex.

4.1 Ion Exchange and Extractable Cations
4.1.1 Cation-Exchange Capacity
4.1.1.4 NH₄OAc Extractable Bases + 1 *N* KCl Aluminum

Effective cation-exchange capacity, NH₄OAc extractable bases + aluminum, application: CEC can be measured by extraction with an unbuffered salt. This method measures the effective cation-exchange capacity (ECEC), i.e., CEC at the normal soil pH (Coleman et al., 1958). Since the unbuffered salt solution, e.g., 1 N KCl, only affects the soil pH one unit or less, the extraction is determined at or near the soil pH and extracts only the cations held at active exchange sites at the particular pH of the soil. Neutral NH₄OAc extracts the same amounts of Ca^{2+}, Mg^{2+}, Na^+, and K^+ as KCl; therefore, extractable bases by NH₄OAc is used at the SSL in place of KCl-extractable bases.

Effective cation-exchange capacity, NH₄OAc extractable bases + aluminum, measurement and calculation: The SSL determines the ECEC by extracting one soil sample with neutral normal NH₄OAc to determine the exchangeable basic cations (Ca^{2+}, Mg^{2+}, Na^+, and K^+) and by extracting another sample of the same soil with 1 N KCl to determine the exchangeable Al. The 1 N KCl extractable Al approximates exchangeable Al and is a measure of "active" acidity present in soils with a pH <5.5. Aluminum is nonexchangeable at pH >5.5 due to hydrolysis, polymerization, and precipitation. The SSL does not analyze for 1 N KCl extractable Al if the 1:2 0.01 M CaCl₂ pH is <5.05. For soils with pH <7.0, the ECEC should be less than the CEC measured with a buffered solution at pH 7.0. The ECEC is not reported for soils with soluble salts.

ECEC is calculated by summing the NH_4OAc bases plus the KCl-extractable Al as follows:

Equation 4.1.1.4.1:

ECEC = NH_4OAc extractable bases + KCl-extractable Al

4.1 Ion Exchange and Extractable Cations
4.1.2 NH_4OAc, pH 7.0 Extractable Bases

Exchangeable and extractable bases, definitions: *Exchangeable cations* have been loosely defined as those removed by neutral salt solutions. Soluble salts, however, can be removed by water alone. The *extractable bases* (Ca^{2+}, Mg^{2+}, K^+, and Na^+) from the extractions by NH_4OAc and NH_4Cl are generally assumed to be those exchangeable bases on the cation-exchange sites of the soil. The term *extractable* rather than *exchangeable* bases is used because any additional source of soluble bases influences the results (Bohn et al., 1979). The most doubtful cation extractions with these kinds of methods are Ca^{2+} in the presence of soluble salts, free carbonates or gypsum, and K^+ in soils that are dominated by mica or vermiculite (Thomas, 1982).

Exchangeable cations, valence and size of hydrated radius: Mineral weathering is a natural source of cations that may potentially be adsorbed as exchangeable cations. The greater the supply of a cation from this weathering, the greater the likelihood that it will be adsorbed, according to the law of mass action (Foth, 1984). The amounts and kinds of cations actually adsorbed are greatly affected by cation valence and hydrated radius. In other words, a cation with greater valence is adsorbed more strongly or efficiently than one of a lower valence, and for a given valence, the cation with the smallest hydrated radius will move closer to the micellar surface and be more strongly adsorbed compared to a cation with a large hydrated radius because the energy of adsorption decreases as the square of the distance increases (Foth, 1984). These differences in the size of hydrated radius and valence are factors in cation adsorption; i.e., Ca^{2+} is more strongly adsorbed (greater valence and smaller hydrated radius) than Na^+. As a result, Ca^2 is preferentially adsorbed and frequently the most abundant exchange cation, whereas Na^+ is readily leached in humid areas (Fig. 4.1.2.1). The order of selectivity or replacement for some of the most common exchangeable cations in soils is as follows: Al^{3+} > Ca^{2+} > Mg^{2+} > K^+ > Na^+. Exchangeable H is difficult to place in this series because of the uncertainties of its hydration properties (Foth, 1984). The distribution of the major exchangeable cations in productive agricultural soils is generally Ca^{2+} > Mg^{2+} > K^+ ≈ $NH4^+$ ≈ Na^+, with this cation abundance and distribution similar to the energy of adsorption sequence (Bohn et al., 1985). Deviation from this usual order signals that some factor or factors, e.g., free $CaCO_3$ or gypsum, serpentine (high Mg^{2+}, Ca/Mg ratios <1), or hydrox material (high Na^+), have altered the soil chemistry. In an arid area, this distribution may be Ca^{2+} > Mg^{2+} > Na^+ > K^+. Other cations can be present on exchange sites under certain suites of minerals; e.g., Ni is an exchangeable cation in serpentine soils (Lee et al., 2001), and Fe is potentially exchangeable under the acid conditions of acid sulfate soils (Claff et al., 2010).

Hydrated ions

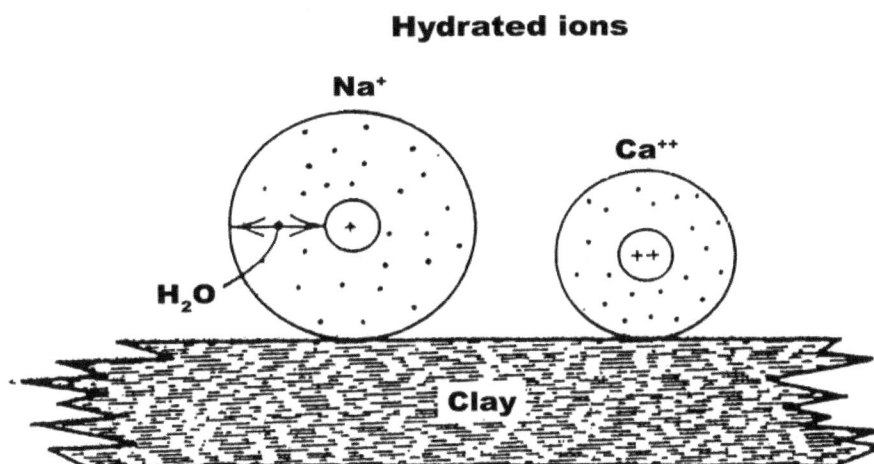

Figure 4.1.2.1.—Calcium ions are more strongly adsorbed by clay than sodium ions because calcium is divalent and has a smaller hydrated radius. After Foth (1984), Fundamentals of Soil Science, 7th edition, reproduced with permission by John Wiley & Sons.

Extractable bases, measurements: The standard SSL method for extractable bases is by NH_4OAc extraction. The NH_4OAc extract is diluted with an ionization suppressant (La_2O_3). The analytes are measured by an atomic absorption spectrophotometer (AAS). An analyte is measured by absorption of the light from a hollow cathode lamp. The AAS converts absorption to analyte concentration. Data are automatically recorded by a microcomputer and printer. In the past, the extractable Ca^{2+} was not reported for soils that contained carbonates or soluble salts ($CaCO_3 > 1.0$ percent). Currently, extractable Ca^{2+} is reported by the SSL for these kinds of soils but is flagged as having such soil components. The SSL less commonly determines extractable bases by NH_4Cl extraction. The analysis of NH_4Cl extractable bases is similar to that of the NH_4OAc extractable bases using atomic absorption spectrophotometry. The SSL reports NH_4OAc and NH_4Cl extractable bases Ca^{2+}, Mg^{2+}, K^+, and Na^+ as cmol(+) kg^{-1}. In the past, the SSL reported extractable bases as meq 100 g^{-1}.

4.1 Ion Exchange and Extractable Cations
4.1.2 NH$_4$OAc, pH 7.0 Extractable Bases
4.1.2.1 Calcium

Calcium, soil-related factors: The calcium present in soils, excluding that in added lime or fertilizer, originates from rocks and minerals in which the soils have developed, e.g., plagioclase minerals (anorthite and impure albite), pyroxenes (augite), amphiboles (hornblende), biotite, epidote, apatite, and certain borosilicates. In semiarid and arid regions, calcite is typically the dominant mineral form of calcium. Sources of calcium include windblown calcareous dust, calcareous ground water, and atmospheric CO_2 (Monger et al., 1991; Kraimer et al., 2005). Another source is calcium sulfate as gypsum and dolomite, often found in association with calcite. Regardless of soil texture, the Ca content of soils in arid regions is typically high because of low rainfall and minimal soil leaching (Tisdale et al., 1985), resulting in secondary deposits of calcium carbonate and calcium sulfate in the soil profiles. In acid, humid regions, Ca occurs largely in the exchangeable form and as undecomposed primary minerals (Tisdale et al., 1985). Typically, very sandy acid soils with low cation-exchange capacity (principally composed of quartz) have inadequate supplies of available Ca for crops.

Calcium is typically the most abundant exchangeable cation in soils. Yields of most agricultural crops are highest when the soil exchange complex is dominated by Ca^{2+}. A Ca-dominated exchange complex usually indicates a near-neutral pH, which is considered optimum for most plants and soil micro-organisms (Bohn et al., 1979). This composition also indicates that the concentrations of other potentially troublesome exchangeable cations are probably low, primarily Al^{3+} in acid soils and Na^+ in sodic soils. Despite the importance of Ca^{2+} as an exchangeable cation, soils derived from limestone can be unproductive; i.e., as the limestone weathers, the Ca^{2+} and HCO_3^- ions are released but are leached out of the system because the soils lack the cation-exchange capacity to retain the Ca^{2+} (Bohn et al., 1979).

Calcium typically occurs in the soil in the same mineral forms as Mg because the chemistry of these elements is very similar. The available forms of Ca and Mg in the soil are present in a Ca:Mg ratio of about 10:1 (Cook and Ellis, 1987). The functions of these two elements within the plant and the way deficiencies affect the plant differ widely.

Calcium, deficiencies in soils: Calcium deficiencies have been reported in soils derived from Mg-rich serpentine rocks and in soils that are highly leached, acidic, and Al saturated. The actual Ca^{2+} content in some soils may be sufficient for plant requirements, but the high concentrations of other cations, e.g., Mg^{2+} and Al^{3+}, may suppress the uptake of Ca^{2+} (Bohn et al., 1979). On the other hand, large quantities of Ca can also induce Mg and K deficiencies. Plant nutrition requires maintenance of a balance between the cations Ca^{2+}, Mg^{2+}, and K^+. The symptoms of Ca and Fe deficiencies in plants are almost identical; however, Fe deficiencies are more common in arid and semiarid regions in soils high in soluble salts, e.g., saline and saline-sodic soils.

Calcium deficiency, serpentine factor: Because of the confusing and often contradictory nature of studies on the serpentine factor as it relates to soil infertility, it is very difficult to assess the evidence and conclusions of these studies in a logical, orderly manner (Brooks, 1987). This infertility has been associated with toxic effects of Ni, Cr, and Co; toxicity of excess Mg; infertility due to low Ca content of serpentine soils; problems arising from an adverse (low) Ca/Mg ratio in the substrate; and infertility arising from low levels of plant nutrients in the soils (Brooks, 1987). In general, it is considered that the content of Cr and Co has little or no influence on vegetation because of the very low abundance of plant-available Cr and the lower toxicity of Co. Nickel, however, continues to be a probable source of some or much of the toxicity in serpentine soils. Calcium appears to play a primary role in the reduction or elimination of the toxic effects of Mg and/or Ni. Simplistically, infertility in serpentinitic soils has been related primarily to the absolute or relative abundance of Mg, Ni, and Ca. The deficiency of plant nutrients is due not so much from absolute concentrations of these elements but rather from the antagonism to their uptake by other constituents, primarily Ni and Mg, the effects of which may or may not be improved by the pH and Ca status of the soil (Brooks, 1987).

Calcium, essential plant element: Calcium is an essential nutrient for plant growth and is absorbed by plants as the ion Ca^{2+}. Calcium has an essential role in cell elongation and division, in cell membrane structure and permeability, in chromosome structure and stability, and in carbohydrate translocation (Tisdale et al., 1985).

Calcium, soil test: The NH_4OAc extractable Ca^{2+} is a common soil test for Ca. Convert Ca^{2+} to kg ha^{-1} for a soil horizon as follows:

Equation 4.1.2.1.1:

$$Ca = Ca^{2+} \times 0.02 \times 1000 \times Hcm \times \rho_{B33} \times Cm$$

where
Ca = Calcium for soil horizon (kg ha^{-1})
Ca^{2+} = NH_4OAc extractable Ca^{2+} (meq 100 g^{-1}) or (cmol(+) kg^{-1})
0.02 = Milliequivalent weight of Ca^{2+} (g meq^{-1})
1000 = Conversion factor to hectares
Hcm = Soil horizon thickness (cm)
ρ_{B33} = Bulk density at 33-kPa water content of <2-mm fraction (g cm^{-3})
Cm = Coarse fragment conversion factor. If no coarse fragments, Cm = 1. If coarse
      fragments are present, calculate Cm using Equations 3.1.2.1.1.3 and 3.1.2.1.1.4.

To convert Ca^{2+} (meq 100 g^{-1}) to lb A^{-1} or kg m^{-3}, replace the conversion factor for hectares (1000) in Equation 4.1.2.1.1 with the factor 2300 or 0.10, respectively. To convert Ca (meq 100 g^{-1}) to $CaCO_3$ (kg ha^{-1}, lb A^{-1}, or kg m^{-3}), replace the milliequivalent weight for Ca (0.02) in Equation 4.1.2.1.1 with the milliequivalent weight for $CaCO_3$ (0.050).

4.1   Ion Exchange and Extractable Cations
4.1.2 NH_4OAc, pH 7.0 Extractable Bases
4.1.2.2   Magnesium

Magnesium, soil-related factors: Magnesium in soils originates from rocks containing primary minerals, e.g., biotite, dolomite, hornblende, olivine, and serpentine, and from secondary clay minerals, e.g., chlorite, illite, smectite, and vermiculite. In arid and semiarid regions, significant amounts of epsomite, hexahydrite, and bloedite may also occur (Buck et al., 2006; Tisdale et al., 1985). In humid regions, Mg deficiency is most often seen in coarse-textured soils.

Magnesium is the second most abundant exchangeable cation in most soils and is absorbed by plants as Mg^{2+}. The concentration of exchangeable Mg^{2+} and other basic cations decreases as soils become leached. An exchange complex with high Mg^{2+} has sometimes been associated with poor physical soil conditions and high pH, e.g., sodic soil conditions. Poor soil structure may be produced by Na during the processes of soil formation under marine conditions. Initially, the soil may have an abundance of Mg and Na; the Na may eventually leach away, leaving the Mg-enriched soil with the inherited structure (Bohn et al., 1979).

Magnesium, deficiencies in soils: Excessive or deficient amounts of Mg are relatively uncommon. Soils associated with Mg deficiencies are acid sandy soils, soils with large amounts of applied calcitic lime, and soils heavily treated with K- or Na-bearing fertilizers (Cook and Ellis, 1987). Liming can usually correct the acidity and the Mg deficiencies in acid soils. Dolomitic limestone and other agricultural limestone typically contain appreciable Mg impurities. On the other hand, excessive amounts of calcium lime can induce Mg deficiencies.

Magnesium, essential plant element: Magnesium is an essential nutrient for plant growth and is absorbed by plants as the ion Mg^{2+}. Magnesium is the mineral constituent of the chlorophyll molecule, which is essential for all autotrophic plants to carry on photosynthesis. It also serves as a structural component of ribosomes, participates in a variety of physiological and biochemical functions, and is associated with transfer reactions involving phosphate-reactive groups (Tisdale et al., 1985). Magnesium in conjunction with sulfur has been related to oil synthesis in plants.

Magnesium, soil test: The NH_4OAc extractable Mg^{2+} is a common soil test for Mg. Convert to kg ha^{-1} for a soil horizon as follows:

Equation 4.1.2.2.1:

$$Mg = Mg^{2+} \times 0.012 \times 1000 \times Hcm \times \rho_{B33} \times Cm$$

where
Mg = Magnesium for soil horizon (kg ha^{-1})
Mg^{2+} = NH_4OAc extractable Mg^{2+} (meq 100 g^{-1}) or (cmol(+) kg^{-1})
0.012 = Milliequivalent weight of Mg^{2+} (g meq^{-1})
1000 = Conversion factor to hectares
Hcm = Soil horizon thickness (cm)
ρ_{B33} = Bulk density at 33-kPa water content of <2-mm fraction (g cm^{-3})
Cm = Coarse fragment conversion factor. If no coarse fragments, Cm = 1. If coarse
      fragments are present, calculate Cm using Equations 3.1.2.1.1.3 and 3.1.2.1.1.4.

To convert Mg^{2+} (meq 100 g^{-1}) to lb A^{-1} or kg m^{-3}, replace the conversion factor for hectares (1000) in Equation 4.1.2.2.1 with the factor 2300 or 0.10, respectively.

4.1 Ion Exchange and Extractable Cations
4.1.2 NH₄OAc, pH 7.0 Extractable Bases
4.1.2.3 Potassium

Potassium, soil-related factors: Potassium in soils, excluding that in added fertilizer, originates from the weathering of rocks containing K-bearing minerals, e.g., potassium feldspars orthoclase and microcline, muscovite, biotite, and phlogopite. The nature and mode of weathering of these K-bearing minerals largely depend on their properties and the environment. As far as the plant response is concerned, the availability (although slight) of K in these minerals is of the order biotite > muscovite > potassium feldspars (Tisdale et al., 1985). Potassium is also found in the form of secondary clay minerals, e.g., illites or hydrous micas, vermiculites, chlorites, and interstratified minerals.

An equilibrium between exchangeable and solution K generally results in some K in soil solution. Therefore, K has the potential to be leached from the system. As small quantities of soluble K exist in soil, many soils of humid and temperate regions may not have sufficient natural reserves to supply sufficient K to agronomic crops. Although potassium (K^+) is monovalent, its concentration in soil solutions is low relative to exchangeable K^+ because of its strong adsorption by many 2:1 layer silicate minerals (Bohn et al., 1979).

Potassium, essential plant element: Potassium is an essential nutrient for plant growth and is the third most important fertilizer element (after N and P). Potassium is absorbed by plants as the ion K^+. Plant requirements for this element are typically high. Potassium is necessary to many plant functions, including carbohydrate metabolism, enzyme activation, osmotic regulation and efficient use of water, nitrogen uptake and protein synthesis, and translocation of assimilates (Tisdale et al., 1985). Potassium also plays a role in minimizing certain plant diseases and in improving plant quality. Potassium deficiencies have been primarily associated with sandy soils because of the scarcity of K-bearing minerals and low clay contents, organic materials low in K, and high lime soils in which K^+ uptake is inhibited by high concentrations of Ca^{2+} (Cook and Ellis, 1987).

Potassium, soil test: The NH₄OAc extractable K^+ is a common soil test for K. Convert to kg ha⁻¹ for a soil horizon as follows:

Equation 4.1.2.3.1:

$$K = K^+ \times 0.039 \times 1000 \times Hcm \times \rho_{B33} \times Cm$$

where
K = Potassium for soil horizon (kg ha⁻¹)
K^+ = NH₄OAc extractable K^+ (meq 100 g⁻¹) or (cmol(+) kg⁻¹)
0.039 = Milliequivalent weight of K^+ (g meq⁻¹)
1000 = Conversion factor to hectares
Hcm = Soil horizon thickness (cm)
ρ_{B33} = Bulk density at 33-kPa water content of <2-mm fraction (g cm⁻³)

Cm = Coarse fragment conversion factor. If no coarse fragments, Cm = 1. If coarse fragments are present, calculate Cm using Equations 3.1.2.1.1.3 and 3.1.2.1.1.4.

To convert K^+ (meq 100 g^{-1}) to lb A^{-1} or kg m^{-3}, replace the conversion factor for hectares (1000) in Equation 4.1.2.3.1 with the factor 2300 or 0.10, respectively.

4.1 Ion Exchange and Extractable Cations
4.1.2 NH$_4$OAc, pH 7.0 Extractable Bases
4.1.2.4 Sodium

Sodium, soil-related factors: Three forms of sodium are typically found in the soil: fixed in insoluble silicates, exchangeable in the structures of other minerals, and soluble in the soil solution (Tisdale et al., 1985). In the majority of soils, most of the Na is present in silicates. In highly leached soils, Na may occur in high-albite plagioclases and in small amounts of perthite, micas, pyroxenes, and amphiboles, which exist mainly in the fine sand and silt fractions (Tisdale et al., 1985). In arid and semiarid soils, Na typically exists in silicates as well as soluble salts, e.g., NaCl, Na_2SO_4, and Na_2CO_3.

Sodium, adverse effects: Sodium is usually a soil chemical concern when it occurs in excess. Sodium has a dispersing action on clay and organic matter, resulting in the breakdown of soil aggregates and reducing permeability to air and water. Because of the loss of large pores, soils with excessive amounts of Na become almost impervious to water and air, root penetration is impeded, clods are hard, seedbed preparation is difficult, and surface crusting results in poor germination and uneven stands. These detrimental effects of excess levels of exchangeable Na^+ are conditioned by soil texture and clay mineralogy (Tisdale et al., 1985).

High concentrations of sodium are toxic to some plants. This toxicity may be relatively insignificant in comparison to the restrictions resulting from the associated physical condition of the soil. Poor physical soil condition normally precedes Na toxicity, and high pH usually accompanies the accumulation of Na in soils; however, these problems are less important than the water and micronutrient problems induced by Na accumulation (Bohn et al., 1979).

Sodium, essential plant element: Sodium is an essential element and is absorbed by plants as Na^+. Halophytic plant species accumulate Na salts in their vacuoles; these salts are necessary to maintain turgor and growth. Sodium can replace part of the K^+ requirement in some plant species. Sodium has been associated with oxalic acid accumulation, K-sparing action, stomatal opening, and regulation of nitrate reductase in plants (Tisdale et al., 1985).

Sodium, soil test: The NH$_4$OAc extractable Na^+ is a common soil test for Na. Convert to kg ha^{-1} for a soil horizon as follows:

Equation 4.1.2.4.1:

Na = Na^+ x 0.023 x 1000 x Hcm x ρ_{B33} x Cm

where

Na = Sodium for soil horizon (kg ha^{-1})

Na$^+$ = NH$_4$OAc extractable Na$^+$ (meq 100 g^{-1}) or (cmol(+) kg^{-1})

0.023 = Milliequivalent weight of Na$^+$ (g meq^{-1})

1000 = Conversion factor to hectares

Hcm = Soil horizon thickness (cm)

ρ_{B33} = Bulk density at 33-kPa water content of <2-mm fraction (g cm^{-3})

Cm = Coarse fragment conversion factor. If no coarse fragments, Cm = 1. If coarse
      fragments are present, calculate Cm using Equations 3.1.2.1.1.3 and 3.1.2.1.1.4.

To convert Na$^+$ (meq 100 g^{-1}) to lb A^{-1} or kg m^{-3}, replace the conversion factor for hectares in Equation 4.1.2.4.1 with the factor 2300 or 0.10, respectively.

4.1   Ion Exchange and Extractable Cations
4.1.3   BaCl$_2$-Triethanolamine, pH 8.2 Extractable Acidity

Soil acidity, definitions: *Soil acidity* is determined largely by soil composition and the ion exchange and hydrolysis reactions associated with the various soil components, which include organic as well as inorganic substances, e.g., layer silicates, oxide minerals (including allophane), and soluble acids (Thomas and Hargrove, 1984). The development or accumulation of soil acidity usually parallels the mineral-weathering sequence in which Al is released and accumulates in the soil (Foth and Ellis, 1988). Hydroxy-Al accumulates as soils become acid, first as interlayer Al and on clay surfaces and as complexes with organic matter and secondly as exchangeable Al^{3+} when soil pH is <5.5 (Foth and Ellis, 1988).

Exchangeable acidity has been defined as the portion of soil acidity that can be replaced with a neutral, unbuffered salt, e.g., 1 N KCl or NaCl. The Al extracted with 1 N KCl approximates the "active" acidity present in soils with a pH <5.5. Exchangeable acidity is due almost entirely to monomeric Al^{3+} ions and is essentially absent at soil pH values >5.5 (Foth and Ellis, 1988). Refer to Figure 4.1.3.1. The SSL uses the 1:2 0.01 M CaCl$_2$ pH <5.05 to determine 1 N KCl extractable Al. The KCl-extractable Al is more related to the immediate lime requirement and existing CEC of the soil (Bohn et al., 1979). *Titratable* or *extractable acidity* is the amount of acid neutralized at a selected pH, commonly pH 8.2, and does not distinguish between exchangeable and virtually nonexchangeable components. Extractable acidity at pH 8.2 is a good measure of the "potential" acidity. Extractable acidity is only a measure of the total acidity present between the initial and final pH levels (Thomas, 1982).

EFFECTS OF ALUMINUM HYDROLYSIS ON SOIL

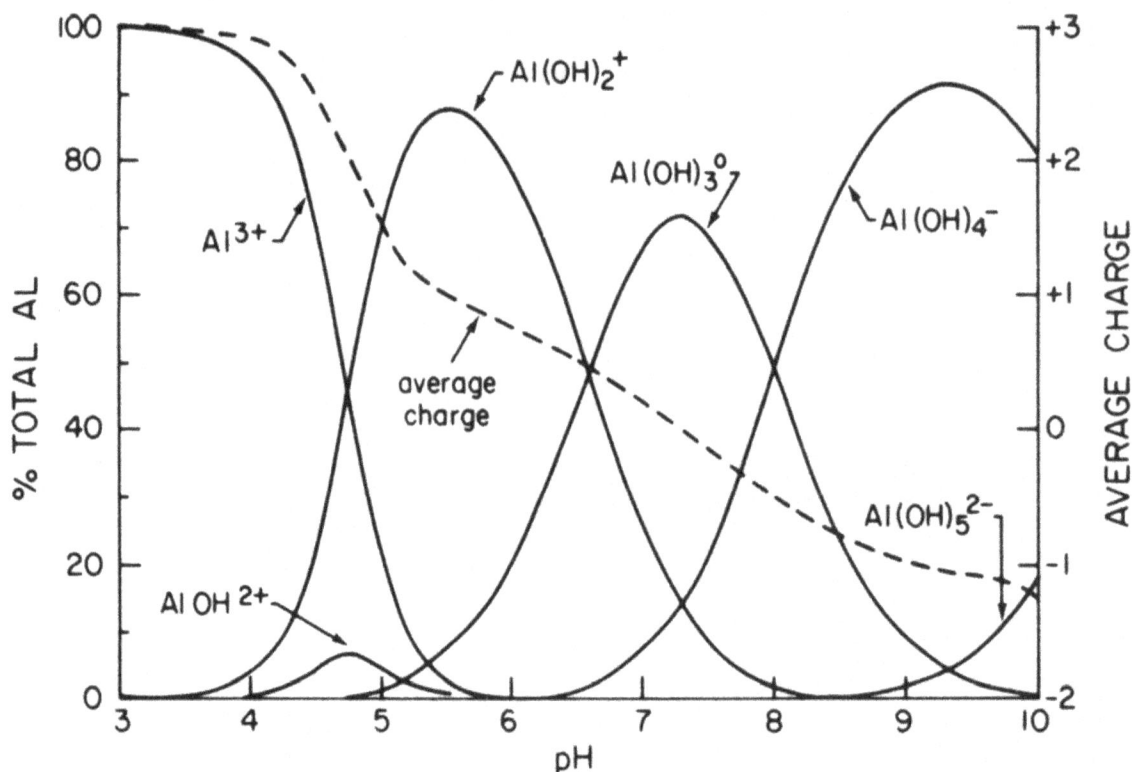

Figure 4.1.3.1.—The relative distribution and average charge on the soluble Al species as a function of pH at $\mu = 0.1$ *M*. After Marion et al. (1976), "Aluminum and Silica Solubility," *Soil Science*, Vol. 121:76-82, and reproduced with permission from Wolters Kluwer, Lippincott Williams & Wilkins, Baltimore, Maryland.

BaCl$_2$-triethanolamine extractable acidity, application: The titratable or extractable acidity released from a soil by a barium chloride-triethanolamine (BaCl$_2$-TEA) solution buffered at pH 8.2 includes all the acidity generated by replacement of H and Al from permanent and pH-dependent exchange sites. Various methods have been used to measure extractable acidity as it may be measured at any pH. The USDA adopted a pH of 8.2 because that pH approximates the calculated pH of a soil containing free CaCO$_3$ in equilibrium with the normal CO$_2$ content (0.03 percent) of the atmosphere. The pH of 8.2 also closely corresponds to the pH of complete neutralization of soil hydroxyl-Al compounds and is conveniently maintained by Mehlich's BaCl$_2$-TEA buffered extraction technique (Bohn et al., 1979). Calcareous soils have little or no acidity to extract by a pH 8.2 solution and are not routinely analyzed by the SSL for extractable acidity. The BaCl$_2$-TEA, pH 8.2, method may not always accurately reflect the nature of soils as they occur in the environment, and other pH values are more valid for some types of soils; however, this method has become a standard reference to which other methods are compared.

BaCl$_2$-triethanolamine extractable acidity, measurement: Since the publication of the original triethanolamine acetate-barium hydroxide method (Mehlich, 1939), several modifications have been proposed in the literature (Peech et al., 1947, 1962; Mehlich, 1948; Pratt and Holowaychuk, 1954). The BaCl$_2$-TEA method as described by Peech et al. (1947) was adopted for use by the SSL. The method evolved from a batch method using Buchner funnels to the mechanical vacuum extraction technique (Holmgren et al., 1977) currently in use at the SSL. A soil sample is leached with a BaCl$_2$-TEA solution buffered at pH 8.2. The sample is allowed to stand overnight and is then extracted using the vacuum extractor. The extract is back-titrated with HCl. The difference between a blank and the extract is the extractable acidity. This vacuum extraction method has increased laboratory production (i.e., number of samples per day) and provided reproducible data for most soils, the main exceptions being soils containing organic or andic material (Seifferlein et al., 2005). Organic soils appeared to resist wetting in the vacuum extraction tube and sometimes floated. Results for highly acidic soils (<5 percent of all soils tested to date at SSL) were variable; the lack of reproducibility was initially attributed to the failure of these soils to wet (Seifferlein et al., 2005). This study found the centrifuge method as an efficient high-volume characterization method, offering advantages over the batch and vacuum extraction methods in providing greater data accuracy and reproducibility for highly acidic soils. The centrifuge method requires an appropriate soil to solution ratio that does not significantly impact the pH of the extraction solution, i.e., 5 g of soil per 40-mL extraction solution for most soils and 0.5 g per 40 mL for highly acidic soils (Seifferlein et al., 2005). The SSL reports extractable acidity in cmol(+) kg^{-1}. In the past, the SSL reported extractable acidity as meq 100 g^{-1}.

4.1   Ion Exchange and Extractable Cations
4.1.4–4.1.5   1 N KCl Extractable Aluminum and Manganese

Aluminum toxicity: Aluminum is not considered an essential nutrient, even though low concentrations have been shown to sometimes increase plant growth or produce other beneficial effects in selected plants (Foy et al., 1978; Foy and Fleming, 1978). Generally, the primary concern with Al is the possible toxic effects of its high concentrations. The critical pH at which Al becomes soluble or exchangeable in toxic concentrations depends on many soil factors, including the predominant clay minerals; organic matter levels; concentrations of other cations, anions, and total salts; and, particularly, the plant species or cultivar (Kamprath and Foy, 1972; Foy, 1974). Such complexity makes it difficult to devise a soil Al test that will accurately predict toxicity under all conditions (Foy, 1984). Among the soil chemical stresses to roots and to the plants as a whole, Al toxicity in strongly acid subsoils and mine spoils can be one of the most serious. The problem is particularly severe where pH is below 5.0, but it may occur where pH is as high as 5.5 in kaolinitic soils (Foy, 1984). Plant sensitivity to Al is typically accentuated in soils low in Ca. Moderate toxicity in the subsoil is usually not readily detectable in the field because the growth of the plant shoots may not be affected in a significant manner while the surface soil is moist and adequate in nutrients (Foy et al., 1978; Alam and Adams, 1979). However, Al toxicity reduces rooting depth and degree of root branching into the subsoil; these conditions are usually more apparent during periods of stress, such as drought (Simpson et al., 1977; Foy, 1984).

Aluminum toxicity, plant effects: The effects of excess Al are frequently cited in the literature. Excess Al can restrict plant root penetration and proliferation in acid subsoils by decreasing water uptake in plants when the soil surface becomes dry. Aluminum toxicity can also damage roots to the extent that they cannot absorb adequate water, even in moist soils (Foy, 1984). There is considerable evidence suggesting that Al toxicity limits microbial breakdown of organic matter in strongly acid soils (Alexander, 1980). High concentrations of Al are also linked to adverse interactions with other elements, e.g., Fe and Ca. At a pH <5.5, the Al x Ca antagonism is probably the most important factor affecting Ca uptake by plants (Foy, 1984). Aluminum toxicity is also linked to P deficiency; conversely, Al tolerance appears to be related to the efficient use of P.

Aluminum toxicity, amelioration: One of the more traditional ways to correct or ameliorate the problem of Al toxicity in field soils is liming. Deep liming of acid subsoils in many instances has been found to be uneconomical or of no significance. In interpreting these data, care is required as percent Al saturation may or may not indicate a problem of Al toxicity. There is evidence that the mechanisms of plant tolerance or sensitivity to Al and acidity may be different. In addition, there is evidence that liming may correct the Al problem by correcting the relative activities of Ca and Al more than by raising the pH or correcting the Ca deficiency. Surface applications of lime have been shown to increase the Ca concentration and base saturation at depths of 180 cm or more below the soil surface (Hartgrove et al., 2006).

Manganese toxicity: Manganese toxicity is probably the second most important growth-limiting factor (after Al toxicity) in acid soils (Foy, 1984). The solubility and, thus, the potential toxicity of Mn to a given crop depend on many soil properties, including total Mn content, pH, organic matter level, aeration, and microbial activity (Foy, 1973; Stahlberg et al., 1976). Manganese toxicity generally occurs in soils with pH values of $\leq$5.5 if the soil contains sufficient total Mn (Foth and Ellis, 1988), but it may also occur at higher soil pH values in poorly drained or compacted soils where reducing conditions favor the production of divalent Mn that plants absorb. This increase in Mn^{2+} occurs as Mn is one of the first elements to undergo reduction in soils with reducing conditions. Some soils do not contain sufficient Mn to produce toxicity, even at pH 5 or below. Manganese has been reported to interact with Fe, Mo, P, Ca, and Si in affecting toxicity symptoms and growth.

Manganese, essential plant element: Manganese is an essential plant nutrient. It is typically involved in the oxidation-reduction processes, decarboxylation, and hydrolysis reactions in photosynthesis as well as in the evolution of oxygen (Tisdale et al., 1985). Manganese has also been associated with maximal activity of many enzyme reactions in the citric acid cycle and can substitute for Mg in many of the phosphorylating and group-transfer reactions. Manganese influences auxin levels in plants. High concentrations of this element favor the breakdown of indoleacetic acid.

1 N KCl extractable Al and Mn, application: Most of the acidity of acid soils is associated with Al. The Al extracted by 1 N KCl approximates *exchangeable Al* and is a measure of the "active" acidity present in soils with a pH <5.5. In soils with a pH above 5.5, Al has undergone hydrolysis and is present in less available forms than Al^{3+}. The SSL analyzes for 1 N KCl extractable Al if the 1:2 0.01 M CaCl$_2$ pH is <5.05. This method does not measure the acidity component of hydronium ions (H_3O^+). If Al is present in measurable amounts, the hydronium is a minor component of the active acidity. Typically, hydroxy-Al accumulates in soils as they become increasingly acid as follows: (1) as interlayer Al, as coatings on clays, and as complexes with organic matter; and (2) exchangeable Al^{3+} when pH is <5.5 (Foth and Ellis, 1988). Because the 1 N KCl extractant is an unbuffered salt and usually affects the soil pH one unit or less, the extraction is determined at or near the soil pH. The KCl-extractable Al is related to the immediate lime requirement and existing CEC of the soil. The use of NH_4Cl in place of KCl is useful where a single extractant for exchangeable bases and Al is preferred since NH_4^+ is as effective as K at displacing Al (Lee et al., 1985; Bertsch and Bloom, 1996).

The Mn extracted by 1 N KCl approximates exchangeable Mn. Mn is an essential trace metal for plant nutrition and is absorbed as the ion Mn^{2+}. Soil analysis for Mn is of interest from the perspectives of both deficiency and toxicity (Gambrell, 1996). The availability of Mn in the field has been difficult to predict. Since Mn mobility is related to oxidation-reduction reactions in the soil, the availability of Mn is closely related to soil moisture and temperature. Cool temperatures may retard organic Mn mineralization. On the other hand, cool temperatures associated with high rainfall levels in early spring may maintain more available Mn through reduction of Mn oxides (Allen and Hajek, 1989; McKenzie, 1989). In general, the soil chemistries of Fe and Mn are similar; i.e., both can exist in more than one oxidation state, both are affected by drainage conditions, both are precipitated as oxides and hydroxides, and both can be complexed with organic matter. Under poorly drained conditions, however, the Mn is more easily reduced and mobilized than Fe (Allen and Hajek, 1989; McKenzie, 1989).

1 N KCl extractable Al and Mn, measurement: A soil sample is leached with 1 N KCl using the mechanical vacuum extractor. The extract is weighed. The KCl-extracted solution is diluted with 0.5 N HCl. The analytes (Al, Mn) are measured by inductively coupled plasma atomic emission spectrophotometer (ICP-AES). The SSL reports Mn and Al in mg kg^{-1} and cmol(+) kg^{-1}, respectively. In the past, Mn and Al were reported as parts per million (ppm) and meq 100 g^{-1}, respectively.

4.1   Ion Exchange and Extractable Cations
4.1.6   Ratios, Estimates, and Calculations Related to Ion Exchange and Extractable Cations
4.1.6.1   Sum of Extractable Bases
4.1.6.1.1   Sum of Extractable Bases by NH₄OAc, pH 7
4.1.6.1.1.1   Sum of Extractable Bases by NH₄OAc, pH 7, Calculated

Sum the NH_4OAc, pH 7 extractable bases (Ca^{2+}, Mg^{2+}, K^+, and Na^+). This value is reported as cmol(+) kg^{-1}.

Equation 4.1.6.1.1.1:

Sum of NH_4OAc extractable bases = $Ca^{2+} + Mg^{2+} + K^+ + Na^+$

141

4.1 Ion Exchange and Extractable Cations
4.1.6 Ratios, Estimates, and Calculations Related to Ion Exchange and Extractable Cations
4.1.6.1 Sum of Extractable Bases
4.1.6.1.2 Sum of Extractable Bases by NH₄Cl
4.1.6.1.2.1 Sum of Extractable Bases by NH₄Cl, Calculated

Equation 4.1.6.1.2:

Sum the NH₄Cl extractable bases (Ca^{2+}, Mg^{2+}, K^+, and Na^+). This value is reported as cmol(+) kg^{-1}.

Sum of NH₄Cl extractable bases = $Ca^{2+} + Mg^{2+} + K^+ + Na^+$

4.1 Ion Exchange and Extractable Cations
4.1.6 Ratios, Estimates, and Calculations Related to Ion Exchange and Extractable Cations
4.1.6.2 Cation-Exchange Capacity (CEC)
4.1.6.2.1 CEC-8.2 (Sum of Cations)
4.1.6.2.1.1 CEC-8.2, Calculated
4.1.6.2.1.2 CEC-8.2, Not Calculated

Calculate the CEC-8.2 by summing the NH₄OAc extractable bases (Ca^{2+}, Mg^{2+}, K^+, and Na^+) plus the BaCl₂-TEA extractable acidity. This value is reported as cmol(+) kg^{-1}. Cation summation is the basis for this method. The CEC-8.2 minus the CEC-7 is considered the pH-dependent charge from pH 7.0 to pH 8.2. The CEC-8.2 is not calculated if significant quantities of soluble salts or carbonates are present in the soil (Soil Survey Staff, 2004). CEC-8.2 is calculated as follows:

Equation 4.1.6.2.1.1:

CEC-8.2 = NH₄OAc Extractable Bases + BaCl₂-TEA Extractable Acidity

4.1 Ion Exchange and Extractable Cations
4.1.6 Ratios, Estimates, and Calculations Related to Ion Exchange and Extractable Cations
4.1.6.2 Cation-Exchange Capacity (CEC)
4.1.6.2.2 Effective Cation-Exchange Capacity (ECEC)
4.1.6.2.2.1 Sum of NH₄OAc Extractable Bases + 1 *N* KCl Extractable Aluminum, Calculated
4.1.6.2.2.2 Sum of NH₄OAc Extractable Bases + 1 *N* KCl Extractable Aluminum, Not Calculated

The CEC can be measured by extraction with an unbuffered salt, which measures the effective cation-exchange capacity (ECEC), i.e., CEC at the normal soil pH

(Coleman et al., 1958). Since the unbuffered salt solution, e.g., 1 N KCl, only affects the soil pH one unit or less, the extraction is determined at or near the soil pH and extracts only the cations held at active exchange sites at the particular pH of the soil. Neutral NH_4OAc extracts the same amounts of Ca^{2+}, Mg^{2+}, Na^+, and K^+ as KCl; therefore, extractable bases by NH_4OAc is used at the SSL in place of KCl-extractable bases.

The ECEC may be determined by extracting one soil sample with neutral normal NH_4OAc to determine the exchangeable basic cations (Ca^{2+}, Mg^{2+}, Na^+, and K^+) and by extracting another sample of the same soil with 1 N KCl to determine the exchangeable Al. The 1 N KCl extractable Al approximates exchangeable Al and is a measure of "active" acidity present in soils with a 1:1 pH <5.5. Aluminum is nonexchangeable at pH >5.5 due to hydrolysis, polymerization, and precipitation. For soils with pH <7.0, the ECEC should be less than the CEC measured with a buffered solution at pH 7.0. The ECEC is not reported for soils with soluble salts. The SSL calculates ECEC by summing the NH_4OAc bases plus the KCl-extractable Al as follows:

Equation 4.1.6.2.2.1:

ECEC = NH_4OAc Extractable Bases + 1 N KCl Extractable Al

4.1 Ion Exchange and Extractable Cations
4.1.6 Ratios, Estimates, and Calculations Related to Ion Exchange and Extractable Cations
4.1.6.3 Base Saturation

Base saturation, historical background: It is important to understand the historical development of *base saturation* and its significance in soil classification and fertility. In early literature on soil acidity, soils were characterized by their percent base saturation values at specified pH levels (Bohn et al., 1979). Soils with low percent base saturation values were considered dominated by kaolinite and hydrous oxide minerals, whereas soils with high percent base saturation were considered dominated by 2:1 type minerals, e.g., montmorillonite, vermiculite, chlorite, and the micas (Bohn et al., 1979).

When work on *Soil Taxonomy* began, base saturation criteria were developed at a time when it was commonplace to determine the CEC at pH 7.0 or 8.2 and when the role of Al and the importance and nature of pH-dependent charge were poorly understood (Foth and Ellis, 1988). The CEC in soils dominated by permanent charge varies less with pH change than it does in soils with pH-dependent charge. Base saturation at pH 8.2 of the subsoil (generally 180 cm below the mineral surface) was used to differentiate the less weathered, fertile Alfisols dominated by permanent charge from the more weathered and less naturally fertile Ultisols dominated by variable charge (Foth and Ellis, 1988). Base saturation at pH 7.0 of the surface horizons was used to distinguish between base-rich mollic epipedons and their base-poor umbric equivalents (Foth and Ellis, 1988). Mollisols typically have more permanent-charge clays, have less leaching of bases, and are more naturally fertile for the more commonly grown crops than Alfisols (Foth and Ellis, 1988). This base saturation criterion, however, was not intended to imply that Ultisols and Alfisols could not become as productive as Mollisols with proper fertilization and liming (Foth and Ellis, 1988). Currently in *Keys to Soil Taxonomy*, base saturation determined by CEC-7 is used in mollic, umbric, and eutro-dystro criteria and base saturation determined by the sum of cations (CEC-8.2) is used in most alfic-ultic criteria (Soil Survey Staff, 2010). Refer to Soil Survey Staff (2010) for additional discussion of these criteria.

Base saturation, index: Percent base saturation is an imprecise index as it is not only a measure of the pH-dependent charge of soils but also of the actual percentage of cation-exchange sites occupied by exchangeable bases (Bohn et al., 1979). The denominator includes any additional charge (CEC) generated by soil organic matter and hydrous oxide-mineral complexes between the actual soil pH and the reference pH (pH 7 or 8.2). Since neither exchangeable Al nor exchangeable H is appreciable in soils with pH >5.5, the ECEC of soils with pH >5.5 typically is essentially 100 percent base saturated (Bohn et al., 1979). If base saturation is based on CEC-7 or CEC-8.2, however, soils in the pH range 5.5 to 7.0 or 8.2 generally still have measured base saturation of <100 percent. These base saturation values are particularly low for weathered soils dominated by such minerals as kaolinite, which has a high proportion of pH-dependent charge (Bohn et al., 1979). Below pH 5.5, exchangeable Al saturation increases, and the exchangeable base saturation decreases with decreasing pH. This phenomenon is expressed as Al saturation (Foth and Ellis, 1988). Refer to the discussion on Al saturation described herein.

Although it is an imprecise index, the percent base saturation is still useful for soil genesis and classification purposes and for empirical liming recommendations (Bohn et al., 1979). From the standpoint of soil chemical properties and reactions, however, base saturation is more correctly an acidity index or liming index, and the degree of nonbase saturation is more meaningful if separated into exchangeable acidity and pH-dependent charge (Bohn et al., 1979). Cation-exchange capacity, and hence the base saturation, is an arbitrary measurement unless the method by which the data are determined is clearly defined (Tisdale et al., 1985).

Base saturation, calculations: In this section, the SSL methods for calculating base saturation are described. These methods include base saturation by NH_4OAc, pH 7 (CEC-7); base saturation by NH_4Cl; base saturation by CEC-8.2 (sum of cations); and base saturation by sum of NH_4OAc extractable bases + 1 N KCl extractable aluminum.

4.1 Ion Exchange and Extractable Cations
4.1.6 Ratios, Estimates, and Calculations Related to Ion Exchange and Extractable Cations
4.1.6.3 Base Saturation
4.1.6.3.1 Base Saturation by NH₄OAc, pH 7 (CEC-7)
4.1.6.3.1.1 Base Saturation by CEC-7, Calculated
4.1.6.3.1.2 Base Saturation by CEC-7, Set to 100 Percent

Calculate the base saturation by dividing by the sum of NH_4OAc extractable bases by CEC-7 and multiplying by 100. This value is reported as percent. If a soil has significant quantities of soluble salts or carbonates, this value is set to 100 percent. Calculate base saturation by CEC-7 as follows:

Equation 4.1.6.3.1.1:

Base Saturation (%) = (NH$_4$OAc Bases/CEC-7) x 100

4.1 Ion Exchange and Extractable Cations
4.1.6 Ratios, Estimates, and Calculations Related to Ion Exchange and Extractable Cations
4.1.6.3 Base Saturation
4.1.6.3.2 Base Saturation by NH$_4$Cl
4.1.6.3.2.1 Base Saturation by NH$_4$Cl, Calculated
4.1.6.3.2.2 Base Saturation by NH$_4$Cl, Set to 100 Percent

Calculate the base saturation by dividing the sum of the NH$_4$Cl extractable bases by CEC by NH$_4$Cl and multiplying by 100. This value is reported as percent. If a soil has significant quantities of soluble salts or carbonates, this value is set to 100 percent. Calculate base saturation by NH$_4$Cl as follows:

Equation 4.1.6.3.2.1:

Base Saturation (%) = (NH$_4$Cl Bases/CEC by NH$_4$Cl) x 100

4.1 Ion Exchange and Extractable Cations
4.1.6 Ratios, Estimates, and Calculations Related to Ion Exchange and Extractable Cations
4.1.6.3 Base Saturation
4.1.6.3.3 Base Saturation by CEC-8.2 (Sum of Cations)
4.1.6.3.3.1 Base Saturation by CEC-8.2, Calculated
4.1.6.3.3.2 Base Saturation by CEC-8.2, Not Calculated

Calculate the base saturation by dividing the sum of the NH$_4$OAc extractable bases by CEC-8.2 and multiplying by 100. This value is reported as percent. In *Keys to Soil Taxonomy*, base saturation determined by the sum of cations (CEC-8.2) is used in most alfic-ultic criteria (Soil Survey Staff, 2010). If a soil has significant quantities of soluble salts or carbonates, this value is not calculated. Calculate base saturation by CEC-8.2 (sum of cations) as follows:

Equation 4.1.6.3.3.1:

Base Saturation (%) = [NH$_4$OAc Bases/(NH$_4$OAc Bases + BaCl$_2$-TEA Acidity)] x 100

4.1 Ion Exchange and Extractable Cations

4.1.6 Ratios, Estimates, and Calculations Related to Ion Exchange and Extractable Cations

4.1.6.3 Base Saturation

4.1.6.3.4 Base Saturation by Effective Cation-Exchange Capacity (ECEC)

4.1.6.3.4.1 Base Saturation by Sum of NH₄OAc Extractable Bases + 1 N KCl Extractable Aluminum, Calculated

4.1.6.3.4.2 Base Saturation by Sum of NH₄OAc Extractable Bases + 1 N KCl Extractable Aluminum, Not Calculated

The base saturation is calculated by dividing the sum of NH₄OAc extractable bases by the ECEC and multiplying by 100. This value is reported as percent. If a soil has significant quantities of soluble salts or carbonates, this value is not calculated. Calculate base saturation by ECEC as follows:

Equation 4.1.6.3.4.1:

$$\text{Base Saturation (\%)} = [\text{NH}_4\text{OAc Bases}/(\text{NH}_4\text{OAc Bases} + 1\ N \text{ KCl Al})] \times 100$$

4.1 Ion Exchange and Extractable Cations

4.1.6 Ratios, Estimates, and Calculations Related to Ion Exchange and Extractable Cations

4.1.6.4 Aluminum Saturation

4.1.6.4.1 Aluminum Saturation by Effective Cation-Exchange Capacity (ECEC)

4.1.6.4.1.1 Aluminum Saturation by Sum of NH₄OAc Extractable Bases + 1 N KCl Extractable Aluminum, Calculated

4.1.6.4.1.2 Aluminum Saturation by Sum of NH₄OAc Extractable Bases + 1 N KCl Extractable Aluminum, Not Calculated

Calculate the Al saturation by dividing the 1 N KCl extractable Al by ECEC and multiplying by 100. This value is reported as percent. If a soil has significant quantities of soluble salts or carbonates, this value is not calculated. Calculate Al saturation as follows:

Equation 4.1.6.4.1.1:

$$\text{Al Saturation (\%)} = [1\ N \text{ KCl Al}/(\text{NH}_4\text{OAc Bases} + 1\ N \text{ KCl Al})]$$

4.1 Ion Exchange and Extractable Cations
4.1.6 Ratios, Estimates, and Calculations Related to Ion Exchange and Extractable Cations
4.1.6.5 Activity
4.1.6.5.1 CEC-7/Clay

Clay activity, data assessments: The CEC of soils is mainly a function of the amount and kind of clay and soil organic matter, their interaction, and pH (Foth and Ellis, 1988). In intensely weathered soils, e.g., Ultisols and Oxisols, the dominant clay minerals are kaolinite, gibbsite, and other oxidic clays; the negative charge properties of kaolinite are modified by these oxidic clays that tend to coat the kaolinite surfaces, masking the kaolinite effects (Foth and Ellis, 1988). These soils have *low-activity clays* (LAC), dominated by variable or pH-dependent charge. These soils have distinctly different fertility characteristics and require distinctly different management practices compared to soils with a much greater CEC dominated by permanent charge. Refer to Soil Survey Staff (2010) for more discussion of Ultisols and Oxisols.

CEC-7/clay ratio, mineralogy assessments: The CEC-7 to clay ratio has been used as auxiliary data to assess clay mineralogy. These data are especially useful when mineralogy data are not available. The CEC-7/clay is an index for clay activity, i.e., probable contribution of clay to the exchange capacity and soil solution chemistry. Clay activity is closely linked to clay mineralogy. The smectites (montmorillonites) and vermiculites are considered high-activity clays; kaolinites and hydroxy-interlayered vermiculites are low-activity clays; and micas (illites) and chlorites are intermediate-activity clays (National Soil Survey Laboratory Staff, 1983). Refer to Soil Survey Staff (2010) for discussion of mineralogy class as a taxonomic criterion of soil families in different particle-size classes. Also refer to Soil Survey Staff (2010) for discussion of the cation-exchange activity classes to help in making interpretations of mineral assemblages and nutrient-holding capacity of soils in mixed and siliceous mineralogy classes of selected particle-size classes. The following guidelines were developed primarily from experience with soil samples from the Central United States and Puerto Rico (National Soil Survey Laboratory Staff, 1983).

CEC-7/Clay	Family mineralogy as assessed by XRD and DTA evidence
>0.7	Smectite
0.5–0.7	Smectite or Mixed
0.3–0.5	Mixed
0.2–0.3	Kaolinite or Mixed
<0.2	Kaolinite

Soils with illitic family mineralogy typically have CEC-7/clay ratios in the range of mixed (lower end of mixed range). Vermiculitic soils typically have CEC-7/clay ratios similar to those of smectitic soils; however, some soil minerals determined as vermiculite by x-ray diffraction analysis (termed hydroxy-interlayered vermiculites) appear more similar in nature to inactive soil chlorites, which have CEC-7/clay ratios in

147

the range of kaolinite (National Soil Survey Laboratory Staff, 1983). The CEC-7/clay ratio is useful both as an internal check of the data and as an estimator of mineralogy when mineralogy data are not available.

A soil with a silt or sand fraction having a significant CEC can have a higher CEC-7/clay ratio than expected. Soils with organic or glassy materials or with a clay fraction that is incompletely dispersed by PSDA or soils with porous silts and sands too coherent to be disaggregated by PSDA also can have a high CEC-7/clay ratio. Users of laboratory data should be alert to any CEC-7/clay ratios >1. In these cases, the 15-bar water to clay ratios also are high (>0.6). In a study of 34 Borolls (National Soil Survey Laboratory Staff, 1990), each percent increase in organic C increases the CEC-7 by 3 meq 100 g^{-1} soil ($r^2 = 0.83$) and the CEC-8.2 by 4 meq 100 g^{-1} soil ($r^2 = 0.85$). A soil with clay-sized materials with little or no CEC, e.g., calcium carbonate, can have a low CEC-7/clay ratio. In this case, the carbonate clay percentage needs to be checked and the CEC-7/clay ratio recalculated based on the noncarbonate clay (National Soil Survey Laboratory Staff, 1983).

CEC-7/clay ratio, calculation: Divide the CEC by NH_4OAc, pH 7 (CEC-7) by total (noncarbonate) clay. This ratio is reported as a dimensionless value. In the past, the ratios of CEC to clay have been reported as meq g^{-1}. For more detailed information on the application of this ratio, refer to Soil Survey Staff (2010).

Equation 4.1.6.5.1:

CEC-7/Clay

4.1   Ion Exchange and Extractable Cations
4.1.7   Nutrient Depletion/Deficiency and Acidification, Processes, Case Studies, and Major Developments

Nutrient depletion/deficiency and acidification, processes: Nutrient depletion, acidification, and loss of organic matter are natural as well as human-induced processes. Normal weathering processes in humid regions, such as the Southeastern United States, result in mineral weathering with release of bases that leach from the soil. This leaching results in the continued acidification of the soil, which in turn contributes to diminished plant growth and the subsequent depletion of organic matter and to soil erosion. In a humid climate, the course of development in a freely drained environment is always toward acidification and leaching of bases (Bache, 2002).

Nutrient depletion is the loss of the capacity of a soil to supply mineral nutrients to plants, commonly assessed by the CEC measurement and its derivative, base saturation. Loss of organic matter typically occurs as a result of oxidation rates (decomposition) or removal of these materials (e.g., crops, erosion) in excess of subsequent accumulation. It is commonly assessed by visual observations and/or laboratory measurements of organic or total C. Soil biology is an important component of soil quality and one that has not received appropriate attention until recently. The biological component is commonly assessed by organic matter content, biomass C, and activity and diversity of soil fauna.

Acidification occurs where there is a net donation of protons to soil components with a loss of bases by leaching or by harvest of plant materials. It is assessed by multiple methods, e.g., pH, acidity, and CEC. Natural acidification is generally the result of processes linked to plants and their ability to assimilate carbon dioxide and to the presence of organic acids from plants, e.g., litter, degradation products of litter, and exudates from plant roots (Bloom et al., 2005). In most cases, soil acidification does not cause serious degradation until pH is <5.5. Below this level, toxic levels of Al and sometimes Mn can be manifested by reduced cropland, forest, or rangeland productivity and in some cases by the transfer of soluble Al to water bodies, which poses a threat to aquatic life (Sumner, 1998).

Nutrient depletion/deficiency and acidification, case studies: Numerous agricultural practices in the United States have induced nutrient depletion/deficiency and soil acidification. Some of these practices were common to many areas in the country, while others were more regionalized. These practices include, but are not limited to, the following: intensive agriculture with inadequate or no return of crop residues; land conversion (e.g., forest or grassland converted to cropland); higher yielding cropping systems with increased nutrient demand and induced micronutrient deficiencies; heavy tillage systems that accelerate organic matter decomposition and increase the release of nutrient elements; inadequate or no implementation of soil and water conservation techniques to constrain soil erosion by wind and water, resulting in accelerated loss of nutrients; volatilization and leaching of nutrients hastened by type of vegetative cover and cropping systems; inadequate, excessive, or inappropriate application of fertilizers and lime, including micronutrients, sometimes resulting in nutrient immobilization or imbalance; application of fertilizer amendments (e.g., ammonium sulfate, ammonium nitrate, anhydrous ammonia, and urea) that produce NH_4^+ (which is nitrified to nitrate, NO_3^-) and acidify soils; nutrient immobilization or imbalance through excessive or inefficient fertilizer applications; inadequate or improper soil testing and plant analysis; and irrigation practices on saline or alkaline soils, resulting in nutrient leaching (e.g., N, P, and B) and/or reduced mineral solubility (e.g., P, Fe, Mn, and Zn) and thus reducing plant absorption.

In the late 1930s to the early 1950s, in response to concerns about nutrient depletion and acidification in U.S. agricultural soils, soil testing laboratories were established at land grant colleges of agriculture. These laboratories have been widely used for decades. They provide statewide testing services to help farmers make decisions about fertilizer and lime applications. It is also important to recognize the growth in the number of commercial laboratories providing these services, many of which were connected to the fertilizer industry. Over the years, both the role and number of extension soil testing programs have changed dramatically. Some of these programs have been closed due to financial cutbacks at the colleges/universities. The remaining programs not only provide analytical services and appropriate nutrient recommendations for successful agriculture but also work with regulatory services to identify best management practices (BMPs) that minimize nutrient-related water-quality impacts—for example, determining appropriate P-based nutrient management plans, such as agronomic soil test P, environmental soil P thresholds, and P indexing of a site (USDA/NRCS and USEPA, 1999). An important development in the 1990s was the increasing pressure on commercial laboratories to participate in proficiency testing programs, e.g., North American Proficiency Testing Program, better ensuring the quality of the soil and plant analysis data and the resulting recommendations to farmers.

Despite the long history of research and development in soil acidity and liming, excessive soil acidity has been recognized as a continuing problem in many agricultural areas of the U.S. and in some cases is considered a yield-limiting factor (Adams, 1984). Significant progress was made in the 1940s (peaking in 1946-47) in addressing soil acidification through increased application of liming materials. This progress was largely due to farm subsidies for the application of these materials. When these subsidies were discontinued by the Federal government, however, the decline in demand suggested that not all farmers were continuing to add lime frequently enough to replace Ca and Mg removed by more intensive cropping and from acids created by the greater use of nitrogen fertilizers after WWII. In 1975, it was estimated that 88 million tons of limestone was required annually, but only 24 million tons had been applied (Tisdale and Nelson, 1975). In more recent years, this trend has been reversed in some parts of the United States (Jackson and Reisenauer, 1984; Lathwell and Reid, 1984; McLean and Brown, 1984).

Even though the beneficial effects of liming were researched as far back as 1906 on major crops in Alabama (Duggar and Funchess, 1911), liming did not become a general practice in the Southern U.S. until the second half of the 20th century (Adams, 1984). In the 1960s and extending into the 1980s, some of the most important research in the yield-limiting effect of subsoil acidity became the focal point at Auburn University (Howard and Adams, 1965; Lyle and Adams, 1971; Adams and Moore, 1983). Nevertheless, in 1979 the National Limestone Institute (unpublished data) estimated that 10 million tons of limestone was applied annually on acid soils in the Southeast and that this amount was a deficit of what was required. Soil acidification is not exclusive to the Southern United States. The pH of soils in the dryland wheat-growing area of the Pacific Northwest dropped an average of one unit between the 1960s and 1980s, and even more dramatic changes have been observed in the surface layers of minimum-tilled fields (Adams, 1984).

In the Midwest during the 1950s, changes in cropping systems and agricultural practices impacted soil nutrients and acidity. There was a decline in, if not abandonment of, the use of legumes as a source of N; legumes typically require large amounts of lime to function properly (McLean and Brown, 1984). There were also significant advances in farm machinery and pesticides, resulting in a shift to greater acreages of cultivated crops, e.g., soybeans and corn, and away from pasture and forage. Higher rates of fertilization, especially N, and removal of large grain yields resulted in a higher rate of depletion of nutrients and lime reserves (McLean and Brown, 1984). While advances in plant breeding and management practices have resulted in higher yielding crops (e.g., 50 bushels per acre in the 1930s compared to over 200 bushels per acre since the 1960s on some productive U.S. soils), these advances have led to an increased nutrient demand (e.g., N, P, and K) and inducement of micronutrient deficiencies. Many U.S. soils are naturally low in available levels of one or more micronutrients, and heavy crop demands over time increase the severity of the deficiency.

While some of the soils in the Western U.S. are naturally or "geologically" acid, there are wide areas of both dryland and irrigated soils that have been made acid through agricultural practices, e.g., fertilization, irrigation, and basic cation removal

(Jackson and Reisenhauer, 1984). Historically, in the West, the use of S-supplying acidifying fertilizers has been preferred, primarily because Western soils are inherently low in available S and responses to soil acidification are well documented (Lorenz and Johnson, 1953; Jackson and Carter, 1976; Jackson and Reisenauer, 1984). In general, the applications of fertilizer amendments (e.g., ammonium sulfate, ammonium nitrate, anhydrous ammonia, and urea) that produce NH_4^+ and are nitrified to nitrate (NO_3^-) can acidify soils. Other practices that have resulted in soil acidification in the West include transfer of basic cations from surface to subsurface layers with percolating waters and the removal of basic cations by crops, especially those crops for which a large fraction of the plant is removed at harvest; such crops are commonly grown in the West and include hay, silage, and pasture (Jackson and Reisenauer, 1984). Pierre and Banwart (1973) reported that an 18 t ha^{-1} crop of alfalfa hay removes an average of 23.2 mmol of basic cations (+) dm^{-2}. This amount represents a significant depletion in soil nutrients as animal feed crops occupy a large amount of cultivated acreage in the West.

Over the years, a number of management practices have been used in the U.S. to ameliorate soil acidity. These include amelioration of topsoil acidity through surface application of lime; amelioration of subsoil acidity through mechanical incorporation of lime to depth, though this method is considered impractical by some (Sumner, 1995); surface applications of gypsum or gypsum plus lime (Sumner, 1970; Reeve and Sumner, 1972; Bradford and Blanchar, 1977; Sumner et al., 1986); and applications of large amounts of animal manure or other organic materials, decreasing subsoil pH (Long, 1979; Lund and Doss, 1980; Wright et al., 1985; Sweeten et al., 1995), although other investigators reported no effect (Sharpley et al., 1991) or subsoil pH decreases (Kingery et al., 1994). In general, the reactions taking place after the application of manure to soil are complex, and the resulting impact on subsoil pH depends on the nature of the manure and the cropping system (Bloom et al., 2005). The benefits of a high content of organic matter in ameliorating the toxic effects of Al in the surface soil have long been recognized in the literature (Evans and Kamprath, 1970; Thomas and Hargrove, 1984). This technique has not been considered feasible in most U.S. agricultural soils, especially under intensive row-cropping. It is worthy to note, however, that a high content of organic matter greatly reduces Al toxicity in the northern temperate forest region, allowing prolific rooting in high organic surface horizons despite pH ≤4.

To a great extent, concern about nutrient removal from U.S. soils has been focused more on private lands in agricultural production than on public lands. Public lands have been leased to ranchers for grazing by livestock for nearly a century. Overgrazing of the public lands can result in a slow deficit in nutrient balance and in soil erosion. Deposition of manure returns some nutrients, but the nutrients captured in weight gain of animals are removed. Fertilization of public lands is typically not practiced.

Major developments in knowledge, science, and technology in soil fertility: At the turn of the 20th century, the "scientific" approach to soil fertility in the United States was gradually developing. This approach has evolved over the last half century to integrate the knowledge and theories about soil weathering and evolution, mineralogy, exchange chemistry, soil taxonomy, fertilizer technology, and plant growth and nutrition. The early work of the agricultural experiment stations, established in 1862 by the USDA Morrill Act, showed the benefits of fertilization and developed a broad

151

outline of the fertility status of soils in the United States. This outline noted the widespread need for P fertilizers, a general lack of K in the coastal plains, deficiencies of N in the South, predominantly acidic soils in Mississippi requiring lime, and the fact that Western soils were generally well supplied with Ca (Tisdale and Nelson, 1975). It soon became apparent, however, that this broad outline, though well defined, could not be used as a basis for blanket fertilizer recommendations (Tisdale and Nelson, 1975). With this evolution in approach and knowledge came a greater understanding of the problems of soil fertility.

At about the same time that the soil testing programs were established at land grant colleges and universities, major soil methods were being developed (e.g., Bray and Kurtz, 1945) that associated available soil nutrients with specific soil types and crops. The development of these methods continued for many years (e.g., Olsen et al., 1954; Mehlich, 1984). The methods encompassed soil test technologies tailored to specific crop nutrient requirements and/or soil amelioration for specific soil types (e.g., acid versus alkaline; fine textured versus coarse textured). The development of tests for soil acidity and lime requirements varied by geographical area and soil types. They include but are not limited to: Woodruff, developed for Mollisols in the Midwest, pH 6.5; Shoemaker, McLean, and Pratt (SMP), developed for Alfisols in the Midwest; and Adams Evan Buffer, developed for acid soils in Alabama, Florida, Georgia, Tennessee, South Carolina, and Virginia.

During the 1970s some of the extension soil testing laboratories began to fulfill a broader mission. North Carolina State University, for example, provided not only soil testing but also nematode assay, plant tissue analysis, waste analysis, solution analysis, and a statewide field services advisory program. During this time there also was renewed emphasis on the efficient use of agricultural inputs, such as fertilizers, largely due to the energy crisis, U.S. grain embargos, and resulting depressed markets and prices for U.S. agricultural products. This emphasis was further enhanced by the increased public concern for the protection of water quality and the prevention of pollution from chemical fertilizers. Some of the soil tests developed in the 1940s are currently being employed in more diverse agronomic and environmental uses (Pierzynski, 2000). With the closure of some of these extension soil testing laboratories due to financial cutbacks at the universities/colleges, the institutional knowledge about analytical methods for soils, water, and plant material is lodged more and more in the U.S. private sector.

A significant development in the last half of the last century (and continuing today) relating to soil, water, and plant analysis for agricultural production is the development (automation, sophistication) of analytical instrumentation (e.g., AAS, ICP-AES, and ICP-MS) for measuring analytes of interest. Aspects of these developments in analytical instrumentation have impacted the accuracy, efficiency, sophistication, and reproducibility of data, i.e., types of analysis and interpretation, speed of analysis, number of analytes, suite of analytes, and detection limits. During the period 1961 to 1986, soil chemists adapted modern technology (including IR, ESR, NMR, and SEM), computer technology, and models to solve complex soil chemical problems and add to our knowledge (Ellis, 1986).

Much of the early work in soil fertility was done when more people on average were either being trained and/or actively working in the area of soil science and agriculture. The prevailing trend of soil science as a discipline is widely perceived to be retrenchment, as evidenced by decreased enrollments in undergraduate soil science curricula at land grant colleges of agriculture. In the aggregate, UC-Davis, University of Florida, University of Nebraska, and Washington State University had 74 students enrolled in such programs in the early 1990s but only 16 in 2001. Ohio State University had 94 students in 1994 compared to 5 in 2004. North Dakota State University had 15 students in the mid 1990s and nearly 5 in 2004 but had plant science enrollments of 115 for 2002 and 120 for 2003 (Prunty, 2004). Over 96 percent of the American public is not involved in agricultural production and thus does not have the training or access to information necessary to draw good conclusions regarding such topics as long-term fertility of U.S. agricultural soils.

4.2   Soil pH

Soil pH, definition: *Soil pH* is one of the most frequently performed determinations and one of the most indicative measurements of soil chemical properties (McLean, 1982). The *pH value* is defined as the negative logarithm to the base of 10 (logarithm of reciprocal) of H-ion activity. Activity is the apparent or effective concentration of an ion in solution. It is affected by various factors, such as the concentration and valence of other ions present in solution. Since pH is logarithmic, H-ion activity in solution increases 10 times when the pH is lowered one unit. The activity of H-ion in soil solution is the intensity factor (index) of soil acidity, whereas exchangeable acidity and lime requirement (quick test), performed by soil testing laboratories, are the capacity factors of soil acidity (McLean, 1982).

Soil pH tells more about a soil than merely indicating whether it is acidic or basic. The availability of essential nutrients and the toxicity of other elements also can be estimated because of their known relationship with pH (Thomas, 1996). The pH of a soil must be determined for an understanding of important chemical processes, such as ion mobility, metal ion equilibria, rate of precipitation and dissolution reactions, nutrient availability, toxicity of trace metals, and the negative response of many plant species to soil acidity (Bloom et al., 2005). Soil pH provides necessary data to help determine liming needs and fertilizer responses. Soil pH can also indicate something about the degree of dissociation of H-ions from cation-exchange sites or the extent of Al hydrolysis (McLean, 1982) and thus can help to develop inferences about many of the chemical processes that have taken place during the genesis of a soil (Buol et al., 1980). Depending on the predominant clay type, the pH may be used as a relative indicator of base saturation (Mehlich, 1943).

Soil pH, related factors: Soils with similar pH can have different levels of acidity; therefore, the quantity of agricultural lime needed to yield the same increase in pH may differ among acidic soils with similar pH values (Bloom et al., 2005). For these reasons, the Soil Science Society of America (2010) defines three measures of the quantity of soil acidity (total, residual, and salt-replaceable). Soil pH is affected by many factors, including the nature and type of inorganic and organic matter, the amount and type of exchangeable cations and anions, the soil:solution ratio, the content of salt or electrolytes, and the CO_2 content (McLean, 1982). The acidity, neutrality, or basicity of a soil influences the solubility of various compounds, the relative ion bonding to exchange sites, and microbial activities. When the pH values of various soils are compared, it is important that the determination be made by the same method (Foth and Ellis, 1988). The pH of an air-dry soil sample in the laboratory will be different from the pH that exists in the same soil in the field during the growing season; i.e., there will be differences in water and salt content, and the roots and micro-organisms will produce CO_2 (Foth and Ellis, 1988). An increase in the soil:water ratio or the presence of salts generally results in a decrease in measured soil pH. The influence of the natural soluble salt content of the soil can be overcome by using dilute salt solutions, e.g., $CaCl_2$ or KCl, instead of distilled water (Foth and Ellis, 1988). The use of dilute salt solutions is a popular method for masking seasonal variation in soil pH. The pH readings are typically lower with dilute salt solutions than with distilled water but may be the same or greater in highly weathered soils with a high sesquioxide content, i.e., soils with a high anion (OH^-) exchange capacity (AEC).

Soil pH, measurements: In this section, the SSL pH methods and their applications are described. The SSL performs several pH determinations. These methods include but are not limited to: NaF (1 N pH 7.5 to 7.8); saturated paste pH; oxidized pH; 1:1 water and 1:2 $CaCl_2$ (final solution: 0.01 M $CaCl_2$); 1 N KCl; and organic materials, $CaCl_2$ (final solution $\approx$ 0.01 M $CaCl_2$). The SSL reports all pH values to the nearest 0.1 pH unit. For detailed descriptions of SSL pH methods which are cross-referenced by method code in the table of contents in this manual, refer to SSIR No. 42 (Soil Survey Staff, 2004), which is available online at http://soils.usda.gov/technical/lmm/. Also refer to the *Soil Survey Field and Laboratory Methods Manual*, SSIR No. 51 (Soil Survey Staff, 2009; available online at http://www.soils.usda.gov/technical/), for detailed descriptions of field methods as used by NRCS soil survey offices.

4.2 Soil pH
4.2.1 1 N NaF, pH 7.5–7.8

1 N NaF, pH 7.5–7.8, application: The action of NaF upon noncrystalline (amorphous) soil material releases hydroxide ions (OH^-) to the soil solution and increases the pH of the solution. The amount of amorphous material in the soil controls the release of OH^- and the subsequent increase in pH (Fields and Perrott, 1966). The following reactions illustrate this action and form the basis of this method.

Equation 4.2.1:

$$Al(OH)_3 + 3\ F^- \longrightarrow AlF_3 + 3\ OH^-$$

Equation 4.2.2:

$$Si(OH)_4 + 4\ F^- \longrightarrow SiF_4 + 4\ OH^-$$

Most soils contain components that react with NaF and release OH⁻, but an NaF pH ≥9.4 is a strong indicator that amorphous material dominates the soil exchange complex (Fields and Perrot, 1966). Amorphous material is generally an early product of weathering of pyroclastic materials in a humid climate. Amorphous material appears to form in spodic horizons in the absence of pyroclastics.

Even though the NaF pH test is one of the simplest and most convenient ways of identifying andic materials, the NaF pH of andic and nonandic materials indicates that there is a continuum of materials that range from clearly andic to marginally andic or nonandic in character (Uehara and Ikawa, 1985). As the glass content of the fine-earth fraction increases and/or the silica/alumina ratio increases, the NaF pH test typically becomes less effective in identifying andic materials (Uehara and Ikawa, 1985). The pH rise with NaF is an intensity rather than a quantity indicator (Bartlett, 1972; Wilson et al., 2002). A small amount of hydroxy Al produces as much pH increase as a large amount, as demonstrated by the constancy of pH in NaF soil suspensions in spite of several-fold increases in solution/soil ratio (Bartlett, 1972). In addition, the NaF pH is not a selective test; e.g., the fluoride in complexing the Al releases OH ions from any form of reactive hydroxy Al, organic or inorganic (Egawa et al., 1960; Birrell, 1961; Bartlett, 1972; Wada, 1977, 1989).

The NaF pH is used as a diagnostic criterion for the isotic family mineralogy class (Soil Survey Staff, 2010). The specific requirements for this family are lack of free carbonates, NaF pH >8.4, and 1500-kPa water retention to clay percentage ratio ≥0.6. For a more detailed discussion of the NaF pH criteria for isotic mineralogy class, refer to Fields and Perrott (1966), Wilson et al. (2002), and Soil Survey Staff (2010).

1 N NaF, pH 7.5–7.8, measurement: A 1-g sample is mixed with 50 mL of 1 N NaF (with an initial pH between 7.5 and 7.8) and stirred for 2 minutes. While the sample is being stirred, the pH is read at exactly 2 minutes in the upper one-third of the suspension. The 1 N NaF pH may be used as an indicator that amorphous material dominates the soil exchange complex and should be comparable to relative acid-oxalate-extractable Fe and Al values. The NaF pH test is based on the ligand exchange between F⁻ and OH⁻ in the noncrystalline materials, which results in a rapid rise in pH when 1.0 g of amorphous soil material is suspended in 1 N NaF solution.

1 N NaF, pH 7.5–7.8, interferences: Soils with a 1:1 water pH >8.2 do not give a reliable NaF pH. Free carbonates in a soil result in a high NaF pH. As high NaF values may be found in soils with large sources of Ca or bases, including carbonates, and in some sesquioxide-rich soils, care is required in the interpretation of these data. In general, soils with a 1:1 water pH <7.0 are not affected.

4.2   Soil pH
4.2.2   Saturated Paste pH

Saturated paste pH, application: When interpretations about the soil are made, the saturated paste pH is usually compared to the 1:1 water pH and the 1:2 $CaCl_2$ pH. The usual pH sequence is as follows: 1:1 water pH > 1:2 $CaCl_2$ pH > saturated paste pH. If saturated paste pH is > 1:2 $CaCl_2$ pH, the soil is not saline. If the saturated paste pH ≥ 1:1 water pH, the soil may be Na saturated and does not have free carbonates.

155

Because of the interrelations among the various soil chemical determinations, the saturated paste pH value may be used as a means of cross-checking salinity data for internal consistency and reliability (U.S. Salinity Laboratory Staff, 1954). Some general rules that apply to the saturated paste pH are as follows:

- Soluble carbonates are present only if the pH is >9.

- Soluble bicarbonate is seldom >3 or 4 meq L^{-1} if the pH is $\leq$7.

- Soluble Ca^{2+} and Mg^{2+} are seldom >2 meq L^{-1} if the pH is >9.

- Soils with gypsum seldom have a pH >8.2.

The saturated paste pH is popular in regions where the soils have soluble salts. The water content of the saturated paste varies with soil water storage characteristics. The saturated paste pH may be more indicative of the saturated, irrigated soil pH than is the soil pH measurement at a constant soil:water ratio. The saturated paste pH is also that pH at which the saturation extract is removed for salt analysis and thus is the pH and the dilution at which the sodium adsorption ratio (SAR) is computed.

Saturated paste pH, measurement: The saturated paste is prepared, and the pH of paste is measured with a calibrated combination electrode/digital pH meter.

4.2 Soil pH
4.2.3 Oxidized pH

Oxidized pH, application: Sulfidic material is waterlogged mineral, organic, or mixed soil material that has a pH of 3.5 or higher, that contains oxidizable sulfur compounds, and that, if incubated as a 1-cm-thick layer under moist, aerobic conditions (field capacity) at room temperature, shows a drop in pH of 0.5 or more units to a pH value of 4.0 or less (1:1 by weight in water or in a minimum of water to permit measurement) within 8 weeks (Van Breemen, 1982; Soil Survey Staff, 2010). The intent of the method described herein is to determine whether known or suspected sulfidic materials will oxidize to form a sulfuric horizon (Soil Survey Staff, 2010). Identification of H_2S in a soil by a "rotten-egg" smell or FeS in a saturated soil by its blue-black color indicates that sulfidic materials may be present. If such soils are drained and oxidized, the soil pH could drop to 3.5 or less, making the soil unsuitable for many uses. A field test for FeS is to add 1 N HCl and note the odor of H_2S.

Oxidized pH, measurement: Samples that will be collected for analysis should be collected in an air-tight container and refrigerated until the analysis is made. Analysis should be initiated as soon as possible (Murray-Darling Basin Authority, 2010). Enough soil is transferred to fill a plastic cup one-half to two-thirds full. A little water should be added if needed to make a slurry. The slurry is stirred thoroughly to introduce air. The pH should be determined immediately. Place the cup in a closed container with openings (inlet and outlet) providing humidified air if possible. Keep the sample at room temperature. After 24 hours, open the container, stir the sample thoroughly, and determine the soil pH. Repeat the procedure for a minimum of 16 weeks until the pH reaches a steady state of ≤0.1 unit over a 2-day period. Daily pH readings are recorded. The initial pH and the oxidized pH (end pH) are reported to the nearest 0.1 pH unit.

4.2 Soil pH
4.2.4 Organic Materials CaCl$_2$ pH, Final Solution ≈ 0.01 M CaCl$_2$

Organic materials CaCl$_2$ pH, final solution ≈ 0.01 M CaCl$_2$, application: The 0.01 M CaCl$_2$ pH is used as a taxonomic criterion to distinguish two family reaction classes in Histosols (Soil Survey Staff, 2010). Dysic families have a pH <4.5 in 0.01 M CaCl$_2$ in all parts of the organic materials in the control section. Euic families have a pH >4.5 in 0.01 M CaCl$_2$ in some part of the control section.

Organic materials CaCl$_2$ pH, final solution ≈ 0.01 M CaCl$_2$, measurement: Place the prepared 2.5-mL sample (2.5 cm^3) in a 30-mL plastic container and add 4 mL of 0.015 M CaCl$_2$, making a final concentration of ≈ 0.01 M CaCl$_2$ with most packed, moist organic materials. Mix, cover, and allow to equilibrate at least 1 hour. Uncover the sample and measure pH with pH paper or a pH meter. This test of organic soil material can be used in field offices. Since it is not practical in the field to base a determination on a dry sample weight, moist soil is used. The specific volume of moist material depends on how the material is packed. Therefore, packing of material must be standardized in order to obtain comparable results by different soil scientists (Soil Survey Staff, 2010).

4.2 Soil pH
4.2.5 1:1 Water pH
4.2.6 1:2 0.01 M CaCl$_2$ pH

1:1 water pH and 1:2 0.01 M CaCl$_2$ pH, application: The 1:1 water pH and 1:2 0.01 M CaCl$_2$ pH determinations are two commonly performed soil pH measurements. A number of general interpretations about soils can be made from these pH measurements, but most are made with more confidence if extractable acidity and bases have been measured (National Soil Survey Laboratory Staff, 1975). Most mineral soils with pH <3.5 or in which pH drops to <3.5 after prolonged drying typically contain acid sulfates. Such soils can be found in coastal marshes or mine spoil areas. They have reduced forms of sulfur, which can oxidize to form H$_2$SO$_4$. Soils with pH of 4.5 to 6.5 have some acidity present as hydroxy-Al and hydronium (H$_3$O$^+$). Exchangeable Al^{+3} is present at pH <5.5. The SSL does not analyze for 1 N KCl Al if the 1:2 0.01 M CaCl$_2$ pH is >5.05. In these soils, base saturation (CEC-8.2) is typically <75 percent and in many cases is <35 percent. Soils with pH of 6.5 to 8 typically have a base saturation (CEC-7) in the range of 75 to 100 percent. Soils with pH of 8 to 8.5 are fully base saturated; they probably contain CaCO$_3$ and some salts, and Ca and Mg typically dominate the exchange sites. Soils with pH >8.5 typically contain significant amounts of exchangeable Na, and soils with pH >10 are highly saturated with Na and have an EC >4 mmhos cm^{-1} (dS m^{-1}) and very low resistivities. Soil pH correlates poorly with corrosion potential, but in general soils with pH <4 and >10 have corrosion potential.

The CaCl$_2$ soil pH is generally less than the 1:1 water pH. The combination of exchange and hydrolysis in salt solutions (0.1 to 1 M) can lower the measured pH from 0.5 unit to 1.5 units, compared to the pH measured in reverse osmosis (RO) water (Foth and Ellis, 1988). The CaCl$_2$ pH measurement estimates the activity of H-ions in a soil suspension in the presence of 0.01 M CaCl$_2$ to approximate a constant ionic strength for all soils regardless of past management, mineral composition, and natural fertility level. The result is a pH measurement that remains somewhat invariable despite the seasonal changes in soil pH. The CaCl$_2$ solution also diminishes the seasonal effect of soluble salt concentration.

Calcium chloride (CaCl$_2$) pH is the standard used in *Keys to Soil Taxonomy* to differentiate the reaction classes used at the family level in mineral and organic soils. The calcium chloride pH 1:2 0.01 M CaCl$_2$ is measured in a mixture, by weight, of one part soil to two parts 0.01 M CaCl$_2$ solution. These pH values are used as taxonomic criteria for the reaction classes (acid and nonacid) in families of mineral soils, such as Entisols, Gelisols (other than Histels), Aquands, Aquepts, and all Gelic suborders and Gelic great groups (Soil Survey Staff, 2010). Soils classified in the acid class have a pH value of <5.0 in 0.01 M CaCl$_2$ (1:2) (about pH 5.5 in water, 1:1). The pH value limit for soils classified in the nonacid class is ≥5.0 in 0.01 M CaCl$_2$ (1:2). The reaction classes used in the families of organic soils (Histosols and Histels) are euic and dysic. Soils classified in the euic reaction class have a pH value of ≥4.5 in 0.01 M CaCl$_2$ (1:2), and soils classified in the dysic reaction class have a pH value of <4.5.

1:1 water pH and 1:2 0.01 M CaCl$_2$ pH, measurement: The pH is measured in soil-water (1:1) and soil-salt (1:2 CaCl$_2$) solutions. For convenience, the pH is initially measured in water and then measured in CaCl$_2$. With the addition of an equal volume of 0.02 M CaCl$_2$ to the soil suspension that was prepared for the water pH, the final soil-solution ratio is 1:2 0.01 M CaCl$_2$. A 20-g soil sample is mixed with 20 mL of reverse osmosis (RO) water (1:1 w:v). The sample is allowed to stand for 1 hour and is stirred occasionally. The sample is then stirred for 30 seconds, and the 1:1 water pH is measured. The 0.02 M CaCl$_2$ (20 mL) is added to the soil suspension, the sample is stirred, and the 1:2 0.01 M CaCl$_2$ pH is measured.

4.2 Soil pH
4.2.7 1 N KCl pH

1 N KCl pH, application: The 1 N KCl pH is an index of soil acidity and is more popular in regions that have extremely acid soils and in which KCl is used as an extractant of exchangeable Al. The KCl pH indicates the pH at which Al is extracted. If the pH is <5.5, significant amounts of Al are expected in the solution. Soils that have pH <4 generally have free acids, such as H$_2$SO$_4$. As with the 1:2 CaCl$_2$ pH, the 1 N KCl pH readings tend to be uniform regardless of time of year.

1 N KCl pH, measurement: A 20-g soil sample is mixed with 20 mL of 1 N KCl. The sample is allowed to stand for 1 hour and is occasionally stirred. The sample is then stirred for 30 seconds; after 1 minute, the KCl pH is read.

4.2 Soil pH
4.2.8 1:5 Water pH

1:5 aqueous extraction, pH, measurement: A 20-g sample of soil is added to 100 mL of water in a 250-mL polyethylene bottle. The soil:water suspension is maintained at room temperature for 23 hours and is then shaken on a reciprocating shaker for 1 hour. The supernatant is filtered into a 100-mL polyethylene bottle. The pH of the extract is measured with a calibrated combination electrode/digital pH meter. The 1:5 extract is also analyzed for EC, cations, anions, nitrate-nitrite, and multielements.

4.3 Soil Test Analyses

For more than 30 years, soil testing has been used as a basis for determining lime and fertilizer needs (Soil and Plant Analysis Council, 1999). In more recent years, some of these tests have been employed in more diverse agronomic and environmental uses (Pierzynski, 2000). For these reasons, the SSL expanded its suite of soil test analyses to more completely characterize the inorganic and organic N fractions and to provide a number of P analyses for a broad spectrum of soil applications. This section describes the suite of SSL test analyses for N, P, and several multielements, along with key definitions and applications of these SSL data. In addition, information is provided on the various soil P and N forms, factors affecting their retention or mobility in soils, and their role as essential plant nutrients and fertilizer components. The SSL P test methods include, but are not limited to, anion-resin extractable, water soluble, Bray P-1, Bray P-2, Olsen sodium-bicarbonate, Mehlich No. 3, citric acid soluble, and New Zealand P Retention. The procedures for total P analysis are described in another section of this manual entitled "Total Analysis." The SSL determines inorganic N (nitrate-nitrite) by KCl extraction. A flow injection automated ion analyzer is used to measure the soluble inorganic nitrate (NO_3^-). The SSL also determines mineralizable N (N as NH_3) by anaerobic incubation. Other N procedures, such as mineralizable N after fumigation incubation and total N, are described in other sections of this manual entitled "Soil Biological and Plant Analysis" and "Total Analysis," respectively. This section also describes water and Mehlich No. 3 extractions for multielement determinations. For detailed descriptions of the SSL methods for N, P, and multielements which are cross-referenced by method code in the table of contents in this manual, refer to SSIR No. 42 (Soil Survey Staff, 2004), which is available online at http://soils.usda.gov/technical/lmm/. Also refer to the *Soil Survey Field and Laboratory Methods Manual*, SSIR No. 51 (Soil Survey Staff, 2009; available online at http://www.soils.usda.gov/technical/), for detailed descriptions of field methods as used by NRCS soil survey offices.

4.3 Soil Test Analyses
4.3.1 Phosphorus, Agronomic, Taxonomic, and Environmental Significance

Total P, organic and inorganic: Phosphorus added to the soil-crop system goes through a series of transformations as it cycles through plants, animals, microbes, soil organic matter, and the soil mineral fraction (National Research Council, 1993). Phosphorus is also an essential plant nutrient and is often related to water-quality problems. Unlike N, most P is tightly bound in the soil, and only a small fraction of the total P found in the soil is available to crop plants. Total P includes both organic and inorganic P forms. Apatite is a common P-bearing mineral.

Organic P levels may range from virtually zero to 0.2 percent, and the inorganic P is typically higher than organic P (Tisdale et al., 1985). The organic P fraction is found in humus and other organic materials. The inorganic P fraction occurs in numerous combinations with Fe, Al, Ca, and other elements. The solubility of these various combinations ranges from soluble to very insoluble (Chang and Jackson, 1957; Lindsay and Vlek, 1977). Phosphates may also react with clays to form generally insoluble clay phosphate complexes (adsorbed P). Refer to Sharpley et al. (1985) for a detailed P characterization of 78 soils representing 7 major soil orders from all regions of the United States. Sharpley et al. (1985) discuss the various soil P forms, e.g., labile, organic, and sorbed; the various pathways of P transformation; the significance of selected soil P test values; and the relationships between soil P and soil test P values. Burt et al. (2002a) published a study of various P extractions and examined correlations with other soil properties of selected benchmark pedons in the United States.

C:N:P ratio: Studies of the mineralization of organic P in relation to the C:N:P ratio have indicated that there is no set ratio for all soils. Some studies have indicated that if the C:inorganic P ratio is 200:1 or less, mineralization of P may occur, and if this ratio is 300:1, immobilization would occur (Tisdale et al., 1985). Other studies have indicated that the N:P ratio is related to the mineralization and immobilization of P and that the decreased supply of one results in the increased mineralization of the other; i.e., if N were limiting, inorganic P may accumulate in the soil and the formation of soil organic matter would be inhibited (Tisdale et al., 1985).

Phosphorus, essential plant element: Phosphorus is an essential nutrient for plant growth and is a primary fertilizer element. Phosphorus is essential in supplying phosphate, which acts as a linkage or binding site in plants. The stability of phosphate enables it to participate in many energy capture, transfer, and recovery reactions, which are important for plant growth (Tisdale et al., 1985). The energy obtained from photosynthesis and metabolism of carbohydrates is stored in phosphate compounds (ATP and ADP) for subsequent use in growth and reproductive processes. In addition to its metabolic role, P also acts as an important structural component of a wide variety of biochemicals, including nucleic acids (DNA and RNA), coenzymes, nucleotides, phosphoproteins, phospholipids, and sugar phosphates (Tisdale et al., 1985). Phosphorus has also been linked to increased root growth and early maturity of crops, particularly grain crops. Plants can absorb P as either the primary $H_2PO_4^-$ ion or smaller amounts of the secondary HPO_4^{2-} orthophosphate ion. The $H_2PO_4^-$ is the principal form absorbed as it is most abundant over the prevailing range in soil pH for most crops. The absorption of $H_2PO_4^-$ is typically greater at low soil pH values, whereas the uptake of HPO_4^{2-} is typically greater at higher pH values (Bidwell, 1979). Some studies have shown that there are 10 times as many absorption sites on plant roots for $H_2PO_4^-$ than for HPO_4^{2-}.

Phosphorus, fertilizer: Fertilizer P is the single most important source of P added to cropland in the United States. Relatively small annual additions of P can cause a soil buildup of P (McCollum, 1991). Phosphorus can be lost from the soil-crop system in soluble form through leaching, subsurface flow, and surface runoff. Particulate P is lost when soil erodes. The fraction of total P lost to erosion and runoff can be substantial. Some of the P added in excess of crop needs remains as residual plant-available P. Without fertilization, the amount of extractable P declines with time because of the slow conversion of P to unavailable forms, e.g., Ca, Al, and Fe-P compounds (Yost et al., 1981; Mendoza and Barrow, 1987; Sharpley et al., 1989; McCollum, 1991). The rate of decline in extractable P (discounting plant uptake) varies with the soil P-level and P-sorption capacity. The P level in the soil is the critical factor in determining actual loads of P to surface water and the relative proportions of P lost in solution and attached to soil particles (National Research Council, 1993). Understanding the relative importance of transport pathways and the processes regulating these transport pathways helps in the design of measures to reduce P losses. When P enters surface waters in substantial amounts, it becomes a pollutant and contributes to the excessive growth of algae and other aquatic vegetation and, thus, to the accelerated eutrophication of lakes and reservoirs (National Research Council, 1993).

Phosphorus, soil testing: Soil chemical tests for estimating soil nutrient pools are relatively rapid. As soil requirements are determined before a crop is planted, these tests have the added advantage over plant analyses and observations of deficiency symptoms during the growing stage. Soil tests usually only measure a part of the total nutrient supplies in the soil. In general, these test values are of no use in themselves and must be calibrated against nutrient rate experiments, i.e., field and greenhouse experiments, before use in the prediction of the nutrient needs of crops. A complete soil testing program includes both the analytical procedures for estimating soil fertility and the appropriate correlation and calibration data for recommending the correct fertilization practices (Corey, 1987; Sabbe and Marx, 1987). Fertilizer recommendations are then based on the interpretation of these calibration data and fertilizer response curves.

Phosphorus, available: The fraction of soil P utilizable for crop growth has been designated as *available P*. This term is also used to refer to the portion of soil P extracted by various solvents, e.g., water, dilute acids or alkalis, and salt solutions (Tisdale et al., 1985). The quantities of total P are much greater than those of the available P, but the available form is of greater importance to plant growth. The term *labile P* has been defined as the fraction that is isotopically exchanged with ^{32}P or that is readily extracted by some chemical extractant or by plants (Foth and Ellis, 1988). Thus, the labile P may include some or all of the adsorbed P in a particular soil, or it may also include some precipitated P (Foth and Ellis, 1988). *Adsorbed P* is generally considered the portion of soil P that is bonded to the surface of other soil compounds when a discrete mineral-phase is not formed; e.g., if soluble P were added to a soil solution, it may be bound to the surface of amorphous Al hydroxide without forming a discrete Al-P mineral (Foth and Ellis, 1988). Labile P has been an important working concept for the soil scientist in relating soil P to plant-available P as it is a measurable fraction, even though it may include P from several of the discrete P fractions held in soils (Foth and Ellis, 1988). Soil tests for P generally try to measure all or part of the labile P (Foth and Ellis, 1988).

Phosphorus retention: Sorption is the removal of P from solution and its retention at soil surfaces. When P is held at the surface of a solid, the P is considered adsorbed, but if the retained P penetrates more or less uniformly into solid phase, it is considered to be absorbed or chemisorbed (Tisdale et al., 1985). The less specific term *sorption* is often preferred because of the difficulty in distinguishing between these two reactions (Tisdale et al., 1985). The reverse reaction, desorption, relates to the release of sorbed P into solution. *Fixation* is a term frequently used to collectively describe both sorption and precipitation reactions of P. There is considerable evidence that supports a wide range of sorption and precipitation mechanisms as causes of P retention but no explicit consensus as to the relative magnitudes of their contributions (Tisdale et al., 1985). Phosphorus retention is viewed by many researchers as a continuous sequence of precipitation, chemisorption, and adsorption. At low P solution concentrations, adsorption may be the dominant mechanism.

Phosphorus retention, soil-related factors: Several of the factors that influence soil P retention are the nature and amount of soil components, pH, cation and anion effects, saturation of sorption complex, organic matter, temperature, and time of reaction. Some of the major soil components that affect P sorption (retention) are hydrous oxides of Fe and Al, type and amount of clay, calcium carbonate, and amorphous colloids. Soils with significant amounts of Fe and Al oxides typically have greater P retention capacity than those soils with more crystalline oxides because of the greater relative surface area and sites of adsorption. Phosphorus is typically retained to a greater extent by 1:1 clays, e.g. kaolinite (low SiO_2/R_2O_3 ratio), than by 2:1 clays. Soils with high amounts of clay typically retain more P than those with small clay contents because of greater surface area. In calcareous soils, calcium carbonate and associated hydrous ferric oxide impurities can function as principal P adsorption sites (Hamad et al., 1992). The active Al in noncrystalline colloids, e.g., allophane, imogolite, and Al humus complexes, is highly reactive with anions, e.g., phosphates, sulfates, and silicates (Van Wambeke, 1992). The affinity of the noncrystalline minerals for P is the result of very high specific surface of these minerals and the density of active Al on colloidal fractions. In addition, fully hydrated gels can deform, partially liquefy, and trap or encapsulate P (occluded P) in voids that are not connected with the soil solution (Uehara and Gillman, 1981).

Phosphorus retention, selected soils: Andisols and other soils that contain large amounts of allophane and other poorly crystalline minerals have capacities for binding P (Gebhardt and Coleman, 1984). The factors that affect P retention in these soils are not well understood; however, allophane and imogolite have been considered as major materials that contribute to P retention in Andisols (Wada, 1985). Phosphate sorption in the surface horizons of 228 acid to neutral soils in Western Australia was found to be more closely related ($r^2 = 0.76$) to the content of oxalate extractable Al than to any other constituent (Gilkes and Hughes, 1994). A phosphate retention of ≥ 85 percent is a taxonomic criterion for andic soil properties (Soil Survey Staff, 2010). The intent of the criterion is diagnostic for the dominance of active Al in amorphous clay minerals that generally synthesize in rapidly weathering volcanic glass (Van Wambeke, 1992). Refer to Soil Survey Staff (2010) for a more detailed discussion of this criterion. The P retention test may not necessarily coincide with soil fertility criteria as the concentration of the P solution in this procedure is higher than current P contents in most soil solutions. Also, in some cases, the method probably overestimates the positive charges on the soil colloids under field conditions (Van Wambeke, 1992).

The P retention test is based on a ligand exchange between HPO_4^{2-} or $H_2PO_4^-$ and OH^-. The NaF pH test is based on a similar ligand exchange, i.e., the exchange between F^- and OH^-. In many ways, the P retention test is a duplication of the information provided by the NaF pH test (Uehara and Ikawa, 1985). Soil pH in NaF was found to be related to P-sorbed ($r^2 = 0.48$), particularly when log(P-sorbed) and log(P-sorbed + 10) were related to pH_{NaF} ($r^2 = 0.66$ and 0.74, respectively) (Gilkes and Hughes, 1994).

Phosphorus, data assessments: Methods development in soil P characterization (Bray and Kurtz, 1945; Olsen et al., 1954; Chang and Jackson, 1957) has been instrumental in developing principles and understanding of the nature and behavior of P in soils (Olsen and Sommers, 1982). Amounts, forms, and distribution of soil P vary with soil-forming factors (Walker, 1974; Stewart and Tiessen, 1987), level and kind of added P (Barrow, 1974; Tisdale et al., 1985; Sharpley and Menzel, 1987; Sharpley, 1996), other soil and land management factors (Haynes, 1982; Sharpley, 1985), and soil P-sorption characteristics (Goldberg and Sposito, 1984; van Riemsdijk et al., 1984; Polyzopoulos et al., 1985; Frossard et al., 1993). Knowledge of these factors and their impact upon the fate and transport of soil P has been used in developing soil P interpretations for such broad and diverse applications as fertility, taxonomic classification, genesis and geomorphology models, and environmental studies (Burt et al., 2002a).

While available (extractable) soil P traditionally has been related to the amount of P available for crop uptake and the probability of crop response to added P, and thereby fertilizer requirements (Pierzynski, 2000), many traditional soil P tests (e.g., Mehlich No. 3, Bray P-1, and water soluble) have also been evaluated for use in environmental studies, e.g., predictive models for P runoff (Tiessen et al., 1984; Gartley and Sims, 1994; Heckrath et al., 1995; Pote et al., 1996; Sims et al., 1998). Comparative studies have shown good correlations between some soil P tests (Wolf and Baker, 1985; Sims, 1989; Tran et al., 1990; Bhiyan and Sedberry, 1995) and between soil P and other soil properties (Jones et al., 1984; Sharpley et al., 1984, 1985, 1989). In a study of 168 U.S. benchmark soils, representing 8 soil orders (Tiessen et al., 1984), the relative proportions of available and stable P and organic and inorganic P were found to be dependent on soil chemical properties and related to the system of soil taxonomic classification. Similarly, predictive P models (labile, organic, sorbed), derived from other soil properties (e.g., total N, clay, and $CaCO_3$) and P data (e.g., extractable P), were improved when 78 soils were divided into groups based on the soil taxonomic classification system (Soil Survey Staff, 2010) and weathering (Sharpley et al., 1984, 1985). The predictive model for potential P runoff (dissolved and bioavailable) in some soils in Oklahoma (Sharpley, 1995) was improved with the integration of the effect of soil type (P-sorption characteristics) over soil P alone, suggesting that relationships between runoff P and soil P need to be soil specific for use in management recommendations.

Soil P-sorption characteristics provide useful information for the assessment of available P (Frossard et al., 1993; Indiati, 2000), determination of pedogenic P pathways (Tiessen et al., 1984), and taxonomic classification (Soil Survey Staff, 2010). Soil P-sorption, expressed as P-saturation indices, has also been investigated for use in risk management of water contamination by P (van der Zee et al., 1987; Breeuwsma and Silva, 1992; Sharpley, 1996; Sims et al., 1998; Beauchemin and Simard, 1999). In a

163

review of these studies, Beauchemin and Simard (1999) suggested that these indices are best developed using homogeneous soil groups to account for their distinctive behavior and characteristics.

Phosphorus-sorption characteristics reflect the chemical and mineralogical properties of the soil, e.g., clay type, Fe and Al oxides, organic matter, pH, and $CaCO_3$ (Loganathan et al., 1987; Owusu-Bennoah and Acguaye, 1989; Solis and Torrent, 1989). Because these soil components are themselves interrelated (Syers et al., 1973), however, it is often difficult to determine which components contribute the most to P-sorption in soils. In a study of intensely weathered acid soils in Kentucky (Mubiru and Karathanasis, 1994), P-sorption was best correlated with extractable Al ($r = 0.93$) and crystalline Fe oxyhydroxides ($r = 0.97$) but had a negative association with organic matter ($r = -0.83$). In a study of P-deficient, highly weathered, lateritic, ironstone gravel soils in Australia (Brennan et al., 1994), Al_2O_3 and organic C explained 45 to 59 percent of the variation in P adsorption, whereas Fe_2O_3 explained less variation ($r^2 = 0.34$), even though it was higher in content than Al_2O_3. The influence of organic matter on P adsorption has been related to the presence of organically chelated Fe and Al oxides (Harter, 1969; Syers et al., 1971). In a study of soils in Italy with low to medium P retention capacity, P adsorption in soils was best correlated with Al_o ($r = 0.94$) but had no significant relationship with organic C, Fe_o, pH, CEC, clay, or exchangeable Ca. In a study of different clay minerals (Oh et al., 1999), P retention was related to pH and P levels (decreasing with increasing pH at greater added P concentrations) and to surface area (maximum adsorption capacities at pH 5.4, decreasing in the order allophane > alumina > goethite > hematite), suggesting that surface site density is greater for Al minerals than for Fe-oxides. However, Goldberg and Sposito (1984) suggest that P-adsorption models need not distinguish between Al and Fe, as similar protonation-dissociation constants and P-surface complexation constants were found for both minerals. Olsen and Watanabe (1957) also found a close relationship between P-sorption and surface area in both acid ($r^2 = 0.92$) and alkaline ($r^2 = 0.96$) soils. Said and Dakermanji (1993) found P adsorption in calcareous soils most significantly related to clay content ($r^2 = 0.69$) but to no other measured soil property, including $CaCO_3$ ($r^2 = 0.106$), which ranged from 1.8 to 49.1 percent. Similarly, Ryan et al. (1984) found no effect by $CaCO_3$ on P adsorption in calcareous soils but a significant relationship with Fe_o. Holford and Mattingly (1975a and 1975b) attributed $CaCO_3$ reactivity to surface area (carbonate particle-size distribution) rather than to total $CaCO_3$.

In a study of 21 benchmark soils of the United States (Burt et al., 2002a), including surface and subsurface horizons, and satellites from the Water Erosion Prediction Project (WEPP), phosphorus (P) was analyzed using methods that included total (TP), water soluble (WP), Bray 1 (BP), Mehlich No. 3 (MP), Olsen (OLP), New Zealand P Retention (NZP), organic (OP), anion-exchange resin (AERP), and acid oxalate (P_o). Objectives of the study were to determine relationships among soil P test values and other soil properties. Knowledge and understanding of these relationships are important to researchers when soil P datasets are evaluated for use in predictive models for agronomic, soil genesis, or environmental purposes. Important relationships were developed, using simple or multiple linear regression models, among P methods and other soil properties, e.g., organic C (OC), total N (TN), dithionite-citrate extractable Fe and Al (Fe_d, Al_d), and clay, as follows:

TP (mg/kg) = 229.02 + 27.76 Al$_d$ (g/kg) + 27.44 (g/kg) + 4.14 Fe$_d$ (g/kg), r^2 = 0.89, P<0.01, n = 263 (all soils)

OP (mg/kg) = 114.07 + 38.07 TN (g/kg) – 14.74 pH + 6.94 OC (g/kg), r^2 = 0.80, p<0.01, n = 262 (all soils)

BP (mg/kg) = -1.82 + 1.11 MP (mg/kg), r^2 = 0.96, P<0.01, n = 268 (all soils)

P$_o$ (mg/kg) = 16.02 – 24.27 Al$_o$ (g/kg) + 25.59 Fe$_o$ (g/kg), r^2 = 0.79, p < 0.01, n = 202 (noncalcareous)

NZP (%) = 16.92 + 1.37 Al$_d$ (g/kg) + 0.28 clay (%), r^2 = 0.91, P < 0.01, n = 203 (noncalcareous)

Phosphorus, measurements: To characterize the P in the soil system requires the selection of an appropriate method of determination. This selection is influenced by many factors, e.g., objectives of the study, soil properties, sample condition or environment, accuracy, and reproducibility (Olsen and Sommers, 1982). Most soil P determinations have two phases, i.e., preparation of a solution that contains the soil P or fraction thereof and the quantitative determination of P in the solution. Most P analyses of soil solutions have been colorimetric procedures, as they are sensitive, reproducible, and lend themselves to automated analysis, accommodating water samples, digest solutions, and extracts (Pierzynski, 2000). The selected colorimetric method for P determination depends on the concentration of solution P, the concentration of interfering substances in the solution to be analyzed, and the particular acid system involved in the analytical procedure (Olsen and Sommers, 1982). Inductively coupled plasma (ICP) spectrophotometry also can be used for P determination. The popularity of this procedure is increasing due to the use of multielement soil extractants (Pierzynski, 2000). Results from colorimetric analyses are not always comparable to those from ICP because ICP measures the total amount of P in solution while the colorimetric procedures measure P that can react with the color developing reagent (Pierzynski, 2000). In the following section, the various SSL P test methods and their respective applications are described.

4.3   Soil Test Analyses
4.3.2   Anion Resin Extraction
4.3.2.1   Two-Point Extraction
4.3.2.1.1   1-h, 24-h, 1 *M* NaCl
4.3.2.1.1.1   Phosphorus

Anion resin P, application: Anion resins remove P from soils without chemical alterations and with only minor pH changes. Amounts of P released from soil and adsorbed by resins have been used as a measure of available P, an assessment of the availability of residual phosphates, an estimation of release characteristics and runoff P for agricultural land (Elrashidi et al., 2003), and a measure of the buffer capacity of soils (Olsen and Sommers, 1982).

Plotting a log of extraction periods (0.25, 0.50, 1, 2, 4, 8, 24, and 48 hours) against amounts of P released (mg kg^{-1}) showed a linear relationship in 24 U.S. benchmark soils (Elrashidi et al., 2003). Two extraction periods (1 hour and 24 hours) are sufficient to develop linear equations that predict P release characteristics (PRC), describing the whole relationship between the extraction time (1 minute to 48 hours) and amount of P released (mg kg^{-1}) for soils (Elrashidi et al., 2003). The SSL method for anion resin P describes a two-point measurement (1-hour and 24-hour extraction).

Anion resin P, measurement and interferences: A 2-g soil sample and 4-g resin bag are shaken with 100 mL of reverse osmosis deionized water for 1 hour. The soil suspension is shaken again with another 4-g resin bag for 23 hours. Phosphorus released from soil during shaking is adsorbed by resin. To remove P from resin, resin bags are shaken for 1 hour in 1 M NaCl solution. Concentrated HCl is added to sample extracts. A 1-mL aliquot is diluted with 4 mL of ascorbic acid molybdate solution. Absorbance of the solution is read using a spectrophotometer at 882 nm. The SSL reports anion resin P as mg P kg^{-1} soil.

The anion resin P analyses similar to the water, Bray, and Mehlich procedures are Mo blue methods, which are very sensitive for P. These methods are based on the principle that in an acid molybdate solution containing orthophosphate ions, a phosphomolybdate complex forms that can be reduced by ascorbic acid, $SnCl_2$, and other reducing agents to a Mo color. The intensity of blue color varies with the P concentration but is also affected by other factors, such as acidity, arsenates, silicates, and substances that influence the oxidation-reduction conditions of the system (Olsen and Sommers, 1982).

4.3   Soil Test Analyses
4.3.3   Aqueous Extraction
4.3.3.1   Single-Point Extraction
4.3.3.1.1   1:10, 30 min
4.3.3.1.1.1   Phosphorus

Water-soluble P, application: Phosphorus occurs in soil in both the solution and solid phases. These forms are well documented, but questions still remain concerning the exact nature of the constituents and ionic forms found in water, soils, and sediments (National Research Council, 1993). These forms influence P availability in relation to root absorption and plant growth, runoff and water-quality problems, and P loadings. Water-soluble P has been defined as P measured in water, dilute salt extracts (e.g., 0.01 M CaCl$_2$), displaced soil solutions, or saturation paste extracts (Olsen and Sommers, 1982). Even though the water-soluble fraction principally consists of inorganic orthophosphate ions, there is evidence that some organic P also is included (Rigler, 1968).

The water or dilute salt extracts represent an attempt to approximate the soil solution P concentration. The objectives of this method, which is an index of P availability, are (1) to determine the P concentration level in the soil extract that limits plant growth (Olsen and Sommers, 1982) and (2) to determine the composition of the soil solution so that the chemical environment of the plant roots may be defined in quantitative terms (Adams, 1974). The sum of water-soluble P and pH 3 extractable P has also been defined as the available P in runoff (Jackson, 1958).

Water-soluble P, measurement: A 2.5-g sample of <2-mm, air-dry soil is mechanically shaken for 30 minutes in 25 mL of reverse osmosis deionized water. The sample is centrifuged until the solution is free of soil mineral particles and then is filtered until clear extracts are obtained. Absorbance of the solution is read using a spectrophotometer at 882 nm. Alternatively, a flow injection automated ion analyzer is used to measure the orthophosphate ion (PO_4^{3-}). Method parameters specific to water-soluble P have been modified from the QuikChem Method 10-115-01-1-A, orthophosphate in waters (U.S. Environmental Protection Agency, 1983; LACHAT Instruments, 1993; U.S. Department of the Interior, Geological Survey, 1993). The ion (PO_4^{3-}) reacts with ammonium molybdate and antimony potassium tartrate under acidic conditions to form a complex. This complex is reduced with ascorbic acid to form a blue complex, which absorbs light at 882 nm. Absorbance is proportional to the concentration of PO_4^{3-} in the sample. The SSL reports water-soluble P as mg P kg^{-1} soil.

4.3 Soil Test Analyses
4.3.3 Aqueous Extraction
4.3.3.1 Single-Point Extraction
4.3.3.1.2 1:5, 23 h, 1 h
4.3.3.1.2.1–22 Aluminum, Arsenic, Barium, Calcium, Cadmium, Cobalt, Chromium, Copper, Iron, Lead, Magnesium, Manganese, Molybdenum, Sodium, Nickel, Phosphorus, Potassium, Selenium, Silicon, Strontium, Vanadium, and Zinc
4.3.3.1.2.23 Boron
4.3.3.1.2.24–30 Bromide, Chloride, Fluoride, Nitrate, Nitrite, Phosphate, and Sulfate

1:5 aqueous extraction, preparation: A 20-g sample of soil is added to 100 mL of water in a 250-mL polyethylene bottle. The soil:water suspension is maintained at room temperature for 23 hours and is then shaken on a reciprocating shaker for 1 hour. The supernatant is filtered into a 100-mL polyethylene bottle. The 1:5 extract is analyzed for pH, EC, cations, anions, nitrate-nitrite, and multielements, including B.

1:5 aqueous extract, anions, measurement: The 1:5 extract is diluted according to its electrical conductivity (EC_s). The diluted sample is injected into the ion chromatograph, and the anions are separated. A conductivity detector is used to measure the anion species and content. Standard anion concentrations are used to calibrate the system. A calibration curve is determined, and the anion concentrations (Br^-, Cl^-, F^-, NO_3^-, NO_2^-, PO_4^{3-}, and SO_4^{2-}, respectively) are calculated. A computer program automates these actions. An aliquot of 1:5 extract is titrated on an automatic titrator to pH 8.25 and pH 4.60 end points. The carbonate and bicarbonate (CO_3^{2-} and HCO_3^-, respectively) are calculated from the titers, aliquot volume, blank titer, and acid normality. The 1:5 extracted anions, Br^-, Cl^-, F^-, NO_3^-, NO_2^-, PO_4^{3-}, SO_4^{2-}, CO_3^{2-}, and HCO_3^-, are reported as mmol (-) L^{-1}. The 1:5 extract is also analyzed for EC and pH.

1:5 aqueous extract, nitrate-nitrite, measurement: An aliquot of the 1:5 aqueous extract is analyzed for soluble inorganic nitrate (NO_3^-) using a flow injection automated ion analyzer. The nitrate is quantitatively reduced to nitrite by passage of the sample through a copperized cadmium column. The nitrite (reduced nitrate plus original nitrite) is then determined by diazotizing with sulfanilamide followed by coupling with N-1-naphthylethylenediamine dihydrochloride. The resulting water-soluble dye has a magenta color, which is read at 520 nm. Absorbance is proportional to the concentration of NO_3^- in the sample. The SSL reports data as mg N kg^{-1} soil as NO_3^- and/or NO_2^-.

1:5 aqueous extract, multielement, measurement: A 20-g sample of soil is added to 100 mL of water in a 250-mL polyethylene bottle. The soil:water suspension is maintained at room temperature for 23 hours and is then shaken on a reciprocating shaker for 1 hour. The supernatant is filtered into a 100-mL polyethylene bottle. The extract is analyzed for Al, As, Ba, Ca, Cd, Co, Cr, Cu, Fe, Pb, Mg, Mn, Mo, Na, Ni, P, K, Se, Si, Sr, V, and Zn by inductively coupled plasma mass spectrophotometry (ICP-MS). The SSL reports water-soluble elements as mg kg^{-1}.

1:5 aqueous extract, boron, measurement: The 1:5 extract is used for B analysis using azomethine-H, which forms a colored complex of H_3BO_3 in aqueous media. A 1-mL aliquot of blank, diluted B standard, or sample, is added to a 50-mL polypropylene centrifuge tube, followed by 2 mL of buffer, and mixed. A 2-mL aliquot of azomethine-H reagent is added to the tube and mixed. After 30 minutes, absorbance is read at 420 nm. Sample concentrations are determined using a standard curve prepared with 0, 2, 4, 6, 8, and 10 B µg B mL^{-1} solutions. The SSL reports B as µg kg^{-1}.

4.3   Soil Test Analyses
4.3.4   Bray P-1 Extraction
4.3.4.1   Phosphorus

Bray P-1, application: The Bray P-1 procedure is widely used as an index of available P in the soil. Bray and Kurtz (1945) originally designed the Bray P-1 extractant to selectively remove a portion of the adsorbed form of P with the weak, acidified ammonium fluoride solution. Adsorbed phosphorus is in the anion form adsorbed by different charged surface functional groups that have varying degrees of adsorption affinity. In general, this method has been most successful on acid soils (Olsen and Sommers, 1982). The acid solubilizes calcium and aluminum phosphates and partially extracts iron phosphate compounds. The NH_4F complexes the aluminum in solution and limits readsorption of P on iron oxides (Kuo, 1996). The Bray P-1 has limited ability to extract P in calcareous soils due to the neutralization of the dilute acid by carbonates. For most soils, Bray P-1 and Mehlich No. 3 are nearly comparable in their abilities to extract native P but exceed the Olsen sodium-bicarbonate method by two- to three-fold, indicating that predictive models for Bray P-1, Mehlich No. 3, and Olsen sodium-bicarbonate are closely associated with pH buffering of extractant (acid versus alkaline) (Burt et al., 2002a).

Bray P-1, measurement: A 2.5-g soil sample is shaken with 25 mL of Bray P-1 extracting solution for 15 minutes. The sample is centrifuged until the solution is free of soil mineral particles and is then filtered until clear extracts are obtained. A 2-mL aliquot is diluted with 8 mL of ascorbic acid molybdate solution. Absorbance of the solution is read using a spectrophotometer at 882 nm. Alternatively, a flow injection automated ion analyzer is used to measure the orthophosphate ion (PO_4^{3}). Method parameters specific to Bray P-1 have been modified from the QuikChem Method 12-115-01-1-A, orthophosphate in soils (U.S. Environmental Protection Agency, 1983; LACHAT Instruments, 1989; U.S. Department of the Interior, Geological Survey, 1993). The ion (PO_4^{3-}) reacts with ammonium molybdate and antimony potassium tartrate under acidic conditions to form a complex. This complex is reduced with ascorbic acid to form a blue complex, which absorbs light at 660 nm. Absorbance is proportional to the concentration of PO_4^{3-} in the sample. The SSL reports Bray P-1 as mg P kg^{-1} soil.

Bray P-1, interferences: Many procedures may be used to determine P. Studies have shown the incomplete or excessive extraction of P to be the most significant contributor to interlaboratory variation. The Bray P-1 procedure is sensitive to the soil/extractant ratio, shaking rate, and time. This extraction uses the ascorbic acid-potassium antimony-tartrate-molybdate method. The Fiske-Subbarrow method is less sensitive but has a wider range before dilution is required (North Central Regional Research Publication No. 221, 1988). For calcareous soils, the Olsen method is preferred. An alternative procedure for calcareous soils uses the Bray P-1 extracting solution at a 1:50 soil:solution ratio. This procedure has been shown to be satisfactory for some calcareous soils (Smith et al., 1957; North Central Regional Research Publication No. 221, 1988).

Silica forms a pale blue complex, which also absorbs light at 660 nm. This interference is generally insignificant as a silica concentration of approximately 4000 mg L^{-1} would be required to produce a 1 mg L^{-1} positive error in orthophosphate (LACHAT Instruments, 1989).

Concentrations of ferric iron greater than 50 mg L^{-1} will cause a negative error due to competition with the complex for the reducing agent ascorbic acid. Pretreating samples high in iron with sodium bisulfite can eliminate this interference. Treatment with bisulfite will also remove the interference due to arsenates (LACHAT Instruments, 1989).

The determination of phosphorus is sensitive to variations in acid concentrations in the sample since there is no buffer. With increasing acidity, the sensitivity of the method is reduced. Samples, standards, and blanks should be prepared in a similar matrix.

4.3   Soil Test Analyses
4.3.5   Bray P-2 Extraction
4.3.5.1   Phosphorus

Bray P-2, application: The Bray P-2 procedure functions to extract a portion of the plant-available P in the soil. The composition is similar to that of the Bray P-1 extraction solution. The difference is a slightly higher concentration of HCl (0.025 *N* compared to 0.1 *N*) in the Bray P-2. The Bray P-2 was originally designated by Bray and Kurtz (1945) to extract the easily acid soluble P as well as a fraction of adsorbed phosphates. The HCl solubilizes calcium and aluminum phosphates and partially extracts iron phosphate compounds. The NH_4F complexes the aluminum in solution and limits readsorption of P on iron oxides (Kuo, 1996). The higher acid concentration of

169

the Bray P-2 should allow greater extraction of P in calcareous soils compared to Bray P-1, but the Bray P-2 is not as widely used by soil testing laboratories as the Bray P-1.

Bray P-2, measurement: A 2.5-g soil sample is shaken with 25 mL of Bray P-2 extracting solution for 15 minutes. The sample is centrifuged until the solution is free of soil mineral particles and then is filtered until clear extracts are obtained. A 2-mL aliquot is diluted with 8 mL of ascorbic acid molybdate solution. Absorbance of the solution is read using a spectrophotometer at 882 nm. The SSL reports Bray P-2 as mg P kg^{-1} soil.

4.3   Soil Test Analyses
4.3.6   Olsen Sodium-Bicarbonate Extraction
4.3.6.1   Phosphorus

Olsen sodium-bicarbonate P, application: The Olsen sodium-bicarbonate extractant is 0.5 M sodium bicarbonate solution at pH 8.5. This extractant is most applicable to neutral to calcareous soils (Buurman et al., 1996). Solubility of Ca-phosphate in calcareous, alkaline, or neutral soils is increased because of the precipitation of Ca^{2+} as $CaCO_3$ (Soil and Plant Analysis Council, 1999). Olsen extractant correlates with Mehlich No. 3 on calcareous soils ($R^2 = 0.918$), even though the quantity of Mehlich No. 3 extractable P is considerably higher (Soil and Plant Analysis Council, 1999). While Mehlich No. 3, Bray P-1, and Olsen sodium-bicarbonate are linearly related, relationships developed between some P tests (e.g., Olsen P and Mehlich No. 3) may have limited predictive capability with increasing soil P content (Burt et al., 2002a).

Olsen sodium-bicarbonate P, measurement: A 1.0-g soil sample is shaken with 20 mL of Olsen sodium-bicarbonate extracting solution for 30 minutes. The sample is centrifuged until the solution is free of soil mineral particles and then is filtered until clear extracts are obtained. Dilute 5 mL of sample extract with 5 mL of color reagent. The absorbance of the solution is read using a spectrophotometer at 882 nm. The SSL reports Olsen sodium-bicarbonate P as mg P kg^{-1} soil.

4.3   Soil Test Analyses
4.3.7   Mehlich No. 3 Extraction
4.3.7.1   Phosphorus

Mehlich No. 3 P, application: Mehlich No. 3 was developed as a multielement soil extraction (Ca, Mg, K, Na, and P) (Mehlich, 1984). In the Mehlich No. 3 procedure, P is extracted by reaction with acetic acid and F compounds. Mehlich No. 3 is used as an index of available P in the soil. Extraction of P by Mehlich No. 3 is designed to be applicable across a wide range of soil properties, ranging in reaction from acid to basic (Mehlich, 1984). Mehlich No. 3 correlates well with Bray P-1 on acid to neutral soils ($R^2 = 0.966$) but does not correlate with Bray P-1 on calcareous soils (Soil and Plant Analysis Council, 1999). Mehlich No. 3 correlates with Olsen extractant on calcareous soils ($R^2 = 0.918$), even though the quantity of Mehlich No. 3 extractable P is considerably higher (Soil and Plant Analysis Council, 1999). The Mehlich No. 3

extractant is neutralized less by carbonate compounds in soil than the double acid (Mehlich No. 1) and the Bray P-1 extractants and is less aggressive towards apatite or other Ca-phosphate than the double acid and Bray P-2 extractants (Tran and Simard, 1993). Mehlich No. 3 can also be used to extract Ca, Mg, K, and Na in a wide range of soils and correlates well with Mehlich No. 1, Mehlich No. 2, and NH_4OAc (Soil and Plant Analysis Council, 1999). The SSL determines Mehlich No. 3 extractable P. Additionally, Mehlich No. 3 can be used to extract multielements (Elrashidi et al., 2003).

Mehlich No. 3 P, measurement: A 2.5-g soil sample is shaken with 25 mL of Mehlich No. 3 extracting solution for 5 minutes. The sample is centrifuged until the solution is free of soil mineral particles and then is filtered until clear extracts are obtained. Dilute 0.5 mL of sample extract with 13.5 mL of working solution. Absorbance of the solution is read using a spectrophotometer at 882 nm. The SSL reports Mehlich No. 3 P as mg kg^{-1} soil.

4.3 Soil Test Analyses
4.3.7 Mehlich No. 3 Extraction
4.3.7.2–22 Aluminum, Arsenic, Barium, Calcium, Cadmium, Cobalt, Chromium, Copper, Iron, Potassium, Magnesium, Manganese, Molybdenum, Sodium, Nickel, Phosphorus, Lead, Selenium, Silicon, Strontium, and Zinc

Mehlich No. 3, multielement, measurement: A 2.5-g soil sample is shaken with 25 mL of Mehlich No. 3 extracting solution for 5 minutes. The sample is centrifuged until the solution is free of soil mineral particles and then is filtered until clear extracts are obtained. Dilute 0.5 mL of sample extract with 13.5 mL of working solution. The extract is analyzed for Al, As, Ba, Ca, Cd, Co, Cr, Cu, Fe, K, Mg, Mn, Mo, Na, Ni, P, Pb, Se, Si, Sr, and Zn by inductively coupled plasma atomic emission spectrophotometer (ICP-AES). The SSL reports Mehlich No. 3 elements as mg kg^{-1} soil.

4.3 Soil Test Analyses
4.3.8 Citric Acid Soluble P_2O_5
4.3.8.1 Phosphorus

Citric acid soluble P_2O_5, application: Citric acid soluble P_2O_5 is used as a taxonomic criterion for distinguishing between mollic (<250 ppm P_2O_5) and anthropic (>250 ppm P_2O_5) epipedons (Soil Survey Staff, 2010). Additional data on anthropic epipedons from several parts of the world may permit improvements in this definition (Soil Survey Staff, 2010).

Phosphorus (citrate-soluble, Method 960.01) and phosphorus (citrate-insoluble, Method 963.03) are recognized methods in the Official Methods of Analysis by the American Association of Analytical Communities (AOAC), International (AOAC, 2000). The AOAC citrate-soluble P method considers the recovery of phosphite source materials as available phosphorus, even though the Association of American Plant Food Control Officials does not recognize phosphite as a source of available phosphorus. The procedure described herein is used by the National Academy of Agricultural Sciences (England and Wales) and is based on the method developed by Dyer (1894).

Citric acid soluble P$_2$O$_5$, measurement: A sample is checked for CaCO$_3$ equivalent. Sufficient citric acid is added to the sample to neutralize the CaCO$_3$ and to bring the solution concentration of citric acid to 1 percent. A 1:10 soil:solution is maintained for all samples. The sample is shaken for 16 hours and filtered. Ammonium molybdate and stannous chloride are added. Absorbance is read using a spectrophotometer at 660 nm. The SSL reports citric acid soluble P$_2$O$_5$ as mg kg^{-1} soil.

Citric acid soluble P$_2$O$_5$, interferences: Unreacted carbonates interfere with the extraction of P$_2$O$_5$; therefore, sufficient citric acid is added to the sample to neutralize the CaCO$_3$. However, a high citrate level in the sample may interfere with the molybdate blue test. If this occurs, the method can be modified by evaporating the extract and ashing in a muffle furnace to destroy the citric acid. Positive interferences in the analytical determination of P$_2$O$_5$ are silica and arsenic, if the sample is heated. Negative interferences in the P$_2$O$_5$ determination are arsenate, fluoride, thorium, bismuth, sulfide, thiosulfate, thiocyanate, and excess molybdate. A concentration of Fe >1000 ppm interferes with P$_2$O$_5$ determination. Refer to Snell and Snell (1949) and Metson (1956) for additional information on interferences in the citric acid extraction of P$_2$O$_5$.

4.3 Soil Test Analyses
4.3.9 New Zealand P Retention
4.3.9.1 Phosphorus

New Zealand P retention, application: Phosphorus retention of soil material is used as a taxonomic criterion for andic soil properties (Soil Survey Staff, 2010). Andisols and other soils that contain large amounts of allophane and other amorphous minerals have capacities for binding P (Gebhardt and Coleman, 1984). The factors that affect soil P retention are not well understood; however, allophane and imogolite have been considered as major materials that contribute to P retention in Andisols (Wada, 1985). Phosphate retention is also called P adsorption, sorption, or fixation.

New Zealand P retention, measurement: A 5-g soil sample is shaken in a 25-mL aliquot of a 1000 mg L^{-1} P solution for 24 hours. The mixture is centrifuged at 2000 rpm for 15 minutes. An aliquot of the supernatant is transferred to a colorimetric tube to which nitric vanadomolybdate acid reagent (NVAR) is added. Absorbance of the solution is read using a spectrophotometer at 466 nm. This absorbance correlates to the concentration of the nonadsorbed P that remains in the sample solution. The New Zealand P retention (Blakemore et al., 1987) is the initial P concentration minus the P remaining in the sample solution and is reported by the SSL as percent P retained.

4.3 Soil Test Analyses
4.3.10 Nitrogen, Agronomic and Environmental Significance

Total nitrogen, organic and inorganic: Nitrogen is ubiquitous in the environment as it is continually cycled among plants, soil organisms, soil organic matter, water, and the atmosphere (National Research Council, 1993). Nitrogen is one of the most important plant nutrients. Because it forms some of the most mobile compounds in the soil-crop system, it is commonly related to water-quality problems. Total N includes both organic and inorganic forms.

Inorganic N in soils is predominantly NO_3^- and NH_4^+. Nitrite is seldom found in detectable amounts except in neutral to alkaline soils receiving NH_4 or NH_4-producing fertilizers (Keeney and Nelson, 1982; Maynard and Kalra, 1993; Mulvaney, 1996). Ammonium ions and nitrate are of particular concern because they are very mobile forms of nitrogen and are most likely to be lost to the environment (National Research Council, 1993). All forms of nitrogen, however, are subject to transformation to ammonium ions and nitrate as part of the nitrogen cycle in agroecosystems, and all can contribute to residual nitrogen and nitrogen losses to the environment (National Research Council, 1993).

Nitrogen, measurements: The SSL determines total N by dry combustion. There is considerable diversity among laboratories in the extraction and determination of NO_3 and NH_4 (Maynard and Kalra, 1993). Nitrate is water soluble, and a number of soil solutions, including water, have been used as extractants; the most common is KCl. Refer to Maynard and Kalra (1993) and Mulvaney (1996) for a review of extractants. The SSL determines inorganic N (nitrate-nitrite) by KCl extraction, in which a flow injection automated ion analyzer is used to measure the soluble inorganic nitrate (NO_3^-). The nitrate is quantitatively reduced to nitrite by passage of the sample through a copperized cadmium column. The nitrite (reduced nitrate plus original nitrite) is then determined by diazotizing with sulfanilamide followed by coupling with N-1-naphthylethylenediamine dihydrochloride. The resulting water-soluble dye has a magenta color, which is read at 520 nm. The concept of an organic N fraction that is readily mineralized has been used to assess soil N availability in cropland, in forestland, and on waste-disposal sites (Campbell et al., 1993). Incubation-leaching techniques have been used to quantify the mineralizable pool of soil organic N. These techniques may be aerobic or anaerobic. The SSL determines mineralizable N (N as NH_3) by anaerobic incubation. In addition, the SSL determines mineralizable N after fumigation incubation.

4.3   Soil Test Analyses
4.3.11   1 *M* KCl Extraction
4.3.11.1–2   Nitrate-Nitrite

1 *M* KCl nitrate-nitrite, application: The inorganic combined N in soils is predominantly NH_4^+ and NO_3^- (Keeney and Nelson, 1982). Ammonium ions and nitrate are of particular concern because they are very mobile forms of nitrogen and are most likely to be lost to the environment (National Research Council, 1993). All forms of nitrogen, however, are subject to transformation to ammonium ions and nitrate as part of the nitrogen cycle in agroecosystems, and all can contribute to residual nitrogen and nitrogen losses to the environment (National Research Council, 1993).

1 *M* KCl nitrate-nitrite, measurement: A 2.5-g soil sample is mechanically shaken for 30 minutes in 25 mL of 1 *M* KCl solution. The sample is then filtered through Whatman No. 42 filter paper. A flow injection automated ion analyzer is used to measure the soluble inorganic nitrate (NO_3^-). Method parameters specific to 1 M KCl have been modified from the QuikChem Method 12-107-04-1-B, nitrate-nitrite in 2 M (1 M) KCl soil extracts (U.S. Environmental Protection Agency, 1983; LACHAT Instruments, 1992; U.S. Department of the Interior, Geological Survey, 1993). The nitrate is quantitatively reduced to nitrite by passage of the sample through a copperized cadmium column. The nitrite (reduced nitrate plus original nitrite) is then determined by diazotizing with sulfanilamide followed by coupling with N-1-naphthylethylenediamine dihydrochloride. The resulting water-soluble dye has a magenta color, which is read at 520 nm. Absorbance is proportional to the concentration of NO_3^- in the sample. The SSL reports data as mg N kg^{-1} soil as NO_3^- and/or NO_2^-.

4.3 Soil Test Analyses
4.3.12 Anaerobic Incubation
4.3.12.1 2 M KCl Extraction
4.3.12.1.1 Nitrogen as NH$_3$

Anaerobic incubation, 2 M KCl nitrogen as NH$_3$, application: The most satisfactory methods currently available for obtaining an index for the availability of soil N are those involving the estimation of the N formed when soil is incubated under conditions that promote mineralization of organic N by soil micro-organisms (U.S. Environmental Protection Agency, 1992). The method described herein for estimating mineralizable N is one of anaerobic incubation and is suitable for routine analysis of soils. This method involves estimation of the ammonium produced by a 1-week period of incubation of soil at 40 °C (Keeney and Bremner, 1966) under anaerobic conditions to provide an index of N availability.

Anaerobic incubation, 2 M KCl nitrogen as NH$_3$, measurement: An aliquot of air-dry homogenized soil is placed in a test tube with water. The test tube is stoppered, and the contents are incubated at 40 °C for 1 week. The contents are rinsed with 2 M KCl. A flow injection automated ion analyzer is used to measure the ammonium produced in the soil after incubation. Absorbance of the solution is read at 660 nm. The SSL reports data as mg N kg^{-1} soil as NH$_3$.

4.4 Carbonate and Gypsum

This section describes the agronomic, taxonomic, and engineering significance of carbonate and gypsum in soils. The SSL methods for the analysis of carbonate and gypsum are described. For detailed descriptions of the SSL methods which are cross-referenced by method code in the table of contents in this manual, refer to SSIR No. 42 (Soil Survey Staff, 2004), which is available online at http://soils.usda.gov/technical/lmm/. For detailed descriptions of other laboratory methods for the quantification of gypsum, refer to U.S. Salinity Laboratory Staff, 1954; Kovalenko, 1972; Sayegh et al., 1978; Lagerwerff et al., 1965; Friedel, 1978; and Nelson et al., 1978. Also refer to SSIR No. 51 (Soil Survey Staff, 2009; available online at http://www.soils.usda.gov/technical/) for detailed descriptions of field methods as used by NRCS soil survey offices.

4.4 Carbonate and Gypsum
4.4.1 Carbonate
4.4.1.1 Agronomic, Taxonomic, and Engineering Significance

Carbonate, soil-related factors: The distribution and amount of $CaCO_3$ have an important effect on fertility, erosion, available water-holding capacity, and genesis of the soil. Calcium carbonate provides a reactive surface for adsorption and precipitation reactions, e.g., phosphate, trace elements, and organic acids (Loeppert and Suarez, 1996; Amer et al., 1985; Talibudeen and Arambarri, 1964; Boischot et al., 1950). The determination of calcium carbonate ($CaCO_3$) equivalent is a taxonomic criterion (Soil Survey Staff, 2010). Carbonate content of a soil is used to define carbonatic, particle-size, and calcareous soil classes and to define calcic and petrocalcic horizons (Soil Survey Staff, 2010). The formation of calcic and petrocalcic horizons has been related to a variety of processes, including translocation and net accumulation of pedogenic carbonates from a variety of sources as well as the alteration of lithogenic (inherited) carbonate to pedogenic carbonate (soil-formed carbonate through *in situ* dissolution and reprecipitation of carbonates) (Rabenhorst et al., 1991).

Carbonate, acid-neutralizing capacity: In agriculture, the term *lime* is defined as the addition of any Ca or Ca- and Mg-containing compound that is capable of reducing soil acidity. Lime correctly refers only to calcium oxide (CaO), but the term almost universally includes such materials as calcium hydroxide, calcium carbonate, calcium-magnesium carbonate, and calcium silicate slags (Tisdale et al., 1985). As used in soil fertility, the term *CaCO₃ equivalent* (CCE) is defined as the acid-neutralizing capacity of an agricultural liming material expressed as a weight percentage of $CaCO_3$. Pure $CaCO_3$ is the standard against which other liming materials are measured, and its neutralizing value is considered to be 100 percent. The molecular constitution is the determining factor in the neutralizing value of chemically pure liming materials (Tisdale et al., 1985). Consider the following discussion and related equations (Tisdale et al., 1985).

Equation 4.4.1.1:

$$CaCO_3 + 2HCl \longrightarrow CaCl_2 + H_2O + CO_2$$

Equation 4.4.1.2:

$$MgCO_3 + 2HCl \longrightarrow MgCl_2 + H_2O + CO_2$$

In both equations, the molecular proportions are the same; i.e., one molecule of either $CaCO_3$ or $MgCO_3$ will neutralize two molecules of acid. The molecular weight of $CaCO_3$ is 100, however, whereas that of $MgCO_3$ is only 84. Therefore, 84 g of $MgCO_3$ will neutralize the same amount of acid as 100 g of $CaCO_3$. The neutralizing value or

CCE of $MgCO_3$ in relation to $CaCO_3$ (CCE = 100) is calculated in the simple proportion as follows:

Equation 4.4.1.3:

$84/100 = 100/x$

$x = 119$ percent

Therefore, on a weight basis, $MgCO_3$ will neutralize 1.19 times as much acid as the same weight of calcium carbonate. This same procedure is used to calculate the neutralizing value of other liming materials, e.g., CaO (CCE = 170); $Ca(OH)_2$ (CCE = 109); and $CaMg(CO_3)_2$ (CCE = 109).

Carbonate, agronomic and engineering significance: In general, crops grown on carbonatic soils may show signs of chlorosis, reflecting nutrient deficiencies (Fe, Zn, or Cu). Alfalfa grown on these soils may indicate symptoms of B deficiency. Carbonatic soils have also been associated with P fixation; hindrance to root ramification; high base status (near pH 8); and lower available water, especially in soils with calcic horizons. Abundant Ca in the soil has a flocculating effect on soil colloids; i.e., clays tend to be coarser. Carbonate particles have a distribution of sizes from coarse clay to gravel. These carbonates affect the soil regardless of the dominant particle size, but the clay-sized carbonate appears to have a stronger influence. Fine carbonates behave like silt and are less coherent than silicate silts and clays. Carbonatic materials are susceptible to frost disruption and to erosion by piping and jugging.

Field test, quantitative calcium carbonate equivalent determination: A field procedure was developed to measure $CaCO_3$ by using a simple volume calimeter from a 50-mL and a 20-mL plastic syringe (Holmgren, 1973). A weighed quantity of soil is placed in the 50-mL syringe, 5 mL 10 percent HCl is added in the small syringe, the syringes are connected, acid is injected into the soil, and the volume of CO_2 produced is measured in either or both syringes, depending on the amount. An array of sample weights to yield 1 mL CO_2 for 1 percent $CaCO_3$ equivalent at various temperatures and elevations is developed. An alternate procedure was developed using a constant weight of 0.33 g of soil (Holmgren, 1973). A simple hand balance is used to weigh this amount in the field, and a monograph is provided to facilitate calculations. Calcium carbonate equivalent by these procedures can be determined within 1 to 2 percent absolute error over the range 0 to 50 percent. These errors can be reduced at lower $CaCO_3$ equivalent values by increasing the sample size. The SSL makes the calimeter kits described above and distributes them to NCSS cooperators upon request.

Field or laboratory staining scheme, differentiation of major carbonate minerals: Carbonate minerals are often difficult to differentiate by typical petrographic procedures. Stains for many carbonate minerals (e.g., calcite, siderite, ankerite and ferroandolomite, dolomite, and witherite) have been described and reviewed (Rodgers, 1940; Hugi, 1945; LeRoy, 1950). Friedman's method (1959) is the most extensively tested, has been found to be the most reliable, and is widely applied (Wolf and Warne, 1960). Warne (1962) presents a more comprehensive diagrammatic scheme for the identification of the major anhydrous carbonate and two sulphate minerals (calcite, aragonite, high magnesian calcite, dolomite, ankerite and ferroandolomite, siderite, magnesite, witherite, rhodochrosite, smithsonite, strontianite, cerussite, anhydrite, and gypsum) in fragments and sections of rocks and coal using five staining solutions (initial classification with alizarin red S first in acid followed in alkaline solution, and final resolution with rhodizonic acid, benzidine, magneson, and Fegl's solution). These tests are easily applied in the field or laboratory. In addition, various methods have been developed to separate calcite from dolomite from different sources of soil carbonates (Evangelou et al., 1984; Kraimer et al., 2005; West et al., 1988).

4.4   Carbonate and Gypsum
4.4.1.2   3 N HCl Treatment
4.4.1.2.1   CO$_2$ Analysis
4.4.1.2.1.1   Carbonate
4.4.1.2.1.1.1   Calcium Carbonate

Calcium carbonate, measurement: The SSL measures the amount of carbonate in the soil by treating the samples with HCl. The evolved CO_2 is measured manometrically. The amount of carbonate is then calculated as percent $CaCO_3$. The SSL most commonly reports the $CaCO_3$ equivalent on the <2-mm basis. In some soils with hard carbonate concretions, however, carbonates are determined on both the <2-mm basis and the 2- to 20-mm basis.

Calcium carbonate, interferences: Chemical interference is the reaction by the acid with other carbonates, e.g., carbonates of Mg, Na, and K, that may be present in the soil sample. The calculated $CaCO_3$ is only a semiquantitative measurement (Nelson, 1982). Analytical interference may be caused by temperature changes within the reaction vessel. When sealing the vessel, the analyst should not hold the vessel any longer than necessary to tighten the cap. The internal pressure must be equalized with the atmosphere. After the septum has been pierced with a needle, ≈ 5 to 10 seconds is required to equalize the internal pressure of the bottle. With extensive use, the septa leak gas under pressure; therefore, they should be replaced at regular intervals. The analyst should not touch the glass of the vessel when reading the pressure.

4.4   Carbonate and Gypsum
4.4.2   Gypsum
4.4.2.1   Agronomic, Taxonomic, and Engineering Significance

Gypsum, soil-related factors: Gypsum is one of the most commonly occurring sulfate minerals. Gypsum occurs as a soil constituent and is frequently associated with gypsiferous geologic deposits, even if the deposits are deep seated or are located some distance away from the site of the gypsum-containing soils (Eswaran and Zi-Tong, 1991). Gypsum may be present as traces in the soil or may dominate the soil system. Generally, gypsum occurs in soils where very little leaching occurs (Lindsay, 1979). Gypsum is frequently found in association with halite (NaCl), the dominant soluble salt in Saliorthids, as well as with some soluble sulfate minerals, e.g., assanite, anhydrite, mirabilite, epsomite, konyaite, hexahydrite, and bloedite. All of these salts are more soluble than gypsum.

Four sources of gypsum can be distinguished (Porta and Herrero, 1990): (1) outcrops of gyprock in which weathering produces a material very rich in gypsum that remains *in situ* or is transported as a mudflow or as calcium and sulfate ions that may be transported over long distances; (2) ions dissolved in water by surface runoff, a flood, or a water table; (3) gypsum dust; and (4) gypsum formed from the oxidation of pyrites in environments rich in calcium carbonate. In all these cases, gypsum can be translocated and accumulated, depending on the characteristics of the soil moisture regime.

Gypsum formation by precipitation of deposits rich in calcium sulfate ($CaSO_4$) is typically highest at the surface layers, whereas gypsum from deposits high in gypsum is typically highest in the lower part of the soil profile; however, leaching may disrupt this sequence. Gypsum accumulations deep in the profile in areas with low annual precipitation may suggest that these soils developed under previously wetter climates. The amount of rainfall and the topographic setting influence the amount and location of gypsum in the soils. The observation that gypsum precipitates near the maximum depth of wetting in some semiarid regions (Nettleton et al., 1982) is supported by the experimental work of Krupkin (1963) with leaching soil columns.

In soils, crystals of gypsum in the solum are generally the size of coarse clay or silt, but deeper in the soil profile the crystals may be sand sized and larger (Nelson, 1982). The crystals are commonly segregated in the soil matrix. In addition, the solubility of gypsum in soils may be reduced in calcareous environments by surface coatings of $CaCO_3$ (Keren and Kauschansky, 1981; Doner and Lynn, 1989).

Pedogenic forms in which gypsum occurs vary widely as follows: as crystal clusters or as single macroscopic crystals (cm) with seemingly little relationship to surrounding components; as macroscopic or microscopic-lenticular crystals in pores, along channels, and along planar voids with no apparent orientation to the associated surfaces or with obvious orientation, especially along planar voids; as interlocking blades or microcrystalline masses; not typically in intimate mixtures with silicate clay (tendency toward segregation of components); commonly in mixtures with calcite but typically in segregated zones (Allen, 1985; Herrero and Porta, 2000).

Gypsum, carbonates, and soluble salts: Gypsum is generally too soluble to persist in soils unless SO_4^{2-} approaches 10^{-2} M. Actual qualitative and quantitative characteristics of soluble salts in a system influence the solubility of gypsum. Gypsum has variable solubility in saline solutions (20 to 50 meq L^{-1}) (U.S. Salinity Laboratory Staff, 1954). The increase in solubility in saline environments tends to impede the formation of gypsum under these conditions.

The solubility of gypsum is not pH dependent, whereas that of calcite is (Lindsay, 1979). With the presence of competing species, such as gypsum and calcite, in a soil system, calcite formation is favored at the higher pH. Gypsum is rarely present in soils if pH >8.2 (Nelson, 1982). Shifts in the dynamics of a system can result in predominating mineral species (calcite versus gypsum). That is, shifts in system inputs, e.g., soluble Ca^{2+}, SO_4^{2-}, acidity, and water content, will ultimately affect system outputs, e.g., calcium and/or gypsum. The window in which these minerals can coexist is relatively small, and the window in which they could possibly exist in equal amounts is even smaller (Fig. 4.4.2.1.1). The use of gypsum in sodic soils is an example of the equilibria between the two mineral species. If gypsum is introduced, naturally or anthropogenically, into a soil system with a pH of 8.5 or higher, the soluble Ca^{2+} is raised above calcite equilibria and leads to the precipitation of calcite. This calcite reaction releases H^+. Over time, the pH will drop to 7.5 to 8.0, which is the pH range in which gypsum and calcite can coexist. In addition, soluble Ca^{2+} is restored to approximately $10^{-2.5}$ M, keeping colloids flocculated and predominantly Ca saturated, with displaced Na^+ leached from the system (Lindsay, 1979; Cresser et al., 1993).

Natural occurrence of soluble salt minerals (including soluble carbonate minerals) in soils requires high evaporation, low rainfall, and a means of concentrating the salts. The main soluble salt minerals reported in soils are some combination of Na, Mg, K, Ca, Cl, SO_4, HCO_3, and CO_3. Salts of NO_3 and IO_3 occur less frequently, and salts of ClO_4 and CrO_3 occur rarely (Doner and Lynn, 1989). The high solubility of $MgSO_4$, $NaSO_4$, and NaCl minerals results in their formation at the furthest extent of soil-water movement. As a result, the crystalline forms of these minerals commonly occur only at the soil surface, under conditions of extreme desiccation (Doner and Lynn, 1989). The formation of Na_2CO_3 minerals can occur in several ways, one of which is through the irrigation of saline-sodic soils. Saline-sodic soils have traditionally been defined as having properties of saline and sodic soils with appreciable contents of soluble salts, ESP $\geq$15, and an EC $\geq$4 dS m^{-1} (mmhos cm^{-1}) (U.S. Salinity Laboratory Staff, 1954). Theoretically, the maximum concentration of Na_2SO_4 in the saturation extract of a medium-textured soil is about 7,000 meq L^{-1}, which is approximately 25 percent salt by weight (Nelson, 1982). The salt content of a soil seldom exceeds 2 percent; hence, all nongypsum sulfate should dissolve if the EC of the saturation extract does not exceed 40 mmhos cm^{-1} (Nelson, 1982). In the Middle East, anhydrite ($CaSO_4$) is a stable mineral species in coastal areas (Shahid et al., 2007; Sanford and Wood, 2001). The loss of waters of hydration (present in gypsum) occurs in areas of extreme salinity and at temperatures that approach and exceed 50 °C.

Gypsum, agronomic significance: Generally, soils with gypsum have an abundance of Ca. These soils are typically associated with high base status, salinity, and possibly more soluble salts than gypsum. When present in excessive amounts, gypsum controls the properties of the soil and may have adverse effects on the agricultural and engineering properties of the soil (Eswaran and Zi-Tong, 1991). Gypsum can also be beneficial. Gypsum is added as a plant nutrient (Ca, S) for improved plant growth in leached Oxisols and Ultisols. Gypsum is also used as an amendment to improve soil structure and permeability in sodic soils.

Soils with gypsum may become impervious to roots and water. Available water content and cation-exchange capacity are generally inversely proportional to gypsum content. The saturated Ca soil solution may result in the fixation of the micronutrients Mn, Zn, and Cu. Gypsum may be used in the reclamation of sodic soils, with calcium replacing the sodium on the exchange complex and sodium sulfate carried out in the drainage water.

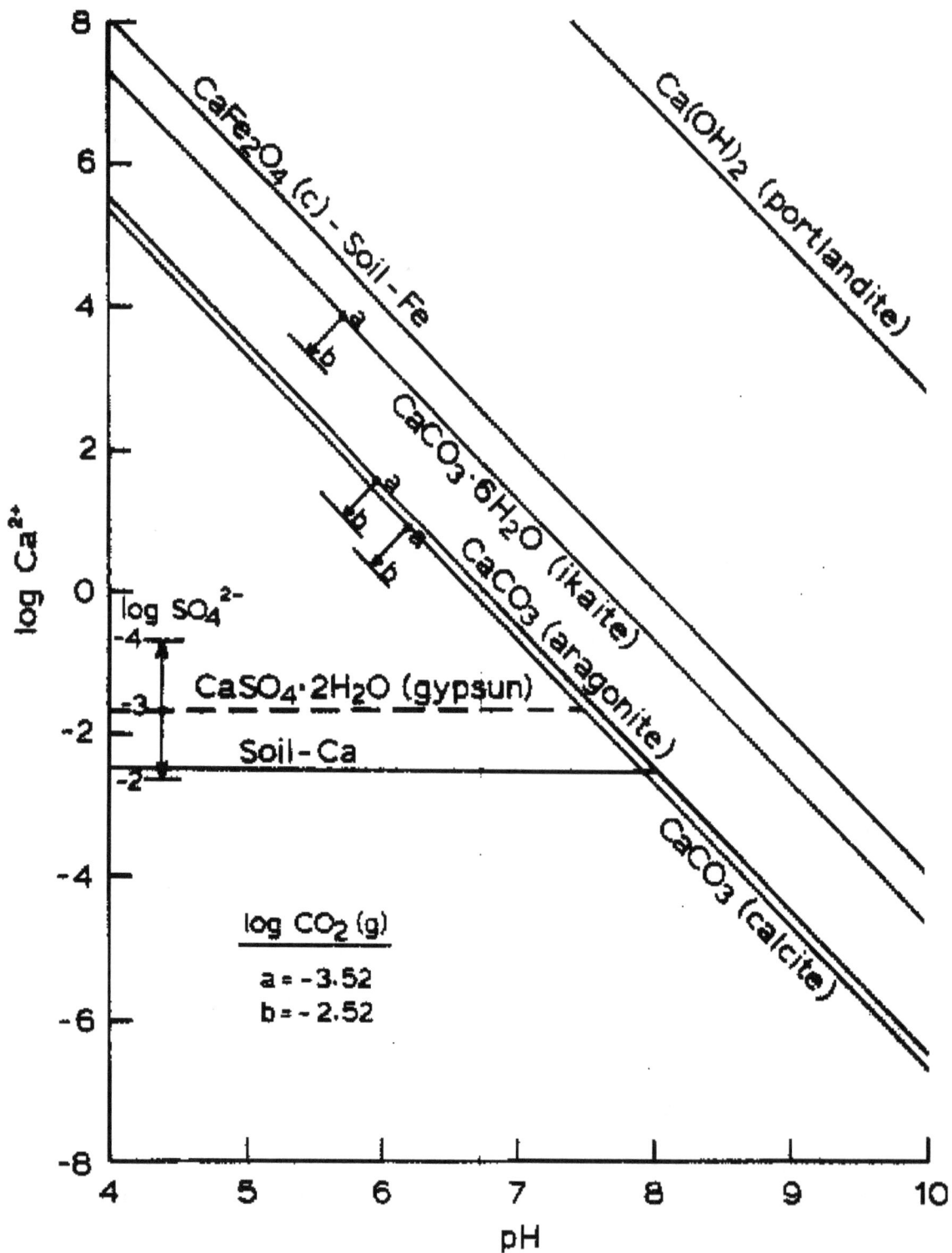

Figure 4.4.2.1.1.—The solubility of various calcium minerals in soils. After Lindsay (1979).

Gypsum, engineering significance: Application of irrigation water on farmland in arid and semiarid areas poses engineering challenges for soils with gypsum (Elrashidi et al., 2007). Soil subsidence through solution and removal of gypsum can crack building foundations, break irrigation canals, and make roads uneven. Failure can be a problem in soils with as little as 1.5 percent gypsum (Nelson, 1982). Typically, soils with gypsum have a number of other water-soluble minerals that are associated with gypsum. Elrashidi et al. (2007) proposed that subsidence should not be solely estimated by gypsum content but also by content of other water-soluble minerals using the Equivalent Gypsum Content (EGC). The EGC is defined as the quantity of both gypsum and other water-soluble minerals and expressed as gypsum percentage (by weight) in soils. Refer to Elrashidi et al. (2007) for the application of EGC to estimate soil subsidence in soils with gypsum. Corrosion of concrete also is associated with soil gypsum. Gypsum content, soil resistivity, and extractable acidity individually or in combination provide a basis for estimating potential corrosivity of soils (USDA/SCS, 1971).

Gypsum, taxonomic significance: The content of gypsum in a soil is a criterion for gypsic and petrogypsic horizons and for mineralogical class at the family level (Soil Survey Staff, 2010). Gypsic horizons typically form abrupt boundaries with overlying cambic or calcic horizons (Allen, 1985). Some characteristics of gypsic horizons are as follows: ≥ 15 cm thick; not cemented or indurated (as in petrogypsic); ≥ 5 percent gypsum and ≥ 1 percent (by volume) secondary visible gypsum; and product thickness (cm) multiplied by gypsum percent by ≥ 150. In a study of Reg soils in southern Israel and Sinai (Dan et al., 1982), gypsic and petrogypsic horizons were associated with older and higher (more stable) geomorphic surfaces that developed over a long period of weathering under extreme arid conditions.

Field test, qualitative, soluble sulfate, calcium, and magnesium: Quantification of gypsum content is important for classification and use and management of soils with gypsum (Soil Survey Staff, 2010). A qualitative field test to identify soluble sulfate in soil material was developed by the SSL for use by NRCS soil survey offices and is available upon request from the NSSC. This test is used conjunctively with other field tests (soluble calcium and magnesium) to identify gypsum.

During a test for sulfate, a soil sample is tested for effervescence with 1 N hydrochloric acid (HCl). Depending on test results, a variable quantity of 0.1 N HCl is added to the sample, followed by barium chromate and a color indicator solution. Development and persistence of a lavender/violet color within 60 seconds represent the presence of sulfate. During a test for calcium and/or magnesium, a soil sample is extracted with water and a portion of the mixture is withdrawn. Half of the mixture is ejected into one test tube, and the other half is ejected into another test tube. Saturated ammonium oxalate solution is added to one test tube. If a cloudy white precipitate forms, calcium is indicated. The amount of precipitate is related to the calcium level. To the other test tube, 0.5 N sodium hydroxide (NaOH) and titan yellow indicator is added. A yellow or brownish yellow color indicates no magnesium. A reddish color indicates magnesium. Red precipitate indicates a high magnesium level.

Other tests, gypsum: Gypsum can also be quantified by low temperature weight loss procedures based on the waters of hydration (Artieda et al., 2006; Lebron et al., 2009; Karathanasis and Harris, 1994).

4.4 Carbonate and Gypsum
4.4.2.2 Aqueous Extraction
4.4.2.2.1 Precipitation in Acetone
4.4.2.2.1.1 Electrical Conductivity
4.4.2.2.1.1.1 Gypsum, Qualitative and Quantitative

Gypsum, measurement: If the electrical conductivity of a soil sample is >0.50 dS cm^{-1}, gypsum content is determined. Additionally, normal amounts of organic matter and a high air-dry/oven-dry ratio (AD/OD) may trigger the determination of gypsum by the SSL. A soil sample is mixed with water to dissolve gypsum. Acetone is added to a portion of the clear extract to precipitate the dissolved gypsum. After the process of centrifuging, the gypsum is redissolved in water. The electrical conductivity (EC) of the solution is read. The EC reading is used to estimate the gypsum content in meq 100 g^{-1}.

Gypsum content (meq 100 g^{-1}) is converted to percent gypsum (uncorrected). The percent gypsum (uncorrected) is used to calculate percent gypsum (corrected). The percent gypsum (corrected) is used to correct the AD/OD. The AD/OD and corrected AD/OD are determined. The corrected AD/OD uses the correction for the crystal water of gypsum. Gypsum content on a <2-mm basis is reported.

Gypsum content may also be determined on the 2- to 20-mm fraction. The gypsum determined on the 2- to 20-mm fraction and the gypsum determined on the fine-earth fraction are combined and converted to a <20-mm soil basis.

Calculate % Gypsum$_{uc}$ (gypsum uncorrected) by using Table 4.4.2.1 to convert EC readings (mmhos/cm) to gypsum content (meq/100 g) and proceeding with the following equation.

Equation 4.4.2.1:

% Gypsum$_{uc}$ = [Gypsum x Water x 0.08609 x AD/OD]/[Sample Weight (g) x 5]

where
% Gypsum$_{uc}$ = % Gypsum in <2-mm fraction or 2- to 20-mm fraction
Gypsum = Gypsum (meq L^{-1}). Refer to Table 4.4.2.1.
Water = Volume RO water (100 mL) to dissolve gypsum
0.08609 = Conversion factor (gypsum % = meq 100 g^{-1} x 0.08609)
AD/OD = Air-dry/oven-dry ratio
5 = Filtrate (5 mL)

Table 4.4.2.1 converts EC (mmhos cm^{-1}) to gypsum (meq L^{-1}) for the above calculations. Enter Table 4.4.2.1 using both the x and y axes for the EC reading to determine gypsum content (meq L^{-1}).

Equation 4.4.2.2:

As an alternative to using Table 4.4.2.1, calculate % Gypsum$_{uc}$ from the following equation:

Result = (Exp (2.420384 + 1.1579713 x Log (EC – blank)) x Water x 0.08609 x ADOD)/(Sample Weight x 5)

Equation 4.4.2.3:

The following equation for calculation of % Gypsum$_c$ (gypsum corrected) assumes that the crystal-water content of gypsum is 19.42 percent (Nelson et al., 1978) as opposed to the theoretical water content (20.21 percent).

% Gypsum$_c$ = [% Gypsum$_{uc}$]/[1 + 0.001942 x % Gypsum$_{uc}$]

Use the % Gypsum$_{uc}$ to recalculate the AD/OD. The corrected AD/OD uses the correction for the crystal-water content of gypsum.

Equation 4.4.2.4:

Calculate gypsum on <20-mm basis as follows:

(%) Gypsum = A x B + [C x (1 - B)]

where
A = Gypsum (%) in <2-mm fraction
B = Weight of the <20-mm fraction minus the 20- to 2-mm fraction divided by the weight of the <20-mm fraction
C = Gypsum (%) in the 20- to 2-mm fraction

Table 4.4.2.1   Convert EC reading (mmhos cm^{-1}) to gypsum content (meq L^{-1})

EC	0.00	0.01	0.02	0.03	0.04	0.05	0.06	0.07	0.08	0.09
0.0						0.40				
0.1	0.80	0.89	0.98	1.10	1.22	1.31	1.40	1.50	1.60	1.70
0.2	1.80	1.90	2.00	2.10	2.20	2.30	2.40	2.50	2.60	2.70
0.3	2.80	2.90	3.00	3.10	3.20	3.30	3.40	3.50	3.60	3.72
0.4	3.85	3.98	4.10	4.22	4.35	4.48	4.60	4.70	4.80	4.90
0.5	5.00	5.12	5.25	5.38	5.50	5.62	5.75	5.88	6.00	6.12
0.6	6.25	6.35	6.45	6.58	6.70	6.82	6.95	7.05	7.15	7.28
0.7	7.40	7.52	7.65	7.78	7.90	8.04	8.18	8.32	8.45	8.58
0.8	8.70	8.82	8.95	9.05	9.15	9.28	9.40	9.55	9.70	9.85
0.9	10.00	10.12	10.25	10.38	10.50	10.62	10.75	10.88	11.00	11.15
1.0	11.30									

Gypsum, interferences: Loss of the precipitated gypsum is the most significant potential error. Care in handling the precipitated gypsum is required. Incomplete dissolution of gypsum also is possible. In soils with large gypsum crystals, use fine-ground samples to reduce sampling errors. When present in sufficiently high concentrations, the sulfates of Na and K are also precipitated by acetone. The concentration limits for sulfates of Na and K are 50 and 10 meq L^{-1}, respectively.

Gypsum, fine grinding: Gypsum ($CaSO_4 \cdot 2H_2O$) solubility in pure water is 2.6 g L^{-1} (25 °C, 0.01 MPa pressure). Soils typically contain gypsum if concentrations of Ca and SO_4 in the saturated paste extract exceed 20 meq L^{-1}. The dissolution rate of gypsum in water increases as the fineness of particles increases (Nelson, 1982). Fine grinding may eliminate carbonate surface coats, which can reduce the solubility of gypsum. In some testing at the SSL, fine grinding of some Arizona soils significantly increased (8 to 19 percent) the gypsum recovery in soils with large amounts of gypsum (e.g., 43 to 50 percent) in association with $CaCO_3$, but fine grinding had little or no effect on the recovery in samples with low amounts of gypsum (e.g., 2 percent). In the past, the SSL did not fine grind soils for gypsum analysis. Currently, this sample preparation is the standard method for samples to be analyzed for gypsum content.

The SSL determines a number of samples for gypsum, primarily from the Western United States. In general, samples with high amounts of gypsum (e.g., 91 percent) pose some complexity as related to appropriate sample size and resulting reproducibility of data. In an effort to verify results, the SSL conducted some methods testing over a period of time. There was reasonably good agreement between the comparative data for a soil gypsum standard, with 7.82 to 7.84 percent by the U.S. Salinity Laboratory multiple dilutions method versus 7.08 to 7.27 percent by the SSL method precipitation in acetone, EC reading. Additionally, soil samples high in gypsum from New Mexico and Arizona were further analyzed by the SSL thermal analysis method. These comparative results also showed reasonably good agreement, e.g., 54 versus 56 percent, 91 versus 93 percent, and 12 versus 13 percent by the standard SSL method for gypsum versus thermal analysis, respectively. Other chemical tests for determination of gypsum are discussed in Nelson (1982).

4.5   Electrical Conductivity and Soluble Salts

This section describes the SSL methods for the preparation of the saturated paste and salinity measurements on extracts derived from this paste (e.g., electrical conductivity and soluble salt composition). Ratios, estimates, and calculations associated with these analyses are described, e.g., exchangeable sodium percentage (ESP) and sodium adsorption ratio (SAR), calculated with and without the saturation paste extract. This section also describes a 1:5 aqueous extraction and provides general information on salt-affected soils, their geographical distribution, soluble salt concentrations, and their effects upon plants (osmotic stress, specific ion effects, and nutritional imbalances). The soil processes of salinization, sodication, and alkalinization also are described, and references to case studies/datasets are presented as evidentiary examples of the actions/practices that have promoted or diminished these processes. In addition, major developments in the research, diagnosis, improvement, and management of salt-affected soils as well as developments in irrigation methods and practices are discussed. For detailed descriptions of the SSL methods which are cross-referenced by method code in the table of contents in this manual, refer to SSIR No. 42 (Soil Survey Staff, 2004), which is available online at http://soils.usda.gov/technical/lmm/. Refer to SSIR No. 51 (Soil Survey Staff, 2009; available online at http://www.soils.usda.gov/technical/) for detailed descriptions of field methods as used in NRCS soil survey offices.

4.5   Electrical Conductivity and Soluble Salts
4.5.1   Salt-Affected Soils

Salt-affected soils, occurrence: *Salt-affected soils*, i.e., excessive amounts of soluble salts and/or exchangeable Na, are common in, though not restricted to, arid and semiarid regions; they also occur in humid areas (Indorante, 2002). The more pronounced salinity problems in arid and semiarid regions have been attributed to annual rainfall amounts that are insufficient for flushing accumulated salts from the crop root zone (Bresler et al., 1982). The main sources of salts in these regions are rainfall, mineral weathering, "fossil" salts, and various surface waters and ground waters that redistribute accumulated salts, often as the result of human activities (Bresler et al., 1982).

Salt-affected soils, soluble salt concentrations: Salt-affected soils are usually described and characterized in terms of the soluble salt concentrations, i.e., major dissolved inorganic solutes (Rhoades, 1982b). Soluble cations and anions have been loosely defined as those removed by water, whereas exchangeable cations are those removed by neutral salt solutions. The aqueous phase outside the electrical double layer of soil colloids is the bulk solution containing the soluble salts. Analyses of water leached in humid regions indicate that the relative cation concentration in the bulk solution is typically $Ca^{2+} > Mg^{2+} > K^+ > Na^+$ (Bohn et al., 1979). There may also be low concentrations of NH_4^+ ions as the result of ammonia fertilization. Sodium may predominate in drainage waters of many irrigated soils. Total salt concentration in the bulk solution of well drained soils in humid and temperate regions is generally in the range 0.001 to 0.01 M (Bohn et al., 1979). In irrigated and arid soils, the soluble salt concentration is typically higher and may be 5 to 10 times as high as in the applied irrigation water because of the salts that remain after the evapotranspiration of water (Bohn et al., 1979). In those areas in which salts (particularly Na salts) accumulate because of improper irrigation, a high water table, or seawater intrusion, the salt concentrations may be as high as 0.1 to 0.5 M (Bohn et al., 1979).

In general, the anion concentration in the aqueous phase of nonsaline soils is less than the cation concentration because most of the negative charge in these soils is from soil colloidal particles (Bohn et al., 1979; Sposito, 1989). The difference between the sum of cation and anion charge narrows as soil salinity increases. Some major anions in soils include NO_3^-, SO_4^{2-}, Cl^-, and HCO_3^-. The relative amounts of these anions vary with fertilizer and management practices, mineralogy, microbial and higher plant activity, saltwater encroachment, irrigation water quality, and atmospheric fallout. In humid regions, the sum of anions rarely exceeds 0.01 M in soil solution (Bohn et al., 1979). In saline soils, the anion concentrations are generally higher because of precipitation of cations and anions as soluble salt with a typical distribution of $Cl^- >$ $SO_4^{2-} > HCO_3^- > NO_3^-$ or $(HCO_3^- + CO_3^{2-}) > Cl^- > SO_4^{2-} > NO_3$ in high-pH sodic soils. The major anions are retained weakly and therefore are mobile in soils if solubilized and leached.

Salt-affected soils, definitions: Traditionally, the classification of salt-affected soils has been based on the soluble salt concentrations in extracted soil solutions and on the exchangeable sodium percentage (ESP) in the associated soil (Bohn et al., 1979). Historically, *saline* soils have been defined as having salt contents >0.1 percent, EC >4 dS m^{-1}, ESP <15 percent, and pH <8.5; *sodic (alkali)* soils typically have an ESP >15 percent, low salt contents, EC <4 dS m^{-1}, and pH 8.5 to 10; and *saline-sodic (saline-alkali)* soils typically have properties of both saline and sodic soils and appreciable contents of soluble salts, ESP >15 percent, and EC >4 dS m^{-1} (U.S. Salinity Laboratory Staff, 1954). The terms *alkali, saline-alkali,* and *saline-sodic* are no longer used in Soil Science Society of America (SSSA) publications (SSSA, 2010). The term *saline* soil as defined by the SSSA (2010) is a nonsodic soil containing sufficient soluble salt to adversely affect the growth of most crop plants with a lower limit saturation extract EC (EC$_e$) conventionally set at 4 dS m^{-1} at 25 °C. Sensitive plants are affected at half this salinity and highly tolerant ones at about twice this salinity. The term *sodic* soil as defined by the SSSA (2010) is a nonsaline soil containing sufficient exchangeable sodium to adversely affect crop production and soil structure under most conditions of soil and plant type. In *Keys to Soil Taxonomy*, the ESP ($\geq$15) and the sodium adsorption ratio (SAR) ($\geq$13) are criteria for natric (*sodium-affected*) horizons, and an EC $\geq$30 ds m^{-1} of a saturated paste is used for salic (salt-affected) horizons (Soil Survey Staff, 2010).

Salt-affected soils, plant effects: Salt composition and distribution in the soil profile affect the plant in various ways, such as osmotic stress, specific ion effects, and nutritional imbalances. Soil texture and plant species also are factors in this plant response to saline soils. The primary effect of soluble salts on plants is osmotic. A high level of salts prevents plants from obtaining water for plant growth. The plant root contains a semipermeable membrane that preferentially permits water to pass but rejects most of the salt. Under increasingly saline conditions, water becomes more difficult to extract osmotically. Plants growing in saline conditions can modify their internal osmotic concentrations by organic acid production or salt uptake, i.e., osmotic adjustment. The term *salt tolerance* can be expressed in terms of *saturation-extract EC levels*. The various saturation-extract EC levels can be associated with relative plant growth, i.e., 10, 25, and 50 percent decreases in yields (U.S. Salinity Laboratory Staff, 1954; Bernstein, 1964; USDA/SCS and U.S. Salinity Laboratory Staff, 1993). In addition to the general osmotic effects of salts, many plants are sensitive to specific ions in irrigation waters or soil solutions. This sensitivity has been termed *specific ion effects*, e.g., Na and B (boron). In many cases, controlling boron toxicity is more difficult than controlling salinity.

Salt-affected soils, specific ion effects: The exchangeable suite of saline soils is highly variable, depending on the amount and kind of salts (Foth and Ellis, 1988). As a soluble salt, Na can have a preferential accumulation in the soil over time. The structure of sodic soils tends to disintegrate because Na is weakly adsorbed and is inefficient in neutralizing the negative charge (Foth and Ellis, 1988). The resulting dispersion of clays and humus reduces soil permeability. Significant amounts of Na in soils severely retard the growth of many plants. Sodic soils that are dominated by active Na exert a detrimental effect on plants in the following ways: (1) caustic influence of the high alkalinity induced by the sodium carbonate and bicarbonate; (2) toxicity of the bicarbonate and other anions; and (3) adverse effects of the active sodium ions on plant metabolism and nutrition (Brady, 1974). Direct sensitivity by plants (specific ion effects) to exchangeable or soluble Na is more apparent at low salt levels and therefore is more difficult to differentiate from the effects of Na on soil permeability.

An important measurement of water quality is the relative amounts of Na or sodicity of the water. Related to the Na hazard of irrigation waters is the bicarbonate concentration. The precipitation of $CaCO_3$ from these waters generally reduces the concentration of dissolved Ca, increases the SAR, and increases the exchangeable Na level in the soil (Bohn et al., 1979). Potential bicarbonate injury or toxicity to plants in saline environments results more from nutritional deficiencies or micronutrient imbalances than from the direct effect of bicarbonate ions or "bicarbonate toxicities"; e.g., in high-bicarbonate, high-pH soils, there are common occurrences of reduced plant-available Fe (Bohn et al., 1979).

Chloride toxicity in saline environments is similar to Na toxicity. Direct sensitivity to soluble chlorine is more apparent at low salt levels. Excessive accumulations in plant tissue near the end of the transpiration stream lead to necrosis, leaf tip and margin burn, and eventual death. Some plants are able to screen out such ions through their root membranes, i.e., selective uptake.

4.5   Electrical Conductivity and Soluble Salts
4.5.2   Salinization, Sodication, and Alkalinization, Processes, Case Studies, and Major Developments

Salinization, sodication, and alkalinization, processes: *Salinization* is a process of accumulation of salts in soils. *Sodication* is a process whereby salt additions to a soil increase the concentration of Na relative to Ca and Mg, resulting in increased exchangeable Na. *Alkalinization* is a process that can occur when a soil solution contains carbonate and bicarbonate in excess of Ca and Mg, the highly soluble Na salts of these anions hydrolyze, and the soil pH typically rises above 8.5 (Derici, 2002).

Salinization, sodication, and alkalinization, agricultural practices: Salinity is a common occurrence in semiarid and arid regions, i.e., where evapotranspiration exceeds rainfall, resulting in accumulation of salts in the root zone (Derici, 2002). The salt problems of greatest importance to agriculture occur when previously productive soils become salinized as a result of agricultural activities, e.g., improper or excessive use of irrigation (high in salts and/or Na relative to Ca and Mg); use of low-quality irrigation water (even good-quality irrigation water that has some salts); inadequate or no drainage system; land conversion from perennial species to annual crops with lower transpiration rates, resulting in the raising of water tables; and actions promoting the formation of saline seeps (Derici, 2002; Rhoades, 2002). Some practices that have been employed in the U.S. to diminish or improve the deleterious effects of salinity, alkalinity, and sodicity include use of high-quality irrigation water, efficient irrigation schemes, site modification using settling basins and alteration in canopy configuration, adequate drainage systems, gypsum application for Na-affected soils, and reduction in irrigated acreage (Rhoades, 2002).

Salinization, sodication, and alkalinization, case studies: In the United States, the area under irrigation doubled between 1949 and 1979 to 21 million hectares; by 1987, this area had more than doubled again (Rhoades, 1990a). In 1989, the Western Region had 15.2 million hectares of irrigated land, approximately 81 percent of the total irrigated land in the United States (U.S. Department of Commerce), using approximately 92 percent of all irrigation water. Of the 33.5 million hectares of arable land in Canada, only 842,000 hectares is irrigated, mostly in Alberta, representing an increase of 19 percent since 1991 (Canadian National Committee on Irrigation and Drainage, 1999).

Artificial elevation of water tables by extensive irrigation, with inadequate or improper drainage, has resulted in salinization of some soils in the Western U.S. (Kapur and Akca, 2002). While salt-affected soils occur naturally in arid and semiarid regions of the West, the problems associated with these soils can be exacerbated under irrigation systems using poor-quality water (e.g., high in Na) and with inadequate delivery and/or drainage systems, resulting in Na accumulation with a dispersing effect on clay and organic matter. This effect can lead to disaggregation, crust formations, and decreased permeability. Approximately 10 million hectares in the West is affected by salinity-related yield reductions, along with very high costs in the Colorado River Basin and the San Joaquin Valley (Barrow, 1994; Kapur and Akca, 2002). Law et al. (1972) estimated that 20 percent of the total water delivered for irrigation in the United States was lost to seepage from conveyance and irrigation canals. These seepage waters typically percolate the underlying strata, dissolving additional salts in the process; flow to lands or waters at lower elevations; and add to the problem of salt-loading associated with on-farm irrigation (Rhoades, 2002).

While primary soil salinity has steadily increased in the prairies of Canada over time because of the increasing ground water levels, there are also major problems with secondary salinity. An estimated 2.2 million hectares of salt-affected land is in Alberta, Saskatchewan, and parts of Manitoba (Kapur and Akca, 2002). Canada has approximately 7 million hectares of sodic soils, and the U.S. has about 2.6 million hectares of sodic soils (Rengasamy, 2002). Farming practices on these soils are commonly under dryland conditions. Sodicity is a latent problem in many salt-affected soils where degradation effects on soil properties are evident only when salts are leached below a threshold level (Rengasamy, 2002).

The Central Valley of California offers one of the best examples of waterlogging and soil salinization associated with irrigation practices. Drainage problems began in this area soon after irrigation began in the 1870s. By 1900, extensive areas had been abandoned because of alkalinity and salinity problems (Nelson and Johnston, 1984). Irrigation continued to expand with some water table control provided by deep wells and open drains as subsurface drainage expanded in the 1950s (Jensen et al., 1990). As a result of the drainage problems in this area, construction began in 1968 on the San Luis Drain. Work was completed to the Kesterson Reservoir, which started receiving irrigation runoff water in 1973. The reservoir received subsurface drainage by 1978, and tile drain water was the sole source by 1981 (Letey et al., 1986). All drainage into the reservoir was terminated in 1985, however, due to bird deformities resulting from selenium in the drainage water (Letey et al., 1986). The Westlands Water District was organized in 1952 to address a lack of adequate drainage water disposal capacity, but by 1983 more than 10,000 hectares had a perched saline water table within 1.5 m of the surface; 47,500 hectares had one between depths of 1.5 and 3.0 m; and 36,800 hectares had one between depths of 3 and 6 m (Letey et al., 1986). Nonuniform and excessive irrigation applications are the main sources of drainage water (Jensen et al., 1990). Because of the closure of the drain to the Kesterson Reservoir, the Westlands farmers were required to greatly improve irrigation management to minimize the quantity of effluent unsuitable for reuse (Jensen et al., 1990).

The center-pivot irrigation system was developed as an alternative to the conventional irrigation systems that were causing the salinity problems. The introduction of this system has caused a decline (about 30 to 50 m) in water table levels in areas north of Lubbock, Texas, reducing the thickness of the Ogallala Aquifer by 50 percent and resulting in subsidence in some areas (Kapur and Akca, 2002). The Ogallala Aquifer, also known as the High Plains Aquifer, is a shallow water table aquifer beneath the Great Plains in the United States. The Ogallala is one of the world's largest aquifers (174,000 mi², or 450,000 km²). It occurs in parts of eight states (South Dakota, Nebraska, Wyoming, Colorado, Kansas, Oklahoma, New Mexico, and Texas) (Glantz, 1989). The use of the aquifer began at the turn of the century and has increased steadily since 1945 (Glantz, 1989). Currently, the withdrawal of this ground water has surpassed the aquifer's rate of natural recharge. In some places the aquifer has been exhausted as a source of irrigation water, whereas other places are less vulnerable because of favorable saturated thicknesses and recharge rates (Glantz, 1989). Irrigated agriculture in western Kansas is expected to decline between 40 and 85 percent by the year 2020 because of a drop in water tables and higher energy costs (Kansas State University, 2006). In the past 70 years, major studies of the Great Plains and the Ogallala Aquifer include the Great Plains Report (Cooke, 1936), the Travelers Insurance studies of the Great Plains (1958-59), and the six-state High Plains study (Banks, 1982). Recent research on salinity shows the influence of rainfall on infiltration of irrigated soils (Suarez et al., 2008).

Erosion also has been related to irrigation practices. Yield reductions in southern Idaho resulting from erosion on undulating irrigated lands were reported by Carter et al. (1985) and Carter (1986) as follows: 75 percent of the fields had a whitish subsoil exposed at the upper ends caused by erosion after 80 seasons of furrow irrigation; some soils had lost all of their topsoil and some of their subsoil near the upper end; most fields had lost about 20 cm of topsoil; topsoil thickness had increased in downslope areas by 60 to 150 cm; and crop yields were estimated to be at 75 percent of what they could have been without erosion.

Soil crusting has been related to irrigation practices. This condition can impact the efficiency of the irrigation system. Crusting has become more of a concern with increased use of center-pivot sprinkling irrigation in the Coastal Plain area (Miller and Radcliffe, 1992). The water application rates of this high energy impact irrigation system are often limited by low infiltration rates due to crust formation. If crusting were prevented, such systems could be made more efficient and higher sprinkling rates with less runoff would be possible. A number of irrigation practices (Singer and Warrington,

1992; Rhoades, 2002) have been used to reduce or ameliorate crusting problems in the West. These practices include using surface mulches to intercept sprinkler drop impact energy; applying chemical amendments (e.g., gypsum and phosphogypsum) to increase electrical conductivity of irrigation water and decrease Na content; using high-quality water and efficient irrigation schemes; using settling basins and alteration in canopy configuration for site modification; and installing adequate irrigation drainage systems.

Major developments in knowledge, science, and technology in the research, diagnosis, improvement, and management of salt-affected soils: In the last half of the last century, extensive work was done in the research, diagnosis, improvement, and management of salt-affected soils in the U.S. and globally, especially in relation to irrigated agricultural lands (U.S. Salinity Laboratory Staff, 1954; Miles, 1977; Moore and Hefner, 1977; Ayers and Westcot, 1985; Hoffman et al., 1990; Rhoades, 1990a, 1990b, 1998, 1999; Tanji, 1990; Rhoades et al., 1992; Umali, 1993; Sinclair, 1994; Rengasamy, 1997; Grattan and Grieve, 1999). Hoffman et al. (1990) reported that high crop productivity was attainable with salt-affected water and soils if management practices were appropriate and environmental conditions were favorable. In the Arkansas River valley of Colorado, sorghum, wheat, and alfalfa were irrigated with water containing 1500 to 5000 mg L^{-1} total dissolved salts (TDS) (Miles, 1977). Moore and Hefner (1977) reported that water averaging 2500 mg L^{-1} had been used for decades in the Pecos River valley of Texas.

Better understanding and employment of new practices to diminish the soil processes of salinization, sodication, and alkalinization on U.S. agricultural lands are evidenced in a paper by Van Doren (1986), one of a series of Golden Anniversary Papers of the Soil Science Society of America (SSSA). Van Doren (1986) reported that since the formation of the SSSA in 1936, subject matter has shown temporal and geographical trends. Related to scientific research and development in the semiarid regions of the U.S., there has been steady increased interest by scientists in the following topical areas: (1) management of saline soils and crop growth and yield on these soils through the maintenance of good soil structure or reclamation efforts at a local level by a variety of regimes (e.g., leaching, water table management, drainage, or irrigation water quality); (2) irrigation (e.g., high frequency or trickle irrigation; maximizing crop use of applied water by limiting excessive evaporation, runoff, or deep drainage losses; and maintenance of good soil structure, especially infiltration under various types of irrigation); and (3) dryland water conservation for crop production (e.g., residue-tillage-weed control management and supplying water via microwatersheds or storing water *in situ* during fallow) (Van Doren, 1986).

Major developments in knowledge, science, and technology in irrigation methods and practices: Irrigation is the process of applying water to soil, primarily to meet the water needs of growing plants. Water from rivers, reservoirs, lakes, or aquifers is pumped or flows by gravity through pipes, canals, ditches, or even natural streams. Irrigation enhances the magnitude, quality, and reliability of crop production. About 14 percent of the 2.13 million farms and ranches in the United States were irrigated in 2002 (USDA, 2004). Although irrigated land is only 18 percent, or 22 million hectares (55 million acres), of the total harvested cropland, farms with irrigated land receive 60 percent of the total market value of crops in the United States (USDA, 2004). Market value of crops on farms where all cropland was irrigated was $3,480/hectare

($1,410/acre) in 2002 compared to $420/hectare ($170/acre) for nonirrigated farms. Irrigation not only increases crop value but also increases efficiency of water use (Howell, 2001) by increasing the mass of crop produced per volume of water.

Various irrigation methods have been developed over time to meet the irrigation needs of certain crops in specific areas. The three main methods of irrigation are surface, sprinkler, and drip/micro. Water flows over the soil by gravity for surface irrigation. Sprinkler irrigation applies water to soil by sprinkling or spraying water droplets from fixed or moving systems. Microirrigation applies frequent, small applications by dripping, bubbling, or spraying. A fourth (and minor) irrigation method is subirrigation, in which the water table is raised to or held near the plant root zone using ditches or subsurface drains to supply the water.

Traditionally, surface irrigation has been the most common method in the United States. Surface irrigation is often considered labor intensive, while sprinkler irrigation is considered capital intensive. The 2003 Farm and Ranch Irrigation Survey showed that sprinkler irrigation exceeded surface irrigation for the first time (USDA, 2005). Sprinklers were used on slightly more than 50 percent of irrigated land, while only 43 percent was irrigated by gravity. The area irrigated by a sprinkler system increased by 8 percent from 1998 to 2003, and the area irrigated by gravity decreased by 15 percent. The increase in sprinkler-irrigated acreage was due almost entirely to an increase in the acreage of center-pivot irrigation, which made up 40 percent of the total irrigated land in 2003. Between 1994 and 2003, the sprinkler-irrigated area steadily increased about 128,000 hectares (315,000 acres) per year and the center-pivot irrigated area increased 148,000 hectares (365,000 acres) per year (USDA, 1995, 2005).

Center pivots are popular because they can uniformly apply water to large fields, typically 50 to 60 hectares (125 to 150 acres). Furthermore, once a circular field has been irrigated, the center pivot is in position to start the next irrigation. Advances in technology also allow center pivots to be monitored and controlled from remote locations. In the future, center-pivot systems may be used to monitor crop and soil conditions in addition to applying water.

A major challenge for irrigated agricultural land is the increasing competition for water, primarily due to increases in population (National Research Council, 1996). Irrigation water cost, or value, will increase with increasing competition for water supplies (Council for Agricultural Science and Technology, 1996). Total water withdrawals in the United States increased steadily from 1950 to 1980. While population continued to increase, water withdrawals were essentially constant from 1985 to 2000 (Hutson et al., 2004). One reason for this trend is that the average irrigation application rate has declined from 1080 ha-mm per hectare (3.55 acre-feet per acre) in 1950 to 756 ha-mm per hectare (2.48 acre-feet per acre) in 2000.

Most of the cropland in the U.S. and globally is classified as dryland; therefore, another major challenge is to improve productivity and stability of production in rainfed areas by searching for technologies and adaptations requiring low external inputs and minimizing crop failure (Rao and Ryan, 2004). Some of the knowledge gained in irrigated agriculture may be used in this endeavor as the search continues for more efficient irrigation systems in a time of dropping water tables and higher energy costs.

4.5   Electrical Conductivity and Soluble Salts
4.5.3   Salinity Measurements and Data Relationships

Saturated paste, definition: *Soil salinity* is conventionally defined and measured on aqueous extracts of saturated soil pastes (U.S. Salinity Laboratory Staff, 1954). The *saturated paste* is operationally defined so that it may be reproduced by a trained analyst using limited equipment. The saturated paste extract derived from the saturated paste is an important aqueous solution because many soil properties have been related to the composition of the saturation extract, e.g., soluble salt composition and electrical conductivity. These soil properties or characteristics are related in turn to the plant response to salinity (U.S. Salinity Laboratory Staff, 1954).

Saturated paste, soil:water ratio: The measurable absolute and relative amounts of various solutes are influenced by the soil:water ratio at which the soil solution extract is made. Therefore, this ratio is standardized to obtain results that can be applied and interpreted universally. This soil:water ratio is used because it is the lowest reproducible ratio at which the extract for analysis can be readily removed from the soil with common laboratory equipment, i.e., pressure or vacuum, and because this soil:water ratio is often related in a predictable manner to field soil water contents (Rhoades, 1982b). Soil solutions obtained at lower soil moisture conditions are more labor intensive and require special equipment.

Saturated paste, data relationships: A means of cross-checking chemical analyses for consistency and reliability is provided by the interrelations that exist among the various soil chemical determinations (U.S. Salinity Laboratory Staff, 1954). The saturated paste pH is the apparent pH of the soil:water mixture and is a key indicator in many of these interrelations. The saturated paste pH is dependent upon the dissolved CO_2 concentration, the moisture content of the mixture, the exchangeable cation composition, the soluble salt composition and concentration, and the presence and amount of gypsum and alkaline-earth carbonates. Some general rules that apply to the saturated paste (U.S. Salinity Laboratory Staff, 1954) are as follows:

Relationship: Total Cation and Anion Concentrations
General Rule: Total cations $\approx$ Total anions, expressed on equivalent basis

Relationship: pH and Ca and Mg Concentrations
General Rule: Concentrations of Ca^{2+} and Mg^{2+} are seldom >2 meq L^{-1} (mmol (-) L^{-1}) at pH >9.

Relationship: pH and Carbonate and Bicarbonate Concentrations
General Rule: Carbonate concentration (meq L^{-1}) is measurable only if pH >9. Bicarbonate concentration is rarely >10 meq L^{-1} (mmol (-) L^{-1}) in absence of carbonates.
Bicarbonate concentration is seldom >3 or 4 meq L^{-1} (mmol (-) L^{-1}) if pH <7.

Relationship:	pH and Gypsum
General Rule:	Gypsum is rarely present if pH >8.2.
	Gypsum has variable solubility in saline solutions (20 to 50 meq L^{-1}, (mmol (-) L^{-1})).
	Check for the presence of gypsum if Ca concentration >20 meq L^{-1} (mmol (-) L^{-1}) and pH $\leq$8.2.

Relationship:	pH, ESP, and Alkaline-Earth Carbonates
General Rule:	Alkaline-earth CO_3^- and ESP $\geq$15 are indicated if pH $\geq$8.5.
	ESP $\leq$15 may or may not be indicated if pH <8.5.
	No alkaline-earth CO_3^- are indicated if pH <7.5.

Relationship:	pH and Exchangeable Acidity
General Rule:	Significant amounts of exchangeable acidity are indicated if pH <7.0.

Saturation percentage, definition and data relationships: The *saturation percentage* (SP) is the amount of water in the saturated paste. An experienced analyst should be able to repeat the saturated paste preparation to an SP within 5 percent. The SP can be related directly to the field moisture range. Measurements on soils, over a considerable textural range (U.S. Salinity Laboratory Staff, 1954), indicate the following general rules:

Equation 4.5.3.1:

$$SP \approx 4 \times 1500\text{-kPa water}$$

Equation 4.5.3.2:

$$SP \approx 2 \times \text{upper end field soil moisture content}$$

Equation 4.5.3.3:

$$AWC \approx SP/4$$

where
SP = Saturation percentage
AWC = Available water capacity

Therefore, at the upper (saturated) and lower (dry) ends of the field moisture range, the salt concentration of the soil solution is $\approx$ 4x and 2x the concentration in the saturation extract, respectively.

If the soil texture is known and the 1500-kPa water content has been measured, the preceding SP relationships may be redefined (U.S. Salinity Laboratory Staff, 1954) as follows:

1500-kPa water %	Texture	Relationship
2.0 to 6.5	Coarse	SP $\approx$ 6 1/3 x 1500 kPa
6.6 to 15	Medium	SP $\approx$ 4 x 1500 kPa
>15	Fine	SP $\approx$ 3 1/4 x 1500 kPa
>15	Organic	SP $\approx$ 3 2/3 x 1500 kPa

Electrical conductivity, definition: The electrical conductivity (EC) measurement is based on the principle that the amount of electrical current transmitted by a salt solution under standardized conditions will increase as the salt concentration of the solution is increased. When electrical conductivity is measured, a sample solution is placed between two electrodes of standardized or known geometry and an electrical potential is imposed across the electrodes. The solution resistance is measured and converted to reciprocal resistance or conductance. The basic unit for resistance measurements is the ohm, and the unit of reciprocal resistance is the mho.

Electrical conductivity, salt index: Electrical conductivity is used as a simplified index to the total concentration of dissolved salts in a given irrigation water and as a water-quality parameter which can be correlated to plant growth (Scofield, 1942). Measuring EC and total dissolved salts (TDS) is not straightforward in soils because salinity is significantly affected by the prevailing moisture content. A primary source of salts is chemical weathering of the minerals present in soils and rocks; the most important reactions include dissolution, hydrolysis, carbonation, acidification, and oxidation-reduction (National Research Council, 1993). All of these reactions contribute to an increase in the dissolved mineral load in the soil solution and in waters.

Electrical conductivity, saturation extract, data relationships: The electrical conductivity of the saturation extract (EC_s) is used as a criterion for classifying a soil as saline. Other uses of this measurement include the estimation of the total cation concentration in the extract, salt percentage in solution (P_{sw}), salt percentage in soil (P_{ss}), and osmotic pressure (OP).

The EC_s may be used to estimate the salt percentage (P_{sw}) in solution (U.S. Salinity Laboratory Staff, 1954) as follows:

Equation 4.5.3.4:

$$P_{sw} \approx 0.064 \text{ x } EC_s$$

where
P_{sw} = Estimated salt percentage in solution
EC_s = Electrical conductivity of saturation extract, mmhos cm^{-1} (dS m^{-1})

The preceding equation may be used to estimate the salt percentage (P_{ss}) in the soil (U.S. Salinity Laboratory Staff, 1954) as follows:

Equation 4.5.3.5:

$$P_{ss} \approx (P_{sw} \times SP)/100$$

where
P_{ss} = Estimated salt percentage in soil
P_{sw} = Estimated salt percentage in solution
SP = Saturated percentage. Water percentage in saturation extract.

The EC_s at 25 °C may be used to estimate the osmotic potential (OP) in atmospheres of a solution (U.S. Salinity Laboratory Staff, 1954) as follows:

Equation 4.5.3.6:

$$OP \approx 0.36 \times EC_s$$

where
OP = Estimated osmotic potential (atmospheres)
EC_s = Electrical conductivity of saturation extract, mmhos cm^{-1} (dS m^{-1})

For solutions with low EC_s, i.e., dilute solutions, the EC_s at 25 °C may be used to estimate the total cation or anion concentration (meq L^{-1}) of the solution (U.S. Salinity Laboratory Staff, 1954) as follows:

Equation 4.5.3.7:

$$\text{Total cations} \approx 10 \times EC_s$$

Equation 4.5.3.8:

$$\text{Total anions} \approx 10 \times EC_s$$

where
EC_s = Electrical conductivity of saturation extract, mmhos cm^{-1} (dS m^{-1})

Electrical conductivity, soil:water ratio: The methods of obtaining soil samples for EC, from the least difficult to the most difficult, are as follows (Corwin, 2007):

Equation 4.5.3.9:

$$EC_p < EC_{1:5} = EC_{1:1} < EC_s < EC_w$$

where
EC_p = EC of saturated paste
$EC_{1:5}$ = EC of 1:5 soil to water extract

$EC_{1:1}$ = EC of 1:1 soil to water extract
EC_s = EC of saturation extract
EC_w = EC of soil:water

General relationships among extracts are as follows (Corwin, 2007):

Equation 4.5.3.10:

$$EC_w = 2EC_s$$

Relationships between extracts >SP, assuming no precipitation-dissolution reactions, are as follows:

Equations 4.5.3.11 and 4.5.3.12:

If SP = 100%, then $EC_e = EC_{1:1} = 5\ EC_{1:5}$ (simple dilution factor)
If SP = 50%, then $EC_e = 2\ EC_{1:1} = 10\ EC_{1:5}$ (simple dilution factor)

The EC of one extract can be converted to another using Suarez and Taber's ExtractChem (v. 0.18) software. Knowledge of major cations and anions is needed.

The relationship between EC_s and EC_p is complex.

The determination of apparent soil EC (EC_a) is a complex measurement influenced by such soil properties as salinity, texture, water content, bulk density, organic matter, clay mineralogy, and temperature. EC_a is determined through geophysical techniques, e.g., electrical resistivity (ER), electromagnetic induction (EMI), and time domain reflectrometry (TDR). Refer to USDA (2007) for a more detailed discussion of these field-scale soil salinity measurement techniques.

Resistivity, definition: *Resistivity* or *specific resistance* has been defined as the resistance in ohms of a conductor, metallic or electrolytic, that is 1 cm long and has a cross-sectional area of 1 cm^2 (U.S. Salinity Laboratory Staff, 1954). The resistance (ohms) is converted to a 60 °F (15.5 °C) basis.

Resistivity, saturated paste: The resistivity of the soil paste (R_s) is used mainly to estimate the salt content in the soil. Soil resistivity, gypsum content, and extractable acidity (individually or in combination) provide a basis for estimating potential corrosivity of soils (USDA/SCS, 1971).

Resistivity and electrical conductivity, data relationships: The R_s has been related to the EC_s (U.S. Salinity Laboratory Staff, 1954). The EC_s is measured and is commonly reported as resistivity (R_s). The EC_s measurement requires more time, i.e., preparation of saturated soil paste, than the R_s measurement. However, the EC_s is the easier measurement from which to make interpretations; i.e., EC_s is more closely related to plant response (U.S. Salinity Laboratory Staff, 1954). There is no simple method to convert electrical conductivity to soil resistivity or vice versa. There is a limited correlation between electrical conductivity and soil resistivity as the relationship is markedly influenced by variations in saturation percentage (SP), salinity, and soil mineral conductivity. The R_s and EC_s have been related by an equation (U.S. Salinity Laboratory Staff, 1954) as follows:

Equation 4.5.3.13:

$$EC_s \approx 0.25/\, R_s$$

where
EC_s = Electrical conductivity of saturation paste, mmhos cm^{-1} (dS m^{-1})
0.25 = Constant for Bureau of Soils electrode cup
R_s = Resistivity of saturation paste (ohms cm^{-1})

Historically, the EC_s is adjusted to a 60 °F (15.5 °C) basis before interpretive use. The EC_s and R_s increase ≈ 2 percent per °C. The unit EC x 10^3 is called the millimho per centimeter (mmhos cm^{-1}).
The following calculation of resistivity is based on a correlation between measurements of EC_s and R_s of the saturated paste ($r^2 = 0.913$, n = 191) (National Soil Survey Laboratory Staff, 1975).

Equation 4.5.3.14:

$$R_s = 1000/[(\text{Log } EC_s/1.1011) - 0.2257] \text{ x } 1.246$$

where
R_s = Resistivity of saturated paste
EC_s = Electrical conductivity of saturated paste

Saturated paste extract, soluble cations and anions: The commonly determined soluble cations in the saturation extract include calcium, magnesium, sodium, and potassium. In soils with a low saturation pH, measurable amounts of Fe and Al may be present. The soluble anions that are commonly determined in saline and alkali soils are carbonate, bicarbonate, sulfate, chloride, nitrate, nitrite, fluoride, phosphate, silicate, bromide, and borate (Khym, 1974; U.S. Salinity Laboratory Staff, 1954). Some less commonly analyzed ions in the saturation extract are manganese, lithium, strontium, rubidium, cesium, hydronium, selenate, selenite, acetate, arsenate, and arsenite. Phosphate, silicate, bromide, and borate are found only occasionally in measurable amounts in soils. In saline and alkali soils, carbonate, bicarbonate, sulfate, and chloride are the anions that occur in the greatest abundance. In general, soluble sulfate is typically more abundant than soluble chloride.
The effect of soluble cations upon the exchangeable cation determination is to increase the cation concentration in the extracting solution, i.e., NH$_4$OAc, buffered at pH 7.0. The dissolution of salts by the extractant necessitates an independent determination of soluble cations and a correction to the exchangeable cations; therefore, in soils with soluble salts or carbonates, the soluble cations must be measured separately and the results subtracted from the extractable bases for determination of exchangeable

bases (Exchangeable = Extractable - Soluble). The presence of alkaline-earth carbonates prevents accurate determination of exchangeable Ca and Mg.

Saturated paste and extract, measurements: This section describes the SSL methods for measuring salinity on aqueous extracts of saturated soil pastes. The saturated paste is prepared and the saturation percentage (SP) determined. The saturated paste extract is obtained with an automatic extractor. Electrical conductivity and soil resistivity of saturated paste are measured. The saturated paste pH is measured. The water-soluble cations of Ca^{2+}, Mg^{2+}, K^+, and Na^+ are measured by atomic absorption spectrophotometry. The water-soluble anions of Br, Cl^-, F^-, NO_3^-, NO_2^-, PO_4^{3-}, and SO_4^{2-} are measured by ion chromatography. The carbonate and bicarbonate concentrations are determined by acid titration. Estimated total salt is calculated. The SSL also performs a salt prediction test which is used not only to predict those soils that have measurable amounts of soluble salts but also to predict the quantity and the appropriate dilutions for salt analyses of those soils. If salt predictions or conductances are >0.25 mmhos cm^{-1}, soils are considered nonsalty and generally no other salt analyses are performed on these soils by the SSL.

4.5   Electrical Conductivity and Soluble Salts
4.5.4   Aqueous Extraction
4.5.4.1   1:2 Extraction
4.5.4.1.1   Electrical Conductivity
4.5.4.1.1.1   Salt Prediction
4.5.4.2   1:5, 23 h, 1 h
4.5.4.2.1   Electrical Conductivity
4.5.5   Saturated Paste
4.5.5.1   Water (Saturation) Percentage
4.5.5.2   Resistivity
4.5.5.3   Saturation Paste Extraction
4.5.5.3.1   Electrical Conductivity
4.5.5.3.2–5   Calcium, Magnesium, Potassium, and Sodium
4.5.5.3.6–12   Bromide, Chloride, Fluoride, Nitrate, Nitrite, Phosphate, and Sulfate
4.5.5.3.13–14   Carbonate and Bicarbonate

Salt prediction, electrical conductivity, measurement: The SSL performs a salt prediction test, which is used not only to predict those soils that have measurable amounts of soluble salts but also to predict the quantity and the appropriate dilutions for salt analyses of those soils. A 5-g soil sample is mixed with 10 mL water (1:2) and allowed to stand overnight. The EC is measured using an electronic bridge. If salt predictions or electrical conductances (EC) are <0.25 mmhos cm^{-1} (dS m^{-1}), the soils are considered nonsalty and generally no other salt analyses are performed on these soils by the SSL; if >0.25 mmhos cm^{-1} (dS m^{-1}), the SSL prepares a saturation paste.

Saturated paste, preparation: The saturated soil paste is a particular mixture of soil and water. This paste is prepared by adding water to a soil sample while stirring the mixture until the soil paste meets the saturation criteria; i.e., the soil paste glistens as it reflects light, flows slightly when the container is tipped, and slides freely and cleanly from a spatula (except in soils that have a high clay content). The mixture is covered and allowed to stand overnight. The saturation criteria are then rechecked. If the mixture fails to meet these criteria, more water or soil is added until the criteria are met. This soil:water ratio is used because it is the lowest reproducible ratio for which enough extract for analysis can be readily removed from the soil with pressure or vacuum and because this ratio is often related in a predictable manner to the field soil water content (U.S. Salinity Laboratory Staff, 1954).

Saturation percentage and extract, measurement: Upon preparation of the saturated paste, a subsample is obtained to determine the water content, i.e., saturation percentage (SP). In addition, an aqueous extract is obtained from the saturated paste by transferring the paste to a plastic filter funnel with filter paper. The funnel is placed on a mechanical vacuum extractor (Holmgren et al., 1977), and the saturated paste is extracted. The SSL uses this extract in a series of chemical analyses, e.g., electrical conductivity of the saturated paste and concentrations of the major solutes as follows:

(1) **Water-soluble cations, saturation extract, measurement:** An aliquot of the extract is diluted with an ionization suppressant (La_2O_3). The analytes are measured by an atomic absorption spectrophotometer (AA). The data are automatically recorded by a computer and printer. Determination of soluble cations is used to obtain the relations between total cation concentration and other properties of saline solutions, such as electrical conductivity and osmotic pressure (U.S. Salinity Laboratory Staff, 1954). The relative concentrations of the various cations in the soil-water extracts also provide information on the composition of the exchangeable cations in the soil. Complete analyses of the soluble ions provide a means to determine total salt content of the soils and salt content at field moisture conditions. The SSL reports the saturation extracted cations, Ca^{2+}, Mg^{2+}, K^+, and Na^+, as mmol (+) L^{-1}. In the past, these cations were reported as meq L^{-1}.

(2) **Water-soluble anions, saturation extract, measurements:** An aliquot of extract is diluted according to its electrical conductivity (EC_s). The diluted sample is injected into the ion chromatograph, and the anions are separated. A conductivity detector is used to measure the anion species and content. Standard anion concentrations are used to calibrate the system. A calibration curve is determined, and the anion concentrations are calculated. A computer program automates these actions. This same method may also be used for water analysis. The SSL reports the saturation extract anions, Br^-, Cl^-, F^-, NO_3^-, NO_2^-, PO_4^{3-}, and SO_4^{2-}, as (mmol (-) L^{-1}). In the past, these anions were reported as meq L^{-1}.

A separate aliquot of the saturation extract is titrated on an automatic titrator to pH 8.25 and pH 4.60 end points. The carbonate and bicarbonate are calculated from the titers, aliquot volume, blank titer, and acid normality. Total dissolved ion amounts generally increase with increasing soil moisture content. While some ions increase, some ions may decrease. Carbonate and bicarbonate anions are among those ions which are most dependent upon soil moisture. Therefore, in making interpretations about carbonate and bicarbonate in soil solution, there must be careful consideration about the chemistry of the soil and soil solution.

The SSL reports the carbonate and bicarbonate as mmol (-) L^{-1}. In the past, these anions were reported as meq L^{-1}.

(3) Electrical conductivity, saturation extract, measurement: The electrical conductivity of the saturation extract (EC$_s$) is measured using a conductivity cell and a direct reading digital bridge. The cell constant is set using a standard solution. The SSL reports the EC$_s$ as dS m^{-1}. In the past, EC was reported as mmhos cm^{-1}.

Resistivity, saturated paste, measurement: The soil resistivity apparatus is simple and rugged, the measurements can be made quickly, and the results are reproducible. Many agencies use the Bureau of Soils electrode cup to estimate the soluble salt content in soils (Davis and Bryan, 1910; Soil Survey Staff, 1951). A saturated paste is placed in an electrode cup, and the resistance is measured. The temperature of the paste is measured. The resistance (ohms) is converted to a 60 °F (15.5 °C) basis using a fourth order equation (Benham, 2003). Resistivity of the saturation extract (R$_s$) is reported in units of ohms at 60 °F (15.5 °C). Use Table 4.5.5.1 to convert measured resistance to specific resistance at 60 °F (15.5 °C).

Equation 4.5.5.1:

Resistivity (ohms cm^{-1}) = ohms @ 60 °F x electrode cup cell factor

Alternatively, the following equation (Benham, 2003) may be used for reducing soil paste resistance readings to values at 60 °F with final results reported to 4 significant figures.

Equation 4.5.5.2:

where B >32 °C

$$A = (-0.013840786 + 0.028627073 \times B - 0.00037976971 \times B^2 + 3.7891593 \times 10^{-6} \times B^3 - 1.2020657 \times 10^{-8} \times B^4) \times C \times D \times E$$

where B ≤32 °C

$$A = (-0.013840786 + 0.028627073 \times (1.8 \times B + 32) - 0.00037976971 \times (1.8 \times B + 32.0)^2 + 3.7891593 \times 10^{-6} \times (1.8 \times B + 32.0)^3 - 1.2020657 \times 10^{-8} \times (1.8 \times B + 32.0)^4 \times C \times D \times E$$

where
A = Resistance (ohms) corrected to 60 °F
B = Temperature (°F) at which the resistance was measured
C = Resistance (ohms) measured at temperature B
D = Electrode cup cell factor
E = Scale (range multiplier)

Table 4.5.5.1   Bureau of soils data for reducing soil paste resistance readings to values at 60 °F (Whitney and Means, 1897)

Temp	Ohms								
°F	1000	2000	3000	4000	5000	6000	7000	8000	9000
40	735	1470	2205	2940	3675	4410	5145	5880	6615
42	763	1526	2289	3052	3815	4578	5341	6104	6867
44	788	1576	2364	3152	3940	4728	5516	6304	7092
46	814	1628	2442	3256	4070	4884	5698	6512	7326
48	843	1686	2529	3372	4215	5058	5901	6744	7587
50	867	1734	2601	3468	4335	5202	6069	6936	7803
52	893	1786	2679	3572	4465	5358	6251	7144	8037
54	917	1834	2751	3668	4585	5502	6419	7336	8253
56	947	1894	2841	3788	4735	5682	6629	7576	8523
58	974	1948	2922	3896	4870	5844	6818	7792	8766
60	1000	2000	3000	4000	5000	6000	7000	8000	9000
62	1027	2054	3081	4108	5135	6162	7189	8216	9243
64	1054	2108	3162	4216	5270	6324	7378	8432	9486
66	1081	2162	3243	4324	5405	6486	7567	8648	9729
68	1110	2220	3330	4440	5550	6660	7770	8880	9990
70	1140	2280	3420	4560	5700	6840	7980	9120	10260
72	1170	2340	3510	4680	5850	7020	8190	9360	10530
74	1201	2402	3603	4804	6005	7206	8407	9608	10809
76	1230	2460	3690	4920	6150	7380	8610	9840	11070
78	1261	2522	3783	5044	6305	7566	8827	10088	11349
80	1294	2588	3882	5176	6470	7764	9058	10352	11646
82	1327	2654	3981	5308	6635	7962	9289	10616	11943
84	1359	2718	4077	5436	6795	8154	9513	10872	12231
86	1393	2786	4179	5572	6965	8358	9751	11144	12537
88	1427	2854	4281	5708	7135	8562	9989	11416	12843
90	1460	2920	4380	5840	7300	8760	10220	11680	13140
92	1495	2990	4485	5980	7475	8970	10465	11960	13455
94	1532	3064	4596	6128	7660	9192	10724	12256	13788
96	1570	3140	4710	6280	7850	9420	10990	12560	14130
98	1611	3222	4833	6444	8055	9666	11277	12888	14499

4.5 Electrical Conductivity and Soluble Salts
4.5.4 Aqueous Extraction
4.5.4.2 1:5, 23 h, 1 h
4.5.4.2.1 Electrical Conductivity

1:5 aqueous extraction, EC, measurement: A 20-g sample of soil is added to 100 mL of water in a 250-mL polyethylene bottle. The soil:water suspension is maintained at room temperature for 23 h and then shaken on a reciprocating shaker for 1 h. The supernatant is filtered into a 100-mL polyethylene bottle. The electrical conductivity (EC) of the 1:5 extract is measured using a conductivity cell and a direct reading digital bridge. The cell constant is set using a standard solution. The SSL reports the EC as dS m^{-1}. The 1:5 extract is also analyzed for pH, cations, anions, nitrate-nitrite, and multielements.

4.5 Electrical Conductivity and Soluble Salts
4.5.6 Ratios, Estimates, and Calculations Related to Soluble Salts
4.5.6.1 Exchangeable Sodium Percentage, NH$_4$OAc, pH 7.0

Exchangeable sodium percentage, application: Historically, the exchangeable sodium percentage (ESP) was used as the main criterion for excessive Na levels in soils. In the last several decades, however, for the classification of Na-affected soils, the emphasis has been on the use of the SAR of the equilibrium soil solution (Bresler et al., 1982). An ESP $\geq$15 percent is a diagnostic criterion for natric soil horizons (Soil Survey Staff, 1999).

Exchangeable sodium percentage, calculations: The U.S. Salinity Laboratory (1954) presented the following relationship for estimating ESP of saturation extract for soils with CEC <50 meq 100 g^{-1}. The SSL does not calculate the ESP using the following equation.

Equation 4.5.6.1.1:

$$ESP = \frac{[100 \, (-0.0126 + 0.01475 \; SAR)]}{[1 + (-0.0126 + 0.01475 \; SAR)]}$$

where
ESP = Exchangeable sodium percentage
SAR = Sodium adsorption ratio

4.5 Electrical Conductivity and Soluble Salts
4.5.6 Ratios, Estimates, and Calculations Related to Soluble Salts
4.5.6.1 Exchangeable Sodium Percentage, NH₄OAc, pH 7.0
4.5.6.1.1 Exchangeable Sodium Percentage, Calculated Without Saturated Paste Extraction

Exchangeable sodium percentage, without saturated paste extract, calculation:
When a saturated paste extraction is not prepared, the exchangeable sodium percentage (ESP) is calculated by dividing the exchangeable sodium (ES) by the CEC by NH₄OAc, pH 7.0 (CEC-7) and multiplying by 100. The ES is calculated by subtracting the water-soluble Na^+ determined from the NH₄OAc extractable Na^+ determined.

When the saturated paste extract is not prepared, the ESP is calculated as follows:

Equation 4.5.6.1.1.1:

$$ESP = (ES/CEC\text{-}7) \times 100$$

where
ESP = Exchangeable sodium percentage
ES = Extractable sodium (NH₄OAc extractable Na^+, (cmol(+) kg^{-1})).
CEC-7 = CEC by NH₄OAc, pH 7.0 (cmol(+) kg^{-1}).

4.5 Electrical Conductivity and Soluble Salts
4.5.6 Ratios, Estimates, and Calculations Related to Soluble Salts
4.5.6.1 Exchangeable Sodium Percentage, NH₄OAc, pH 7.0
4.5.6.1.2 Exchangeable Sodium Percentage, Calculated With Saturated Paste Extraction

Exchangeable sodium percentage, with saturated paste extract, calculation:
When a saturation paste extract is prepared, the SSL calculates the exchangeable sodium percentage (ESP) as follows:

Equation 4.5.6.1.2.1:

$$ESP = 100 \times [Na_{ex} - (Na_{ws} \times (H_2O_{ws}/1000)))]/CEC\text{-}7$$

where
ESP = Exchangeable sodium percentage
Na_{ex} = Extractable Na (NH₄OAc extractable Na^+, (cmol(+) kg^{-1}))
Na_{ws} = Water-soluble Na (mmol (+) L^{-1})
H_2O_{ws} = Water saturation percentage
CEC-7 = CEC by NH₄OAc, pH 7.0 (cmol(+) kg^{-1})
1000 = Conversion factor to (cmol(+) kg^{-1})
100 = Conversion factor to percent

Exchangeable sodium is computed with acceptable accuracy unless salt content is >20 dS m^{-1} at 25 °C. Exchangeable Na equals extractable Na minus saturation extract Na multiplied by saturation percentage. Saturation percentage is the water percentage in the saturated paste divided by 1000. Exchangeable Na can be determined with greater accuracy than the other cations in the presence of gypsum or carbonates. If exchangeable K is negligible compared to exchangeable Ca and Mg, then exchangeable Ca plus Mg equals CEC (NH$_4$OAc, pH 7.0) minus exchangeable Na. This approximation is suitably reproducible for comparison between soils and for soil classification. Exchangeable Ca can be computed in the same manner as exchangeable Na. Results are not so satisfactory for exchangeable Ca when computed in the presence of carbonates or large amounts of gypsum.

4.5 Electrical Conductivity and Soluble Salts
4.5.6 Ratios, Estimates, and Calculations Related to Soluble Salts
4.5.6.2 Sodium Adsorption Ratio (SAR)

Sodium adsorption ratio, application: In addition to the total salinity (osmotic) hazard of irrigation water on plants, the tendency of the solution to produce excessive exchangeable Na must also be considered (Bresler et al., 1982). Significant amounts of Na in soils severely retard the growth of many plants. The SAR was developed as a measurement of the quality of irrigation water, particularly when the irrigation water is salt or Na affected (U.S. Salinity Laboratory Staff, 1954). The SAR is used as an indirect estimate of the equilibrium relation between soluble Na in the salt solution and exchangeable Na adsorbed by the soil. Measurements of analytes for this calculation are fewer and simpler relative to the more complex ESP calculation. An SAR $\geq$13 percent is used as a diagnostic criterion for natric soil horizons (Soil Survey Staff, 2010).

Unlike most theoretical ion "ratios," the SAR is expressed in terms of ion concentration (Gapon-type exchange) rather than ion activity. However, the use of concentration is valid as the activity coefficient decreases more rapidly with increasing salt concentration for divalent cations, e.g., Ca^{2+} and Mg^{2+}, than it does for monovalent cations, e.g., Na$^+$ (Sposito, 1989; Cresser et al., 1993). In addition, this ratio is relatively constant over a broad range of concentrations.

Sodium adsorption ratio, calculation: Compute the sodium adsorption ratio (SAR) by dividing the molar concentration of the monovalent cation Na$^+$ by the square root of the molar concentration of the divalent cations Ca^{2+} and Mg^{2+} (U.S. Salinity Laboratory Staff, 1954). The water-soluble Ca^{2+}, Mg^{2+}, and Na$^+$ are determined. The SSL calculates the SAR as follows:

Equation 4.5.6.2.1:

$$SAR = \frac{\left[Na^+\right]}{\sqrt{\dfrac{\left[Ca^{++}\right]+\left[Mg^{++}\right]}{2}}}$$

SAR = Sodium adsorption ratio
Na^+ = Water-soluble Na^+ (mmol (+) L^{-1})
Ca^{2+} = Water-soluble Ca^{2+} (mmol (+) L^{-1})
Mg^{2+} = Water-soluble Mg^{2+} (mmol (+) L^{-1})

An ESP determination requires (1) measurement of all soluble and exchangeable Na from the soil; (2) subsequent subtraction of any soluble Na in the saturation extract; and (3) a soil CEC determination. The SAR can be readily determined on the saturation extract if that extraction has been made for an EC determination in the soil salinity analysis. The ESP analysis is less precise than the SAR determination.

4.5   Electrical Conductivity and Soluble Salts
4.5.6   Ratios, Estimates, and Calculations Related to Soluble Salts
4.5.6.3   Estimated Total Salt

Total salt concentration is an important water-quality parameter from the standpoint of salinity as it may be used to estimate the osmotic potential of the solution (Bresler et al., 1982). Historically, this parameter was determined as the total dissolved salts (TDS) by evaporating a known volume of water to dryness and then weighing the quantity of dissolved materials contained therein. This measurement has its ambiguities and limitations (Bresler et al., 1982). The various salts exist in different dehydration states, depending on the degree of drying employed. The TDS method fails to account for variations in composition of the water under analysis. In addition, the TDS is a more laborious measurement than EC, which is the currently preferred measure of salinity (Bresler et al., 1982).

Use the charts and graphs available in U.S. Salinity Laboratory Staff (1954) to estimate total salt content from the electrical conductivity (EC_s) of the saturation extract. The essential relations are summarized in the following equations.

Equation 4.5.6.3.1:

Total Salt in soil (%) = (12.2347 X Ec_s + (0.058 x Ec_s^2) – (0.0003 x Ec_s^3) - 4.2333) x 0.000064 x SP

where
EC_s = Electrical conductivity of saturation extract (dS m^{-1})
SP = Saturation percentage of saturation paste (% wt)

Previous equations that have been used to estimate total salt content are as follows:

Equation 4.5.6.3.2:

Log total salt in soil (ppm) = 0.81 + 1.08 x Log EC_s (mmhos cm^{-1}) + Log SP

205

where
EC_s = Electrical conductivity of saturation extract
SP = Saturation percentage of saturation paste

Equation 4.5.6.3.3:

Total salt in soil (%) = Total salt (ppm) x 10^{-4}

These equations are applicable to saturation extracts with an EC_s <20 mmhos cm^{-1}. Deviations occur at higher salt concentrations.

At one time, estimated total salt was also reported on the Supplementary Characterization Data Sheet. This value was calculated using soluble cations (Ca^{2+}, Mg^{2+}, Na^+, and K^+) as follows:

Equation 4.5.6.3.4:

Total salt (%) = (Soluble cations x 0.058 x SP)/1000

where
Total salt = Total salt percentage on a <2-mm basis
Soluble cations = Sum of soluble cations (Ca^{2+}, Mg^{2+}, Na^+, and K^+ in saturation extract, meq L^{-1})
SP = Saturation percentage. Water percentage in saturation paste.

The amounts of salts were reported in cumulative amounts; i.e., each horizon included the salt from the overlying horizons in addition to the amount measured in the horizon. The calculation assumed that chloride was the only anion in the saturation extract (National Soil Survey Laboratory Staff, 1990). To the extent to which this is not so, adjustments need to be made. A good general factor is 0.064 for a mixture of sulfate and chloride (National Soil Survey Laboratory Staff, 1990). This value is no longer reported by the SSL.

4.5 Electrical Conductivity and Soluble Salts
4.5.7 Case Study

Some of the data relationships discussed in Section 4.5 of this manual are graphically presented for selected saline soils in Utah as follows:

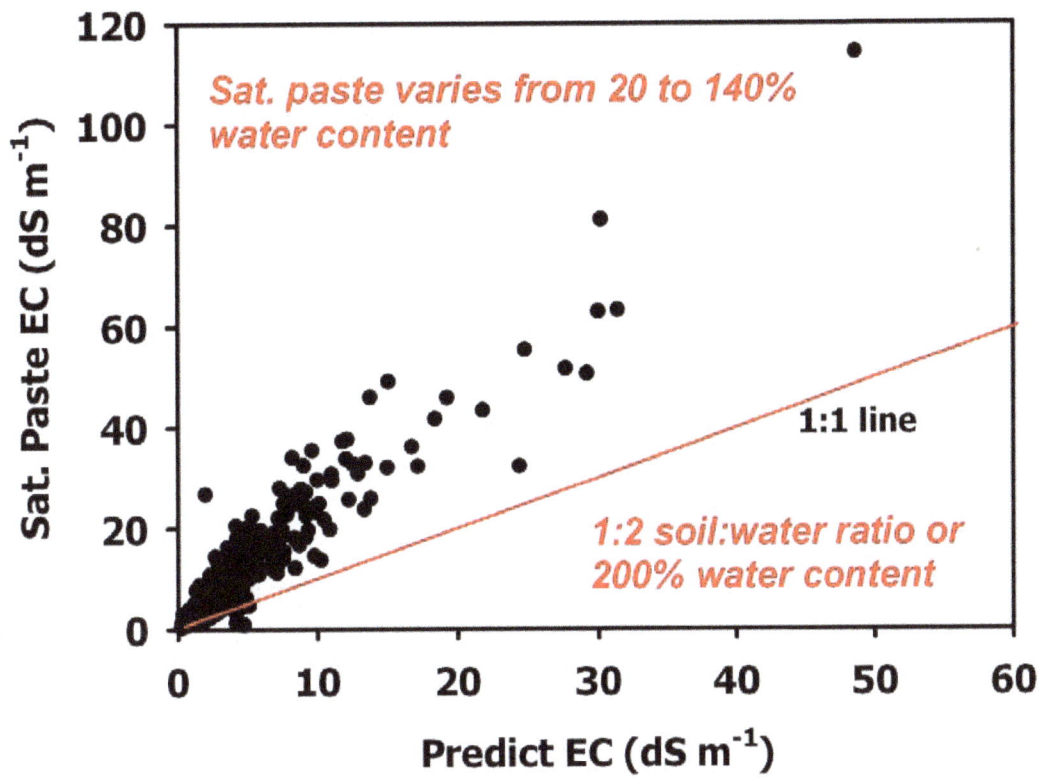

Sat. paste varies from 20 to 140% water content

1:1 line

1:2 soil:water ratio or 200% water content

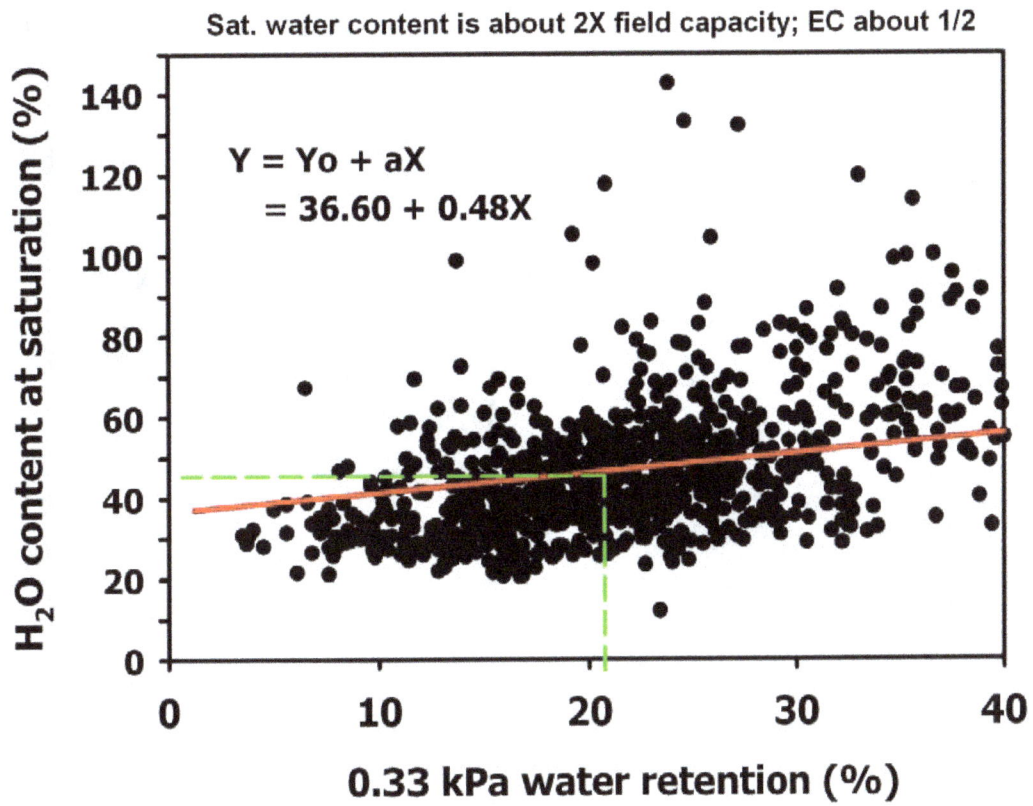

Sat. water content is about 2X field capacity; EC about 1/2

$Y = Yo + aX$
$= 36.60 + 0.48X$

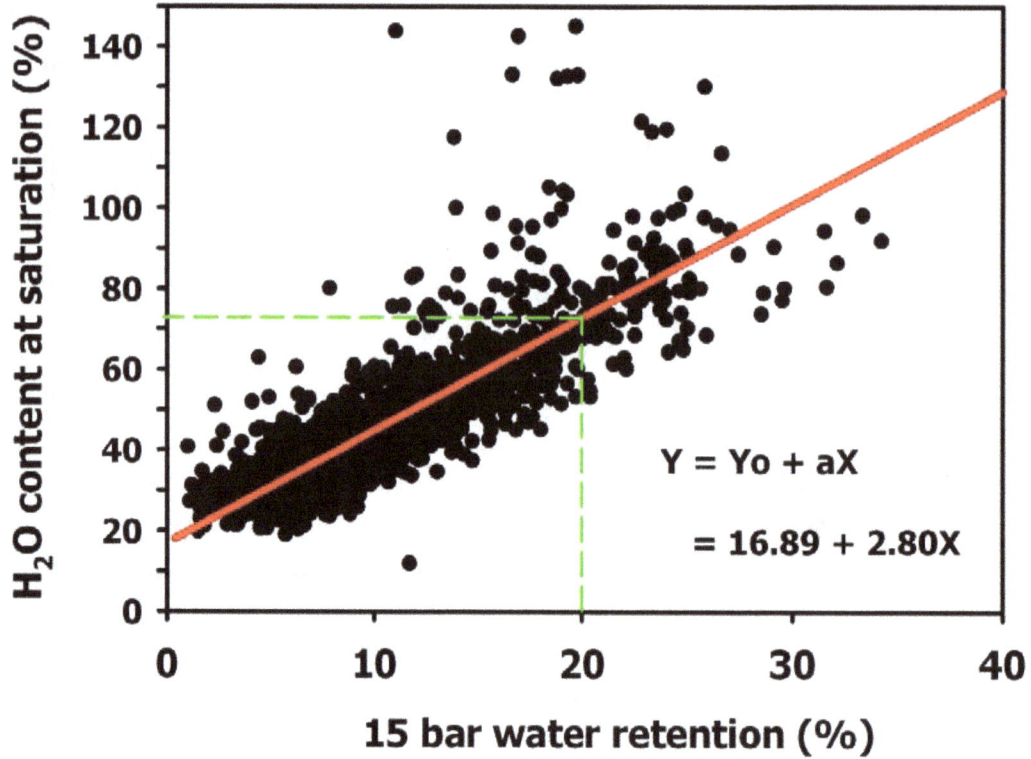
Sat. water content is about 4X wilting point; EC is about 1/4

$Y = Y_o + aX$

$= 16.89 + 2.80X$

(top figure axes: H$_2$O content at saturation (%) vs 15 bar water retention (%))

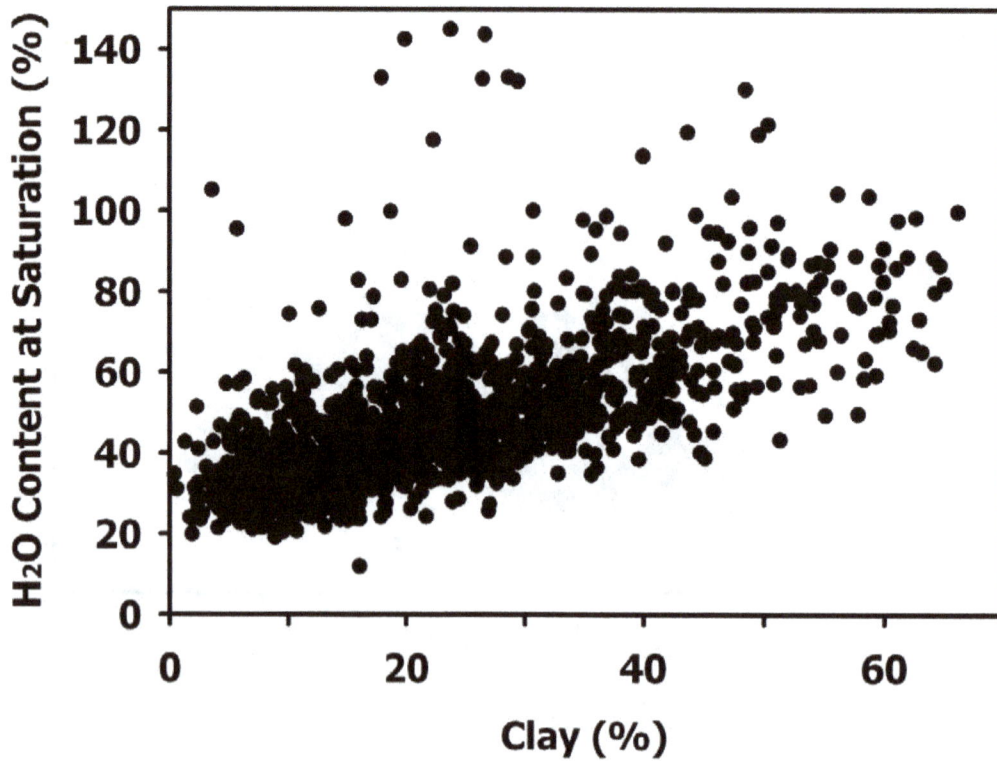

(bottom figure axes: H$_2$O Content at Saturation (%) vs Clay (%))

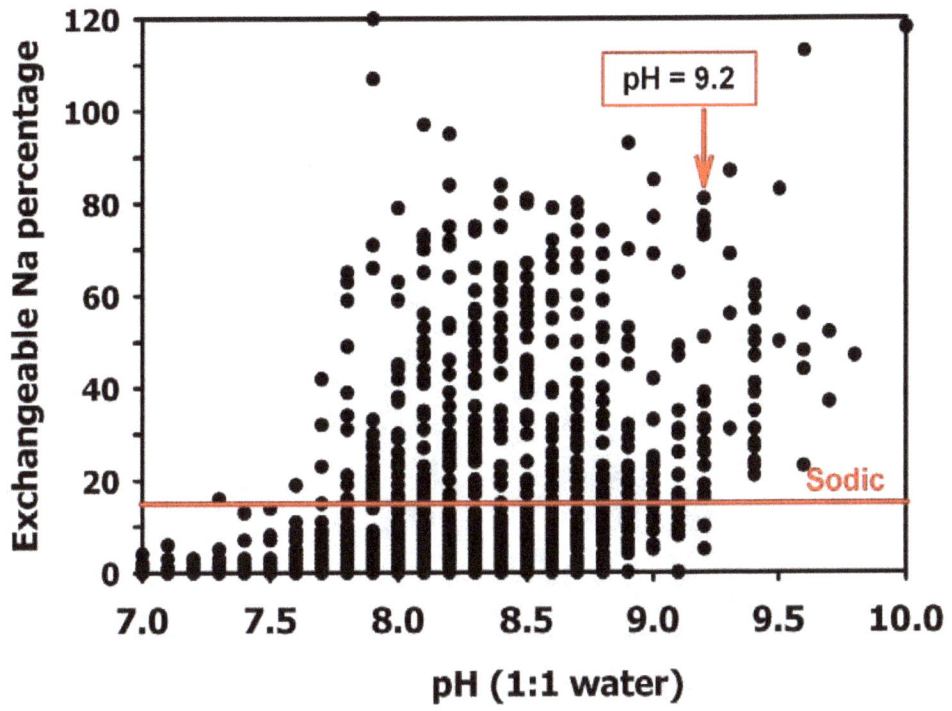

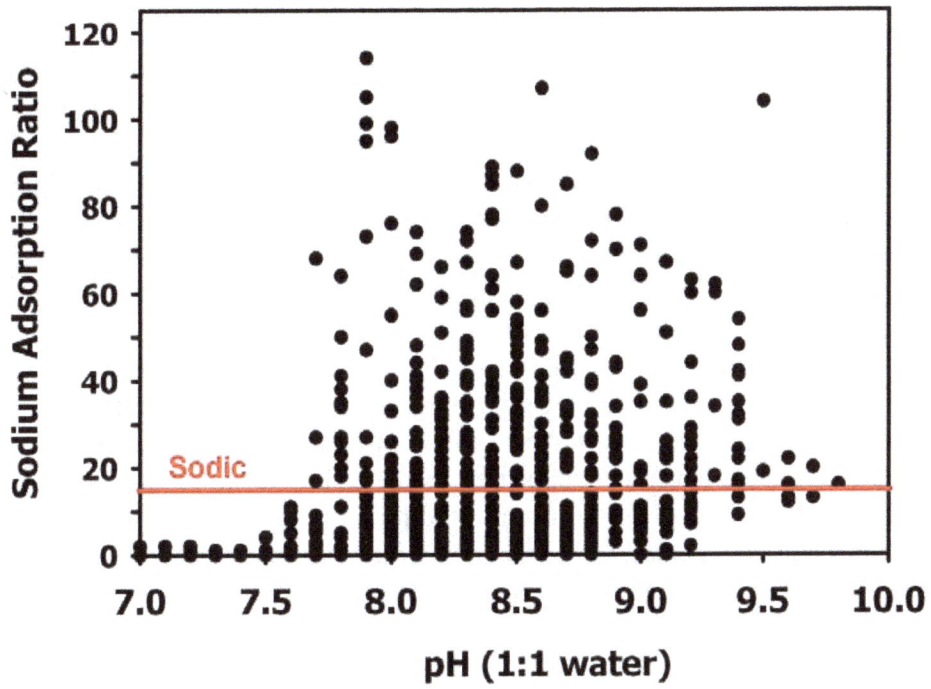

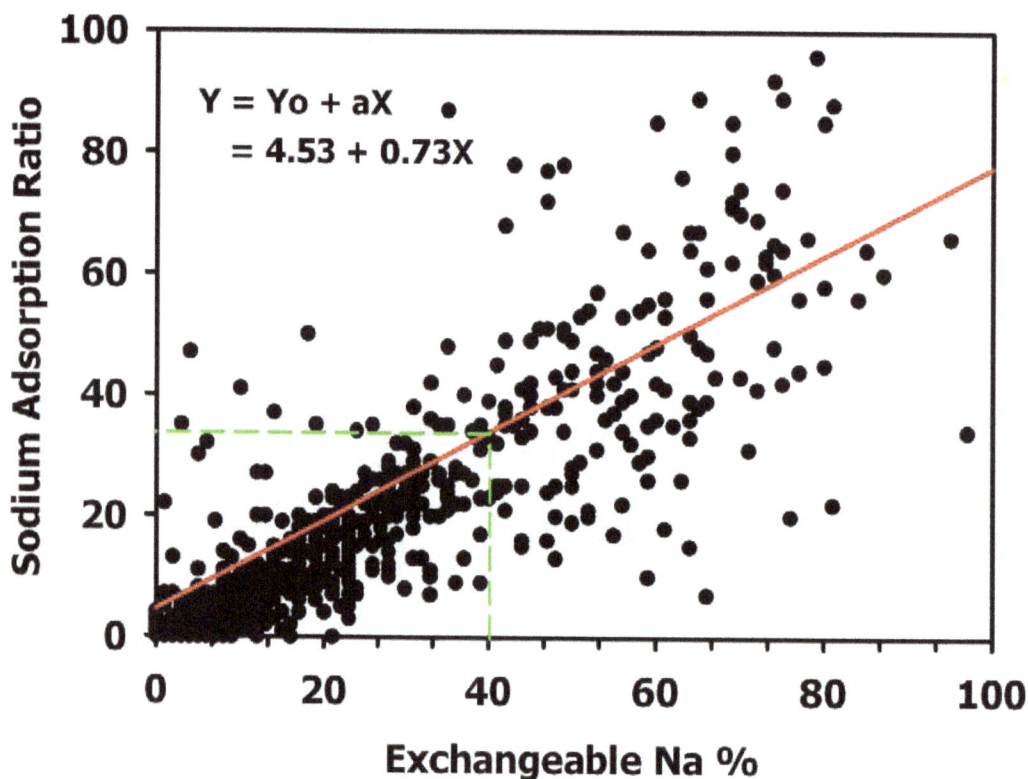

Y = Yo + aX

= 4.53 + 0.73X

4.6   Selective Dissolutions

This section provides definitions for some broad groupings of soil components, such as crystalline phyllosilicates, amorphous, poorly crystalline, paracrystalline, noncrystalline, allophane, imogolite, and short-range-order minerals (SROMs). These groupings have been related in part to various laboratory analyses and therefore have been operationally defined quantitatively and semiquantitatively by these analyses. Information is provided on the limitations as well as applications of these dissolution methods, e.g., taxonomic classification, genesis and geomorphology models, and environmental studies. The SSL selective dissolution methods include dithionite-citrate extractable Al, Fe, and Mn; ammonium oxalate extractable Al, Fe, Mn, P, and Si; and sodium pyrophosphate extractable Al, Fe, and Mn. Ratios, estimates, and calculations associated with these analyses also are described. Some of these ratios, estimates, and calculations appear on the SSL data sheets, while other ratios described herein were former data elements or have never appeared on the SSL data sheets but can be calculated using SSL data. These ratios are used for a variety of purposes, e.g., soil genesis studies and taxonomic criteria for spodic materials, andic soil properties, and mineralogy classes. For detailed descriptions of the SSL selective dissolution methods which are cross-referenced by method code in the table of contents in this manual, refer to SSIR No. 42 (Soil Survey Staff, 2004), which is available online at http://soils.usda.gov/technical/lmm/. Also refer to SSIR No. 51 (Soil Survey Staff, 2009; available online at http://www.soils.usda.gov/technical/) for detailed descriptions of field methods as used in NRCS soil survey offices.

4.6   Selective Dissolutions
4.6.1   Soil Components

Soil components, broad groupings: Over the years, various terms have been used to describe broad groupings of soil components, e.g., *crystalline phyllosilicates, amorphous, poorly crystalline, paracrystalline, noncrystalline, allophane, imogolite,* and *short-range-order minerals* (SROMs). These groupings have been related in part to various laboratory analyses and therefore have been operationally defined quantitatively and semiquantitatively by these analyses. Some of these analytical procedures include x-ray diffraction analysis and selective chemical dissolutions, e.g., dithionite-citrate, sodium pyrophosphate, and ammonium oxalate extractions. These terms have not been used consistently in the literature. In addition, there is not always a clear delineation between dissolution data, either conceptually or operationally. This discussion on terminology is pertinent not only to the data for dithionite-citrate extractable Fe, Al, and Mn but also to sodium pyrophosphate and ammonium oxalate extractions.

Soil aluminosilicates and crystalline phyllosilicates, definitions: *Soil aluminosilicates* include a broad spectrum of constituents, ranging from noncrystalline materials (exhibiting local and nonrepetitive short-range order) to paracrystalline materials (intermediate-range order) to crystalline phyllosilicates (layer silicates) characterized by three-dimensional periodicity over appreciable distances (long-range order) (Jackson et al., 1986). *Crystalline phyllosilicates* have been defined by Bailey (1980) as follows: "containing two-dimensional tetrahedral sheets of composition T_2O_5 (T = Si, Al, Be ...) with tetrahedra linked by sharing three corners of each, and with the fourth corner pointing in any direction. The tetrahedral sheets are linked in the unit structure to octahedral sheets, or to groups of coordinated cations, or individual cations." In soils and sediments, the phyllosilicates of common interest include the 1:1 layer types (kaolinite and halloysite) and the 2:1 layer types (smectite, vermiculite, mica, and kaolinite).

Amorphous, SROMs, poorly crystalline, noncrystalline, and paracrystalline, definitions: The term *amorphous* has been used to refer to an array of materials that are amorphous to x-rays and have no more than short-range order, e.g., allophane and imogolite (Kimble et al., 1984). The term *SROMs* has been used interchangeably with the terms amorphous, poorly crystalline, and noncrystalline. The term SROMs is preferred by some investigators because as the resolving power of mineralogical instruments improves, the possibility of determining and distinguishing among the various kinds of SROMs also improves (Uehara and Ikawa, 1985). Allophane and imogolite are two SROMs that would have been called amorphous 25 years ago. There are other soil materials with even shorter range order than allophane and imogolite, e.g., allophanelike constituents (Wada, 1980) and surface coatings of SROMs on crystalline minerals (Jones and Uehara, 1973; Uehara and Jones, 1974). The term *noncrystalline* is used by some investigators (Wada, 1977, 1989; Jackson et al., 1986) in preference to the more commonly used term *amorphous*. The term *paracrystalline* includes the somewhat ordered (short-range-order) materials, e.g., imogolite. Hence, the term *noncrystalline aluminosilicates* has been cited in the literature to address the whole

211

spectrum of short-range-order minerals in weathered parent materials and soils (Jackson et al., 1986).

Although most soils consist primarily of crystalline minerals, many contain appreciable amounts of noncrystalline or SROMs (Jones et al., 2000). Soils derived from volcanic ash and pumice may consist primarily of allophane and imogolite or other noncrystalline Al, Fe, or Si materials (Thorp and Smith, 1949; Mitchell et al., 1964; Jones and Uehara, 1973; Wada and Wada, 1976); however, these noncrystalline materials are not limited to soils derived from volcanic materials. Even small amounts of these noncrystalline materials can contribute significantly to the physical and chemical properties of soils (Fey and LeRoux, 1976); therefore, it is helpful to quantify the relative amounts of noncrystalline as well as crystalline components in these soils. In general, some of the more notable properties of these kinds of soils are high variable charge, high surface area, high reactivity with phosphate and organics (high anion retention), high water retention, and low bulk density (Wada, 1985).

Allophane and imogolite, definitions: Allophanes are associated mainly with weathered volcanic ash (Jackson et al., 1986). A definition of the term *allophane* was proposed by van Olphen (1971), in accordance with Ross and Kerr (1934), as follows: "Allophanes are members of a series of naturally occurring minerals which are hydrous aluminum silicates of widely varying chemical composition, characterized by short-range-order, by the presence of Si-O-Al bonds, and by a differential thermal analysis curve displaying a low temperature endotherm and a high temperature exotherm with no intermediate endotherm." By these criteria, allophane is limited to a small sector of the total spectrum of noncrystalline and paracrystalline aluminosilicates developed by weathering of volcanic ash and pumice and other materials of soils and deposits (Jackson et al., 1986). *Imogolite*, a mineral closely associated with allophane, is a hydrated aluminosilicate with a threadlike morphology that consists of paracrystalline cylindrical assemblies of a one-dimensional structure unit (Cradwick et al., 1972). *Allophanelike* constituents are difficult to identify or have more defective structures than allophane. Opaline silica occurs as amorphous spheres. There are also amorphous Al hydroxide gels and Fe oxides (ferrihydrite). In some studies, ferrihydrite has been considered the primary factor responsible for the irreversible drying of some soils into hard aggregates (Kubota, 1972; Espinoza et al., 1975; Maeda et al., 1977; Parfitt and Childs, 1988).

Allophane and imogolite, spot test, toluidine: Some soils with predominantly allophane and/or imogolite on the exchange complex have a small negative charge in the condition under which they occur in the field (Okamura and Wada, 1983), and large organic cations, such as tetramethyl- or tetraethyl-ammonium ion, are excluded from these negative charge sites (Wada and Tange, 1984). Toluidine blue, $(CH_3)_2$ $N^+C_6H_3NSC_6H_2(CH_3)NH_2$, is adsorbed from aqueous solution on negatively charged colloids and exhibits a characteristic color change from blue to purplish red (metachromasis). From these observations, a hypothesis was tested; i.e., the absence of metachromasis of toluidine blue (large organic cation) can be used for a test of allophane and/or imogolite (Wada and Kakuto, 1985a). In this study, some soils in Chile and Ecuador derived from volcanic ash with predominantly allophane and/or imogolite on the exchange complex were analyzed; results showed the absence of metachromasis. Though this test is not specific for imogolite and allophane, it is a test

212

for the absence or near-absence of negative charge sites in the soil or for negative charge sites with no access to a large organic cation. When this test is used in conjunction with NaF pH or the ammonium oxalate extractable Al data, the toluidine blue test can provide key information on the soil-forming processes from volcanic ash and may be useful for taxonomic classification.

Selective dissolutions, general applications: Selective dissolution data have been used extensively in the study of the noncrystalline material content of soils and sediments; however, there are limitations affecting the use of these data. In general, there is a continuum of crystalline order, ranging from no long-range order to paracrystalline to poorly crystalline to well crystalline (Follet et al., 1965). Selective dissolution data are necessary for independent determinations of various inorganic constituents of soils because of the difficulty with many physical analytical methods in estimating or even recognizing the presence of noncrystalline and paracrystalline free oxides or aluminosilicates mixed with crystalline soil components (Jackson et al., 1986). In general, the crystalline free oxides and phyllosilicates of soils can be identified qualitatively and estimated semiquantitatively by x-ray diffraction analysis. Those soils containing hydroxyls (-OH groups), e.g., kaolinite, gibbsite, and goethite, can sometimes be determined quantitatively by differential thermal analysis (DTA), differential scanning calorimetry (DSC), and thermogravimetric analysis (TGA).

Selective dissolutions, measurements and limitations: With selective chemical dissolution data, there are difficulties in the adequate assessment of the portion that is extracted by particular reagents, e.g., dithionite-citrate, sodium pyrophosphate, and ammonium oxalate. In principle, it cannot be expected that chemical methods are able to perfectly distinguish the degrees of crystallinity, and some caution is required in the interpretation of these analytical data (Van Wambeke, 1992).

Refer to Table 4.6.1–3 (Wada, 1989) for the dissolution of Al, Fe, and Si in various clay constituents and organic complexes by treatment with different reagents. These reagents include sodium pyrophosphate ($Na_4P_2O_7$), dithionite-citrate, Na_2CO_3, oxalate-oxalic (ammonium oxalate), and NaOH.

Table 4.6.1–3 Dissolution of Al, Fe, and Si in various clay constituents and organic complexes by treatment with different reagents[1]

Element in specified component and complex	Treatment with				
	0.1 M $Na_4P_2O_7$[2]	Dithionite-citrate[3]	20 g L^{-1} Na_2CO_3[4]	0.15-0.2 M oxalate-oxalic acid (pH 3.0-3.5)[5]	0.5 M NaOH[6]
Al and Si in					
Allophane	Poor	Poor	Poor	Good	Good
Imogolite	Poor	Poor	Poor	Good-fair	Good
"Allophanelike"	Poor	Good	Good	Good	Good
Layer Silicates	No	No-poor	No	No-poor	Poor-fair
Al in					
Organic complexes	Good	Good	Good	Good	Good
Hydrous oxides					
Noncrystalline	Poor	Good	Good	Good	Good
Crystalline	No	Poor	Poor	No	Good
Si in					
Opaline silica	No	No	Poor	No	Good
Crystalline silica	No	No	No	No	Poor
Fe in					
Organic complexes	Good	Good	No	Good	No
Hydrous oxides					
Noncrystalline[7]	Poor	Good	No	Good	No
Crystalline	No	Good	No	No-poor	No

[1] After Wada, 1989; used by permission from SSSA.
[2] McKeague et al., 1971; Higashi and Wada, 1977.
[3] Mehra and Jackson, 1960; Tokashiki and Wada, 1975.
[4] Jackson, 1956; Tokashiki and Wada, 1975.
[5] Schwertmann, 1964; Higashi and Ikeda, 1974; Wada and Kakuto, 1985b.
[6] Hashimoto and Jackson, 1960; Tokashiki and Wada, 1975.
[7] Includes ferrihydrite.

The SSL routinely performs the above-described dithionite-citrate, sodium pyrophosphate, and ammonium oxalate extractions. These selective dissolutions are discussed in greater detail in the following section.

4.6   Selective Dissolutions
4.6.2   Dithionite-Citrate Extraction
4.6.2.1–3   Aluminum, Iron, and Manganese

Dithionite-citrate extraction, background: The original objectives of the dithionite-citrate extraction were to determine the free Fe oxides and to remove the amorphous coatings and crystals of free Fe oxide, acting as cementing agents, for subsequent physical and chemical analyses of soils, sediments, and clay minerals (Weaver et al., 1968; Jackson, 1969; Jackson et al., 1986). Dithionite-citrate extractable Fe (Fe_d) is considered a measure of "free iron" in soils; as such, it is pedogenically significant. Dithionite-citrate extractable Fe data are of interest in soil genesis-classification studies because of increasing concentration with increasing weathering and the effect of Fe on soil colors (Schwertmann, 1992). "Free iron" is also considered an important factor in P fixation and soil aggregate stability.

This extraction is also sometimes referred to as citrate-dithionite or sodium citrate-dithionite. The method described herein is after Soil Survey Staff (2004) and is not the same extraction as that described by Soil Survey Staff (1972), which is an obsolete method. This retired method, commonly referred to as the CBD method, incorporated sodium bicarbonate as a buffer (pH 7.3) in the dithionite-citrate method, resulting in a buffered neutral citrate-bicarbonate-dithionite (Aguilera and Jackson, 1953; Mehra and Jackson, 1960; Jackson, 1969).

Dithionite-citrate extraction, application: Dithionite-citrate (CD) is used as a selective dissolution extractant for organically complexed Fe and Al, noncrystalline hydrous oxides of Fe and Al, and amorphous aluminosilicates (Wada, 1989). The CD solution is a poor extractant of crystalline hydrous oxides of Al, allophane, and imogolite. The CD solution does not extract opal, Si, or other constituents of crystalline silicate minerals (Wada, 1989).

Dithionite-citrate extraction, measurement: A soil sample is mixed with sodium dithionite, sodium citrate, and reverse osmosis deionized (RODI) water and is shaken overnight. The solution is centrifuged, and a clear extract is obtained. The CD extract is diluted with RODI water. The analytes are measured by an atomic absorption spectrophotometer (AAS). The data are automatically recorded by a computer and printer. The AAS converts absorption to analyte concentration. The CD extractable Al_d, Fe_d, and Mn_d are reported as percent.

Dithionite-citrate extraction, interferences: The redox potential of the extractant is dependent upon the pH of the extracting solution and the soil system. Sodium citrate complexes the reduced Fe and usually buffers the system to a pH of 6.5 to 7.3. In some soils the pH may be lowered, resulting in the precipitation of Fe sulfides. Filtered extracts can yield different recoveries of Fe, Mn, and Al relative to unfiltered extracts.

4.6 Selective Dissolutions
4.6.3 Ammonium Oxalate Extraction
4.6.3.1–5 Aluminum, Iron, Manganese, Phosphorus, and Silicon
4.6.3.6 Optical Density, Ammonium Oxalate Extract (ODOE)

Ammonium oxalate extraction, background: The intent of the ammonium oxalate procedure is to measure the quantities of poorly crystalline materials in the soil. At the present time, the ammonium oxalate extraction is considered the most precise chemical method for measuring these soil components. In principle, however, it cannot be expected that chemical methods are able to perfectly distinguish the degrees of crystallinity, and some caution is needed in the interpretation of the analytical data (Van Wambeke, 1992). A more reliable and accurate estimation of soil properties and a better understanding of noncrystallinity are provided when ammonium oxalate extraction is used in conjunction with other selective dissolution procedures, thermal techniques, and chemical tests (Jackson et al., 1986).

Ammonium oxalate extraction, application: In general, ammonium oxalate allowed to react in darkness has been considered to be a selective dissolution for noncrystalline materials (McKeague and Day, 1966; Higashi and Ikeda, 1974; Fey and LeRoux, 1976; Schwertmann and Taylor, 1989; Hodges and Zelazny, 1980). The ammonium oxalate procedure removes most noncrystalline and paracrystalline materials (allophane and imogolite) from soils (Higashi and Ikeda, 1974; Hodges and Zelazny, 1980) as well as short-range-ordered oxides and hydroxides of Al, Fe, and Mn (Schwertmann, 1959, 1964; McKeague and Day, 1966; McKeague et al., 1971; Fey and LeRoux, 1976). In addition, this method is assumed to extract Al + Fe humus complexes. Ammonium oxalate does not extract opaline silica, layer silicates, crystalline hydrous oxides of Fe and Al, or crystalline silicate (Wada, 1977, 1989). This procedure has been reported to dissolve very little hematite and goethite and small amounts of magnetite (Baril and Bitton, 1969; McKeague et al., 1971; Walker, 1983). There have been conflicting data on the effect of this procedure on clay minerals, but in general the ammonium oxalate treatment is considered to have very little effect on phyllosilicates (kaolinite, montmorillonite, illite) or gibbsite. The ammonium oxalate extraction is assumed to dissolve selectively "active" Al and Fe components that are present in noncrystalline materials as well as associated or independent poorly crystalline silica. The method also extracts allophane, imogolite, Al + Fe humus complexes, and amorphous or poorly crystallized oxides and hydroxides. In the system of soil taxonomic classification, ammonium oxalate extractable Fe and Al are criteria for andic soil properties and spodic materials (Soil Survey Staff, 2010). This extraction is sometimes referred to as acid ammonium oxalate, acid oxalate, oxalate-oxalic acid, or oxalic acid-acid ammonium oxalate.

Ammonium oxalate extraction, measurement: A soil sample is extracted with a mechanical vacuum extractor (Holmgren et al., 1977) in a 0.2 M ammonium oxalate solution buffered at pH 3.0 under darkness. The ammonium oxalate extract is weighed. The ammonium oxalate extract is diluted with reverse osmosis deionized water. The analytes are measured by an inductively coupled plasma atomic emission spectrophotometer (ICP-AES). Data are automatically recorded by a computer and printer. The SSL reports ammonium oxalate extractable Al_o, Fe_o, and Si_o as percent and the Mn_o and P_o as mg kg^{-1}. The optical density of the ammonium oxalate extract (ODOE) is measured with a UV spectrophotometer at 430 nm and reported to the nearest 0.01 unit.

Ammonium oxalate extraction, interferences: The ammonium oxalate buffer extraction is sensitive to light, especially UV light. The exclusion of light reduces the dissolution effect of crystalline oxides and clay minerals. The dissolution of magnetite during the extraction may give erroneous high results due to the dissolution of this mineral (Walker, 1983). If the sample contains large amounts of amorphous material (>2 percent Al), an alternate method should be used, i.e., shaking with 0.275 *M* ammonium oxalate, pH 3.25, 1:100 soil:extractant.

Optical density, ammonium oxalate extract, background: There is a similarity between optical density and absorbance. In early spectrophotometry, what is now called absorbance was often termed optical density (Skoog, 1985). Optical density or absorbance of a solution or solid is defined as follows:

Equation 4.6.3.1:

$$D = \log_{10} I_o/I = \log_{10} 1/T = -\log_{10} T$$

where
D = Optical density
I_o = Intensity of incident light
I = Light intensity after passage through solution
T = Transmission

Transmittance of a solution is the fraction of incident radiation transmitted by the solution, i.e., = I/I_o, and is often expressed as a percentage, i.e., $\%T = I/I_o$ x 100. These relationships are derived from Beer's Law.

Optical density, ammonium oxalate extract, application: The ODOE is used as a taxonomic criterion to help identify spodic materials (Soil Survey Staff, 2010). Soils with spodic materials show evidence that organic materials and aluminum, with or without Fe, have been moved from an eluvial horizon to an illuvial horizon, and an increasing ODOE indicates an accumulation of translocated organic materials in an illuvial horizon (Soil Survey Staff, 2010). Refer to Soil Survey Staff (2010) for a more detailed discussion of ODOE as a taxonomic criterion.

4.6   Selective Dissolutions
4.6.4   Sodium Pyrophosphate Extraction
4.6.4.1–4   Aluminum, Iron, Manganese, and Organic Carbon

Sodium pyrophosphate extraction, background: Sodium pyrophosphate extractable Al, Fe, and C were former criteria for spodic materials in *Soil Taxonomy* (Soil Survey Staff, 1975). These sodium pyrophosphate data were used in conjunction with dithionite-citrate data to help identify translocated Fe and Al humus complexes in spodic horizons (Soil Survey Staff, 1975). At one time, these data were referred to as spodic horizon criteria on the SSL data sheets, but they have since been replaced by other taxonomic criteria, such as ammonium oxalate extractable Fe and Al (Soil Survey Staff, 2010). They are still useful as an approximate measure of the organically chelated Fe and Al in some soils.

Sodium pyrophosphate extraction, application: Sodium pyrophosphate (0.2 M $Na_4P_2O_7$) is used as a selective dissolution extractant for organically complexed Fe and Al (Wada, 1989), while the dithionite-citrate extraction tends to extract these compounds plus the free oxides (McKeague et al., 1971). The sodium pyrophosphate solution is a poor extractant for allophane, imogolite, amorphous aluminosilicates, and noncrystalline hydrous oxides of Fe and Al. The sodium pyrophosphate solution does not extract opal, crystalline silicates, layer silicates, and crystalline hydrous oxides of Fe and Al (Wada, 1989).

Sodium pyrophosphate extraction, measurement: A soil sample is mixed with 0.1 M $Na_4P_2O_7$ and shaken overnight. The solution is then allowed to settle overnight before it is centrifuged and filtered to obtain a clear extract. The analytes (Al, Fe, Mn) are measured by an atomic absorption spectrophotometer (AAS). The data are automatically recorded by a computer and printer. The AAS converts absorption to analyte concentration. The SSL reports sodium pyrophosphate extractable Al_p, Fe_p, and Mn_p as percent. The organic C in the sodium pyrophosphate extract (C_p) is wet oxidized in a fume hood and gravimetrically measured.

Sodium pyrophosphate extraction, interferences: Sodium pyrophosphate extraction can result in the peptization and dispersion of microcrystalline iron oxide (Jeanroy and Guillet, 1981). The quantity of Fe extracted with pyrophosphate decreases with increasing centrifugation (McKeague and Schuppli, 1982); therefore, uniform high-speed centrifugation or micropore filtration treatments are required (Schuppli et al., 1983; Loveland and Digby, 1984). Sodium pyrophosphate extraction works best at pH 10 (Loeppert and Inskeep, 1996). The concentration of $Na_4P_2O_7$ solution must be close to 0.1 M. Variable amounts of Fe, Al, Mn, and organic C may be extracted by varying the pyrophosphate concentration.

4.6   Selective Dissolutions
4.6.5   Ratios, Estimates, and Calculations Associated With Selective Dissolutions
4.6.5.1 Fe_d, Al_d, Mn_d, Al_o, and Fe_o, Pedogenic Significance

The Fe_d and Fe_o are pedogenically significant. In a general way, the Fe_d is considered to be a measure of the total pedogenic Fe, e.g., goethite, hematite, lepidocrocite, and ferrihydrite, and the Fe_o (probably ferrihydrite) is a measure of the paracrystalline Fe (Schwertmann, 1973; Birkeland et al., 1989), but the Fe_p does not necessarily correlate with organic-bound Fe (Birkeland et al., 1989) as commonly thought (Schwertmann and Taylor, 1977; Parfitt and Childs, 1988). Ammonium oxalate analysis appears to release some Fe from magnetite grains (Walker, 1983). The Fe_o/Fe_d ratio is often calculated because it is considered an approximation of the relative proportion of noncrystalline Fe oxide or ferrihydrite in soils (Schwertmann, 1985). The Fe_d can be compared to total Fe in the horizon to evaluate the degree of weathering that has occurred (Wilson et al., 1996). The Al_d and Al_o also are pedogenically significant. The Al_d represents the Al substituted in Fe oxides, which can have an upper limit of 33 mole percent substitution (Schwertmann et al., 1977; Schwertmann and Taylor, 1989). The Al_o is generally an estimate of the total pedogenic Al in soils dominated by allophane, imogolite, and organically bound Al (Wada, 1977, 1989; Childs et al., 1983). Unlike Fe_d, the Al_d extract is commonly less than the Al_o (Childs et al., 1983; Birkeland et al., 1989) and therefore does not necessarily represent the total pedogenic Al (Wada, 1977, 1989). The Mn_d is considered the "easily reducible Mn." The Fe_d/Fe_o ratio is not reported on the Primary Characterization Data Sheet.

4.6 Selective Dissolutions
4.6.5 Ratios, Estimates, and Calculations Associated With Selective Dissolutions
4.6.5.2 Al_p, Fe_p, C_p, Al_d, and Fe_d, Spodic Horizon, Former Laboratory Criteria
4.6.5.3 $Al_o + \frac{1}{2} Fe_o$, Spodic Horizon, Current Laboratory Criteria

Spodic horizon, field and laboratory criteria: Historically, the spodic horizon has been identified by both field and laboratory criteria, and these criteria have changed over time. In this section, the former and current laboratory criteria for spodic horizons are discussed. The intent of this discussion is to provide some historical perspective as well as current understanding of the appropriate applications and limitations of the various selective dissolutions. Refer to Soil Survey Staff (2010) for a more detailed discussion of the past and current taxonomic criteria for spodic horizons as well as the distinctions between spodic horizons and andic soil materials and other diagnostic subsurface horizons.

Spodic horizon, former laboratory criteria: The original concept of the Podzol in Russia emphasized the ashy gray eluvial horizon, and the name was applied to soils having such a horizon regardless of the nature of the underlying illuvial horizon (McKeague et al., 1983). In general, as this term was carried into German and later into English, it became associated with soils having, in addition to a bleached layer, an underlying reddish to dark brownish or black illuvial horizon typical of Podzols (McKeague et al., 1983). These illuvial horizons typically consist largely of humus and sesquioxides. At one time, the Al_p, Fe_p, and C_p were used in conjunction with Fe_d and Al_d to help identify translocated Al and Fe humus complexes in spodic horizons and thus served as chemical requirements for spodic horizons (Soil Survey Staff, 1975). These chemical requirements included the ratios $(Fe_p+Al_p)/(Fe_d+Al_d)$, $(Fe_p+Al_p)/clay$, and $(Al_p+C_p)/clay$. In the past, these ratios, estimates, and calculations were reported as spodic horizon criteria on the SSL data sheets, but they have since been replaced by other taxonomic criteria (Soil Survey Staff, 2010).

The original intent of these chemical determinations was to define some of the more weakly defined spodic horizons (McKeague et al., 1983). These requirements were based on the chemical composition and activity of the illuvial material. Sodium pyrophosphate extracting solutions tend to selectively extract mainly Fe and Al associated with organic compounds, while the dithionite-citrate extractions tend to extract these compounds plus the free oxides (McKeague et al., 1971). To emphasize the organic-sesquioxide complexes, the taxonomic definition for a spodic horizon required the following: (1) a high amount of Fe and Al extracted by pyrophosphate relative to the amount extracted by dithionite-citrate; (2) a high pyrophosphate-extractable Al + Fe or Al + C percentage relative to the percentage of clay in the horizon in order to eliminate horizons dominated by silicate clay; and (3) relatively large CEC from nonsilicate clay sources so as to eliminate weakly developed soils (McKeague et al., 1983). Numerous evaluations of pyrophosphate extracts have indicated that the pyrophosphate extraction does not necessarily correlate with organic-bound Fe and Al (Schuppli et al., 1983; Kassim et al., 1984; Parfitt and Childs, 1988; Birkeland et al., 1989) as commonly thought (Schwertmann and Taylor, 1977; Parfitt and Childs, 1988). Pyrophosphate not only extracts organic-bound Fe but also peptizes solid particles of ferrihydrite and in some instances even goethite (Yuan et al., 1993).

Spodic horizon, current laboratory criteria: Organic substances are believed to be of the first importance in the development of spodic horizons because of their dominant role in the processes of mobilization, migration, and accumulation (De Connick, 1980). If enough polyvalent cations, especially Al and Fe, are available, mobile organic substances are immobilized in place, but if there are insufficient amounts of Al and/or Fe to completely immobilize the mobile compounds, these cations are complexed by the mobile compounds and are transported downward (De Connick, 1980). Other theories of deposition include microbial decomposition of organic ligands (Lundstrom et al., 1995, 2000) and eluviations and precipitation or *in situ* formation of proto-imogolite sols in Bs horizons followed by adsorption of mobile humus (Farmer et al., 1980; Anderson et al., 1982; Ugolini and Dahlgren, 1987).

In *Keys to Soil Taxonomy*, spodic horizons are defined as illuvial horizons with ≥85 percent spodic materials (Soil Survey Staff, 2010). Spodic materials contain illuvial "active amorphous" materials composed of organic matter and Al, with or without Fe. The term "active" is used here to describe materials that typically express high pH-dependent CEC, large surface area, and high water retention (Soil Survey Staff, 2010). Chemical criteria for spodic placement include: (1) $Al_o + \frac{1}{2} Fe_o$ percentage totaling ≥0.50 and half that amount or less in an overlying horizon and (2) an ODOE value ≥0.25 and a value half as high or lower in an overlying horizon (Soil Survey Staff, 2010). The intent of these criteria is to show evidence that organic materials and Al, with or without Fe, have been moved from an eluvial horizon to an illuvial horizon. The sensitivity and consistency of ODOE and $Al_o + \frac{1}{2} Fe_o$ as chemical indicators of spodic development are evidenced in a chronosequence of soil development on moraines of the Mendenhall Glacier in Alaska. Spodic development was achieved within a relatively short period of time, with the accumulation and mineralization of organic matter as the dominant factor. The $Al_o + \frac{1}{2} Fe_o$ and ODOE are reported on the Primary Characterization Data Sheets.

4.6 Selective Dissolutions
4.6.5 Ratios, Estimates, and Calculations Associated With Selective Dissolutions
4.6.5.4 $Al_o + \frac{1}{2} Fe_o$, Andic Soil Properties

Andic soil properties result mainly from the presence of significant amounts of allophane, imogolite, ferrihydrite, or aluminum-humus complexes in soils (Soil Survey Staff, 2010). The $Al_o + \frac{1}{2} Fe_o$ is used in conjunction with other laboratory data (ρ_{B33}, phosphate retention, and volcanic glass) as a taxonomic characteristic to identify andic soil properties. The use of $\frac{1}{2} Fe_o$ in the equation is because the weight of Fe atoms is approximately twice that of Al atoms. In evaluating the relative proportion of Fe and Al atoms solubilized by ammonium oxalate, the weight percent of Fe must be divided by two, i.e., $Al_o + \frac{1}{2} Fe_o$. The $Al_o + \frac{1}{2} Fe_o$ is reported on the SSL taxonomic data sheets. Refer to Soil Survey Staff (2010) for a more detailed discussion of these chemical criteria for andic soil properties.

4.6 Selective Dissolutions
4.6.5 Ratios, Estimates, and Calculations Associated With Selective Dissolutions
4.6.5.5 Al_o, Si_o, and Al_p, Allophane Estimation

Allophane is a mineral series that has an Al/Si molar ratio ranging from 1.0 to 3.0 (Parfitt and Kimble, 1989), but allophanes with an Al/Si molar ratio >2.0 have not been isolated (Dahlgren et al., 1993). Allophanes with Al/Si = 2.0 are thought to be the most readily formed and stable in soils (Parfitt and Kimble, 1989). Allophane in soils has been estimated from the Al_o, Si_o, and Al_p (Parfitt and Henmi, 1982; Parfitt and Wilson, 1985). The Al_o represents the Al dissolved from allophane, imogolite, and Al-humus complexes, and the Al_p is the Al from the Al-humus complexes alone (Parfitt and Kimble, 1989). The Al_o minus the Al_p gives an estimate of the Al in allophane and imogolite, whereas the Si_o gives an estimate of the Si in allophane and imogolite. The $(Al_o - Al_p)/Si_o$ times the molar ratio (28/27) is an estimate of the Al/Si ratio of allophane and imogolite in the soil. The values of 28 and 27 represent the atomic weights of Si and Al, respectively. Parfitt and Henmi (1982) characterized an allophane that had a disordered imogolite structure, an Al/Si ratio of 2.0, and Si_o content of 14.1 percent. Using this allophane and additional samples as references, Parfitt and Henmi (1982) and Parfitt and Wilson (1985) developed an algorithm to estimate the percent allophane in a sample using the calculated Al/Si molar ratio and Si_o content. The amount of allophane is estimated by multiplying the Si_o by the appropriate factor provided in Table 4.6.5.5.1 (Parfitt, 1990). The $(Al_o - Al_p)/Si_o$ is not reported on the SSL data sheets.

Table 4.6.5.5.1   Al:Si atomic ratios of allophane and factor used to estimate allophane[1]

Al:Si	Factor[2]	Al:Si	Factor[2]
1.0	5	2.5	10
1.5	6	3.0	12
2.0	7	3.5	16

[1] From Parfitt (1990); used by permission from the Australian Journal of Soil Research. Copyright CSIRO 1990. Published by CSIRO PUBLISHING, Melbourne, Australia; available at http://www.publish.csiro.au/nid/84/paper/SR9900343.htm.
[2] Factor to use with Si_o.

4.6   Selective Dissolutions

4.6.5   Ratios, Estimates, and Calculations Associated With Selective Dissolutions

4.6.5.6 Fe_o, Si_o, and Fe_d, Mineralogy Classes

Selective chemical dissolution data are used as taxonomic criteria for mineralogy classes. The Fe_o and Si_o are used for the amorphic and ferrihydritic mineralogy classes. The Fe_d is used for the ferritic, ferruginous, and parasesquic mineralogy classes. Refer to Soil Survey Staff (2010) for a more detailed discussion of these mineralogy classes.

4.7   Total Analysis

This section describes the SSL methods for total analysis of major and trace elements. It also provides geologic and environmental information relevant to specific elements. The soil processes of anthropogenic pollution and contamination are described, and references to case studies/datasets are presented as evidentiary examples of the actions/practices that have promoted or diminished these processes. In addition, major developments in knowledge and scientific technology related to soil organic and inorganic components are discussed. Major elements (Al, Ca, Fe, K, Mg, Mn, Na, P, Si, Sr, Ti, and Zr) are analyzed by digesting 100 mg of dried clay suspension, the fine-earth (<2-mm) fraction, or other particle-size separate with HF + HNO_3 + HCl. Trace elements (Ag, As, Ba, Be, Cd, Co, Cr, Cu, Hg, Mn, Mo, Ni, P, Pb, Sb, Se, Sn, Sr, V, W, and Zn) are analyzed by digesting 500 mg of the fine-earth (<2-mm) fraction with HNO_3 + HCl. Samples are placed in Teflon digestion vessels and heated in a microwave. Elemental concentrations of digestates are determined by inductively coupled plasma atomic emission spectrophotometry (ICP-AES) for major elements and ICP mass spectrometry (ICP-MS) for trace elements. The SSL major and trace element methods follow U.S. Environmental Protection Agency (EPA) methods 3051A and 3052, respectively. Refer to Appendix 7 (Newport Pedon) for example SSL reports for major and trace element data.

This section also describes the SSL methods for total C, N, and S by dry combustion and specific method applications and interferences. Included in this discussion is information about agronomic ratios (e.g., C:N), organic and inorganic forms, and their role in nutrient cycling and plant nutrition. Information on the estimation of organic matter and organic C using total C data also is provided. For detailed descriptions of the SSL total analysis methods which are cross-referenced by method code in the table of contents in this manual, refer to SSIR No. 42 (Soil Survey Staff, 2004), which is available online at http://soils.usda.gov/technical/lmm/.

4.7   Total Analysis
4.7.1–2   Major and Trace Elements
4.7.1–2.1   Elemental Evaluations

Elemental analysis, general applications: Prior to the development of modern analytical techniques (e.g., x-ray diffraction and thermal analysis), the identification of rocks and minerals was historically based on elemental analysis and optical properties (Washington, 1930; Bain and Smith, 1994). Chemical analysis is still essential to determine mineral structural formulas and to identify and quantify specific mineral species through elemental allocation to minerals. Many clay mineral groups are subdivided based on composition. Elemental analysis of the fine-earth (<2-mm) fraction or specific particle-size separates has also been used in the study of soil properties on a pedon, landscape, or ecosystem basis by providing information on geologic processes and parent material uniformity (Chapman and Horn, 1968; Kaup and Carter, 1987); pedogenesis (Brimhall et al., 1991; Wilson et al., 1996); or mineral weathering, composition, and phase quantification (Nettleton et al., 1970; Dubbin et al., 1993; Amonette and Sanders, 1994). Elemental analysis has also been used in soil fertility evaluations and environmental studies. Some of these applications of elemental data are discussed in more detail in the following sections.

Elemental analysis, clay fraction components: Total elemental analysis data may be used to estimate various components of the clay fraction. The percent Fe_2O_3 and Al_2O_3 in the clay fraction of soils alone is not generally used for mineral calculations; however, these data are useful to compare horizons within or between pedons as indicators of depletion or concentration by weathering or other pedogenic processes. If percent Al_2O_3 is <10, there should be a reason; e.g., calcite may represent a large proportion of the clay fraction. If kaolinite = 5 by x-ray and/or gibbsite is present, percent Al_2O_3 should approach or exceed 30 percent. The percent K_2O may be used to estimate mica (illite) in soils that do not contain feldspars in the clay fraction. Some general rules are as follows:

In clay derived from a soil developed in igneous parent materials, the soil mica is assumed to contain 10 percent K_2O, which is less than found in theoretically ideal micas (11.8 percent) (Alexiades and Jackson, 1966; Fanning et al., 1989).

Equation 4.7.1.1.1:

% Mica = % K_2O x 10

In clay derived from marine sedimentary materials, the mica is assumed to contain only 8.0 percent K_2O (Berdanier et al., 1978). The 8.0 percent K_2O value, which represents 100 percent mica, is used as a taxonomic criterion to help define the illitic family mineralogy class (Soil Survey Staff, 2010). Illitic mineralogy is defined as more than one-half (by weight) illite (hydrous mica) and commonly more than 4 percent K_2O in the <0.002-mm fraction (Soil Survey Staff, 2010).

Equation 4.7.1.1.2:

% Mica (illite) = % K_2O x 12.5

Some adjustment to these rules may be needed in dryland areas (Nelson and Nettleton, 1975) and in other source materials of illites. In some dryland areas, clays have been found with mica as the only mineral (by x-ray diffraction) and having 4 to 6 percent K_2O.

Equation 4.7.1.1.3:

% Mica (illite) = % K_2O x 20

The percent K_2O may also be used as a check for internal consistency in laboratory data. The relative amounts of mica by x-ray analysis are adjusted in accordance with the percent K_2O in total clay as follows:

X-ray	% K_2O
1 or 2	0–1
2 or 3	1–2
3 or 4	2–4
5	>4

Elemental analysis, pedogenic evaluations: Mineral weathering during pedogenesis results in translocation and/or accumulation of major or trace elements via chemical and biological processes, such as leaching, podzolization, and oxidation-reduction (Davies, 1980; Kabata-Pendias and Pendias, 1992; Mausbach and Richardson, 1994; Pierzynski et al., 1994; Wilson et al., 1996). The determination of soil processes and resulting pedogenic changes often involves elemental evaluations of horizons within the solum, between the solum and parent material, and along bioclimatic gradients. Examples of pedogenic evaluations using elemental data include the ratios of Si/Al and Si/Al+Fe as soil weathering indexes (Jackson, 1979), ratios of alkali cations (e.g., Ca and K) and the relatively stable elements Ti/Zr as measures of pedogenic changes and geologic uniformity (Sudom and St. Arnaud, 1971; Muhs et al., 2001), total and fractionated forms of P (e.g., inorganic, organic, and residual) as indices of soil development (Syers et al., 1968; Walker, 1974; Walker and Syers, 1976; Tiessen et al., 1984; Singleton and Lavkulich, 1987; Guzel and Ibrikci, 1994; Cross and Schlesinger, 1995; Burt and Alexander, 1996), and the relative accumulation of Fe and Al and depletion of P as a function of climate indicative of pedogenic gradients (Tedrow, 1968; Jenny, 1980; Bockheim, 1980; Birkeland et al., 1989). Geochemical studies have also used trace elements to document such soil processes as element release and transport (weathering and profile leaching) in Spodosols (Jersak et al., 1997) and elemental immobilization (oxidation-reduction and associated pH changes) in wetlands (Gambrell, 1994).

Elemental analysis, soil fertility: Elemental analysis has also been used in fertility evaluations. Total P and fractionated forms (e.g., organic P) have served as indicators of fertility (Walker and Adams, 1958) as well as long-term changes in soil P as affected by changes in short-term processes (e.g., soil management) (Hedley et al., 1982; Sharpley and Smith, 1985). Phosphorus fertility of a soil and potential water-quality problems can be better understood by measurements of total P, especially when compared to other P measurements, such as water-soluble or Bray-1 extractable P. The inherent fertility of a soil derived from its parent material can be examined by determination of the basic cations relative to the Si or Al content. Elemental analysis has been important in fertility studies of serpentinitic soils documenting nutrient deficiencies and imbalances (low N, P, and K; adverse Ca/Mg ratios) and nonanthropogenic metal enrichments (Ni, Cr, and Co) (Johnston and Proctor, 1979; Alexander et al., 1985; Brooks, 1987). Studies have shown Ca/Mg ratios <0.7, which are considered unfavorable for the growth of most plants (Proctor and Woodell, 1975; Woodell et al., 1975; Woolhouse, 1983). A Ca/Mg ratio <0.1 is used in the Magnesic Great Group in the Australian Soil Classification System (Isbell, 1996), but this ratio (neither total nor available Ca, Mg) has yet to be incorporated into the U.S. soil taxonomy system (Soil Survey Staff, 2010).

Elemental analysis, environmental studies: Elemental data are useful information in environmental studies of soil and water, e.g., metal contamination through atmospheric deposition (Storm et al., 1994; Burt et al., 2000) or transport in surface or ground waters (Mesuere et al., 1991; Jones et al., 1997; Martens and Suarez, 1997; Whatmuff, 2002). For example, total P provides useful baseline data for environmental studies. Soils with naturally high P levels (660 to 3600 mg kg-1) and high P sorption in the Tualatin River Basin in Oregon were found to serve as potential sources of nonpoint P pollution (Abrams and Jarrell, 1995). A common procedure used in studies to assess environmental levels or background amounts of trace elements in soils is the quantification of the total or the total extractable pool of trace elements in soils (Shacklette and Boerngen, 1984; Shuman, 1985; Holmgren et al., 1993; Mermut et al., 1996; Chen et al., 1999; Burt et al., 2003). This pool is routinely determined not only because these data are important to any overall assessment of the fate, bioavailability, and transport of trace elements (Shacklette and Boerngen, 1984; Holmgren et al., 1993; Wilson et al., 2001) but also because of the lack of a widely applicable method to assess the bioavailable fraction.

Elemental analysis, data relationships: Many studies have examined relationships among elements (major and trace) and between elemental concentrations and other soil properties in noncontaminated soils. Al and Fe were strongly related to total P, with Al (r^2 = 0.83) explaining most of the variation in 22 U.S. benchmark soils (Burt et al., 2002a). Total Al and Fe explained 29 percent of the variation in baseline concentrations of trace elements in Florida surface soils (Chen et al., 1999), and total Al strongly correlated with selected trace elements (Pb, Cd, and Cr with r^2 = 0.88, 0.81, and 0.45, respectively) in Louisiana coastal wetlands (Pardue et al., 1992). Trace element concentrations have also been related to clay and CEC in 40 mineral soils in Florida (Ma et al., 1997) and to physical surface area (e.g., clay, organic matter, and CEC) in the Southeastern U.S. (Shuman, 1985). Other studies have related elemental extractability (speciation) or potential reactivity of trace elements to other soil properties, such as total concentrations and/or soil type, pH, texture, organic matter, and carbonates (Chlopecka et al., 1996; Ma and Rao, 1997; Kabala and Singh, 2001; Andersen et al., 2002).

Most studies that have developed correlations between trace elements and other soil properties have generally focused on nonanthropogenic sources of trace elements or have not evaluated natural and anthropogenic sources separately. Elements derived from anthropogenic sources likely have different correlations with other soil properties (Burt et al., 2003). Differences in correlation have been used to speculate on possible anthropogenic additions (Andersen et al., 2002). The study by Burt et al. (2003) was initiated to examine a wide variety of soils from across the U.S. with anthropogenic additions (AD) and without anthropogenic additions (NAD). Results showed that total Al, Fe, CEC, organic C, pH, and clay exhibited significant correlations (0.56, 0.74, 0.50, 0.31, 0.16, and 0.30, respectively) with total trace element concentrations of all horizons of the NAD and that Mn showed the best interelement correlation (0.33). Total Fe was shown to have one of the strongest relationships, explaining 55 percent of the variation in these associated total concentrations for all horizons in the NAD and 30 percent for all horizons in the AD.

Trace elements: The term *trace elements* is widely applied to a variety of elements that are normally present in relatively low concentrations in plants, soils, and natural waters (or what is termed background levels) and that may or may not be essential for the growth and development of plants, animals, or humans. Knowledge of these levels is important in understanding the consequences of increasing levels of trace elements in ecosystems (Tiller, 1989; Holmgren et al., 1993). Concentrations of these elements may become elevated due to natural processes (e.g., magmatic activity, mineral weathering, and translocation through the soil or landscape) or through human-induced activities (e.g., pesticides, mining, smelting, and manufacturing). The relative reactivity or bioavailability of these elements in soils is governed by a variety of chemical factors, such as pH, redox potential, organic concentrations, and oxides (Pierzynski and Schwab, 1993; Gambrell, 1994; Keller and Vedy, 1994; Burt et al., 2002b). Uses of elemental data in soil survey applications are broad and diverse (Wilson et al., 2008). They include understanding natural distributions (Wilcke and Amelung, 1996; Jersak et al., 1997) and human-induced distributions (Wilcke et al., 1998). Knowledge of the elemental amounts and distribution in soils and their relationships with other soil properties can enhance the understanding of the fate and transport of anthropogenic elements, thereby expanding the utility and application of soil survey knowledge in areas of environmental concern, such as urban development, mine spoil reclamation, smelter emissions, and agricultural waste applications (Burt et al., 2003).

The following section provides information related to specific trace elements (As, Cd, Cr, Cu, Hg, Ni, Pb, Se, and Zn). Elements described are typically of particular interest to scientists as well as to the general public, especially in relation to their impacts upon the environment and human health. The following information from Wilson et al. (2008) covers the topics of ionic form, general reactivity, mobility in soils, sources of pollution, and parent materials with high concentrations of trace elements.

Arsenic is an extremely toxic element that occurs in soils in both organic and inorganic forms. Arsenic is typically found in low concentrations in the soil but has been widely applied to soils as a component in pesticides and herbicides and also via industrial pollution and smelting operations. Parent materials with high concentrations of As include sedimentary material rich in sulfides (e.g., coal containing arsenopyrite and organic-rich coastal sediments) and sulfides of ores, such as Ag, Pb, Cu, and Ni (Fio et al., 1991; Liao et al., 2005). Arsenic is used in drugs, soaps, dyes, and metals; 90 percent of industrial As in the U.S. is used in wood preservatives (Pinsker, 2001; Stilwell and Gorny, 1997). Concerns exist for both short-term (acute) and long-term (chronic) soil exposure. The primary route for exposure is via soil ingestion or inhalation of air-borne particles. Data from As measurements in ground water by the U.S. Geological Survey and the Environmental Protection Agency suggest that most As in ground water is related to natural sources (Ryker, 2001), i.e., from mineral dissolution in geologic formations and soils. In Bangladesh, for example, As released via pyrite oxidation is in part responsible for ground-water As levels ranging between 50 and 2,500 ug L^{-1} in many wells. Soil-applied As is generally immobile and has soil chemistry similar to that of phosphorus. The element occurs as arsenate (As^{5+}) and arsenite (As^{3+}) and in soils in the form of the oxyanion AsO_4^{3-}. The weathering of limestone and the biological accumulation of As by aquatic organisms are responsible for the high levels in wetland soils of Florida (Chen et al., 2002).

Cadmium is extremely toxic and has no known function in biological processes (Wilson et al., 2008). The geochemistry of cadmium is similar to that of Zn, and cadmium is often found in association with Zn deposits (Wilson et al., 2008). Cadmium occurs in low concentrations in magmatic and most sedimentary rocks, but higher concentrations are found in argillaceous or shale materials (Kabata-Pendias and Pendias, 2001). Solution activity for Cd is strongly related to pH (Kabata-Pendias and Pendias, 2001). Adsorption increases with increases in organic matter, sesquioxides, clay, and pH (Romkens and Salomons, 1998). Adsorption to carbonates (Renella et al., 2004) forms soluble complexes of $CdSO_4^0$ and $CdCl^+$ in saline soils (Sposito, 1989; Suave and Parker, 2005).

Chromium is found in ultramafic rocks or those formations that have undergone low-grade metamorphism or in serpentinites (Burt et al., 2001a). Chromium exists in soils as both valence states but mostly as Cr^{3+}, whereas in water the predominant forms are Cr^{6+} and colloidal, organically bound Cr^{3+} (Bartlett and James, 1996). Cr^{3+} remains in cationic form at most soil pH levels and has very low solubility. Cr^{3+} precipitates as hydroxides at pH >4.5, e.g., Cr(OH) or Cr(OH)$_3$ or Cr$_2$O$_3$, and forms both soluble and insoluble complexes with organics and minerals (Wilson et al., 2008). Cr^{6+} exists in soils as an anionic form similar to orthophosphate or sulfate, and the chromate anion may be adsorbed by oxides or precipitated by cations (Bartlett and James, 1996). Cr^{3+} is generally considered immobile as it tightly binds to both organic and inorganic materials, whereas Cr^{6+} is considered more mobile (Suave and Parker, 2005). Anthropogenic sources of Cr include industrial wastes, metal finishing/plating, electronics, and wood treatment (Forstner, 1995).

Copper is most abundant in mafic and intermediate rocks and is excluded from carbonate rocks (Kabata-Pendias and Pendias, 2001). Copper is found in geologic ore deposits containing minerals with Cu in association with Fe and S. Soils forming from Cu mineral deposits are rare (Wilson et al., 2008). Copper is specifically adsorbed by carbonates, clays, oxides, and organic matter (Romkens and Salomons, 1998; Reed and Martens, 1996). Copper may be present as an exchangeable cation (Reed and Martens, 1996), i.e., adsorption involving exchange sites in acid soil conditions and chemisorption with organic ligands in alkaline conditions (Kabata-Pendias and Pendias, 2001). Copper is generally considered immobile in soils due to adsorption on organic and mineral surfaces, but biocycling results in increasing surface concentrations (Reed and Martens, 1996; Kabata-Pendias and Pendias, 2001). Anthropogenic sources of Cu include municipal sludge, waste from smelting, and poultry and swine manure (Reed and Martens, 1996).

Lead is typically found in low concentrations in most geologic materials, but higher concentrations occur where sulfides are present in the rocks, e.g., ore deposits high in sulfide minerals (PbS, galena) (Wilson et al., 2008). Lead can form dissolved and colloidal organometallic compounds, organic complexes, or soluble complexes with sulfate, bicarbonate, and carbonate (Sposito, 1989). Lead is generally considered immobile, but limited mobility of lead has been shown in Norway (Steinnes et al., 2005). Primary anthropogenic sources are automotive exhaust, paint, and long-range atmospheric transport (Steinnes et al., 2005). Lead is a major contaminant in urban areas, and children are especially vulnerable (Hamel et al., 2010).

Mercury is typically found in low concentrations in most geologic materials, but higher concentrations occur where sulfides are present in the rocks, e.g., shales. Mercury is present as a sulfide mineral, cinnabar (HgS), and is associated with other sulfide minerals containing As, Se, Ag, Au, Zn, and Pb (Crock, 1996). Mercury is highly toxic to both plants and animals. It enters the food chain primarily through atmospheric deposition (smelting, coal combustion, and volcanic activity) and pesticide usage (Pais and Jones, 1997). Because of Hg absorption by both organic and inorganic soil components, many studies have examined soil-Hg interactions (MacNaughton and James, 1974; Barrow and Cox, 1992; Yin et al., 1996) and ecosystem distributions (Hall et al., 1987; Inacio et al., 1998).

Nickel is found in high amounts in ultramafic rocks or those formations that have undergone low-grade metamorphism or in serpentinites (Burt et al., 2001a). Nickel can form organic complexes or soluble complexes with sulfate, bicarbonate, or carbonate (Sposito, 1989). Nickel functions as an exchangeable cation and is bound by oxides (Wilson et al., 2008). Nickel is mobile under reducing conditions (Lee et al., 2001). Common anthropogenic sources include mining, smelting, and industrial activities.

Selenium is a naturally occurring element in rocks and occurs in many sedimentary formations. It is especially concentrated in certain marine formations of Tertiary and Upper Cretaceous age and in soils derived from these formations, such as Mancos Shale in Colorado and Wyoming and in the shales of the Moreno and Kreyenhagen

Formations of California (Martens and Suarez, 1997). Selenium is a major problem in soils of certain areas in the Western United States. It can impact entire ecosystems, affecting the quality of life at all levels of the food chain. The problem has been exacerbated by mining and agriculture, especially drainage/irrigation of soils (Goldberg et al., 2006). Irrigation mobilizes the Se in sediments, and this element becomes concentrated in streams, rivers, and lakes as well as in the associated alluvial soils and sediments. The impacts are not restricted to rural areas but can affect life in urban areas as well. Selenium occurs in four species (related to valence states): selenate (Se^{6+}), selenite (Se^{4+}), elemental Se (Se^0), and selenide (Se^{2-}). The bioavailability and toxicity of Se are related to speciation. The oxidized species are the more mobile and predominant species in soils and natural waters (Huang and Fujii, 1996). Selenium is important due to both deficiency (forages for animals) and toxicity (bioaccumulation) concerns (Huang and Fujii, 1996).

Zinc generally occurs in low concentrations that are relatively similar in most geologic materials, but higher concentrations occur where sulfides are present in the rocks, e.g., ore deposits of Zn, sulfide, and sphalerite (Wilson et al., 2008). Zinc is readily adsorbed by carbonates, clays, oxides, and organic matter (Romkens and Salomons, 1988; Reed and Martens, 1996). Zinc may also be present as an exchangeable cation (Reed and Martens, 1996), i.e., adsorption involving exchange sites in acid soil conditions and chemisorption with organic ligands in alkaline conditions (Kabata-Pendias and Pendias, 2001). Zinc is generally considered immobile in soils due to adsorption on organic and mineral surfaces, but biocycling results in increasing surface concentrations (Reed and Martens, 1996; Kabata-Pendias and Pendias, 2001). Anthropogenic sources of Zn include municipal sludge, waste from smelting, and poultry and swine manure (Reed and Martens, 1996).

4.7   Total Analysis
4.7.1–2   Major and Trace Elements
4.7.1–2.2   Anthropogenic Pollution and Contamination, Processes, Case Studies, and Major Developments

Concentrations of organic and inorganic soil components: Soils are made up of organic and inorganic components in a variety of forms as well as soluble ions in the solution phase. The inorganic components are generally the mineral matter that is derived from the weathered rock and/or sediment in which a soil formed (parent material). The inorganic component is composed of a fine-earth fraction with particle sizes of clay (<0.002 mm), silt (0.002 to 0.05 mm), sand (0.05 to 2.0 mm), and larger rock fragments (gravel, stones, and boulders that are >2 mm). The finer (<2-mm) fractions provide important land use functions.

Organic constituents in soils are carbon-based compounds that generally occur as dead and decomposing plant materials or as humus, the stable end product. The organic materials in a soil have beneficial characteristics, such as water and nutrient retention. Organic components, such as manures, can also be added as amendments to soils to provide a source of plant-available nutrients or for the beneficial properties mentioned above. Another source of organics added to soils is in the form of manmade pesticides (synthetic organic chemicals) to control weeds, insects, or diseases.

A third component in soils consists of ions located in the soil liquid phase (soil solution) or adsorbed to the clay fraction. These ions are available for plant absorption or a variety of precipitation/dissolution/exchange/complexation reactions with organic and inorganic components. These ions are of concern in terms of soil quality as they can become concentrated in soil or in plant materials, which can then become toxic to plants or animals. Organic and inorganic soil amendments are commonly added to provide a source of plant-available elements.

Anthropogenic pollution and contamination of soils, processes: Organic and inorganic constituents are added for improvement of agricultural productivity. These compounds contain the elements of interest, but they may also have accessory components or elements that can cause environmental/land use problems if allowed to accumulate via application or redistribution. These added materials are from manures and other biosolids, inorganic fertilizers, or pesticides. They solubilize or break down to elemental or molecular form that can be either valuable or detrimental to agriculture. It is important to understand the distribution of these compounds or elements as these materials enter the food chain or drinking water or concentrate in water bodies (rivers, lakes, and wetlands). Any element or molecule must be considered through the entire biogeochemical cycle. This cycle is composed of both the abiotic, defined as the inorganic phase that exists in the atmosphere, lithosphere (rocks, soils), or hydrosphere (soluble ions in solution), and the biotic, the organic (generally having plant or animal origins) phase. In general, cycling or transformations are much slower in the abiotic phase than in the biotic phase. The result is an accumulation of elemental constituents in soils as inorganic ions.

Many elements (both macronutrients, such as N, P, K, Ca, and Mg, and micronutrients, such as Fe, Mn, Mo, Cu, Bo, and Zn) are required for the normal growth and function of both plants and animals, although the elements may become toxic in elevated concentrations. Trace elements in soils may be elevated naturally when derived from specific geologic materials, such as serpentine rocks (Burt et al., 2001a; Lee et al., 2001), or they may be elevated by industrial or other urban causes. Regardless of the source, higher concentrations of these constituents may move out of the soil and into rivers or lakes. The effects of this movement are undesirable. Ecological impacts on humans, animals, and plants resulting from this redistribution are a concern. Both organic and inorganic components used for agriculture have variable form and fates based on reactivity with soil constituents (functional groups), degradation products of the contaminant, and various soil reactions (e.g., adsorption, translocation, precipitation, runoff, and leaching).

Pesticides have multiple fates when added to the soil. They may be absorbed and accumulate in the soil, may leach or be transported by surface runoff, or may break down into multiple reaction products. Absorption is a function of both the pesticide and soil properties (content of clay and organic matter, pH, permeability, surface charge characteristics). Breakdown or degradation occurs by biological, chemical, or photochemical processes. Generally, degradation is beneficial as these reaction products are generally nontoxic to the environment. It is the mobile, persistent pesticides that are the most harmful.

Anthropogenic pollution and contamination of soils, case studies: Agricultural experiment stations associated with U.S. universities demonstrated the value of fertilizer additions at the beginning of the 20th century, but the most dramatic improvements in crop yields have been made since 1945 because of the development of readily available, inexpensive fertilizers and improvements in crop varieties. Several inorganic elements have anthropogenically accumulated in selected U.S. soils as a result of agricultural practices, mostly from inorganic or organic materials added for fertilizers. Use of these fertilizers created concerns for anthropogenic pollution both onsite and offsite.

For example, cadmium is a natural component of the raw phosphate-rich deposits that are used to make P fertilizers. While P fertilizers have low concentrations of Cd and other selected trace elements, these elements may accumulate in the soil over time and be absorbed by crops. Nitrogen inputs into soil are from rainfall, fertilizer application, N fixation by leguminous plants (e.g., soybeans and clovers), animal/human waste, and urban runoff. Typically, the issue with N is not its accumulation in soils (N is in deficient supply for many crops) but rather its runoff into and accumulation in water systems (e.g., ground water, bays) that are used for human consumption. Production and distribution of cheap, readily available sources of N (anhydrous ammonia and ammonium nitrate) in the 1950s led to increased use of N fertilizers. As one example of N movement, Keeney and DeLuca (1993) studied water quality/flow rate data on the Des Moines River and concluded that agricultural use of N was a principal component of the increased amounts of nitrates in the river since 1945.

Certain elements become a problem as a result of their natural concentration in soils, not because of their application via fertilizers. Selenium naturally occurs in elevated concentrations in Upper Cretaceous and Tertiary marine sedimentary deposits in Colorado, California, Utah, Arizona, Nevada, North Dakota, South Dakota, Nebraska, and New Mexico (e.g., Mancos Shale). Widespread irrigation of these arid, seleniferous soils has resulted in mobilization of Se and deposition downstream or offsite in potentially hazardous concentrations.

Municipal wastewater treatment plants concentrate biosolids that are used for land application. Metals or organic compounds can be a concern in biosolids that originate from industrial inputs. For example, application of municipal wastes as amendments can result in accumulations of As, Zn, and Cu in soils as these elements are accessory constituents from industrial additions. Also, agricultural biosolids (manures from poultry or swine) may have elevated concentrations of these elements via feed additives.

A wide variety of synthetic organic and inorganic chemicals may be used in agriculture to control a variety of plant, insect, or animal pests. The persistence (decomposition), transport, and fate of these applied products are of concern, especially as many of these pesticides and their decomposition products end up in surface water and ground water (Thurman et al., 1992). The use of pesticides is widespread across agriculture. Over 0.5 million Mg of pesticides are used in the U.S. annually (Pierzynski et al., 2000). For example, soils in the Cotton Belt in the Southern U.S. receive large quantities of both herbicides and insecticides. Another widespread problem is accumulation of Pb and As in soils from long-term use of lead arsenate as an insecticide in orchards or in soils managed for vegetable production (Codling and Richie, 2005;

Renshaw et al., 2006). Copper compounds were used in Florida for many years as a fungicide in orchard crops, resulting in the buildup of this element in soils.

Major developments in knowledge, science, and technology in organic and inorganic soil components: Soil testing methodologies and laboratories across the U.S. have increased since the 1940s. This increase in laboratory evaluation of soils for nutrient application is parallel to the increasing application of organic and inorganic fertilizers. It has resulted in a more quantitative application of nutrients, thereby avoiding wasteful accumulation. With the development of higher yielding crop varieties and new irrigation technologies, improved fertilizers that increase efficiency have been developed. Examples include foliar application, slow-release nitrogen fertilizers, high-analysis polyphosphates, and fluid fertilizers for use in irrigation systems. With the growing number of soil testing laboratories, research in this area has been progressing. For example, there have been developments in soil test correlations to understanding the growth curve (rate) of specific crops in relation to nutrient absorption and use efficiency (Melsted and Peck, 1977).

Management practices that maximize the absorption of N in soils via crop uptake can minimize the loss of nitrates. These practices can include crop rotation, buffer strips, prairie restoration, soil testing, improved livestock manure management, and fine tuning fertilization practices (Keeney and DeLuca, 1993). The increasing use of irrigation, especially center-pivot systems, has greatly expanded the amount of land available for agriculture. Irrigation also provides benefits related to soil nutrients as there is a greater efficiency of the fertilizer applied (Tisdale et al., 1985).

The value of manures and sludges for providing both nutrients and organic matter to soils used for crop production is becoming more recognized. Proper application rates have been increasingly understood to minimize movement of nutrients offsite. The nutrient content of the waste material (solid or liquid) must be analyzed, and uniform application can be obtained via broadcasting on the surface or injecting into the subsurface.

The Uncompahgre Project was initiated in the early 1900s to develop irrigation for crop production in western Colorado. Irrigation of the saline and seleniferous soils in this area has resulted in an increase in salt (including Se) load of the Colorado River. A public and multiagency governmental task force (including the USDA/NRCS) was formed in the Gunnison Basin of Colorado to assess the problem of Se from irrigated soils in the Uncompahgre and Gunnison River Basins of the Upper Colorado River (Gunnison Basin & Grand Valley Selenium Task Forces Web page). A report on the reclamation was completed in 1984. The Colorado River Basin Salinity Control Program of the Lower Gunnison Basin Unit, Colorado, suggested irrigation water management alternatives to lower the total salt load to the river from these soils. It was found that much of the salt load occurred when irrigation waters leave cultivated fields and pass through highly saline soils and the underlying Mancos Shale Formation. Thus, reducing the ground-water flow through these soils was an objective used to reduce salt content. In addition, "off-farm" irrigation delivery systems were improved by lining canals or piping of irrigation laterals. The domestic water system was used to water livestock in winter months rather than deriving water from canals. This effort reduced seepage in nonirrigation seasons and reduced salt loadings by approximately 30 percent. Success of this project was stimulated by Federal funding for installation costs and annual operation and maintenance. Success of this project was also produced by upgrades to onfarm irrigation systems (water-control structures, gated pipes, underground piping, and land leveling) and by improvement of irrigation management, both with the assistance of the Federal government (USDA/NRCS) though incentives and cost sharing.

Knowledge of the concentration of trace elements was originally derived from the elemental analysis of rocks and soils. Development of instrumentation and general improvements in analytical techniques have allowed detection of lower concentrations of these elements. This analytical ability has changed the focus from merely characterization to environmental assessment. This assessment can be addressed by several pathways. Elemental concentrations in soils are governed by contributions of parent materials (geological materials), soil-forming factors, and anthropogenic additions. Land use planners or reclamation specialists may compare two sites with similar soil types to determine whether one site may have anthropogenic contributions of concern. This comparison requires the use of soil surveys and onsite examination of the location by trained pedologists. Availability of knowledgeable soil scientists is the result of training from a widespread, state-based university system as well as mapping activities of the National Cooperative Soil Survey. Extensive mapping of the soils in the U.S. was initiated following World War II; this effort has evolved into a comprehensive digital map of the soils (publicly available on the Internet). A second pathway for evaluation that has developed is evaluation of the reactivity of an element by fractionation, or sequential extraction of an element in question by a series of solutions with increasing strength (Tessier et al., 1979; Burt et al., 2003). Direct analytical methods of speciation also are available, often in association with computer modeling (D'Amore et al., 2005). A third pathway is the direct measure of the bioavailability of trace elements, regardless of natural or anthropogenic additions. Measures of bioavailability are designed to use chemical extractants that remove a portion of the element that may be absorbed by plants or animals. This technique is very similar to soil fertility tests. To date, no single chemical extractant has been found to model the bioavailability for all plants or animals (Houba et al., 1996; McBride et al., 2003; Darmondy et al., 2004). Bioavailability of an element within the human stomach also has been tested, measuring the solubility of an element under chemical conditions mimicking the human digestive system (Yang et al., 2002; Stewart et al., 2003; Fendorf et al., 2004).

Certain plant species have been discovered to have both a tolerance and a large capacity for elemental absorption (Kukier and Chaney, 2001). This degree of absorption, termed phytoextraction, is being evaluated as an efficient and environmentally sound soil remediation technique (Chen et al., 2004). This remediation is being performed by plants, known as "hyperaccumulators," capable of absorbing large amounts of trace elements (>1000 mg/kg above-ground biomass) in contaminated soils (Lombi et al., 2001; Cong and Ma, 2002).

A program for registration of new pesticides was initiated in 1947 by the USDA and is currently under the guidance of the U.S. Environmental Protection Agency (Pierzynski et al., 2000). Part of this regulatory strategy is ground-water assessments (Federal Register, 1996). Currently, there is a large amount of research by universities and Federal agencies, such as the U.S. Geological Survey, in an effort to better understand the fate and transport of pesticides applied to crops (Schraer et al., 2000; Thurman and Aga, 2001). There have been many improvements in the development of pesticides in recent years, including changes in formulations to lower amounts of active ingredients and to contain more readily degradable materials. Also, user education programs have been initiated to address changes in the application of these materials and the timing of irrigation to increase the effectiveness and limit transport to ground water (Spalding et al., 2003). In 1987, Canada instituted a government program called Food Systems 2002. This program was intended to reduce the use of pesticides in agriculture by 50 percent by the year 2002 (Gallivan et al., 2001). Based on data compiled from 1983 through 1998, the use of pesticides declined by 38.5 percent, due in part to a reduction in crop area but principally resulting from a reduction in mean application rates. Many factors influenced this reduction, but the results are a positive step toward improving the environment.

4.7   Total Analysis
4.7.1   Major Elements
4.7.1.3   HF + HCl + HNO$_3$ Acid Digestion
4.7.1.3.1–12   Aluminum, Calcium, Iron, Potassium, Magnesium, Manganese, Sodium, Phosphorus, Silicon, Strontium, Titanium, and Zirconium

Major elements, acid digestion, rationale: Hydrofluoric acid (HF) is efficient in the digestion and dissolution of silicate minerals for elemental dissolution (Bernas, 1968; Sawhney and Stilwell, 1994). The addition of HNO$_3$ + HCl aids in the digestion of soil components, especially the organic fraction (Wilson et al., 1997). Insoluble fluorides of various metals may form. Formation of SiF$_4$ results in gaseous losses of Si, but additions of H$_3$BO$_3$ retard formation of this molecule and dissolve other metal fluorides (Lim and Jackson, 1982). Heating of H$_3$BO$_3$ with the digested sample has been found to be important for initiating the complexation forming HBF$_4$ (Wilson et al., 2006). Heating 20 mL 4.5 percent H$_3$BO$_3$ with digested standard reference soil produced recoveries of 94, 98, and 99 percent for Al, Ca, and Mg, respectively, compared to 46 percent recovery for Al and Mg and 37 percent recovery for Ca in extracts where H$_3$BO$_3$ was added but not heated (Wilson et al., 2006).

Major elements, measurements: A 250-mg <2-mm or other particle-size soil separate that has been oven dried and ground to <200 mesh (75 μm) is weighed into a 100-ml Teflon (PFA) sample digestion vessel. In addition, dried clay (<0.002 mm) may be used, or a clay suspension containing approximately 250 mg of clay material is pipetted into a digestion container and dried at 110 °C. An equal amount of suspension is pipetted into a tared aluminum-weighing dish and dried at 110 °C to obtain a dried sample weight. The P and Na content of the clay fraction is not measurable when the soil is dispersed in sodium hexametaphosphate. To the vessel, 9.0 mL HNO$_3$, 3.0 mL HCl, and 4 mL HF are added. The vessel is inserted into a protection shield and covered and is then placed into a rotor with temperature control. Following microwave digestion, the rotor and samples are cooled and 20 ml of 4.5 percent boric acid solution is added. The samples are then covered and heated in the microwave. The digestate is then quantitatively transferred onto a 100-ml polypropylene volumetric with boric acid solution to achieve a final boric acid concentration of 2.1 percent. The volumetrics are allowed to stand overnight and are then filled to volume. Approximately 60 mL is saved for analysis. The concentrations of Al, Ca, Fe, K, Mg, Mn, Na, P, Si, Sr, Ti, and Zr are determined by ICP-AES. All major elements are reported as mg kg^{-1}.

4.7 Total Analysis
4.7.2 Trace Elements
4.7.2.3 HNO$_3$ + HCl Acid Digestion
4.7.2.3.1–21 Silver, Arsenic, Barium, Beryllium, Cadmium, Cobalt, Chromium, Copper, Mercury, Manganese, Molybdenum, Nickel, Phosphorus, Lead, Antimony, Selenium, Tin, Strontium, Vanadium, Tungsten, and Zinc

Trace elements, acid digestion, rationale: The approach of the HNO$_3$ + HCl digestion methodology is to maximize the extractable concentration of elements in digested soils while minimizing the matrix interferences found in digestion procedures that use HF acid. Organic constituents may contain metals and are difficult to digest if present in high concentrations. Certain elements are subject to volatile losses during digestion and transfer. Certain soil minerals (e.g., quartz and feldspars) are not soluble in HNO$_3$ + HCl.

Trace elements, measurements: A 500-mg <2-mm soil separate that has been air dried and ground to <200 mesh (75 μm) is weighed into a 100-ml Teflon (PFA) sample digestion vessel. To the vessel, 9.0 mL HNO$_3$ and 3.0 mL HCl are added. The vessel is inserted into a protection shield and covered and is then placed into a rotor with temperature control. Following microwave digestion, the rotor and samples are cooled and the digestate is quantitatively transferred into a 50-ml glass volumetric of high purity reverse osmosis deionized water. The volumetrics are allowed to stand overnight and are then filled to volume. The samples are transferred into appropriate acid-washed polypropylene containers for analysis. The concentrations of Ag, As, Ba, Be, Cd, Co, Cr, Cu, Hg, Mn, Mo, Ni, P, Pb, Sb, Se, Sn, Sr, V, W, and Zn have previously been determined using ICP-AES in axial mode. Mercury had been analyzed by a cold-vapor atomic absorption spectrophotometer (CVAAS) and As and Se by flow through hydride-generation and atomic absorption spectrophotometery (HGAAS). Currently, all trace elements are analyzed by ICP-MS. All trace element data are reported as mg kg^{-1}, except Hg and Se, which are reported as μg kg^{-1}.

4.7 Total Analysis
4.7.3–5 Carbon, Nitrogen, and Sulfur
4.7.3–5.1 Elemental Evaluations

Total C, organic and inorganic: Total C is the sum of organic and inorganic C. Most of the organic C is associated with the organic matter fraction, and the inorganic C is generally found with carbonate minerals. The organic C in mineral soils generally ranges from 0 to 12 percent (Nelson and Sommers, 1996). In humid regions in which there has been extensive leaching of the soil profiles, the organic C is typically the dominant form present. In arid and semiarid regions, however, carbonate minerals (e.g., calcite and dolomite) along with soluble carbonate and bicarbonate salts typically constitute a significant portion of the total C (Nelson and Sommers, 1982).

Total C, significance and soil-related factors: Total C is used to estimate organic matter and organic C, which serve as taxonomic criteria to help classify a wide variety of soils. Soil and sediments contain a large variety of organic materials, the characteristics of which include their ability to form water-soluble and water-insoluble complexes with metal ions and hydrous oxides, interact with clay minerals and bind particles together, sorb and desorb both naturally occurring and anthropogenically introduced organic compounds, absorb and release plant nutrients, and hold water in the soil environment (Schumacher, 2002). Because of these characteristics, total organic C (TOC) is often used in contaminant analyses as part of an ecological risk assessment package.

Carbon cycle: The role of soils in global C cycling is diverse. Soil and detritus contain 1200 Gt of C, which is one of the larger pools of C in this cycle (Pierzynski et al., 2000). Plants fix atmospheric CO_2 during photosynthesis. Soil micro-organisms decompose dead plants, returning a portion of the C to the atmosphere as CO_2 and retaining a portion of the C as soil organic matter; thus, soil plays a significant role in C storage (Pierzynski et al., 2000). This flow of C from plant roots through the microbial biomass is one of the key processes in terrestrial ecosystems (Van Veen et al., 1989). When soils are taken out of natural vegetation and used for agricultural production, they will typically lose 25 to 50 percent of their organic matter unless they are properly managed. Also, when soil erosion occurs, there is a significant C loss in the eroded material (Pierzynski et al., 2000). A form of carbon called "black C" or biochar was initially investigated in the Amazon basin, where this form of C is found to be very stable and resistant to degradation (Lima et al., 2002; Sombroek et al., 2003). Currently, the production of black C or biochar (from the process of pyrolysis) is believed to be not only a beneficial additive to soils (Liang et al., 2006) but also a mechanism of long-term C storage (Lehmann, 2007).

Total N, organic and inorganic: Total soil N includes both the organic and inorganic forms and may make up 0.06 to 0.5 percent of the surface layer in many cultivated soils, <0.02 percent of the subsoil, and 2.5 percent of peats (Bremner and Mulvaney, 1982). Organically complexed N makes up over 90 percent of the total N in the surface layer of most soils and is an important factor in soil fertility (Stevenson, 1982). Inorganic forms of N were once considered to constitute only a few percent of the total soil N pool (Young and Aldag, 1982). More recently, however, many soils have been found to contain appreciable amounts of N in the form of fixed or nonexchangeable NH_4^+, particularly in the lower horizons. Soils with large amounts of illites or vermiculites can "fix" significant amounts of N compared to those soils dominated by smectite or kaolinite (Bower, 1950; Nommik and Vahtras, 1982; Young and Aldag, 1982).

Total N, significance and soil-related factors: The uses of total N data include, but are not limited to, the determination of the N distribution in the soil profile, the soil C:N ratio, and the potential of the soil to supply N for plant growth. Factors associated with N accumulation in soils include those that favor the growth of plants, which are the major source of soil organic matter (SOM), and those that inhibit organic matter decomposition (Foth and Ellis, 1997). Histosols have some of the highest amounts of N, whereas Aridisols have the lowest. Typically, grassland soils have more OM than nearby forested soils that formed under similar conditions. Other factors favoring organic N accumulation in soils include a high clay content and low temperatures (Foth and Ellis, 1997). Generally, the OM and N contents of soils decrease with increasing soil depth. In some soils, there is a secondary N increase in subsoils because of the protection of OM in Bt horizons with high clay content (Foth and Ellis, 1997). Organic matter in subsoils is generally older and more decomposed than that in A horizons and has a lower C:N ratio.

Nitrogen cycle: Nitrogen is ubiquitous in the environment as it is continually cycled among plants, soil organisms, soil organic matter, water, and the atmosphere (National Research Council, 1993). Nitrogen is one of the most important plant nutrients and forms some of the most mobile compounds in the soil-crop system; thus it is commonly related to water-quality problems (Bremner, 1996). Nitrogen undergoes a wide variety of transformations in the soil, most of which involve the organic fraction. An internal "N cycle" exists in the soil distinct from the overall cycle of N in nature. Even if gains and losses in N are equal, as may be the case in some mature ecosystems, the N cycle is not static. Continuous turnover of N occurs through mineralization-immobilization with transfer of biological decay products into stable humus forms (Stevenson, 1982). This N cycle is critical to crop growth. The balance between the inputs and outputs and the various interactive transformations (mineralization, nitrification, immobilization, and denitrification) in the N cycle determine how much N is available for plant growth and how much may be lost to the atmosphere, surface water, or ground water (National Research Council, 1993). Nitrogen inputs to a particular agricultural field include rainfall; fertilizers; mineralization from soil organic-N, crop residues, and manure; N-fixation by micro-organisms; and even delivery of N from irrigation waters (National Research Council, 1993). The primary desired output is N uptake in harvested crops and crop residues (National Research Council, 1993). Nitrogen is applied to soils as NH_4^+ (ammonium) or NO_3^- (nitrate) ions. Generally, NH_4^+ ions rapidly undergo nitrification, forming NO_3^- in warm, aerobic soils. Ammonium can be adsorbed by soil particles and lost by fixation or erosion, while NO_3^- remains in soil solution and is subject to leaching or denitrification. Denitrification is the chemical reduction of NO_3^- to gaseous nitric oxide (NO), nitrous oxide (N_2O), or dinitrogen (N_2) forms. Volatilization of these forms of N represents atmospheric N losses (Bremner et al., 1981; Nelson, 1982; Meisinger and Randall, 1991). Some nutrients may be removed by weeds or immobilized by microbes and thus can enter the organic-N storage pool. These minor outputs are secondary factors and typically have been implicitly included in nutrient-crop yield response models (National Research Council, 1993).

Total S, organic and inorganic: Total S is composed of both organic and inorganic S forms. The organic S fraction accounts for more than 95 percent of the total S in most soils in humid and semihumid regions (Tabatabai, 1982). The proportion of organic and inorganic S in a soil sample varies widely according to soil properties (pH, moisture status, organic matter, clay contents, and depth of sampling) (Tabatabai, 1996; Pierzynski et al., 2000). Mineralization of organic S and its conversion to sulfate by chemical and biological activity may provide a source of plant-available S. Total S typically ranges from 0.01 to 0.05 percent in most mineral soils but can be greater in organic soils. In arid regions, soils that contain gypsum can contain higher amounts of measurable S. In well drained soils, most of the inorganic S normally occurs as sulfate. Large amounts of reduced S compounds occur in marine tidal flats, other anaerobic marine sediments, and mine spoils and can oxidize to sulfur acid upon exposure to the air. Significant amounts of inorganic S are found as sulfates, e.g., gypsum and barite in arid regions (Tabatabai, 1996).

Total S, significance: Total S has been used as an index of the total reserves of this element that can be converted to plant-available S. Extractable sulfate S (SO_4^2-S) is an index of readily plant-available S. Extractable SO_4^2-S does not include the labile fraction of soil organic S that is mineralized during the growing season (Tabatabai, 1982).

Sulfur cycle: In many parts of the United States and in agricultural areas around the world, it has not been necessary to supplement the natural S sources to meet plant requirements (Johnson, 1987). The chemistry of S favors its conservation in soils either as the sparingly soluble sulfate salts, e.g., gypsum and barite, or as a component of organic matter (Johnson, 1987). The recycling of S, along with the significant additions of S from rainfall and irrigation waters, ensures adequate S for crops in most environments. Crop deficiencies of S in desert and arid regions are extremely rare as soils in these areas typically developed under conditions in which S was not leached from within the rooting depth of most crops (Johnson, 1987). In those areas with S deficiencies for specific crops, the proper sampling, testing, and use of appropriate calibration tables are factors in the efficient use of S fertilizers.

Essential plant elements: The presence of an element in a plant is not, in and of itself, a valid basis upon which to assess its essentiality to plant life (Noggle and Fritz, 1976). Of the many elements that have been detected in plant tissues, only 20 are essential to the growth of some plant or plants. In the absence of each of the essential elements that are characteristic of the elements, symptoms develop resulting in reduced plant growth and yields (Arnon and Stout, 1939; Noggle and Fritz, 1976). It was proposed that if elements were metabolically active but are not essential, then these elements are better termed functional or metabolic elements rather than essential elements (Bollard and Butler, 1966; Nicholas, 1969). An example would be if one element could be substituted for another, e.g., Br for Cl.

Macroelements (C, H, O, N, P, K, Ca, Mg, and S) are those elements required in relatively large amounts by plants, whereas microelements (B, Fe, Mn, Cu, Zn, Mo, Cl, Co, V, Na, and Si) are those required in relatively small or trace amounts. The above-named macroelements, with the exception of C, H, O, and N, are known as mineral elements. These mineral elements typically constitute what is known as the plant ash or the mineral remaining after "burning off" C, H, O, and N. Carbon, hydrogen, and oxygen in plants are obtained from carbon dioxide and water and are converted to simple carbohydrates by photosynthesis; when combined with N, they are converted to amino acids, proteins, and protoplasm.

There exists in nature a soil-plant continuum, i.e., the soil with its properties and reactions that affect plant-available elements; the root with its growth, distribution, and response to environmental factors; and the plant with its requirements, absorption, and utilization of elements. The interaction of all these components is the continuum that is more critical than any one single component. This continuum can be extended to include the microbial component, as the physical and chemical soil characteristics determine the nature of the environment in which micro-organisms are found (Alexander, 1977).

Nitrogen, essential plant element: Nitrogen is an essential plant nutrient that is used in protein formation and serves as an integral part of the chlorophyll molecule, the primary light energy absorber for photosynthesis. Nitrogen has been related to carbohydrate utilization and associated with vigorous vegetative growth and dark green color. Many proteins are enzymes, and the role of N in plant growth is considered as both structural and metabolic. An imbalance of N or an excess of N in relation to other nutrients, e.g., P, K, and S, can prolong the growing period and delay crop maturity (Bidwell, 1979). Plants absorb N in the form of ammonium, urea, and nitrate. The NO_3^- is typically the dominant form in moist, warm, well-aerated soils (Tisdale et al., 1985).

Sulfur, essential plant element: Sulfur is an essential nutrient for plant growth. Sulfur plays an important role in protein formation, in the functioning of several enzyme systems, in chlorophyll synthesis, and in the activity of nitrate reductase (Tisdale et al., 1985). Sulfur is absorbed by plant roots almost exclusively as the sulfate ion SO_4^{2-}. Low levels may also be absorbed through plant leaves and utilized within plants. High levels of the gaseous form (SO_2) are toxic. Symptoms of S deficiencies in plants are commonly very similar to those of N deficiencies.

Total C, measurements and estimates: Total C is quantified by two basic methods, i.e., wet or dry combustion. The SSL uses dry combustion. In total C determinations, all forms of C in a soil are converted to CO_2 followed by a quantification of the evolved CO_2. Organic C is determined by either wet or dry combustion. The SSL formerly used the wet combustion, Walkley-Black modified acid-dichromate digestion, $FeSO_4$ titration. The Walkley-Black organic C method is an obsolete SSL method. Total C can be used to estimate the soil organic C content. The difference between total and inorganic C is an estimate of the organic C. The inorganic C should be approximately equivalent to carbonate values measured by CO_2 evolution with strong acid (Nelson and Sommers, 1996). The SSL determines the amount of carbonate in a soil by treating a sample with HCl and then manometrically measuring the evolved CO_2. The carbonate amount is then calculated on a $CaCO_3$ equivalent basis.

Total N, measurements: Two methods of analysis of total N have gained acceptance for the determination of total N in soils. These are the Kjeldahl (1883) method, which is essentially a wet oxidation procedure, and the Dumas (1831) method, which is fundamentally a dry oxidation (i.e., combustion) procedure (Bremner, 1996). The SSL currently uses the dry combustion technique for the analysis of total N.

Total S, measurements: The SSL uses dry combustion for the analysis of total S. The SSL does not analyze for extractable sulfate S (SO_4^{2-}S). Reagents that have been used for measuring SO_4^{2-}S include water, hot water, ammonium acetate, sodium carbonate and other carbonates, ammonium chloride and other chlorides, potassium phosphate and other phosphate, and ammonium fluoride (Bray-1). Extractable SO_4^{2-}S does not include the labile fraction of soil organic S that is mineralized during the growing season (Tabatabai, 1996). Extraction reagents for organic S include hydrogen peroxide, sodium bicarbonate, sodium hydroxide, sodium oxalate, sodium peroxide, and sodium pyrophosphate. Other methods are available for determination of S, especially for total S and SO_4^{2-}S. The investigator may refer to the review by Beaton et al. (1968).

The following section describes the SSL total analysis method for C, N, and S. These elements are analyzed in succession with a thermal conductivity detector. In the past, the SSL determined total N by wet combustion and determined total C and S separately.

4.7   Total Analysis
4.7.3–5   Carbon, Nitrogen, and Sulfur
4.7.3–5.2   Dry Combustion

The SSL method uses an air-dry (80 mesh, <180 μm) sample packed in tinfoil, weighed, and analyzed for total C, N, and S by an elemental analyzer. The elemental analyzer works according to the principle of catalytic tube combustion in an oxygenated CO_2 atmosphere and high temperature. The combustion gases are freed from foreign gases. The desired measuring components (N_2, CO_2, and SO_2) are separated from each other with the help of specific adsorption columns and are determined in succession with a thermal conductivity detector; helium is used as the flushing and carrier gas. The SSL reports percent total C, N, and S.

4.8   Analysis of Ground Water and Surface Water

This section describes the SSL methods for the analysis of ground water and surface water. These methods include pH, EC, specific cations (Ca^{2+}, Mg^{2+}, K^+, and Na^+) and anions (Br^-, Cl^-, F^-, NO_3^-, NO_2^-, PO_4^{3-}, and SO_4^{2-}), carbonate and bicarbonate, and total elemental analysis. For detailed descriptions of the SSL methods which are cross-referenced by method code in the table of contents in this manual, refer to SSIR No. 42 (Soil Survey Staff, 2004), which is available online at http://soils.usda.gov/technical/lmm/. Also refer to SSIR No. 51 (Soil Survey Staff, 2009; available online at http://www.soils.usda.gov/technical/) for detailed descriptions of field methods as used by NRCS soil survey offices.

4.8   Analysis of Ground Water and Surface Water
4.8.1   pH
4.8.2   Electrical Conductivity and Salts
4.8.2.1–4   Calcium, Magnesium, Potassium, and Sodium
4.8.2.5–11   Bromide, Chloride, Fluoride, Nitrate, Nitrite, Phosphate, and Sulfate
4.8.2.12–13   Carbonate and Bicarbonate
4.8.3   Total Analysis

Analysis of ground water and surface water, application: The pH of a water sample is a commonly performed determination and one of the most indicative measurements of water chemical properties. The acidity, neutrality, or basicity is a key factor in the evaluation of water quality. Nutrients (nitrogen and phosphorus), sediments, pesticides, salts, or trace elements in ground water or surface water also affect water quality (National Research Council, 1993).

240

pH and EC, measurement: The SSL measures the pH of the water sample with a calibrated combination electrode/digital pH meter. The EC of the water sample is measured using an electronic bridge. The SSL reports the EC as dS m^{-1}.

Cations, measurement: For determination of cations, the SSL filters the water sample and dilutes it with an ionization suppressant (La_2O_3). The analytes are measured by an atomic absorption spectrophotometer (AAS). The data are automatically recorded by a computer and printer. The SSL reports water cations, Ca^{2+}, Mg^{2+}, K^+, and Na^+, as mmol (+) L^{-1}.

Anions, measurement: The SSL determines the anions of the water sample by filtering and diluting according to its electrical conductivity (EC_s). The diluted sample is then injected into the ion chromatograph, and the anions are separated. A conductivity detector is used to measure the anion species and content. Standard anion concentrations are used to calibrate the system. A calibration curve is determined, and the anion concentrations (Br^-, Cl^-, F^-, NO_3^-, NO_2^-, PO_4^{3-}, and SO_4^{2-}) are calculated. A computer program automates these actions. Some water samples contain suspended solids and require filtering. Low molecular weight organic anions will co-elute with inorganic anions from the column. The SSL analyzes the carbonate (CO_3^{2-}) and bicarbonate (HCO_3^-) by filtering the water sample with an aliquot titrated on an automatic titrator to pH 8.25 and pH 4.60 end points. The carbonate and bicarbonate are calculated from the titers, aliquot volume, blank titer, and acid normality. The SSL reports water anions, Br^-, Cl^-, F^-, NO_3^-, NO_2^-, PO_4^{3-}, SO_4^{2-}, CO_3^{2-}, and HCO_3^-, as mmol (-) L^{-1}.

Total analysis, measurement: The SSL determines the major elements (Al, Ca, Fe, K, Mg, Mn, Na, P, Si, Sr, Ti, and Zr) and trace elements (Ag, As, Ba, Be, Cd, Co, Cr, Cu, Hg, Mn, Mo, Ni, P, Pb, Sb, Se, Sn, Sr, Ti, V, W, and Zn) by filtering and acidifying the water sample with HCl. Two calibration standards plus a blank are prepared for elemental analysis. Elemental concentrations are determined using an ICP-AES or ICP-MS. The SSL reports all elements as mg L^{-1}, except Hg (μ L^{-1}).

5 Analysis of Organic Soils or Materials

This section describes the various SSL methods used to characterize organic soils or materials. The process of the loss of organic matter also is described, and references to case studies/datasets are presented as evidentiary examples of the actions/practices that have promoted or diminished this soil process. The SSL methods include mineral content, fiber volume, sodium pyrophosphate color, and melanic index. The fiber volume and sodium pyrophosphate color are usually performed in conjunction. Ratios, estimates, and calculations that have been traditionally used to characterize organic soils or materials also are described. Included in this discussion are the description and application of the obsolete SSL Walkley-Black method for soil organic carbon (SOC). While the C:N ratio has been a long-standing data element on the SSL data sheets, the basis of its calculation changed upon the retirement of the Walkley-Black method. Other estimates discussed herein include some values that have been added (e.g., estimated organic matter and organic C) or that do not appear on the SSL data sheets (e.g., C:N:S and organic C accumulation index). These ratios, estimates, and calculations are not described or assigned method codes in SSIR No. 42 (Soil Survey Staff, 2004). For detailed descriptions of the SSL methods for organic soils or materials which are cross-referenced by method code in the table of contents in this manual, refer to SSIR No. 42 (Soil Survey Staff, 2004), which is available online at http://soils.usda.gov/technical/lmm/. Also refer to SSIR No. 51 (Soil Survey Staff, 2009; available online at http://www.soils.usda.gov/technical/) for detailed descriptions of field methods as used in NRCS soil survey offices.

5.1   Mineral Content

Mineral content, application: The mineral content is the plant ash and soil particles that remain after organic matter removal. The percentage of organic matter lost on ignition (LOI) can be used to define organic soils in place of organic matter estimates by the Walkley-Black organic C method (6A1c, method obsolete, Soil Survey Staff, 1996). Organic C data by Walkley-Black are generally considered invalid if organic C is >8 percent. The determination of organic matter by LOI is a taxonomic criterion for limnic organic materials (coprogenous and diatomaceous earth), i.e., requiring CEC <240 cmol(+) kg^{-1} organic matter (measured by LOI). The standard SSL method to determine organic matter by LOI calls for heating the sample overnight (16 hours) at 400 °C. Other standard methods heat the sample to 360 °C for 2 hours (Schulte and Hopkins, 1996).

Mineral content, measurement: The mineral content is determined by drying a sample overnight at 110 °C in a moisture can. The sample is cooled and weighed and then placed in a cold muffle furnace. The temperature is raised to 400 °C. The sample is heated overnight (16 hours), cooled, and weighed. The SSL reports the ratio of the weights (400 °C/110 °C) as the mineral content percentage.

Equation 5.1.1:

Mineral Content (%) = (R_W/OD_W) x 100

where
R_W = Residue weight after ignition
OD_W = Oven-dry soil weight

Organic matter percent can then be calculated as follows:

Equation 5.1.2:

Organic Content (%) = 100 - Mineral Content (%)

5.2   Pyrophosphate Color

Pyrophosphate color, application: Decomposed organic materials are soluble in sodium pyrophosphate. The combination of organic matter and sodium pyrophosphate forms a solution color that correlates with the decomposition state of the organic materials. Dark colors are associated with sapric (strongly decomposed) materials and light colors with fibric (weakly decomposed) materials (Soil Survey Staff, 2010). Refer to Soil Survey Staff (2010) for a discussion of the use of sodium pyrophosphate solubility data as taxonomic criteria for organic soils as they relate to various kinds of organic materials, e.g., fibric, sapric, coprogenous, and diatomaceous.

Pyrophosphate color, measurement: Organic material is combined with sodium pyrophosphate. After the material is allowed to stand, the color is evaluated by moistening a chromatographic strip in the solution and comparing the color with standard Munsell color charts. The SSL reports the color using Munsell color notation.

Pyrophosphate color, interferences: These tests of organic soil material can be used in field offices. Since it is not practical in the field to base a determination on a dry sample weight, moist soil is used. The specific volume of moist material depends on how it is packed; therefore, packing of material must be standardized in order to obtain comparable results by different soil scientists.

5.3 Fiber Volume

Fiber volume, application: The water-dispersed fiber volume is a method of characterizing the physical decomposition state of organic materials. The decomposition state of organic matter is used as a taxonomic criterion to define sapric, hemic, and fibric organic materials (Soil Survey Staff, 2010). Sapric material passes through a 100-mesh sieve (0.15-mm openings). Fibers are retained on the sieve. As defined in *Keys to Soil Taxonomy*, organic materials that are >2 cm in cross section and too firm to be readily crushed between thumb and fingers are excluded from fiber. Refer to Soil Survey Staff (2010) for a discussion of the use of sodium pyrophosphate solubility data as taxonomic criteria for organic soils as they relate to various kinds of organic materials, e.g., fibric, sapric, coprogenous, and diatomaceous.

Fiber volume, measurement: The sample is prepared to a standard gravimetric content. The unrubbed fiber content is determined in a series of three steps designed to remove the sapric material by increasingly vigorous treatments. The rubbed fiber content is determined by rubbing the sample between the thumb and fingers. The SSL reports the percent unrubbed fiber after each step and the final unrubbed and rubbed fiber.

Equation 5.3.1:

Fiber volume (%) = Reading on half-syringe (mL) x 20

where
Fiber volume = Rubbed + unrubbed fiber

Fiber volume, interferences: These tests of organic soil material can be used in field offices. Since it is not practical in the field to base a determination on a dry sample weight, moist soil is used. The specific volume of moist material depends on how it is packed; therefore, packing of material must be standardized in order to obtain comparable results by different soil scientists.

5.4   Melanic Index

Humic acid types and extractions: Extractable humus is traditionally fractionated into humic acids and fulvic acids based on differential solubility at low pH values (Shoji et al., 1993). In Japan, humic acids are further fractionated into A-, B-, P-, and RP-types according to relative color intensity (RF) and color coefficient ($\Delta \log K$) (Kumada, 1987). The A-type can be distinguished from other humic acid types based on the distinctive features of its absorption spectrum (Shoji et al., 1993). P-type humic acid is found in the extract of organic matter from Spodosols (Watanabe et al., 1996). Pg absorption (optical characteristics observed in absorption spectra of humic acid) indicates the presence of Perylenequinone compounds (green pigments derived from species of litter-composing fungi, or fungal sclerotia and hyphae) (Watanabe et al., 1996). The name Pg originates from green humic acid found in P-type humic acid.

There are various methods to extract humus. Solutions of dilute NaOH and sodium pyrophosphate are the most common. Pyrophosphate solution preferentially extracts metal-humus complexes, whereas hot NaOH typically dissolves >60 percent of the humus (Adachi, 1973). The 0.5 percent NaOH solution is considered the standard for humus extraction to determine the melanic index in Andisols (Shoji et al., 1993).

Melanic index, application: Melanic and Fulvic Andisols have a high content of humus, which is related to soil color and indicative of prevailing pedogenic processes (Honna et al., 1988). Typically, Melanic Andisols formed under grassland ecosystems and have humus dominated by A-type humic acid (highest degree of humification), whereas Fulvic Andisols formed under forest ecosystems and have humus characterized by the high ratio of fulvic acid to humic acid (low degree of humification, e.g., P- or B-type humic acid) (Kumada and Hurst, 1967; Orlov, 1968; Honna et al., 1988). The organic matter thought to result from large amounts of gramineous vegetation can be distinguished from organic matter formed under forest vegetation by the melanic index (Soil Survey Staff, 2010). The melanic index (≤ 1.70) is used as a taxonomic criterion to identify the melanic epipedon (Soil Survey Staff, 2010). Refer to Soil Survey Staff (2010) for a more detailed discussion of the melanic index as a taxonomic criterion.

Melanic index, measurement: A 0.5-g soil sample is mechanically shaken for 1 hour in 25 mL of 0.5 percent NaOH solution. One drop of 0.2 percent superfloc solution (flocculation aid) is added to the sample, and the sample is then mechanically shaken for 10 minutes. Either a 1- or 0.5-mL extract (<10 percent or >10 percent organic C, respectively) is pipetted into a test tube, 20 mL of 0.1 percent NaOH solution is added, and the sample is mixed thoroughly. Absorbance of the solution is read using a spectrophotometer at 450 and 520 nm, respectively, within 3 hours after extraction. The SSL calculates the melanic index by dividing the absorbance at 450 nm by the absorbance at 520 nm.

5.5 Ratios, Estimates, and Calculations Related to Organic Matter
5.5.1 Organic Matter
5.5.1.1 Organic Matter, Estimated
5.5.2 Organic Carbon
5.5.2.1 Organic Carbon, Walkley-Black Modified Acid-Dichromate Digestion, $FeSO_4$ Titration, Automatic Titrator (Method Obsolete)
5.5.2.2 Organic Carbon, Estimated
5.5.3 Carbon, Nitrogen, and Sulfur Ratios
5.5.3.1 C:N Ratio
5.5.3.2 C:N:S Ratio
5.5.3.3 N:S Ratio
5.5.3.4 C:S Ratio
5.5.3.5 C:P Ratio

Organic matter, definition: The principal feature that separates soil from rock is *organic matter*. *Soil organic matter* (SOM), also referred to as *soil humus*, is defined as the organic fraction of soil exclusive of undecayed plant and animal residues (Soil Science Society of America, 2010). For laboratory analysis, however, the SOM estimates generally include only those organic materials that pass through a 2-mm sieve, and thus it is difficult to quantitatively estimate the organic matter content of a soil (Nelson and Sommers, 1982). Soil organic matter may be partitioned into humic and nonhumic substances, and the major portion in most soils and waters consists of humic substances (Schnitzer and Khan, 1978). Humic materials include humic acid (HA), fulvic acid (FA), and humin. Humic substances are partitioned into these three main fractions based on their solubility in alkali and acid. The HA fraction is soluble in dilute alkali but is precipitated by acidification of the alkaline extract; FA is the fraction that remains in solution when the alkaline extract is acidified, i.e., soluble in both dilute alkali and dilute acid; and humin is the fraction that cannot be extracted from the soil or sediment by dilute base and acid (Schnitzer and Khan, 1978). Nonhumic substances include those substances with still-recognizable physical and chemical characteristics, e.g., carbohydrates, proteins, peptides, amino acids, fats, waxes, alkanes, and low molecular weight organic acids (Schnitzer, 1982). Most of these substances have a short survival period in the soil as they are readily metabolized by micro-organisms.

Organic matter, significance: The quantity and properties of organic matter help to determine the direction of soil formation processes as well as the biochemical, chemical, physical, and soil fertility properties (Kononova, 1966). Organic matter affects the composition and mobility of adsorbed cations as well as soil color, energy balance, volume weight, consistency, and specific gravity of the solid phase. The organic matter content influences many soil properties, e.g., water retention capacity, extractable bases, aggregate stability, soil aeration, and capacity to supply N, P, and micronutrients (Nelson and Sommers, 1996). The overall influence of accumulating organic matter typically leads to higher soil fertility, and the resultant higher humus content often serves as the first indication of a fertile soil (Orlov, 1985). In addition to changes in soil properties, components of humus and the level of productivity can have a direct physiological influence on plants as well as on the biological activity of the soil (Orlov, 1985).

The loss of SOM resulting from agricultural and other land management practices is intimately tied to biological degradation of soils. Mechanical effects of land management, e.g., erosion, compaction, and changes in patterns of drainage, can affect soil biology (Sims, 1990). Agricultural practices include prescribed burning; plowing; irrigation; and long-term applications of N, P, micronutrients, herbicides, nematicides, and pesticides, which impact biological activity and diversity (Miller et al., 1995; Miller, 2000). High concentrations of heavy metals have also been shown to adversely affect the size, diversity, and activity of microbial populations in the soil, and total arbuscular mycorrhizal fungi (AMF) spore numbers decreased with increasing amounts of heavy metals in the soil (Del Val et al., 1999).

Organic matter as determined by LOI is used as a taxonomic criterion for limnic organic materials, such as coprogenous and diatomaceous earth (Soil Survey Staff, 2010). Organic matter, as determined by percent organic C multiplied by 1.724, is also used to define the *n* value (Pons and Zonneveld, 1965), which characterizes the relation between the percentage of water in a soil under field conditions and its percentages of inorganic clay and humus (Soil Survey Staff, 2010).

Organic soil materials, definition: *Organic soil materials* have been defined by the Soil Science Society of America (2010) as soil materials that are saturated with water and have 174 g kg^{-1} or more SOC if the mineral fraction has 500 g kg^{-1} or more clay, or 116 g kg^{-1} SOC if the mineral fraction has no clay or has proportional intermediate contents, or if never saturated with water, have 203 g kg^{-1} or more organic C.

Organic C, significance: *Organic C* is a major component of soil organic matter. Studies of organic matter and nutrient cycling (N, P, and S) emphasize the central role of C (Stevenson, 1982). Carbon is important as a major source of CO_2, and humus is a C reservoir sensitive to changes in climate and atmospheric CO_2 concentrations (Schnitzer and Khan, 1978; Schnitzer, 1982). Organic C consists of the cells of micro-organisms; plant and animal residues at various stages of decomposition; stable "humus" synthesized from residues; and nearly inert and highly carbonized compounds, e.g., charcoal, graphite, and coal (Nelson and Sommers, 1982). Organic C content defines mineral and organic soils as well as lower taxonomic levels, e.g., Ustollic and Fluventic subgroups (Soil Survey Staff, 2010). Organic C is also used as a taxonomic criterion for a number of diagnostic horizons, e.g., mollic, melanic, anthropic, and plaggen epipedons (Soil Survey Staff, 2010).

Organic C, general rules: Some general rules about the properties of organic C (National Soil Survey Laboratory Staff, 1975) are as follows:

Equation 5.5.1:

$$1 \text{ g organic C} \approx 3 \text{ to } 4 \text{ meq CEC (NH}_4\text{OAc, pH 7.0)}$$

Equation 5.5.2:

$$1 \text{ g organic C} \approx 1.5 \text{ g H}_2\text{O (1500 kPa)}$$

Equation 5.5.3:

$$1 \text{ g organic C} \approx 3.5 \text{ g } H_2O \ (33 \text{ kPa})$$

Organic carbon, "Van Bemmelen Factor": Organic C can serve as an indirect determination of organic matter through the use of an approximate correction factor. The "Van Bemmelen factor" of 1.724 has been used for many years and is based on the assumption that organic matter contains 58 percent organic C. The literature indicates that the proportion of organic C in soil organic matter for a range of soils is highly variable. Any constant factor that is selected is only an approximation. Studies have indicated that subsoils have a higher factor than surface soils (Broadbent, 1953). Surface soils rarely have a factor <1.8 and commonly have a factor ranging from 1.8 to 2.0. The subsoil factor may average ≈ 2.5 (Broadbent, 1953). In the past, the preference of the SSL was to determine and report organic C concentration in a soil rather than to convert the analytically determined organic C value to organic matter content through the use of an approximate correction factor.

Organic C, measurement: Organic C is determined by either wet or dry combustion. In the past, the SSL used the wet combustion, Walkley-Black modified acid-dichromate digestion, $FeSO_4$ titration, automatic titrator (6A1c, method obsolete, Soil Survey Staff, 2004). This organic C value represents decomposed soil organic matter and normally excludes relatively fresh plant residues, roots, charcoal, and C of carbonates. Even though the Walkley-Black method converts the most active forms of organic C in soils, it does not yield complete oxidation of these compounds. Walkley and Black (1934) determined that ≈ 76 percent of organic C was recovered by their method and therefore proposed a correction factor of 1.32 to account for unrecovered organic C. Allison (1960) found that the percent recovery of organic C by Walkley-Black procedure ranged from 63 to 86 percent in a wide variety of soils and that the correction factor ranged from 1.16 to 1.59. The SSL used the Walkley-Black correction factor for data generated using this method. The Walkley-Black method and similar procedures provide approximate or semiquantitative estimates of organic C in soils because of the lack of an appropriate correction factor for each soil analyzed (Nelson and Sommers, 1982). Organic C data by Walkley-Black are generally considered invalid if organic C is >8 percent. In these cases, the SSL used a more direct determination of soil organic matter. The organic matter is destroyed on ignition (400 °C), and the soil weight loss is used as a measure of the organic matter content, commonly referred to as loss on ignition (LOI).

C:N ratio, application: The C:N ratio relates to fertility and organic matter decomposition. In many soils, the level of "fixed" N typically remains constant or increases with depth while organic C typically diminishes with depth, resulting in a C:N ratio that narrows with depth (Young and Aldag, 1982). The potential to "fix" N has important fertility implications as the "fixed" N is slowly available for plant growth. In cultivated, agriculturally important soils of the temperate regions, the C:N ratio of surface soil horizons, e.g., mollic epipedons, typically falls within the narrow limits of about 10 to 12 (commonly a few units higher in forest soils) (Young and Aldag, 1982). Higher ratios in soils may suggest low decomposition levels or low N levels in plant residues and soils. In many cases, the C:N ratio narrows in the subsoil, partly because of the higher content of NH_4^+-N and the generally lower amounts of C. Variations in the C:N ratio may serve as an indicator of the amount of inorganic soil N. The C:N ratios in uncultivated soils are generally higher than in cultivated soils. When C or N values are very low, particularly when both C and N contents are low, ratios are most likely unrealistic and care is required in interpreting the data. The C:N ratio can be calculated using measured or estimated organic C data; it is important to ensure that both C and N measurements are in the same units (e.g., wt% or mg kg^{-1}).

C:N:S, N:S, C:S, and C:P ratios, application: There have been many investigations of the soil C:N:S ratios, and results have been mixed. Some studies have indicated that these ratios are very similar for different groups of soils. Other studies (Stewart and Bettany, 1982a, 1982b) have found significant differences in the mean C:N:S ratios among and within types of world soils; the differences are attributed to variations in parent material and other soil-forming factors, e.g., climate, vegetation, leaching intensity, and drainage. However, a close association generally exists between the N and S constituents of soil organic matter (Tisdale et al., 1985). Total N, which is principally organic, and the organic S are often more closely correlated than organic fractions of C and S. The N:S ratio in many soils falls within the narrow range of 6 to 8:1 (Tisdale et al., 1985). Rates of S mineralization are not proportional to the total amount of S in organic matter due to the variety of S-containing organic compounds in soils that have different decomposition rates; the type of plant and animal residues that affect mineralization-immobilization rates and release; and the formation of S-containing precipitates, e.g., $CaSO_4$ and $Al_2(SO_4)_3$, that can influence the amount of plant-available S. The C:S ratio of these materials is important because, in general, net mineralization occurs when the C:S ratio is <200:1, net immobilization occurs when the C:S ratio is >400:1, and a steady state results when the C:S ratio is between 200:1 and 400:1 (Pierzynski et al., 2000). Fresh plant residues may rapidly release P into the soil solution, whereas more stable forms of organic matter, e.g., soil humus, animal manures, municipal biosolids, or composts, generally act as long-term, slow-release sources of P (Pierzynski et al., 2000). Mineralization (the conversion of organic P to inorganic P) usually occurs rapidly if the C:P ratio of the organic matter is <200:1, while immobilization (the incorporation of P into microbial biomass) occurs if C:P ratios are >300:1 (Pierzynski et al., 2000).

Organic C, organic matter, and C:N ratio, calculations: Calculations with and without Walkley-Black organic C data are as follows: (1) estimated organic C, (2) estimated organic matter, and (3) C:N ratio. These estimated values may differ slightly, depending on whether organic C by Walkley-Black or total C by dry combustion is used in the calculations.

(1) Estimated organic C

Equation 5.5.4:

$$OC_e (\%) = [TC - (CaCO_3 \times 0.12)]$$

where
OC_e = Organic C estimated (%)

TC = Total carbon (%) by dry combustion
0.12 = Carbon is 12% of $CaCO_3$ (see Equation 5.5.5)
$CaCO_3$ = Calcium carbonate measured by CO_2 evolution with HCl treatment

Equation 5.5.5:

Atomic weights, $CaCO_3$
Ca = 40
O = 16
C = 12

C = [40 + (3 x 16) + 12] = 12% of $CaCO_3$

Example: Refer to Appendix 4 (Sverdrup Pedon, S87MN051001, C horizon, 102-127 cm)

TC = 2.49% measured by dry combustion
$CaCO_3$ = 16% measured by CO_2 evolution with HCl treatment
OC_m = 0.13% measured by Walkley-Black procedure

OC_e (%) = [TC – ($CaCO_3$ x 0.12)]

OC_e (%) = [2.49 – (16 x 0.12)] = 2.49 – 1.92 = 0.57%

(2) Estimated organic matter

Equation 5.5.6:

$$OM_{e1} \ (\%) = 1.724 \ x \ OC_m$$

where
OM_{e1} = Organic matter estimated (%)
1.724 = "Van Bemmelen Factor"
OC_m = Organic carbon (%) by Walkley-Black procedure

OR (alternatively)

Equation 5.5.7:

$$OM_{e2} \ (\%) \ = 1.724 \ x \ [TC – (CaCO_3 \ x \ 0.12)]$$

where
OM_{e2} = Organic matter estimated (%)
1.724 = "Van Bemmelen Factor"
TC = Total carbon (%) by dry combustion

0.12 = Carbon is 12% of $CaCO_3$ (see Equation 5.5.5)
$CaCO_3$ = Calcium carbonate measured by CO_2 evolution with HCl treatment

Example: Refer to Appendix 4 (Sverdrup Pedon, S87MN051001, C horizon, 102-127 cm)

TC = 2.49% measured by dry combustion
$CaCO_3$ = 16% measured by CO_2 evolution with HCl treatment
OC_m = 0.13% measured by Walkley-Black procedure

OM_{e1} (%) = 1.724 x OC

OM_{e1} (%) = 1.724 x 0.13% = 0.22%

OM_{e2} (%) = 1.724 x [TC – ($CaCO_3$ x 0.12)]

OM_{e2} (%) = 1.724 x [2.49 – (16 x 0.12)] = 0.98%

(3) C:N Ratio

Equation 5.5.9:

$$C:N \text{ Ratio} = OC/TN$$

where
C:N = Carbon to nitrogen ratio
OC = Organic C (%) measured by Walkley-Black procedure
TN = Total nitrogen measured by dry combustion

OR (alternatively)

Equation 5.5.9:

$$C:N \text{ Ratio} = [TC – (CaCO_3 \times 0.12)]/TN$$

where
C:N = Carbon to nitrogen ratio
TC = Total carbon (%) by dry combustion
0.12 = Carbon is 12% of $CaCO_3$ (see Equation 5.5.5)
$CaCO_3$ = Calcium carbonate measured by CO_2 evolution with HCl treatment
TN = Total nitrogen (%)

Example: Refer to Appendix 5 (Caribou Pedon, S88ME003001, Ap1 horizon, 0-11 cm)

TC = 1.98% by dry combustion
OC_m = 2.28% by Walkley-Black procedure
TN = 0.193% by dry combustion
$CaCO_3$ = 0%

C:N Ratio = OC/TN

C:N Ratio = 2.28/0.193 = 11.18 or 12

C:N Ratio = [TC – ($CaCO_3$ x 0.12)]/TN

C:N Ratio = [1.98 – (0 x 0.12)]/0.193 = 10.26 or 10

Organic C, accumulation index (C stocks), calculation: An accumulation index may be calculated for organic C or other data. Another term for this calculation is "C stocks," which is defined as a quantity of C contained in a "pool," meaning a reservoir or system that has the capacity to accumulate or release C.

An example calculation of organic C accumulation index (C stocks) to a depth of 1 meter (kg/m^{2-m}) is as follows:

Equation 5.5.10:

$$\text{Product } (kg/m^{2-m}) = Wt_{oc} \times 0.1 \times \rho_{B33} \times Hcm \times Cm$$

Equation 5.5.11:

$$\text{Accumulation Index} = \text{Sum of Products to 1 m } (kg/m^{2-m})$$

where
Wt_{oc} = Weight percentage of organic C on <2-mm basis by Walkley-Black procedure
　　　or, alternatively, estimated weight percentage of organic C (see Equation 5.5.5)
0.1 = Conversion factor, constant
ρ_{B33} = Bulk density at 33-kPa water content on <2-mm basis (g cm^{-3}). Alternatively, use
　　　ρ_{Bf} at field-state water content (e.g., soil cores at field-state).
Hcm = Horizon thickness (cm)
Cm = Coarse fragment conversion factor. If no coarse fragments, Cm = 1. If coarse
　　　fragments are present, calculate Cm using Equations 3.1.2.1.1.3 and 3.1.2.1.1.4.

Example: Refer to Appendix 2 (Wildmesa Pedon, S89CA027004)

Horizon	Depth	OC	Factor	ρ_{B33}	Hcm	Cm	Product
	cm	%		g cm^{-3}	cm		
A	0-8	0.45	0.1	1.47	8	.84	0.44
AB	8-15	0.15	0.1	1.60	7	.94	0.16
2Bt	15-46	0.19	0.1	1.45	31	.99	0.85
2Btk	46-74	0.15	0.1	1.38	28	1.00	0.58
2Btk	74-109	0.12	0.1	1.26	26	.99	0.39
ACCUMULATION INDEX					100 cm		2.42 kg/m^{2-m}

Organic matter, weight to volume basis, calculation: Organic carbon is routinely reported as a weight percent. Convert organic C on a weight basis to a volume basis as follows:

Equation 5.5.12:

$$V_{om} = (Wt_{oc} \times 1.72 \times \rho_{B33})/1.1$$

where
V_{om} = Volume percentage of organic matter on <2-mm basis
Wt_{oc} = Weight percentage of organic C on <2-mm basis by the Walkley-Black procedure or, alternatively, estimated weight percentage of organic C (see Equation 5.5.5)
1.724 = "Van Bemmelen Factor"
ρ_{B33} = Bulk density at 33-kPa water content on <2-mm basis (g cm^{-3})
1.1 = Assumed particle density of organic matter (g cm^{-3})

5.6 Loss of Organic Matter, Processes and Case Studies

Loss of organic matter, processes: Organic matter is typically lost as a result of oxidation processes (decomposition) or by removal of organic materials (e.g., by crops or erosion) in excess of subsequent accumulation. The loss of organic matter is commonly assessed by visual observations and/or by laboratory measurements of organic or total C. Soil biology is an important component of soil quality and one that has not received appropriate attention until recently. The biological component is commonly assessed by organic matter content, biomass C, and activity and diversity of soil fauna.

Loss of organic matter, case studies: Over long periods of time, SOM is the result of climatic, biological, and geological factors, whereas over shorter periods SOM varies with disturbance of vegetative communities and changes in land use patterns, which affect the rate of organic matter inputs and the mineralization of organic matter. Losses of 50 percent in the top 20 cm and 30 percent in the surface 100 cm are average (West and Post, 2002). Soil organic carbon (SOC) was found to be the only soil quality indicator that consistently showed significant differences between land uses in both the Central and Southern High Plains (Brejda et al., 2000). Several common practices resulting in SOM loss over the last half-century on agricultural soils in North America are heavy tillage systems that accelerate organic matter decomposition and increase the release of nutrients and intensive agriculture with inadequate or no return of crop residues. Other practices include land conversions (e.g., converting forestland or grassland to cropland); overgrazing or uncontrolled grazing; and drainage and cropping of productive organic soils (e.g., Histosols), resulting in SOM loss and subsidence.

In general, tillage-induced loss of SOM occurs very rapidly, more so in hot humid climates than in cool wetter climates (Stewart et al., 1991; Zobeck et al., 1995; Robinson et al., 1996; Reeves, 1997). Janzen et al. (1998) estimated that approximately 25 percent of the SOC originally present in the surface layer of Canadian agricultural soils was lost to the atmosphere upon conversion to arable agriculture. Sparrow (1984) reported that the wide use of mechanical tillage for weed control and seedbed preparation had caused deterioration in the quality of Canadian soils. Organic matter is considered an important measure of soil quality. Reductions in SOM due to cultivation have been reported in western and eastern Canada (Martel and MacKenzie, 1980; McGill et al., 1981). Reduced levels of SOM have been related to nutrient depletion and increased susceptibility to erosion in some Canadian soils (Janzen, 1987; Monreal et al., 1995a, 1995b). Janzen et al. (1998) suggested that these trends could be influenced to favor larger amounts of crop residue resulting from crop yields that have shown increases in recent decades due to improved cultivars, enhanced nutrient application, and more intensive management practices.

Conservation tillage has been a major part of the conservation program in the U.S. since the 1970s. This management practice became feasible with the advances in herbicide developments during the 1960s (Baeumer and Bakermans, 1970; Cannell and Hawes, 1994). The use of conservation tillage to sustain or increase SOC has been continually reevaluated and adapted over the years because of the need for and effectiveness of this practice under different climates on different U.S. soils (Bruce et al., 1990; Havlin et al., 1990; Wood et al., 1991; Reeves and Wood, 1994; Franzluebbers et al., 1994; Aase and Pikul, 1995). More importantly, conservation tillage, which encompasses a range of tillage practices, is currently considered as one aspect of the soil management system, inclusive of a wide array of practices, e.g., crop rotations, adequate and appropriate fertilization, residue inputs, and manures (Reeves, 1997).

In the past 25 years, there has been important work relating SOC not only to agronomic productivity, soil quality, and economic sustainability but also to the C cycle and the role of soils in sequestering C (Lal et al., 1998b, 1998c). Quantifying SOC stocks is critical to understanding C dynamics and potential storage capacity; therefore, C storage is an important component in cycling processes. Waltman and Bliss (1997) estimated SOC in the lower 48 States at 59.4 Pg (or about 4 percent total SOC in world soils), with an additional 13.5 Pg in Alaska. Estimates by Kern (1994) for the 48 States were 78 to 84.5 Pg. In comparison, SOC was estimated at 262 Pg for Canada and 11.5 Pg for Mexico (Waltman and Bliss, 1997). Erosion and mineralization caused by the conversion of forest to natural ecosystems has resulted in loss of SOC. The magnitude of the historic loss is more difficult to estimate than that of the current SOC pool (Lal and Kimble, 2006). Kern (1994) estimated the loss of SOC pool at 1 to 1.7 Pg, and Lal et al. (1998a) estimated the loss of SOC pool from U.S. cropland at about 5 Pg. Lal and Kimble (2006) estimated that U.S. cropland has a potential to sequester about 5 Pg C through agricultural intensification and the adoption of recommended management practices.

6   Soil Biological and Plant Analysis

This section describes the SSL biological and plant analysis procedures. It also describes key definitions and applications of these data. Typically, biological and plant samples are collected for SSL analysis in conjunction with pedon sampling or for specific research projects or agency initiatives, e.g., soil quality and dynamic soil properties. The long history of research and development in soil biology and plant analysis precedes the incorporation of these important components in the concept of soil quality. Soil biology was integrated into the soil quality initiative by the USDA/NRCS soil survey program in the 1990s and continues to be important (http://soils.usda.gov/sqi/).

Traditionally, *plant analysis* has been defined as the diagnosis and correction of plant growth limitations imposed by relative insufficiencies of inorganic nutrients (Mills and Jones, 1996). Despite decades of research and practice, this nutritional diagnosis is still an inexact science that uses a variety of techniques, e.g., leaf analysis, tissue testing, and combined soil and crop nutrient element status (which forms the basis for assessing lime and fertilizer needs) (Mills and Jones, 1996). In the latter half of the 20th century, there was increased interest in the importance of the role of soil micro-organisms in the retention and release of nutrients and energy. In any attempts to assess nutrient and energy flow in soil systems, the role of microbial biomass must be recognized (Parkinson and Paul, 1982). In more recent years, soil biology and plant analysis have been employed in more diverse agronomic and environmental uses. For these reasons, the SSL expanded its suite of analyses to include selected biological and plant procedures so as to more completely characterize sampling sites and pedons for a broad spectrum of soil survey applications.

The SSL biological and plant analysis methods include root and plant (above-ground) biomass and nutrient cycling, organic carbon extractions, separation and total analysis of soil organic matter fraction, microbial biomass characterization, and β-glucosidase C cycle assay. Carbonates are determined by acid decomposition and CO_2 analysis by gas chromatography. This method is more commonly used in soil biochemical and biology studies, where organic C in soils with carbonates may be more precisely determined by subtracting the total carbonates (inorganic C) from total C. For detailed descriptions of the SSL methods which are cross-referenced by method code in the table of contents in this manual, refer to SSIR No. 42 (Soil Survey Staff, 2004), which is available online at http://soils.usda.gov/technical/lmm/. Also refer to SSIR No. 51 (Soil Survey Staff, 2009; available online at http://www.soils.usda.gov/technical/) for field methods as used by NRCS soil survey offices.

6.1   Soil Biological and Plant Analysis as Assessment Tools

Soil is one of the most basic of natural resources. It serves as a critical link between agricultural productivity, economic progress, and environmental quality (Lal, 1998). In the 1950s, pedologists were among the first to use the concept of soil quality in the development of methodology and criteria for land evaluation and soil capacity assessment (Lal, 1998). Since the late 1980s, soil scientists have defined soil quality in terms of the capacity of soil to perform specific functions (Soil Science Society of America, 2010; Larson and Pierce, 1991; Papendick and Parr, 1992; Lal, 1993; Doran and Parkin, 1994). While the concept of *soil quality* is still being developed by soil scientists (Pierzynski et al., 2000), the definition used by the Soil Science Society of America (2010) is used herein as follows: "The capacity of a soil to function within ecosystem boundaries to sustain biological productivity, maintain environmental quality, and promote plant and animal health."

Soil quality has both an inherent or natural component determined by geology, landscape, climate, and native soil characteristics (soil formation processes) and a dynamic component; therefore, it cannot be evaluated in isolation without relating it to productivity under different land uses and management scenarios (e.g., land conversion, crop rotation, fertilization, and conservation measures) (Lal, 1998; Eswaran et al., 2001; Gregorich, 2002). Natural processes can degrade soil quality (e.g., erosion resulting in acidification, compaction, and nutrient depletion), and anthropogenic activities (e.g., agriculture) can accelerate these degradation processes; however, some agricultural land uses and practices, e.g., various tillage methods, cropping systems, and nutrient management plans, can help to stabilize or improve soil quality (Gregorich, 2002).

Factors impacting soil quality and its assessment are complex and interactive. In order to achieve a logical approach to the assessment of soil quality as impacted by changes in agriculture or land management systems, researchers commonly use three broad categories (physical, chemical, and biological) as indicators of soil quality. Soil properties are characteristics described by measurements and are important in determining the limitations and practical uses for a unit of land. Within these three categories are specific soil properties (e.g., structure, pH, salinity, biota, and biomass) that are commonly measured in soil quality assessments and are indicative of certain soil processes. The following section describes the SSL biological and plant analysis measurements that typically have been used in soil survey assessments of soil quality.

6.2   Hot Water Extractable Organic Carbon

Hot water extractable organic carbon, application: Hot water soluble soil carbohydrates are thought to be primarily extracellular polysaccharides of microbial origin. They help bind soil particles together into stable aggregates. Water-stable aggregates reduce the hazard of soil erosion (reducing the loss of organic matter and nutrients and improving infiltration). They also occur as part of the fast or labile SOC pool in soils. This labile pool contains the most available carbon for plant, animal, and microbial use. The hot water soluble organic C makes up 4 to 10 percent of the microbial biomass C determined by chloroform fumigation. It also makes up about 6 to 8 percent of the total carbohydrate content in the soil. This pool is the most easily depleted of the three organic C pools (Joergensen et al., 1996; Haynes and Francis, 1993).

Hot water extractable organic carbon, measurement: Water is added to a 10-g soil sample and autoclaved for 1 hour at 121 °C. Extractable carbohydrates are measured by adding disodium bicinchoninic acid (BCA) reagent to a 0.5 M K_2SO_4 soil extract, heating to 60 °C for 2 hours, and then cooling. The absorbance is read at 562 nm using a spectrophotometer. Glucose is used as a standard, and results are expressed as glucose-C. Data are reported as mg glucose equivalent-carbon kg^{-1} soil.

6.3 Active Carbon

Active carbon, application: This method is designed to be a quick and easy field test for the assessment of easily oxidized or "active" soil organic C (Blair et al., 1995; Weil et al., 2003). Following the principle of bleaching chemistry, a weak solution of potassium permanganate ($KMnO_4$) is used to oxidize organic matter present in soil. An active soil carbon index can be expressed as the quotient of active carbon to SOC (Blair et al., 2001). The stability of this index over time is considered to be a useful measure of soil quality (Islam and Weil, 2000).

Active carbon, measurement: A 5-g sample is oxidized with 0.02 M potassium permanganate diluted with reverse osmosis water. The sample is shaken for 2 minutes and allowed to stand undisturbed for 5 to 10 minutes. A small aliquot of the supernatant is diluted with reverse osmosis water, and the absorbance of the solution is read at 550 nm using a spectrophotometer. Extractable carbon by 0.02 M $KMnO_4$ is reported as mg C kg^{-1} soil.

Active carbon, interferences: Chemical oxidation methods for the determination of labile soil carbon have a number of limitations. Because different soil samples may have variable amounts of readily oxidizable fractions, standardization of any method can be difficult; results may be influenced by the amount of C in the sample, MnO_4^- concentration, and contact time (Blair et al., 1995).

6.4 Carbonates

Carbonates, application: Methods involving determination of CO_2 have usually been preferred for measuring soil carbonate (Loeppert and Suarez, 1996). CO_2 released can be measured gravimetrically (Allison, 1960; Allison and Moodie, 1965), titrimetrically (Bundy and Bremner, 1972), manometrically (Martin and Reeve, 1955; Presley, 1975), volumetrically (Dreimanis, 1962), spectrophotometrically by infrared spectroscopy, or by gas chromatography (Loeppert and Suarez, 1996). The SSL routinely determines the amount of carbonate in the soil by treating the $CaCO_3$ with HCl, with the evolved CO_2 measured manometrically for <2-mm and 2- to 20-mm bases. The SSL method herein describes soil carbonate by acid decomposition and CO_2 analysis by gas chromatography. This method is more commonly used in soil biochemical and biology studies, where organic C in soils with carbonates may be more precisely determined by subtracting the total carbonates (inorganic C) from total C. These data can be used to estimate SOC by subtracting (CO$_2$-C x 0.2727) from total carbon.

Carbonates, measurement and interferences: Soil carbonate is determined by chromatographic analysis of CO$_2$ evolved upon acidification of soil in a closed system of known headspace. Ferrous iron (FeCl$_2$) is added to the acid as an antioxidant, and the dilute acid solution (1N HCl) is chilled before addition to soil to minimize the decarboxylation of organic matter by the acid. Data are reported as mg CO$_2$-C per g of soil to the nearest 0.1 g.

Carbonates, interferences: It is essential that precautions be taken to ensure that there is no interference from organic matter oxidation (Loeppert and Suarez, 1996). This procedure may be more appropriate for soils with relatively low amounts of carbonates (<15 percent).

6.5 Particulate Organic Matter

Particulate organic matter, application: Particulate organic matter (POM) is a physical fraction of the soil >53µm in diameter (Elliott and Cambardella, 1991; Cambardella and Elliott, 1992; Follett and Pruessner, 1997). Some researchers combine this fraction with the fast or labile pool. Others have described this pool as slow, decomposable, or stabilized organic matter (Cambardella and Elliott, 1992). To avoid confusion, this fraction may best be described as representing an intermediate pool with regards to decomposition. This fraction is similar to various sieved and physical fractions, such as the resistant plant material (RPM) (Jenkinson and Rayner, 1977), size fractions (Gregorich et al., 1988), and variously determined light fractions of the soil organic matter (Strickland and Sollins, 1987; Hassink, 1995).

Under tillage, the POM fraction becomes depleted (Jenkinson and Rayner, 1977; Cambardella and Elliott, 1992). Reductions of more than 50 percent have been found in long-term (20-year) tillage plots. Measurable reductions are believed to occur in the range of 1 to 5 years (Cambardella and Elliott, 1992).

When paired samples are selected, a comparison can be made, either in time or between two tillage treatments, to determine the impact of the tillage practice. POM can be used in soil organic matter modeling as a soil quality indicator and as an indicator of the SOM that can move into the active C pool.

Since the late 1970s, several models have been developed to estimate the dynamics of organic matter in the soil. All of these models have at least two phases—slow and rapid. In measuring these two phases, chemical fractionation (humic and fulvic acids) has been found to be less useful than physical fractionation (Hassink, 1995). Examples of these models can be found in Jenkinson and Rayner (1977) and in tests of the CENTURY Soil Organic Model (Parton et al., 1987; Metherell et al., 1993; Montavalli et al., 1994). A minimum dataset for SOC proposed by Gregorich et al. (1994) includes POM as one of the primary parameters.

Particulate organic matter, measurement: The Hyper POM procedure is primarily the physical separation of <2-mm sieved soil into two fractions: (1) ≥53-µm, GT53, and (2) <53 µm, LT53 (Cambardella and Elliot, 1992; Follett and Pruessner,

1997). Typically, this procedure is determined on the A horizon (Soil Survey Staff, 2010) because detectable levels of both C and N are most likely to occur in this horizon. The GT53 fraction is retained, oven dried, and analyzed by dry combustion for total C and N. While POM-C and POM-N of the GT53 fraction are measured values, the MIN-C and MIN-N are calculated by subtracting the POM-C and POM-N from the total C and N in the <2-mm oven-dry soil. In the past, POM-C and POM-N were measured directly.

Particulate organic matter, interferences: In some weathered soils, approximately the same amount of C is in both fractions. To date, no research has been done to establish the interpretation of this result. Charcoal in native sod that has been historically burned does not affect the POM determination and C and N analysis, if residence time were to be determined from the two fractions.

6.6   β-glucosidase C Cycle Assay

β-glucosidase C cycle assay, application: β-glucosidases and galactosidases are widely distributed in nature and are important in the C cycle (Bandick and Dick, 1999). These enzymes have been detected in soils and fungi (Skujins, 1967, 1976; Jermyn, 1958; Viebel, 1950). β-glucosidase C cycle assay can be used to detect treatment effects that are related to C inputs and loss of organic matter in soils. β-glucosidase is one of the more stable assays, showing less seasonable variability than other assays.

β-glucosidase C cycle assay, measurement: Air-dry soils are mixed with a pH-buffered solution with substrate p-nitrophenyl-β-D-glucoside. Activity is determined by spectrometrically quantifying the amount of p-nitrophenol enzymatically released from substrate during 1-hour reaction time with the soil. While activity decreases with air drying of soil, the ranking of soil treatments within a soil type or the ranking across different soil types stays constant (Bandick and Dick, 1999).

6.7   Microbial Biomass
6.8   Mineralizable Nitrogen

Microbial biomass and mineralizable N, application: Soil is an ecosystem that contains a broad spectrum of biological components representing many physiological types (Germida, 1993). Soil biota (e.g., fungi, bacteria, earthworms, protozoa, arthropods, and nematodes) is critical to soil quality. It affects nutrient cycling, soil stability and erosion, water quality and quantity, and plant health (USDA/NRCS, 2004). Soil micro-organisms are an important component of soil organic matter. One of their functions is to break down nonliving organic matter in the soil. There are various methods of measuring the biomass of living soil microbes. The method described herein is a two-step procedure that includes chloroform fumigation incubation (Jenkinson and Powlson, 1976), with modifications, and measurement of CO_2 evolution by gas chromatography. Mineralizable N may also be determined on microbial biomass (Anderson and Domsch, 1978). Biota and its relationship to soil health are discussed in the USDA/NRCS "Soil Biology Primer" (Tugel and Lewandowski, 2001). Refer to Reeder et al. (2001) for additional information on microbial biomass.

Microbial biomass and mineralizable N, measurement: A freshly collected soil sample is weighed into two separate vials. One sample is fumigated using chloroform, and the other is used as a control (nonfumigated). After fumigation, both the fumigated and nonfumigated samples are brought up to 55 percent water-filled pore space (WFPS) (Horwath and Paul, 1994). Both samples are placed in a sealed container and aerobically incubated for 20 days. During this incubation period, it is assumed that normal respiration occurs in the control sample container. The fumigated sample, having a large C source for food (supplied from the dead micro-organisms), has a higher CO_2 production. At the end of 10 days, respiration readings are taken on both the control sample and the fumigated sample to determine the amount of CO_2 evolved by gas chromatography. The CO_2 level of the control sample is measured at the end of 20 days. CO_2 produced by biomass flush (g CO_2-C/g of soil) and soil microbial biomass (kg C/ha for a given depth interval) are reported (Soil Survey Staff, 2004). Mineralizable N by 2 M KCl extraction may also be determined on microbial biomass using a flow injection automated ion analyzer. The CO_2 produced by biomass flush (g CO_2-C/g of soil) and soil microbial biomass (kg C ha^{-1} for a given depth interval) are reported. The difference between mineralizable N of fumigated and nonfumigated samples is reported as mg N kg^{-1} soil as NH_3.

Microbial biomass, interferences: The determination of CO_2 evolution by gas chromatography gives a rapid and accurate measurement and can be used in acidic soils; however, this technique is prone to error in neutral and alkaline soils (Martens, 1987) as accumulation of carbonate species in the soil solution can lead to lowered CO_2 determinations (Horwath and Paul, 1994).

6.9   Root Biomass
6.10   Plant (Above-Ground) Biomass

Root and plant biomass, application: Root biomass in the upper 4 inches of the soil is an input value for the Revised Universal Soil Loss Equation (RUSLE) (Renard et al., 1997). The mass, size, and distribution of roots in the near surface are among the most important factors in determining the resistance of the topsoil to water erosion and wind erosion. Root biomass is also one of the major carbon pools found in soil. Above-ground biomass (production) represents annual yield and can be measured following the protocols in the "National Range and Pasture Handbook" (USDA/NRCS, 2009a) and in Sosebee (1997). For more information on root biomass and microbial biomass, refer to Reeder et al. (2001), Harwood et al. (1998), Sosebee (1997), Bedunah and Sosebee (1995), and Paul and Clark (1989).

The development of new roots and ultimately the decomposition of roots within the soil are major contributors to the soil organic carbon (SOC) pool. In this way, plant roots also contribute to the fertility of soils by slowly releasing macronutrients and micronutrients back into the soil. Root biomass and SOC help bind the soil together by forming aggregates and granular structure. As a result, tilth and the resistance of soil to erosion are improved. Depending upon the root turnover rate (known for some species), climate, and residue decomposition rate (known for some areas, based on climate and soil moisture status), the amount of carbon stored in the soil can be determined from the root biomass, plant residue, and SOC.

Root biomass is frequently used to calculate root/shoot ratios in order to evaluate the health and vigor of plants and to determine the success of establishment of seeded plants at the four-leaf stage (Thornley, 1995). Dried roots can be fine ground, and total C, N, P, and S can be determined. The C:N ratio also can be determined. This ratio is typically different from the C:N ratio of the above-ground plant material. Low levels of N in the soil will promote root growth over top growth (Bedunah and Sosebee, 1995). The C:N ratio of roots, plant residue in the soil, and SOC contribute to the residue decomposition rate for soils. Low C:N values lead to more rapid decomposition; high C:N levels slow decomposition. The C:N ratio required for decomposition of plant residue, without a net tie-up of N, is approximately 25:1 (Franks, 1998). Plant residue from young legumes commonly has a C:N ratio of 15:1; the ratio for plant residue from woody materials commonly is 400:1 (Harwood et al., 1998). The C:N ratio of soil microbes is quite variable but commonly falls between 15:1 and 3:1 (Paul and Clark, 1989).

Root biomass/soil horizon can be paired with the description of roots in each soil horizon (i.e., few fine, many very fine, etc.) in the pedon description; thus, a qualitative estimate can be made of the mass in each size fraction of roots. Refer to the *Field Book for Describing and Sampling Soils* (Schoeneberger et al., 2002) for detailed instructions for describing the quantity, size, and location of roots in soil horizons.

The automated method for determining root biomass also includes some plant residue. Woody material is removed and weighed separately. Because root biomass determined in this manner includes plant residue, it can be used to estimate the soil plant residue pool in most models (Jenkinson and Rayner, 1977; Metherell et al., 1993). The SSL method for measuring root biomass is after Soil Survey Staff (2004), Lauenroth and Whitman (1971), and Fribourg (1953). The automated root washer employed in the SSL method is after Brown and Thilenius (1976) and was developed and modified at a relatively low cost. Other, more expensive root washers include, but are not limited to, the one described in Carlson and Donald (1986) and the commercially available one developed by Gillison's Variety Fabrication, Inc., after Smucker et al. (1982).

Root and plant biomass, measurements: The procedural steps encompass the physical separation of roots and plant residue from a soil sample using an automated root washer (Brown and Thilenius, 1976). Weights are recorded for root and plant biomass, and these fractions are analyzed for total C, N, and S. The SSL reports root biomass as kg ha^{-1} at a given depth interval (cm). If plant residue was separated from roots, the SSL reports the root biomass and plant residue separately. The SSL also reports percent total C, N, and S for roots and plant material, using fine-grind samples ($\approx$ 180 µm).

Root biomass, interferences: The soil must be dispersed for successful separation of the roots and plant residue from the soil sample. Tapwater rather than distilled water should be used to help avoid puddling and dispersion problems.

6.11 Ratios, Estimates, and Calculations Related to Soil Biological and Plant Analysis

Estimates or calculated values associated with soil biological and plant analyses are described within the respective methods. Additional information on the reporting of these calculated values can be obtained from the SSL upon request.

7 Mineralogy

This section describes the SSL mineralogy methods and their applications. It includes a simplified key to family mineralogy classes, a table showing the land area of these classes in the U.S., and a discussion of the agronomic, taxonomic, and engineering significance of the classes. There are two broad groups of SSL mineralogy methods: instrumental analyses and optical analyses. Instrumental analyses include x-ray diffraction, thermogravimetric analysis, differential scanning calorimetry, and the measurement of specific surface area by N_2-BET. Optical analyses include grain studies and platy minerals. Grain studies include the preparation of grain mounts using epoxy with a specified sand or silt fraction. On occasion, less abundant minerals with a specific gravity >2.8 or 2.9 are separated with heavy liquids and analyzed. Platy minerals can be analyzed after separation by magnetic separation, static tube separation, or froth flotation. For detailed description of the SSL methods which are cross-referenced by method code in the table of contents in this manual, refer to SSIR No. 42 (Soil Survey Staff, 2004), which is available online at http://soils.usda.gov/technical/lmm/. Also refer to SSIR No. 51 (Soil Survey Staff, 2009; available online at http://www.soils.usda.gov/technical/) for detailed descriptions of field methods as used by NRCS soil survey offices.

7.1 Soil Mineralogy Classes

Differentiae at the family level of soil taxonomy are guides for the practical uses of soil. Mineralogy classes are a component of the family classification that is used in all mineral soils, except for Quartzipsamments. Quartzipsamments are a taxon defined by their siliceous mineralogy at the great group level. Accurate classification of a pedon into a family mineralogy class requires quantitative mineralogy data on particular horizons from the *mineralogy control section* (Lynn et al., 2002). Refer to Soil Survey Staff (2004) for descriptions of the analytical methods. Refer to the "Key to Mineralogy Classes" in *Keys to Soil Taxonomy* (Soil Survey Staff, 2010 or latest version) for the appropriate data requirements for these family mineralogy classes. Table 7.1.1 provides a simplified key to the family mineralogy classes, and table 7.1.2 shows their land area in the United States.

Table 7.1.1   Simplified key to mineralogy classes. Start at top of key and work down. The mineralogy family is the first class for which a pedon meets all of the criteria. After Lynn et al. (2002) and reproduced with permission by Soil Science Society of America, Madison, Wisconsin, with modifications based on the 11th edition of *Keys to Soil Taxonomy* (Soil Survey Staff, 2010).

A. For Oxisols and highly weathered Ultisols and Alfisols with:
1. >40% Fe oxides, fine-earth fraction[1]   Ferritic
2. >40% gibbsite, fine-earth fraction   Gibbsitic
3. 18–40% Fe oxides and 18–40% gibbsite, fine-earth fraction   Sesquic
4. 18–40% Fe oxides, fine-earth fraction   Ferruginous
5. 18–40% gibbsite, fine-earth fraction   Allitic
6. >50% (kaolinite + gibbsite + other nonexpending minerals), clay fraction   Kaolinitic
7. >50% (halloysite + kaolinite + allophane), clay fraction   Halloysitic
8. All other soils   Mixed

B. For soils with a substitute class, replacing particle-size class, other than fragmental:
1. ≥40% gypsum, fine-earth fraction or <20-mm fraction   Hypergypsic
2. ≥5% noncrystalline components (allophane > ferrihydrite), fine-earth fraction   Amorphic
3. ≥5% noncrystalline components (ferrihydrite > allophane), fine-earth fraction   Ferrihydritic
4. ≥30% volcanic glass, 0.02- to 2.0-mm fraction   Glassy
5. All other soils   Mixed

C. For all other soils with:
1. ≥15% gypsum, fine-earth fraction or <20-mm fraction   Gypsic
2. >40% (carbonates + gypsum), fine-earth or <20-mm fraction   Carbonatic
3. >40% Fe oxides minerals, fine-earth fraction   Ferritic
4. >40% (gibbsite + boehmite), fine-earth fraction   Gibbsitic
5. >40% Mg silicate minerals (serpentines and others), fine-earth fraction   Magnesic
6. >20% glauconitic pellets, fine-earth fraction   Glauconitic

D. For all other clayey soils (≥35% clay):
1. >10% (Fe oxide minerals + gibbsite), fine-earth fraction   Parasesquic
2. >50% (halloysitic + kaolinite + allophane) and more halloysite   Halloysitic
   than others, clay fraction[2]
3. >50% (kaolinite + gibbsite + other nonexpandables)   Kaolinitic
   and <10% smectite, clay fraction
4. More smectite than any other clay mineral, clay fraction   Smectitic
5. >50% illite (usually >4% K in clay fraction), clay fraction   Illitic
6. More vermiculite than any other clay mineral, clay fraction   Vermiculitic
7. Poorly crystalline aluminosilicates (e.g., allophane), clay fraction   Isotic
8. All other soils, clay fraction   Mixed

E. For other mineral soils with:
1. >40% (mica + mica pseudomorphs) in the 0.02- to 0.25-mm fraction   Micaceous
2. >10% (Fe oxides minerals + gibbsite), fine-earth fraction   Parasesquic
3. Poorly crystalline aluminosilicates (e.g., allophane), fine-earth fraction   Isotic
4. >90% (quartz + other resistant minerals) in 0.02- to 2.0-mm fraction   Siliceous
5. All other soils   Mixed

---

[1] Fine-earth fraction = material <2.0 mm in diameter.

[2] Clay fraction = material <2.0 μm in diameter.

Table 7.1.2   Land area in the United States in family mineralogy classes. After Lynn et al. (2002) and reproduced with permission by Soil Science Society of America, Madison, Wisconsin.

Mineralogy family class	ha x 1000	acres x 1000[1]	%
Kaolinitic	11,929	29,465	2.17
Halloysitic	46	113	<0.01
Illitic	5,818	14,371	1.06
Illitic (calcareous)	8	21	<0.01
Smectitic	73,756	182,184	13.40
Smectitic (calcareous)	5,507	13,603	1.00
Vermiculitic	63	156	0.01
Siliceous	56,572	140,177	10.31
Micaceous	1,084	2,678	0.20
Magnesic	226	558	0.04
Glauconite	24	59	<0.01
Carbonatic	15,881	39,225	2.89
Calcareous	10	25	<0.01
Hypergypsic and Gypsic	1,014	2,504	0.18
Mixed (clayey, loamy, sandy families)	349,059	862,176	63.44
Mixed (calcareous)	29,051	71,156	5.28
Ferritic	6	14	<0.01
Ferruginous	1	4	<0.01
Sesquic	1	2	<0.01
Total	550,237	1,359,084	100.00

[1] Multiply the reported numbers by 1,000 to approximate the actual numbers.

7.2 Agronomic, Taxonomic, Engineering, and Environmental Significance

Soil mineralogies: Soil mineralogy has important agronomic, taxonomic, engineering, and environmental implications for many soils and can be related and described in broad concepts (National Soil Survey Laboratory Staff, 1983; Lynn et al., 2002). The following descriptions include those mineralogies discussed earlier in relation to the CEC-7 to clay ratio as well as some mineralogies not previously discussed. The CEC-7 to clay ratio is used as auxiliary data to assess clay mineralogy. These data are especially useful when mineralogy data are not available. Refer to Soil Survey Staff (2010) for a discussion of mineralogy class as a taxonomic criterion for soil families in different particle-size classes.

Amorphic, ferrihydritic, glassy, and isotic (noncrystallinity): The colloidal fractions of many Andisols and Spodosols are mixtures that contain predominantly noncrystalline minerals with small amounts of layer lattice clays (1:1 and 2:1 phyllosilicates) and that also include Al and Fe humus complexes. In general, noncrystalline inorganic components (e.g., allophane and imogolite) predominate in well drained soils with andic properties in the udic soil moisture regime, whereas organo-chelated metals (spodic materials) predominate where soil pH is <5.0 (Parfitt and Kimble, 1989; Wilson and Righi, 2010). Additionally, some of the more notable properties of these soils are high variable charge; high surface area; high reactivity with sulfate, phosphate, and organics (high anion retention); low nutrient retention (e.g., Ca, Mg, NH_4^+ and K^+); high water retention; and low bulk density (Wada, 1985; McDaniel and Wilson, 2007). Because of their open structure and their capacity to accept rainwater, Andisols have a reputation for being more resistant to water erosion than many other soils; however, demographic pressure and steep slopes very often lead to compaction and severe losses of soil, and "thixotropic" properties of some Andisols make them very susceptible to landslides (Van Wambeke, 1992). Soils classified with the glassy mineralogy class have volcanic glass contents of 30 percent or more (by grain count) in the 0.02- to 2.0-mm fraction (coarse silt through very coarse sand). For glassy soils, the "pumiceous" or "ashy" substitutes for particle-size class are commonly used, although the "medial" substitute classes are used for some. Soils in the glass mineralogy class may or may not possess andic soil properties or be classified as Andisols.

Carbonatic: A carbonatic mineralogy class indicates that the soil pH is nearly always above 7 and most likely about 8.3 (Lynn et al., 2002). Carbonate minerals tend to precipitate P and to remove divalent cations (e.g., Ca, Mg, and Fe) from the soil solution. In general, crops grown on carbonatic soils may show signs of chlorosis, which could reflect nutrient deficiencies. Carbonatic soils have also been associated with P fixation; hindrance to root ramification; high base status; and both lower cation-exchange capacity and available water capacity, especially in soils with calcic horizons. Abundant Ca in the soil has a flocculating effect on soil colloids; i.e., clays tend to be coarser. Carbonates are considerably less soluble in water than gypsum, and the problems of dissolution noted for the hypergypsic and gypsic mineralogy classes are not often serious for soils in a carbonatic family (Lynn et al., 2002).

Ferritic, sesquic, ferruginous, and parasesquic: In general, soils with ferritic, sesquic, ferruginous, and parasesquic mineralogy classes are low in nutrients from mineral components and recycle nutrients from organic components. These soils have also been associated with fixation of plant nutrients, such as P and B. Ferritic soils are considered a good source of roadbed material and a possible source of Fe ore. In ferritic soils, the Fe typically adsorbs on clay surfaces and/or concentrates in nodular forms, and if the Fe cements, these soils may have a low amount of available water. In soils with sesquic, ferruginous, and parasesquic mineralogy classes, the clay contents, Atterberg limits, and other engineering tests are often difficult to measure and interpret. Soils in ferritic, sesquic, ferruginous, and parasesquic mineralogy classes are usually considered stable and permeable, from an engineering standpoint, if not disturbed. Dithionite-citrate extractable Fe (Fe_d) is used as a taxonomic criterion for the ferritic, sesquic, ferruginous, and parasesquic mineralogy classes. The sesquic and ferruginous mineralogy classes are used only in Oxisols and in "kandi" and "kanhap" great groups of Alfisols and Ultisols. The ferritic and parasesquic mineralogy classes are also used in these taxa as well as in other taxa with higher CEC and/or a higher content of weatherable minerals.

Gibbsitic and allitic: In general, soils with a significant content of gibbsite have low nutrient supplies from mineral components and recycle nutrients from organic matter. Gibbsitic soils have also been associated with fixation of plant nutrients, such as P. Atterberg limit tests and other engineering tests on these soils are often difficult to read and interpret. The gibbsitic and allitic mineralogy classes are incorporated into soil taxonomy because gibbsite is reported in the literature as part of the weathering cycle of soil clays in tropical regions. The gibbsitic mineralogy class requires more than 40 percent gibbsite (by weight), and the allitic mineralogy class requires 18 to 40 percent gibbsite (by weight) in the fine-earth fraction.

Glauconitic: Glauconitic soils are typically high in Fe, P, and Mg. These soils have been associated with unusual 1500-kPa water to clay ratios because the clay is held in glauconite pellets (greenish color), structural conditions are favorable for root growth, and structural stability is apparently not much of a problem. Also, if clay is high in smectite, these soils exhibit shrink-swell properties similar to those of smectitic soils. The content of K in glauconite is similar to that in illite, i.e., 6 to 8 percent.

Gypsic and hypergypsic: Soils with gypsic and hypergypsic mineralogy classes are recognized by their unique chemical and physical properties. The hypergypsic mineralogy class is intended to separate the gypsum-bearing soils with 40 percent or more (by weight) gypsum from those having at least 15 percent but less than 40 percent gypsum, by weighted average. Gypsum-bearing soils may become impervious to roots and water if they contain cemented horizons (i.e., petrogypsic horizons). Available water content and CEC are generally inversely proportional to gypsum content and are significantly lower when gypsum content is 40 percent or more. The saturated Ca soil solution may result in the fixation of the micronutrients Mn, Zn, and Cu. Soil subsidence through dissolution and removal of gypsum can crack building foundations, break irrigation canals, and make roads uneven. Failure can be a problem in soils with as little as 1.5 percent gypsum (Nelson, 1982). Corrosion of concrete also is associated with soils that contain gypsum. Accurate determination of soil texture classes is difficult in gypsum-bearing soils, both in the field and after laboratory analysis. This difficulty is due to the softness of gypsum, its solubility in water, and its flocculating effect on soil colloids. Gypsum content, soil resistivity, and extractable acidity singly or in

265

combination provide a basis for estimating potential corrosivity of soils (USDA/SCS, 1971).

Halloysitic: Halloysitic soils typically contain significant amounts of amorphous materials (Lynn et al., 2002). Halloysitic soils are commonly found in forested areas and receive rainfall so frequently that they rarely dry out. Trees obtain nutrients from underlying saprolite. Halloysitic soils are typically unstable because of moisture and are prone to slumping and solifluction. On sloping land, the soils are more likely to fail if the forest is replaced by housing.

Illitic: Illitic soils can be acid or alkaline, depending on the depositional environment of the parent sediments. These soils typically have a moderate to high natural fertility and base status. Illitic soils commonly contain intergrade illite-smectite clays. Illitic materials are typically stable for construction or foundation purposes but tend to be less stable as the clay content increases. Illites typically contain 6 to 8 percent K, which becomes available to plants as the soil weathers (Lynn et al., 2002).

Kaolinitic: Kaolinitic clays are low-activity clays with a low nutrient supply. Kaolinite forms in soils and is resistant to breakdown. Kaolinitic soils commonly have lost nearly all of their weatherable minerals (Lynn et al., 2002). They may have argillic horizons or clayey B horizons that tend to perch water and cause problems with drain fields for septic tanks. These soils can provide a stable construction base and stable construction material. Ponds may not seal well if they are floored or lined with kaolinitic clays. Small quantities of high-activity clays have a large impact on the properties of these soils.

Magnesic: The magnesic mineralogy class is intended to classify soils with a high content of serpentine minerals. Serpentinite is a rock consisting of primarily serpentine minerals, e.g., chrysotile, lizardite, and antigorite, with the generalized chemical formula [$Mg_3Si_2O_5(OH)_4$] derived from the alteration (serpentinization) of such minerals as olivines, pyroxenes, and amphiboles in dunites, peridotites, pyroxenites, and other ultramafic rocks (Bates and Jackson, 1987). Lizardite and antigorite have platy morphology, whereas chrysotile has tubular or fibrous morphology. Fibrous chrysotile is a mineral of the asbestos group (Bates and Jackson, 1987) and is known to cause serious human-health problems (Schreier et al., 1987; Skinner et al., 1988).

Currently, serpentinitic soils at the family category level are identified in the magnesic mineralogy class (Soil Survey Staff, 2010). The magnesic mineralogy class is defined as "Any particle-size class, except for fragmental, and more than 40% (by weight) magnesium-silicate minerals, such as serpentine minerals (antigorite, chrysotile, and lizardite) plus talc, olivines, Mg-rich pyroxenes, and Mg-rich amphiboles, in the fine-earth fraction." With little substitution of Fe for Mg in serpentine, Fe in peridotite minerals is incorporated into common accessory minerals, such as magnetite and hematite (Deer et al., 1992).

Serpentinitic soils have frequently been associated with infertility related to the toxic effects of Ni, Cr, and Co; toxicity of excess Mg; low Ca content; and adverse Ca/Mg ratio in the substrate (Burt et al., 2001a; Burt and Wilson, 2006). The physical nature of these soils, particularly droughtiness, is also a factor in the poor vegetative response. In addition, the physical properties of these soils make them structurally unstable, prone to landslides, and highly erodible. Controlling erosion with terraces, diversions, and waterways is difficult. These soils may also serve as a ready source of sediment in watersheds.

Micaceous: Muscovite and biotite are the most common mica minerals in soils in the micaceous mineralogy class. Muscovite is rather resistant and does not release K readily, whereas biotite is highly weatherable and readily releases K into the soil solution (Lynn et al., 2002).

Mica content of the 0.02- to 0.25-mm fraction is used as a taxonomic criterion for the micaceous mineralogy class. This size range uses the three fractions commonly quantified at the SSL (coarse silt, very fine sand, and fine sand) and is a modification to the earlier criterion (Soil Survey Staff, 2010) that required mica content in the 0.02- to 2.0-mm fractions. There are concerns that the taxonomic criteria (size range quantified, taxonomic limits, and depth of the control section) do not consistently identify soils with interpretatively significant mica content. Also, there are questions regarding the best procedure to use for mica quantification. The SSL uses grain counts for quantification for the micaceous family mineralogy class, as specified in *Keys to Soil Taxonomy* (Soil Survey Staff, 2010). This method appears the most reliable (quantitative, easily obtainable) for providing information about mica in soils (Rebertus and Buol, 1989). Weight or volume percent mica in soils is desired for many engineering interpretations, and there is no suitable conversion from grain counts to weight percent available for all soils. Refer to Table 7.2.1 (Kelley, 2006).

Such properties as size, shape, and density all limit determination of a conversion factor. Refer to additional discussion about the problems and issues related to grain counts versus weight percentages (Harris and Zelazny, 1985). Refer to Soil Survey Staff (2010) for additional discussion of micaceous mineralogy. The taxonomic criterion for the micaceous mineralogy class was revised in 2010 to be more than 45 percent, by grain count, mica and stable mica pseudomorphs. This revision incorporated the mica content range of the now-obsolete paramicaceous mineralogy class.

Many factors, e.g., mineralogy and particle size, collectively influence engineering properties of soils. Micaceous soils have been associated with stability problems, especially on slopes. Piping and jugging may occur on embankments. Micaceous soils are also susceptible to frost action. In studies of mica content and soil engineering properties, a high mica content has been generally associated with low strength, poor compactability, and high compressibility. Refer to equations and associated discussion below for more information on load-carrying capacity, soil strength, and compaction. Many of these studies have used artificial mixtures in order to control particle size and systematically vary mineral weight fractions (Gilroy, 1928; Harris et al., 1984a; Moore, 1971). While this technique is useful in isolating relationships, its use is uncertain in natural soils (most artificial mixtures are coarse textured, poorly graded, and unrepresentative of natural soils) (Harris et al., 1984b).

Table 7.2.1   Methods used to convert grain count values to percent mica by weight (Kelley, 2006)

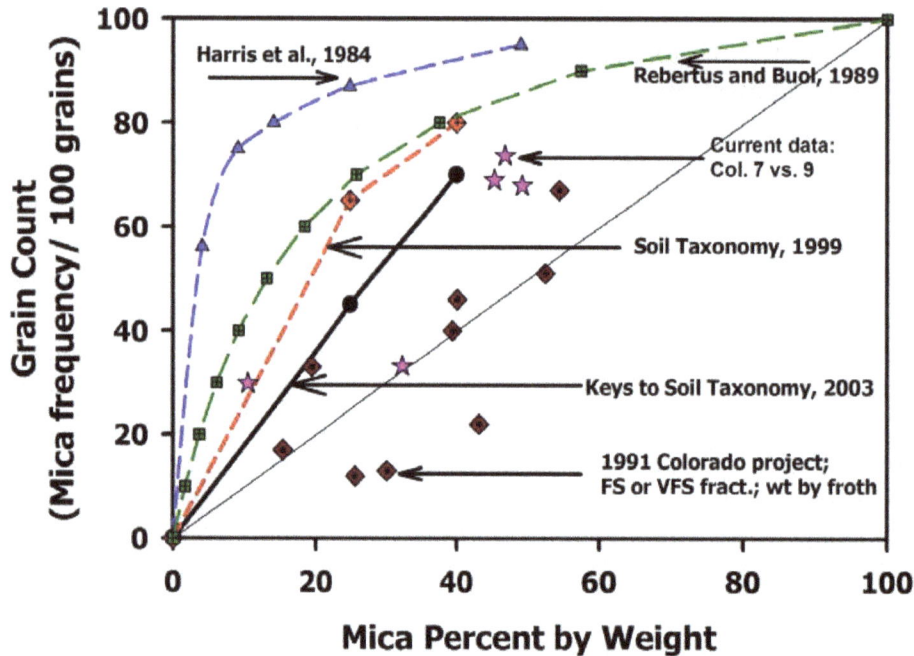

The objectives of the studies by Harris et al. (1984a, 1984b) were to determine relationships between mica content and engineering properties of sand in which the effects of mica are most likely to be most observable and least modified by external factors and to explore relationships between these engineering properties of natural soils and mineralogical and particle-size variables. Some of the conclusions of these studies (Harris et al., 1984a, 1984b) are as follows:

- Soil strength decreases and compressibility increases with decreasing particle size and increasing phyllosilicate content.
- Mica amounts as low as 10 percent can significantly reduce the strength and bearing capacity and increase the compressibility of sandy materials. Platy-mineral (mica) effects may be less in soils where nonplaty grains are coarse enough to form a skeletal network.
- The relative effect of mica content on shear strength, bearing capacity, and California Bearing Ratio (Goodwin, 1965) is most pronounced at lower weight percentages and tapers off at about 10 to 15 percent.

The study by Harris et al. (1984b) presents competing multiple linear regression (MLR) models relating engineering variables (internal friction, California Bearing Ratio, and compression index) and selected soil variables (e.g., coarse sand, mica, and smectite). Two variables explained a significant portion of the variability when used in these multiple regression models despite their low individual correlations with engineering variables. These variables were (1) estimated smectite weight fraction in the clay and (2) count frequency of 0.05- to 0.42-mm phyllosilicates. Harris et al. (1984b) concluded that the study soils probably contained too little smectite for the mineral effects to be clearly evident (masked by multiple interactions with natural soil systems). Other studies (Kenney, 1967; Olsen, 1974) have shown that smectite reduces soil strength to a greater extent than other phyllosilicates. In general, these study results indicate that regression models should be developed and interpreted using site-specific datasets.

The influence of mica on soil strength has not been well investigated. For example, the Virginia Department of Transportation has a rating for soil load support characteristics (strength) for highways. This rating identifies mica as one of the ratings criteria, thus acknowledging its impact, but does not provide a method to quantify mica. The mica content is determined visually, and the quantity is placed in three general classes (Kelley, 2006).

The American Association of State Highway and Transportation Officials (AASHTO) system uses a group index (GI) as a parameter indicative of the *load-carrying capacity* within an AASHTO soil group. The GI is numerically equal to the equation as follows:

Equation 7.2.1:

$$GI = (F - 35) [0.2 + 0.005 (LL - 40)] + 0.01 (F - 15) (PI - 10)$$

where
GI = Group index
LL = Liquid limit
F = Percentage passing 0.075 sieve, expressed as whole percent. This percentage is based only on the material passing the 75-mm sieve.
PI = Plastic index

The GI is considered an inverse indicator of load-carrying capacity; i.e., as GI increases, the load-carrying capacity decreases. The NRCS uses the AASHTO GI to identify low strength (high GI) as a soil limitation. There are three classes:

Group index (GI)	Degree of limitation
GI <5	not limiting
GI >5 but <8	somewhat limiting
GI ≥8	limiting

In a study of 26 representative high-mica soils from the Southern Appalachian Mountains and the Western Piedmont (Kelley, 2006), data from the USDA Soil Mechanics Laboratory, Ft. Worth, Texas, showed that 20 of the samples (77 percent) had a nonlimiting GI, 2 samples had a somewhat limiting GI, and 4 samples had a limiting GI. The soils with limiting GI had clay contents >40 percent, and those with somewhat limiting had 18 to 30 percent clay. In general, mica is only one of the factors determining soil strength; clay percentage, liquid limit, and plasticity index also are important factors. Mica content is not the overriding factor in determining soil strength, but the size of mica can play an important role (Kelley, 2006).

Refer to ASTM D 4555-01 (ASTM, 2008g), "Standard Test Method for Determining Deformability and Strength of Weak Rock by an In Situ Uniaxial Compressive Test." Since there is no reliable method of predicting overall *soil strength* and deformation data of a rock mass from the results of laboratory tests on small specimens, *in situ* tests on large specimens are necessary. In such tests, the rock specimen is tested under environmental conditions similar to those prevailing for the rock mass. Refer to ASTM D 1883-99, "Standard Test Method for CBR (California Bearing Ratio) of Laboratory-Compacted Soils" (ASTM, 2008e). Also refer to ASTM D 4429-04, "Standard Test Method for CBR (California Bearing Ratio) of Soils in Place" (ASTM, 2008f).

Refer to ASTM D 1557-07, "Standard Test Methods for Laboratory Compaction Characteristics of Soil Using Modified Effort (56,000 ft-lbf/ft^3 (2,700 kN-m/m^3))" (ASTM, 2008j). This test method (Proctor Test) covers laboratory *soil compaction* procedures used to determine the relationship between water content and dry unit weight of soils (compaction curve) compacted in a 4-inch or 6-inch (101.6- or 152.4-mm) diameter mold with a 10-lbf (44.5-N) rammer dropped from a height of 18 inches (457 mm), producing a compactive effort of 56,000 ft-lbf/ft^3 (2,700 kN-m/m^3).

Mixed: Mixed mineralogy classes are the most prevalent in soil taxonomy, and soils with mixed mineralogy cover the largest land area in the United States. These soils contain mixtures of several minerals, and no particular mineral has an overriding influence on soil properties or uses of the soil (Lynn et al., 2002).

Siliceous: Soils with siliceous mineralogy have >90 percent resistant minerals in the 0.02- to 2.0-mm fraction (sand plus coarse silt fractions). The major constituent is quartz, which is essentially inert chemically and physically and provides no plant nutrients (Lynn et al., 2002). Siliceous soils commonly require fertilizer for agronomic crop growth, and conifers grow better than hardwood tree species without applications of fertilizer. Siliceous soil materials are typically stable for construction, e.g., embankments, or as base or subbase for road construction. The siliceous mineralogy class is not applied to Quartzipsamments (sandy soils inherently defined by a high content of resistant, siliceous minerals), but these soils have chemical properties similar to those of soils to which the siliceous mineralogy class is applied. Quartzipsamments have very low CEC activity and cannot prevent soluble components from fertilizers, pesticides, or liquid wastes from passing through the soils and into the ground water (Lynn et al., 2002). In most sandy soils, the sand grains are coated with silicate clays, oxides, or organic matter. This coating provides some interaction with the soil solution. In some cases, however, the sand grains are uncoated and serve as poor filters for soil pollutants (Lynn et al., 2002). *Soil Taxonomy* recognizes coated and uncoated classes of coatings on sands for Quartzipsamments (Soil Survey Staff, 1999). Most soils in the siliceous mineralogy class are classified in "loamy" particle-size classes (e.g., fine-loamy), and they commonly have significant evidence of pedogenesis in the profile (e.g., argillic horizons). Soils in the siliceous mineralogy class are the third most extensive in the U.S. in terms of land area.

Smectitic: Smectitic clays are high-activity clays that are associated with high base status and natural fertility. These soils can be acidic if associated with weathered acidic volcanic materials. Smectitic soils typically have a high shrink-swell potential (indicated by linear extensibility) and can cause problems when used in foundations and septic drain fields and as construction material. Soils classified in the smectitic mineralogy class are widespread in the U.S. and are second only to soils with the mixed mineralogy class in terms of land area.

Vermiculitic: Vermiculitic clays are P fixing, and thus high levels of P fertilization are necessary. Vermiculitic soils have a tendency towards landslide and solifluction because of strong cleavage. This instability is more pronounced on sloping terrain.

Summary, mineralogy classes: Summary information about some soil mineralogy classes (from Lynn et al., 2002) is as follows:

- Mineralogy classes with low CEC activity include *ferritic, gibbsitic, sesquic, ferruginous, allitic, kaolinitic, hypergypsic, gypsic,* and *carbonatic*. Soils in ferritic, gibbsitic, sesquic, ferruginous, and allitic classes contain enough Fe oxide or oxyhydroxide minerals and Al hydroxide minerals to make P fixation a concern. The parasesquic class requires less Fe oxides plus gibbsite than the sesquic class, and contributions to and concerns about soil behavior are correspondingly less.
- Mineralogy classes with relatively higher CEC include *smectitic, illitic, vermiculitic,* and *mixed*. Soils high in smectite and vermiculite are some of the world's most fertile soils because of their high CEC and greater AWC.
- Mineralogy classes rich in mica minerals are *micaceous, illitic,* and *glauconitic* families.
- Mineralogy classes with predominantly short-range-order minerals (SROMs) and/or amorphous components include *amorphic, ferrihydritic, glassy,* and *isotic* families.

Summary mineralogies, soil orders: Even though mineralogy is not explicitly used to define soil orders, definitions of these orders are designed so that they contain a distinct suite of minerals (Lynn et al., 2002). General mineralogical information as it relates to soil orders (from Lynn et al., 2002) is as follows:

- *Oxisols* have a subsurface oxic or kandic horizon that has low CEC activity and is nearly devoid of weatherable primary minerals (<10 percent in the 50- to 200-μm fraction). The clay fraction is primarily kaolinite and lesser amounts of gibbsite, goethite, hematite, anatase, and rutile. Fe oxides facilitate development of a fine granular, low bulk density soil with good internal drainage and physical stability if the soil is not disturbed. Most of the sand fraction consists of quartz.
- *Ultisols* have a subsurface argillic or kandic horizon in which clay has accumulated relative to the surface horizon. Ultisols are deeply weathered and have low base saturation in the lower part of the subsoil, but they are less weathered than Oxisols and more weathered than Alfisols. High amounts of kaolinite and low CEC are common characteristics of Ultisols. Significant amounts of hydroxy-interlayered vermiculite and smectite occur

271

in some Ultisols. In the clay fraction, common accessory minerals are goethite, hematite, and gibbsite. In the sand and silt fractions, potassium feldspars and muscovite mica are common weatherable primary minerals. Ultisols commonly occur on old stable landscapes.

- *Alfisols* have an argillic, kandic, or natric horizon and are less deeply weathered than Ultisols. The higher base saturation deep in the profile is used to separate Alfisols from Ultisols. Generally, Alfisols contain a mixture of layer silicate minerals (e.g., illite, smectite, vermiculite, and kaolinite) with no one mineral dominating. Hydroxy-interlayered minerals are not common in Alfisols but can occur in the more acid Alfisols. In the sand and silt fractions, weatherable primary minerals (e.g., feldspars, muscovite mica, and some amphiboles) are common. Most Alfisols have weathered in temperate climates. Carbonates have been leached in humid climates but commonly remain in the profile in subhumid climates. Alfisols are extensive on older glacial landscapes.

- *Mollisols* have a dark diagnostic surface horizon (i.e., mollic epipedon) rich in organic matter and have high base saturation throughout the profile. Mollisols occur in practically any climate. They can contain a variety of minerals, but smectite or illite typically predominates in the clay fraction. Formation of hydroxy-interlayered vermiculite and smectite is usually inhibited because of the high base saturation and relatively high pH. Sands and silts typically include quartz, feldspars, and micas and have small amounts of other minerals. As with Alfisols, carbonates have been leached from the soils in humid climates but remain in the subsoil in the drier climates and may accumulate.

- *Aridisols* are defined primarily by the presence of an aridic moisture regime. They occur in a desert or semidesert climate, whether it is hot, temperate, or cold. Agronomic crops cannot be grown on these soils without irrigation. Weathering and formation of a deep soil profile are retarded under an aridic soil moisture regime. Illuviation of clay into diagnostic subsoil horizons (e.g., argillic and nitric) does occur but at a relatively shallow depth, and horizon formation is often aided by the presence of Na. Clays commonly contain a mixture of smectite, kaolinite, and illite. More than any other soil order, Aridisols contain very soluble minerals (e.g., halite and gypsum). Accumulations of calcite and gypsum commonly occur in Aridisols. Aridisols with kaolinitic clay that occur in present deserts are assumed to have formed under previous climates that were more moist and are called paleosols. Some Aridisols simply inherited kaolinite clay from parent materials that were derived from ancient geologic formations and have little evidence of prior pedogenic development in a wetter paleoclimate.

- *Spodosols* have a subsurface spodic horizon where organic matter and/or Fe and Al oxides have accumulated. These horizons are typically thin and contorted. Refer to Soil Survey Staff (2010) for the definition of a spodic horizon. Spodic horizons are best expressed in cool to cold climates in areas of sandy parent materials. Spodosols commonly occur on young glaciated land or in alpine areas as well as in wet, sandy coastal areas and typically have a minimal Fe oxide component. Spodic horizons typically contain amorphous materials and some short-range-order minerals (e.g., ferrihydrite and allophane). Well crystallized minerals in the clay, silt, and sand are inherited from the parent materials, as mineral weathering after deposition is negligible. Sand and silt are primarily quartz, and clays typically have small amounts of illite, vermiculite, and kaolinite.

- *Andisols* have andic soil properties in the surface horizon. Refer to Soil Survey Staff (2010) for the definition of andic soil properties. Volcanic ash deposits are the most extensive parent material. Andisols can also develop in mafic lava flow or deposits from cinder cones. Parent materials are generally amorphous with short-range-order minerals, e.g., allophane, imogolite, ferrihydrite, and in some cases, halloysite. Short-range-order materials commonly weather to smectite in drier climates.

- *Vertisols* have at least 30 percent clay in surface horizons and evidence of shear failure (slickensides) or wedge-shaped peds due to swelling clays in subsurface horizons. Reversible cracks, which prograde toward the soil surface and extend across horizon boundaries, are commonly induced by dewatering of subsoil horizons. The clay content is typically >50 percent in the subsurface and almost always is dominated by smectite. In a few Vertisols, clays contain a mixture of layer silicate minerals, including smectite, illite, vermiculite, and kaolinite, with none dominant. Vertisols occur in all except the coldest climates but in most cases occur on clayey geologic deposits or on clayey Quaternary alluvial or deltaic deposits. Shrink-swell properties are commonly well expressed in subhumid climates.

- *Inceptisols* are young to middle-aged, developed soils that can form in any climate except the desert (i.e., aridic/torric soil moisture regime). They must show evidence that parent materials have been altered enough by pedogenic processes to form diagnostic surface and/or subsurface horizons (e.g., umbric, cambic, and petrocalcic horizons), and most do meet the requirements of any of the other soil orders. Most Inceptisols form on young land surfaces. Minerals are inherited from the parent material; therefore, there is no well defined mineralogy associated with Inceptisols.

- *Entisols* show little evidence of weathering, and mineralogy is diverse because most minerals are inherited from parent materials. Entisols occur in any climate or in areas of any soil moisture status. They form in very young alluvial, glacial, or eolian deposits, and some form in colluvial or mass-wasting deposits on steep and/or rocky terrain. Entisols can form in very sandy deposits on geomorphic surfaces that may not be young.

- *Gelisols* have permafrost and/or gelic materials within 200 cm of the soil surface. They form in very cold climates and express minimal weathering of primary minerals.

- *Histosols* are primarily composed of organic soil materials and contain little mineral material. They commonly occur on young geomorphic surfaces in cool climates.

7.3 Instrumental Analyses

Mineral identification: The physical and chemical properties of a soil are controlled to a very large degree by the soil minerals, especially by those minerals constituting the clay fraction (McBride, 1989; Whittig and Allardice, 1986). Positive identification of mineral species and quantitative estimation of their proportions in soils usually require the application of several complementary qualitative and quantitative analyses (Whittig and Allardice, 1986). Some of the semiquantitative and quantitative procedures that have been performed by the SSL include x-ray diffraction and thermal analysis. Thermal procedures include thermogravimetric analysis (TGA) and differential scanning calorimetry (DSC). While the SSL no longer performs DSC analysis, discussion about this method and application of resulting data are included herein as these data are in the SSL database. Refer to Amonette and Zelazny (1994) for a more detailed discussion of quantitative methods in soil mineralogy. Other indirect, ancillary procedures that infer mineral composition include linear extensibility, elemental analysis, and CEC/clay ratios.

Analysis by x-ray diffraction facilitates the identification of crystalline mineral components of soil and semiquantitative estimates of relative amounts. This analysis is commonly applied by the SSL to the clay fraction in soils and to layer silicate (phyllosilicates) minerals in particular. Identification is by the d-spacings (spatial distance between repeating planes of atoms) that are characteristic of a mineral, according to Bragg's Law. Because some layer silicate structures are similar from one mineral to another, several treatments (cation saturation and heating) must be used to correctly identify the several minerals. In x-ray analysis of soils or clay samples, there are difficulties in evaluation of and compensation for the variations in chemical composition, crystal perfection, amorphous substances, and particle size (Whittig and Allardice, 1986; Hughes et al., 1994). A more reliable and accurate estimation of mineral percentages is provided when x-ray diffraction analysis is used in conjunction with other methods, e.g., differential-thermal, surface-area, elemental analysis, and other species-specific chemical methods (Alexiades and Jackson, 1966; Karathanasis and Hajek, 1982a).

Many soil constituents undergo thermal reactions upon heating, and these reactions serve as diagnostic properties for qualitative and quantitative identification of these substances (Tan et al., 1986; Karathanasis and Harris, 1994). Thermogravimetric analysis (TGA) is a technique for determining weight loss of a sample when it is heated at a constant rate. The TGA is an outgrowth of dehydration curves that were used in early studies of various phyllosilicate clay minerals (Jackson, 1956); however, with TGA the sample weight is monitored continuously rather than measured at discrete intervals after periods of heating at a constant temperature (Wendlandt, 1986). The TGA measures only reactions that involve weight loss of the sample.

Differential scanning calorimetry (DSC) is a calorimetric technique that theoretically measures the amount of energy required to establish zero temperature difference between sample and reference material as the two are heated side by side at a controlled rate (Tan et al., 1986; Karathanasis and Harris, 1994). Most common DSC instruments have sample and reference pans heated in a single furnace, and the difference in temperature is measured during various endothermic and exothermic reactions. This difference in temperature is then converted to a value equivalent to an enthalpy change (expressed in calories) using instrumental calibrations (Karathanasis and Harris, 1994).

The TGA and DSC are complementary methods available to the analyst. The SSL is currently equipped to perform TGA. Many of the same clay mineral reactions that are studied by DSC, e.g., dehydroxylation, loss of surface adsorbed water, decomposition of carbonates, and oxidation, can also be studied by TGA; however, some transformation reactions, e.g., melting or structural reorganization (quartz alpha-beta transition), cannot be measured by TGA because no weight loss is involved (Karathanasis and Harris, 1994). The DSC procedures provide information about energy relationships in the structures and reactions of the solid phase, whereas TGA provides quantitative information about substances gained or lost by the solid phase during certain thermally driven reactions.

7.3 Instrumental Analyses
7.3.1 X-Ray Diffraction Analysis

X-ray diffraction, application: Clay fractions of soils are commonly composed of mixtures of one or more phyllosilicate minerals together with primary minerals inherited directly from the parent material (Olson et al., 1999). Positive identification of mineral species and quantitative estimation of their proportions in these polycomponent systems usually require the application of several complementary qualitative and quantitative analyses (Whittig and Allardice, 1986; Amonette and Zelazny, 1994; Wilson, 1994; Moore and Reynolds, 1997; Harris and White, 2008). One of the most useful methods of identifying and making semiquantitative estimates of the crystalline mineral components of soil is x-ray diffraction analysis (Schulze, 1989; Hughes et al., 1994; Kahle et al., 2002). Quantification of a mineral by x-ray diffraction requires attention to many details, including sample (slide) size relative to the incident x-ray beam, thickness and particle-size uniformity of the sample, and beam-sample orientation (Moore and Reynolds, 1997). More complex quantification procedures include using standard additions, full pattern fitting, and determining mineral intensity factors (Kahle et al., 2002). At best, quantification can approach a precision of ± 5 percent and an accuracy of ± 10 to 20 percent (Moore and Reynolds, 1997).

The operational strategy at the SSL has been to base mineral quantification on first-order peak intensities. Semiquantitative interpretations have been held consistent over time (1964 to the present) by adjusting instrumental parameters (e.g., scan speed) to maintain a constant peak intensity for an in-house reference clay standard and, subsequently, soil samples. The intent is to keep interpretations consistent from sample to sample.

X-ray diffraction, measurement: Soils are dispersed and separated into fractions of interest. Sands and silts are mounted on glass slides as slurries or placed into a well of a sample holder (powder mount), on a smear of Vaseline, or on double sticky tape for analysis. Clay suspensions are placed on glass slides to dry and to preferentially orient clay minerals. Most samples of soil clays contain fewer than seven minerals that require identification. The soil clay minerals of greatest interest are phyllosilicates, e.g., kaolinite, mica (illite), smectite, vermiculite, hydroxy-interlayered vermiculite, hydroxy-interlayered smectite, and chlorite.

Diffraction maxima (peaks) develop from the interaction of x-rays with planes of elements that repeat at a constant distance (d-spacing) through the crystal structure. Generally, no two minerals have exactly the same d-spacings in three dimensions, and the angles at which diffraction occurs are distinctive for a particular mineral (Whittig and Allardice, 1986; Moore and Reynolds, 1997). Many phyllosilicates (or layer silicate minerals) have very similar structures, except in the direction perpendicular to the layers (c-dimension). Several treatments are needed to sort out which minerals are present. Glycerol is added to expand smectites. Ionic saturation and/or heat treatments are used to collapse some 2:1 layer silicates. Heating a sample will dehydroxylate (destroy) kaolinite, gibbsite, and goethite, thus eliminating characteristic peaks.

The crystal d-spacings of minerals, i.e., the interval between repeating planes of atoms, can be calculated by Bragg's Law as follows:

Equation 7.3.1:

$$n\lambda = 2d \sin \theta$$

where
n = integer that denotes order of diffraction
λ = x-radiation wavelength (Angstroms, Å or nm)
d = crystal d-spacing (Å or nm)
θ = angle of incidence

When $n = 1$, diffraction is of the first order. The wavelength of radiation from an x-ray tube is constant and characteristic for the target metal in the tube. Copper radiation (CuKα) with a wavelength of 1.54 Å (0.154 nm) is used at the SSL. Because of the similar structure of layer silicates commonly present in soil clays, several treatments that characteristically affect the d-spacings are necessary to identify the clay components. At the SSL, four treatments are used: Mg^{2+} (room temperature), Mg^{2+}-glycerol (room temperature), K^+ (300 °C), and K^+ (500 °C).

Standard tables to convert θ or 2θ angles to crystal d-spacings are published in the U.S. Geological Survey Circular 29 (Switzer et al., 1948) and in other publications (Brown, 1980). Through the years, hardware has been updated and the recording of data has evolved from a strip chart recorder through several kinds of electronic software.

From the "Detected Peaks File" and graphics chart, the minerals present are identified according to the registered d-spacings. For a more complete list of d-spacings for confirmation or identification of a mineral, consult the Mineral Powder Diffraction File—Data Book (JCPDS, 1980). As a first approximation, the following peak intensities are used, i.e., peak heights above background in counts s[-1], to assign each layer silicate mineral to one of the five semiquantitative classes.

Class	Approximate mineral percentage	Peak height above background (counts sec^{-1})
5 (Very large)	>50	>1800
4 (Large)	30-50	1120 to 1800
3 (Medium)	10-30	360 to 1120
2 (Small)	3-10	110 to 360
1 (Very small)	<3	<110

Class placement is adjusted to reflect area under the curve if peak is broad relative to peak height or if thermal, elemental, or clay activity data or other evidence warrants class adjustment. If there are no peaks or evidence of crystalline components, the sample is placed in the NX class (noncrystalline). If there are only one to three very small (class 1) peaks, NX is also indicated to imply a major noncrystalline component. Sometimes when slides are of poor quality due to high organic matter, curling of clay, or other slide preparation problems, the NX designation is not used.

X-ray diffraction, interferences: Interstratification of phyllosilicate minerals causes problems in identification. These interstratified mixtures, differences in crystal size, purity, chemical composition, atomic unit cell positions, and background or matrix interferences affect quantification (Moore and Reynolds, 1997; Kahle et al., 2002). In the SSL, no pretreatments other than ionic saturation and dispersion with sodium hexametaphosphate are used for separation and isolation of the clay fraction in the routine procedure. Impurities, such as organic matter, carbonates, and iron oxides, may act as matrix interferences and cause peak attenuation during x-ray analysis, or they may interfere with clay dispersion and separation. Pretreatments to remove these impurities serve to concentrate the crystalline clay fraction and may increase accuracy, but they may also result in degradation of certain mineral species (e.g., smectites) and loss of precision (Hughes et al., 1994).

The separation (centrifuge) procedure used to isolate the clay fraction from the other size fractions of the soil skews the <2-μm clay suspension toward the fine clay, but it minimizes the inclusion of fine silt in the fraction. Sedimentation of the clay slurry on a glass slide tends to cause differential settling by particle size (i.e., increasing the relative intensity of finer clay minerals).

Dried clay may peel from the XRD slide. One remedy is to rewet the peeled clay on the slide with one drop of glue-water mixture (1:7). Other remedies are:

a. Place double sticky tape on the slide before rewetting the dried clay with the glue-water mixture.

b. Dilute the suspension if it is thick.

c. Crush with ethanol and dry, and then add water to make a slurry slide.

d. Roughen the slide surface with a fine-grit sandpaper.

An optimum amount of glycerol on the slides is required to solvate the clay, i.e., to expand smectites to 18 Å. X-ray analysis should be performed 1 to 2 days after glycerol addition. If excess glycerol is applied to the slide and free glycerol remains on the surface, XRD peaks are attenuated. Suggestions for drying the slides and achieving optimum glycerol solvation are as follows:

a. Use a chamber, such as a desiccator (with no desiccant), to dry the slide, especially when the clay is thin.

b. If the center of the slide is whitish and dry (commonly with thick clay), brush the slide with glycerol or add an extra drop of glycerol.

7.3 Instrumental Analyses
7.3.2 Thermogravimetric Analysis
7.3.3 Differential Thermal Analysis

Thermal analysis, application: Thermal analysis defines a group of analyses that determine some physical parameter, e.g., energy, weight, or evolved substances, as a dynamic function of temperature (Tan et al., 1986; Karathanasis and Harris, 1994; Karathanasis, 2008). Thermogravimetric analysis (TGA) is a technique for determining weight loss of a sample as it is being heated at a controlled rate. The weight changes are recorded as a function of temperature, i.e., a thermogravimetric curve, and provide quantitative information about substances under investigation, e.g., gibbsite ($Al(OH)_3$), kaolinite ($Al_2Si_2O_5(OH)_4$), goethite (FeOOH), and 2:1 expandable minerals (smectite and vermiculite).

Thermal analysis, measurement: A 5- to 10-mg sample of soil clay or fine earth (finely ground) is placed in a platinum sample pan, and the pan is placed in the TGA balance. The instrument records the initial sample weight. The sample is then heated from a temperature of 30 °C to 900 °C at a variable rate (using a slower rate during mineral reactions) in a flowing N_2 atmosphere. The computer collects weight changes as a function of temperature and records a thermogravimetric curve. Gibbsite and kaolinite are quantified by calculating the weight loss between approximately 250 to 350 °C and 450 to 550 °C, respectively, and then relating these data to the theoretical weight loss of pure gibbsite or kaolinite. The weight loss is due to dehydroxylation, i.e., loss of crystal lattice hydroxyl ions. Though not presently performed by the SSL, quantification of the 2:1 expandable minerals (smectite + vermiculite) is related to weight loss at <250 °C, i.e., loss of adsorbed water (Karathanasis and Hajek, 1982a, 1982b; Tan et al., 1986). At this low temperature, adsorbed water is proportional to the specific surface area of the sample (Jackson, 1956; Mackenzie, 1970; Tan and Hajek, 1977; Karathanasis and Hajek, 1982b). In the absence of gibbsite, goethite (Fe oxyhydroxide) can be quantified based on the characteristic weight loss of 10.1 to 11.2 percent between 300 and 400 °C (Karathanasis and Harris, 1994). Recent work in the SSL has found good agreement between gypsum quantification using dissolution procedures and thermal analysis. Gypsum has a weight loss of 20.9 percent between 100 and 350 °C (Karathanasis and Harris, 1994). The TGA method is especially useful for soils with a large percentage (>20 percent) of gypsum. Burt et al. (2001a) found good agreement between total Mg analysis and TGA quantification (12.9 percent weight loss between 600 and 650 °C) of serpentine minerals in ultramafic-derived soils in Oregon. The percentages of gibbsite, kaolinite, gypsum, and antigorite are reported.

Equation 7.3.2.1:

% Kaolinite = {[(Δ sample weight % 450-550 °C)]/14} x 100

or (Δ sample weight % 450-550 °C) x 7.14

where
Δ sample weight = Total change in sample weight expressed as relative percent
14 = Percent weight of hydroxyl water lost from pure kaolinite during dehydroxylation

Equation 7.3.2.2:

% Gibbsite = {[(Δ sample weight % 250-350 °C)]/34.6} x 100

or (Δ sample weight % 250-350 °C) x 2.89

where
Δ sample weight = Total change in sample weight expressed as relative percent of the 110 °C base weight
34.6 = Percent weight of hydroxyl water lost from pure gibbsite during dehydroxylation

If Fe oxides are removed prior to analysis to prevent the interference with gibbsite determination, the calculation is modified to account for weight loss due to deferration as follows:

Equation 7.3.2.3:

% Gibbsite ={[Δ Sample weight % 250-350 °C x (Wt_2/Wt_1)]/34.6} x 100

where
Wt_1 = Weight before deferration
Wt_2 = Weight after deferration

The percent weights of hydroxyl water lost from kaolinite and gibbsite are derived from the following assumed dehydroxylation reactions.

Equations 7.3.2.4 and 7.3.2.5:

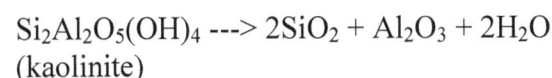

$Si_2Al_2O_5(OH)_4$ ---> $2SiO_2 + Al_2O_3 + 2H_2O$
(kaolinite)

$2Al(OH)_3 \longrightarrow Al_2O_3 + 3H_2O$
(gibbsite)

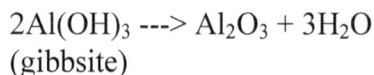

Using kaolinite as an example, percent weight of hydroxyl water lost is calculated from the following formula weights:

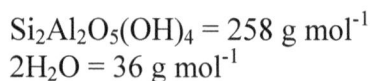

$Si_2Al_2O_5(OH)_4 = 258$ g mol^{-1}
$2H_2O = 36$ g mol^{-1}

Percent weight of hydroxyl water lost = (36/258) x 100 = 14%

If serpentine minerals are present in the sample, TGA can be used to quantify these minerals (Burt et al., 2001a) based on the onset temperature of 600 to 650 °C (Karathanasis and Harris, 1994) and a weight loss from 600 to 900 °C (12.9 percent) based on the mineral structure [$Mg_3Si_2O_5(OH)_4$]:

Equation 7.3.2.6:

% Serpentine minerals = {[(Δ sample weight % 600-900 °C)]/12.9} x 100

or (Δ sample weight % 600-900 °C) x 7.75

Gypsum can be quantified based on a loss of 20.9 percent (Karathanasis and Harris, 1994) based on the weight loss in the region of 100 to 350 °C:

Equation 7.3.2.7:

% Gypsum = {[(Δ sample weight % 100-350 °C)]/20.9} x 100

or (Δ sample weight % 100-350 °C) x 4.78

Thermal analysis, interferences: Organic matter is objectionable because it has a weight loss by dehydrogenation and by oxidation to CO_2 between 300 and 900 °C (Tan et al., 1986). Analysis in an inert N_2 atmosphere helps to alleviate this problem, but samples with significant organic matter should be pretreated with H_2O_2. Mineral salts that contain water of crystallization also may interfere. Samples should be washed free of any soluble salts. In some cases, weight loss from gibbsite and goethite overlap and prevent quantitative interpretation. These samples can be deferrated to eliminate goethite.

A representative soil sample is important, as sample size is small (<10 mg). Large aggregates in the sample should be avoided because they may cause thermal interferences, i.e., differential kinetics of gas diffusion through the sample and physical movement of the sample in a reaction.

In general, the same reactions that interfere with DSC/DTA also interfere with TGA determinations of kaolinite, gibbsite, and 2:1 expandable minerals; however, TGA is more sensitive to small water losses at slow rates, whereas DSC/DTA is more sensitive to large water losses at rapid rates (Tan et al., 1986). This sensitivity difference may help to explain why quantification of kaolinite and gibbsite in TGA vs. DSC/DTA often is not equivalent; i.e., TGA estimates tend to be greater than the corresponding DSC/DTA estimates. With TGA, there is a greater probability of measuring water losses in specific temperature regimes that are not specifically associated with dehydroxylation reactions of interest. This problem is particularly apparent with illitic samples, which characteristically contain more "structural" water than ideal structural formulae would indicate (Rouston et al., 1972; Weaver and Pollard, 1973).

Even though it is well established that various minerals lose the major portion of their crystal lattice water in different temperature ranges (Tan et al., 1986), there are overlaps in weight loss regions (WLR) of minerals that interfere in the identification and measurement of the minerals of interest. The goethite WLR (250 to 400 °C) overlaps the gibbsite WLR (250 to 350 °C) (Mackenzie and Berggen, 1970). The illite WLR (550 to 600 °C) overlaps the high end of the kaolinite WLR (450 to 550 °C) (Mackenzie and Caillere, 1975). The WLR of hydroxy-Al interlayers in hydroxy-Al interlayered vermiculite (HIV) (400 to 450 °C) overlaps the low end of the kaolinite WLR (450 to 550 °C), especially in the poorly crystalline kaolinites (Mackenzie and Caillere, 1975). Similarly, the dehydroxylation of nontronites, which are Fe-rich dioctahedral smectites (450 to 500 °C), may interfere with kaolinite identification and measurement (Mackenzie and Caillere, 1975).

7.3 Instrumental Analyses
7.3.4 Differential Scanning Calorimetry

Differential scanning calorimetry, application: Calorimetry measures specific heat or thermal capacity of a substance. Two separate types of differential scanning calorimetry (DSC) instruments have evolved over time. The term "DSC" is most appropriate for the power-compensated type of instrument, in which the difference in the rate of heat flow between a sample and a reference pan is measured as the materials are held isothermal to one another using separate furnaces (Karathanasis and Harris, 1994; Karathanasis, 2008). The DSC therefore directly measures the magnitude of an energy change (ΔH, enthalpy or heat content) in a material undergoing an exothermic or endothermic reaction. Heat-flow DSC instruments are more common than other types and are similar in principle to differential thermal analyzers (DTA). The heat-flow instruments have the sample and reference pans in a single furnace and monitor pan temperature from the conducting base. The difference in pan temperatures (ΔT) results from clay mineral decomposition reactions in the sample as the furnace temperature is increased. The configuration of this instrument results in a signal that is independent of the thermal properties of the sample, and ΔT can be converted to a calorimetric value via instrument calibration (Karathanasis and Harris, 1994). DSC is commonly used to quantify gibbsite ($Al(OH)_3$) and kaolinite ($Al_2Si_2O_5(OH)_4$) in soils and clays by measuring the magnitude of their dehydroxylation endotherms, which are between approximately 250 to 350 °C and 450 to 550 °C, respectively (Jackson, 1956; Mackenzie, 1970; Mackenzie and Berggen, 1970; Karathanasis and Hajek, 1982a).

Differential scanning calorimetry, measurement: An 8-mg sample of soil clay is weighed into an aluminum sample pan and placed in the DSC sample holder. The sample and reference pans are heated under flowing N_2 atmosphere from a temperature of 30 to 600 °C at a rate of 10 °C min^{-1}. Data are collected by the computer, and a thermograph is plotted. Gibbsite and kaolinite are quantified by measuring the peak area of any endothermic reactions between 250 to 350 °C and 450 to 550 °C, respectively, and by calculating the ΔH of the reaction. These values are related to the measured enthalpies of standard mineral specimens (gibbsite and kaolinite). Percent kaolinite and gibbsite are reported.

Differential scanning calorimetry, interferences: Organic matter is objectionable because it produces irregular exothermic peaks in air or O_2, commonly between 300 and 500 °C, which may obscure important reactions from the inorganic components of interest (Schnitzer and Kodama, 1977). Analysis in an inert N_2 atmosphere helps to alleviate this problem, although thermal decomposition of organic matter is still observed. Pretreatment with H_2O_2 may be necessary for soils with significant amounts of organic matter. Mineral salts that contain water of crystallization also may be interferences. Samples should be washed free of any soluble salts.

A representative soil sample should be used, as sample size is small (<10 mg). Large aggregates in the sample should be avoided because they may cause thermal interferences, i.e., differential kinetics of gas diffusion through the sample and physical movement of the sample in a reaction.

The dehydroxylation of goethite is between 250 and 400 °C and may interfere with the identification and integration of the gibbsite endotherm (250 to 350 °C) (Mackenzie and Berggen, 1970). The dehydroxylation of illite is between 550 and 600 °C and partially overlaps the high end of the kaolinite endotherm (450 to 550 °C), resulting in possible peak integrations (Mackenzie and Caillere, 1975). The dehydroxylation of hydroxy-Al interlayers in hydroxy-Al interlayered vermiculite (HIV) is between 400 and 450 °C and may interfere with the low end of the kaolinite endotherm (450 to 550 °C), especially in the poorly crystalline kaolinites (Mackenzie and Mitchell, 1972). Similarly, the dehydroxylation of nontronites, which are Fe-rich dioctahedral smectites, is between 450 and 500 °C and may interfere with kaolinite identification and measurement (Mackenzie and Caillere, 1975).

7.3   Instrumental Analyses
7.3.5   Surface Area by N_2-BET

Surface area, application: Surface area influences many physical and chemical properties of materials, e.g., physical adsorption of molecules and the heat loss or gain that results from this adsorption, shrink-swell capacity, water retention and movement, cation-exchange capacity, pesticide adsorption, and soil aggregation (Carter et al., 1986). In addition, many biological processes are closely related to specific surface area. Soils vary widely in their relative surface area because of differences in mineralogical and organic composition and in their particle-size distribution. Specific surface area (SSA) is an operationally defined concept, dependent upon the measurement technique and sample preparation (Pennell, 2002).

The most common approach used to measure SSA is considered indirect, based on measurements of the adsorption or retention of probe molecules on a solid surface at monolayer coverage (Pennell, 2002). Two common methods of measuring SSA are by ethylene glycol monoethyl ether (EGME) and N_2-sorption, using the theory of Brunauer, Emmett, and Teller (N_2-BET). N_2 as a nonpolar gas does not interact with or have access to interlayer crystallographic planes of expandable clay minerals and thus is considered to provide a measure of external surface area, whereas polar molecules, such as EGME, are known to penetrate the interlayer surfaces of expandable clay minerals and therefore have been used to provide a measure of total surface areas (internal + external surface area) (Pennell, 2002). Significant differences between these methods are most apparent in soils that contain expandable clay minerals and soil organic matter (Chlou and Rutherford, 1993; Pennell et al., 1995; de Jong, 1999; Quirk and Murray, 1999). In the past, the SSL determined surface area by glycerol retention (7D1, method obsolete, Soil Survey Staff, 1996) or EGME retention (7D2, method obsolete, Soil Survey Staff, 1996). The current method described herein is N_2-BET (multipoint).

Surface area, N_2 adsorption, multipoint and single point, measurement: A <2-mm, air-dry soil sample is ground to pass a 0.25-mm (60-mesh) sieve and is oven dried (24 hours, 110 °C). Enough soil (typically 0.5 to 1 g) is added to the weighed sample cell to achieve a total area of 2 to 50 m^2. The soil is cleaned of contaminants, e.g., water and oils, by vacuum degassing at 10 millitorr for a minimum of 3 hours at 110 °C and then is reweighed to obtain the degassed sample weight. The sample is brought to a constant temperature by means of an external bath (77 °K), and then small amounts of gas (N_2), called the absorbate, are admitted in steps to evacuate the sample chamber. Gas molecules that stick to the surface of the solid (adsorbent) are said to be adsorbed and tend to form a thin layer that covers the entire adsorbate surface. The number of molecules required to cover the adsorbent surface with a monolayer of adsorbed molecules, N_m, can be estimated based on the BET theory. Multiplying N_m by the cross-sectional area of an adsorbate molecule yields the sample's surface area. Specific surface area is reported in $m^2 g^{-1}$.

Surface area, interferences: Organic material can coat or cover mineral surfaces, generally reducing SSA by N_2-BET. Removal of organic matter prior to analysis will typically increase these values. Freeze-drying may provide SSA values that are more representative of field conditions; air-drying may result in the collapse and shrinkage of soil humic acid, whereas freeze-drying maintains an intricate structural network more characteristic of a natural state (Pennell, 2002). Sample size will vary, depending on the SSA of the solid.

7.4 Optical Analyses[1]
7.4.1 Grain Studies
7.4.1.1 Analysis and Interpretation

Minerals

Identification criteria: Important properties in grain identification are listed below in approximate order of ease and convenience of determination. Estimates of several of these properties may allow identification of a grain so that detailed or extremely accurate measurements are seldom necessary (Lynn et al., 2008). In the finer soil separates, grain identification may be impossible because the grains may be too small or not in the right position to permit measurement of some properties, e.g., optic angle (2V) or optic sign. A process to help practice estimating properties is to crush, sieve, and mount a set of known minerals and compare these known standards to unknowns.

Refractive index is the ratio of the speed of light in the medium (mineral) to the speed of light in a vacuum. It can be estimated by relief or can be accurately determined by using calibrated immersion liquids. When relief is used to estimate refractive index, the grain shape, color, and surface texture are considered; i.e., thin platy grains may be estimated low, whereas colored grains and grains with rough, hackly surface texture may be estimated high. Estimation is aided by comparing an unknown with known minerals.

Relief is an expression of the difference in refractive index between the grain and the mounting medium. The greater the difference, the greater the relief. The analogy is to topographic relief. When viewed through the microscope, grains with high relief are distinct, whereas grains with low relief tend to fade into the background. The SSL selects a mounting medium with an index of refraction close to that of quartz; i.e., quartz has low relief. Most other minerals are identified by comparison.

Becke line is a bright halo of light that forms near the contact of the grain and the mounting medium because of the difference in refractive index between the two. As the plane of focus is moved upward through the grain, the Becke line appears to move into the component with the higher refractive index. In Petropoxy 154$_{TM}$, the Becke line moves away from potassium feldspar (index of refraction <1.54) but moves into mica (index of refraction >1.54).

Birefringence is the difference between the highest and lowest refractive index of the mineral. Accounting for grain thickness and orientation, the birefringence is estimated by interference color. Interference color is observed when an anisotropic mineral is viewed between cross-polarized light. Several grains of the same species must be observed because the grains may not all lie in positions that show the extremes

---

[1] The discussion of the identification and significance of minerals, microcrystalline aggregates, and amorphous substances in optical studies of grain mounts is from material after John G. Cady (1965), used with permission by Soil Science Society of America, Madison, Wisconsin, and modified by Warren C. Lynn, Research Soil Scientist, NRCS, Lincoln, Nebraska.

of refractive index. For example, birefringence of mica is high but appears low when the platy mineral grain is perpendicular to the microscope axis because the refractive indices of the two crystallographic directions in the plane are similar; however, a mica grain viewed on edge in a thin section shows a high interference color. The carbonate minerals have extremely high birefringence (0.17 to 0.24). Birefringence of most of the ferromagnesian minerals is intermediate (0.015 to 0.08), and that of orthoclase feldspar and apatite is low (0.008) and very low (0.005), respectively.

Color helps to discriminate among the heavy minerals. Pleochroism is the change in color or light absorption with stage rotation when the polarizer is inserted. Pleochroism is a good diagnostic characteristic for many colored minerals. Tourmaline, green beryl, and staurolite are examples of pleochroic minerals.

Shape, cleavage, and crystal form are characteristic or possibly unique for many minerals. Cleavage may be reflected in the external form of the grain or may appear as cracks within the grain that show as regularly repeated straight parallel lines or as sets of lines that intersect at definite repeated angles. The crystal shape may be different from the shape of the cleavage fragment. Plagioclase feldspars, kyanite, and the pyroxenes have strong cleavage. Zircon and rutile typically appear in crystal forms.

Extinction angle and character of extinction observed between cross-polarized light are important criteria for some groups of minerals. For extinction angles to be measured, the grain must show its cleavage or crystal form. These angles may be different along different crystallographic axes. Some minerals have sharp, quick total extinction, whereas other minerals have more gradual extinction. In some minerals with high light dispersion, the interference color dims and changes at the extinction position.

Optic sign, optic angle, and sign of elongation are useful, if not essential, determinations but are often difficult, unless grains are large or in favorable orientation. Determination of optic sign requires that the grains show dim, low-order interference colors or show no extinction. Grains with bright colors and with sharp, quick extinction rarely provide usable interference figures.

Particular mineral species: The following are the outstanding diagnostic characteristics of the most commonly occurring minerals and single-particle grains in the sand and silt fractions of soils. The refractive indices that are provided are the intermediate values.

Quartz has irregular shapes. The refractive index of quartz (1.54) approximates that of the epoxy (Petropoxy 154$_{TM}$) mounting medium. The Becke line may be split into yellow and blue components. The interference colors are low order but are bright and warm. There is sharp extinction with a small angle of rotation, i.e., "blink extinction." Crystal forms are sometimes observed and usually indicate derivation from limestone or other low-temperature secondary origin.

Potassium feldspars. Orthoclase may resemble quartz, but the refractive index (1.52) and birefringence are lower than those of quartz. In addition, orthoclase may show cleavage. *Microcline* has a refractive index of 1.53. The Becke line moves away from the grain with upward focus. A twinning intergrowth produces a plaid or grid effect between cross-polarized light that is characteristic of microcline. *Sanidine* has the same refractive index and birefringence as other potassium feldspars. Grains are typically clear, and twinning is not evident. In sanidine, the 2V angle is low (12°) and

285

characteristic. The 2V angle is the acute angle between two optic axes, or more simply, the optical axial angle.

Plagioclase feldspars have refractive indices that increase with an increase in the proportion of calcium. The refractive index of the sodium end-member *albite* (1.53) is lower than that of quartz, but the refractive index of the calcium end-member *anorthite* (1.58) is noticeably higher than that of quartz. Some *oligoclase* has the same refractive index as quartz, which prevents distinctions by the Becke line. Plagioclase feldspars commonly show a type of twinning (defined as albite twinning) that appears as multiple alternating dark and light bands in cross-polarized light. Cleavage is good in two directions parallel to (001) and (010), often producing lathlike or prismatic shapes.

Micas occur as platy grains that are commonly very thin. The plate view shows very low birefringence, whereas the edge view shows a very high birefringence. Plates are commonly equidimensional and may appear as hexagons or may have some 60° angles. *Biotite* is green to dark brown. Green grains may be confused with chlorite. Paler colors, a lowering of refractive index, and a distortion of the extinction and interference figure indicate weathering to *hydrobiotite, kaolinite,* or *vermiculite.* *Muscovite* is colorless. It has a moderate refractive index (1.59) in the plate view and an interference figure that shows a characteristic 2V angle of 30 to 40°, which can be used as a standard for comparing 2V angles of other minerals.

Amphiboles are fibrous to platy or prismatic minerals with slightly inclined extinction or, in some cases, with parallel extinction. Color and refractive index increase as the Fe content increases. Amphiboles have good cleavage at angles of ≈ 56 and 124°. Refractive index of the group ranges from 1.61 to 1.73. *Hornblende* is the most common member of the amphiboles. It is slightly pleochroic, typically has a distinctive color close to olive-green, has inclined extinction, and is often used as an indicator of weathering.

Pyroxenes. Enstatite and *aegerine-augite* are prismatic and have parallel extinction. Aegerine-augite has unique and striking green-pink pleochroism. *Augite* and *diopside* have good cleavage at angles close to 90° and large extinction angles. Colors typically are shades of green with interference colors of reds and blues. Refractive indices in the pyroxenes (1.65 to 1.79) are higher than in amphiboles.

Olivine is colorless to very pale green and is typically irregular in shape (weak cleavage). It has vivid, warm interference colors. Olivine is an easily weathered mineral and may have cracks or seams filled with serpentine or goethite. It is seldom identified in soils but has been observed in certain soils from Hawaii.

Staurolite is pleochroic yellow to pale brown and in some cases contains holes, i.e., the "Swiss cheese" effect. The refractive index is ≈ 1.74. Grains may have a foggy or milky appearance, which may be caused by colloidal inclusions.

Epidote is a common heavy mineral, but the forms that occur in soils may be difficult to identify positively. Typical epidote is unmistakable with its high refractive index (1.72 to 1.76), strong birefringence, and a pleochroism that includes pistachio green. It has typical interference colors of reds and yellows. Commonly, grains show an optic axis interference figure with a 2V angle that is nearly 90°; however, epidote is modified by weathering or metamorphism to colorless forms with lower birefringence and refractive index. *Zoisite* and *clinozoisite* in the epidote group are more common than some of the literature indicates. These minerals of the epidote group commonly appear as colorless, pale-green, or bluish-green, irregularly shaped or roughly platy grains with high refractive index (1.70 to 1.73). Most of these minerals show anomalous interference colors (bright pale blue) and no complete extinction and can be confused with several other minerals, e.g., kyanite and diopside. Zoisite has a distinctive deep blue interference color. Identification usually depends on determination of properties for many grains.

Kyanite is a common mineral but is generally not abundant. The pale blue color, the platy, angular cleavage flakes, the large cleavage angles, and the large extinction angles (30° extinction) usually can be observed and make identification easy.

Sillimanite and *andalusite* resemble each other. These minerals are fibrous to prismatic with parallel extinction. Their signs of elongation, however, are different. Also, sillimanite is colorless, and andalusite commonly has a pink color.

Garnet is found in irregularly shaped, equidimensional grains that are isotropic and have a high refractive index ($\geq$1.77). Garnet of the fine sand and silt size is commonly colorless. Pale pink or green colors are diagnostic in the larger grains.

Tourmaline has a refractive index of 1.62 to 1.66. Prismatic shape, strong pleochroism, and parallel extinction are characteristic. Some tourmaline is almost opaque when at right angles to the vibration plate of the polarizer.

Zircon occurs as tetragonal prisms with pyramidal ends. It has a very high refractive index (>1.9), parallel extinction, and bright, strong interference colors. Broken and rounded crystals are common. Zircon crystals and grains are characteristically clear and fresh appearing.

Sphene, in some forms, resembles zircon, but the crystal forms have oblique extinction. The common form of sphene, a rounded or subrounded grain, has a color change through ultrablue with crossed polarizers instead of extinction because of its high dispersion. Sphene is the only pale or colorless high-index mineral that provides this effect. It appears amber in reflected light. The refractive index of sphene is slightly lower than that of zircon, and the grains are commonly cloudy or rough surfaced.

Rutile grains are prismatic. The refractive index and birefringence are extremely high (2.6 to 2.9). The interference colors typically are obscured by the brown, reddish brown, or yellow colors of the mineral. Other TiO_2 minerals, such as *anatase* and *brookite*, also have very high refractive indices and brown colors and may be difficult to distinguish in small grains. The anatase and brookite typically occur as tabular or equidimensional grains.

Apatite is common in youthful soil materials. Apatite has a refractive index slightly <1.63 and a very low birefringence. Crystal shapes are common and may appear as prisms, but apatite can also be bullet shaped. Rounding by solution produces ovoid forms. Apatite is easily attacked by acid and may be lost in pretreatments.

Carbonates. Calcite, dolomite, and *siderite,* in their typical rhombohedral cleavage forms, are easily identified by their extremely high birefringence. In soils, these minerals have other forms, e.g., scales and chips, cements in aggregates, microcrystalline coatings or aggregates, and other fine-grained masses that are commonly mixed with clay and other minerals. The extreme birefringence is always the identification clue and is shown by the bright colors between cross-polarized light and by the marked change in relief when the stage is rotated with one polarizer in. The microcrystalline aggregates produce a twinkling effect when rotated between cross-polarized light. These three minerals have different refractive indices, which may be used to distinguish them. Siderite is the only one with both indices >Petropoxy 154$_{TM}$. It is more difficult to distinguish calcite from dolomite, and additional techniques, such as staining or x-ray diffraction, may be used.

Gypsum occurs in platy or prismatic flat grains with a refractive index approximately equal to that of orthoclase. Gypsum typically has a brushed or "dirty" surface.

Opaque minerals, of which *magnetite* and *ilmenite* are the most common, are difficult to identify, especially when they are worn by transportation or otherwise affected by weathering. Observations of color and luster by reflected light, aided by crystal form if visible, are the best procedures. Magnetic separations help to confirm the presence of magnetite and ilmenite. Many grains that appear opaque by plain light can appear translucent if viewed between strong cross-polarized light. Most grains that behave in this way are altered grains or aggregates and are not opaque minerals.

Microcrystalline Aggregates and Amorphous Substances

Identification criteria: Most microcrystalline aggregates have one striking characteristic feature; i.e., they show birefringence but do not have definite, sharp, complete extinction in cross-polarized light. Extinction may occur as dark bands that sweep through the grain or parts of the grain when the stage is turned or may occur in patches of irregular size and shape. With a few exceptions, e.g., well-oriented mineral pseudomorphs and certain clay-skin fragments, some part of the grain is bright in all positions. Aggregates and altered grains should be examined with a variety of combinations of illumination and magnification in both plain and polarized lights. The principal properties that can be used to identify or at least characterize aggregates are discussed below.

Color, if brown to bright red, is typically related to Fe content and oxidation. Organic matter and Mn may contribute black and grayish brown colors.

Refractive index is influenced by a number of factors, including elemental composition, atom packing, water content, porosity, and crystallinity. Amorphous (noncrystalline) substances have a single index of refraction, which may vary depending on chemical composition. For example, allophane has a refractive index of 1.47 to 1.49, but the apparent refractive index increases with increasing inclusion of ferrihydrite (noncrystalline Fe oxide) in the mineral.

Strength of birefringence is a clue to the identity of the minerals. Even though the individual units of the aggregate are small, birefringence can be estimated by interference color and brightness. Amorphous substances, having only a single index of refraction, exhibit no birefringence and are isotropic between cross-polarized light.

Morphology may provide clues to the composition or origin of the aggregate. Some aggregates are pseudomorphs of primary mineral grains. Characteristics of the original minerals, i.e., cleavage traces, twinning, or crystal form, can still be observed. Morphology can sometimes be observed in completely altered grains, even in volcanic ash shards and basalt fragments. Other morphological characteristics may be observed in the individual units or in the overall structure; e.g., the units may be plates or needles, or there may be banding.

Particular species of microcrystalline aggregates and amorphous substances:
For purposes of soil genesis studies, the aggregates that are present in sand or silt fractions are not of equal significance. Some are nuisances but must be accounted for, and others are particles with important diagnostic value. Useful differentiating criteria for some of the commonly occurring aggregate types are discussed below.

Rock fragments include chips of shale, schist, and fine-gained igneous rocks, e.g., rhyolite. Identification depends on the recognition of structure and individual components and the consideration of possible sources. Rock fragments are common in mountainous regions and are commonly hydrothermally altered in the Western United States.

Clay aggregates may be present in a wide variety of forms. Silt and sand that are bound together into larger grains by a nearly isotropic brownish material usually indicate incomplete dispersion. Clay skins may resist dispersion and consequently may appear as fragments in grain mounts. Such fragments are typically brown or red and translucent with wavy extinction bands. Care is required to distinguish these fragments from weathered biotite. Clay aggregates may be mineral pseudomorphs. Kaolin pseudomorphs of feldspar commonly are found. Montmorillonite aggregates, pseudomorphic of basic rock minerals, have been observed. In this form, montmorillonite shows high birefringence and an extinction that is mottled or patchy on a small scale. Coarse kaolinite flakes, books, and vermicular aggregates resist dispersion and may be abundant in sand and silt. These particles may resemble muscovite, but they are cloudy, show no definite extinction, and have very low birefringence. Many cases of anomalously high cation-exchange capacity (CEC) of sand and silt fractions that are calculated from whole soil CEC and from clay CEC and percent content can be accounted for by the occurrence of these aggregates in the sand and silt fractions.

Volcanic glass is isotropic and has a low refractive index—lower than most of the silicate minerals. The refractive index ranges from 1.48 in the colorless siliceous glasses to as high as 1.56 in the green or brown glasses of basalt composition. Shapes vary, but the elongated, curved shard forms, in many cases with bubbles, are common. This glassy material may adhere to or envelop other minerals. Particles may contain small crystals of feldspar or incipient crystals with needles and dendritic forms. The colorless siliceous types (acidic, pumiceous) are more common in soils as the basic glasses weather easily. Acidic glasses are more commonly part of "ash falls" and pyroclastic flows (i.e., ash flows) as the magma typically is gaseous and explosive when pressure is released. Basic glasses are more commonly associated with volcanic flow (e.g., basaltic lava) rocks, which are typically not as gaseous or as viscous during eruption as the more acidic magmas.

Allophane occurs in many soils that are derived from the weathering of volcanic ash. Allophane seldom can be identified directly, but its presence can be inferred when sand and silt are cemented into aggregates by isotropic material with low refractive index, especially if volcanic ash shards also are present.

Opal, an isotropic material, occurs as a cementing material and in separate grains, some of which are of organic origin, i.e., plant opal, sponge spicules, and diatoms. The refractive index is very low (<1.45); this value is lower than the value for volcanic ash. Identification may depend in part on form and occurrence.

Iron oxides may occur as separate grains or as coatings, cementing agents, and mixtures with other minerals. Iron oxides impart brown and red colors and raise the refractive index in the mixtures. *Goethite* is yellow to brown. Associated red areas may be hematite. These red varieties have a refractive index and birefringence that are higher and seem to be better crystallized, commonly having a prismatic or fibrous habit. Aggregates have parallel extinction. In oriented aggregates, the interference colors commonly have a greenish cast. *Hematite* has higher refractive index than goethite and is granular rather than prismatic. Large grains of hematite are nearly opaque.

Gibbsite commonly occurs as separate, pure crystal aggregates, either alone or inside altered mineral grains. The grains may appear to be well crystallized single crystals, but close inspection in cross-polarized light shows patchy, banded extinction, indicating intergrown aggregates. Gibbsite is colorless. The refractive index (1.56 to 1.58) and the birefringence are higher for gibbsite than the corresponding values for quartz. The bright interference colors and aggregate extinction are characteristic of gibbsite.

Chalcedony is a microcrystalline form of quartz that was formerly considered a distinct species. Chalcedony occurs as minute quartz crystals and exhibits aggregate structure with patchy extinction between cross-polarized light. It may occur in nodules of limestone deposits and may be a pseudomorphic replacement in calcareous fossils. The refractive index is slightly lower than that of quartz, and the birefringence is lower than that of gibbsite. *Chert* is a massive form of chalcedony.

Glauconite occurs in aggregates of small micaceous grains with high birefringence. When fresh, glauconite is dark green and almost opaque, but it weathers to brown and more translucent forms. Glauconite is difficult to identify by optical evidence alone. Knowledge of source area or history is helpful in identification.

Titanium oxide aggregates have been tentatively identified in the heavy mineral separates of many soils. These bodies have an extremely high refractive index and high birefringence (similar to those of rutile). The yellow to gray colors are similar to those of anatase. The TiO_2 aggregates are granular and rough surfaced. This growth habit with the little spurs and projections suggests that TiO_2 aggregates may be secondary.

7.4   Optical Analyses
7.4.1   Grain Studies
7.4.1.1   Analysis and Interpretation
7.4.1.1.1   Separation by Heavy Liquids

Grain studies, separation by heavy liquids, application: The sand and silt fractions of most soils are dominated by quartz or by quartz and feldspars (Cady, 1965). These minerals have a relatively low specific gravity (2.57 to 2.76). The large numbers of "heavy" mineral grains (specific gravity >2.8 or 2.9) with a wide range of weatherability and diagnostic significance may be only a small percentage of the grains (Cady, 1965); however, these "heavy" minerals are often indicative of provenance, weathering intensities, and parent material uniformity (Cady et al., 1986).

Grain studies, separation by heavy liquids, measurement: To study "heavy" minerals, a common practice is to concentrate these grains by specific-gravity separations in a heavy liquid. This procedure is rarely used at the SSL but is used on occasion for special studies.

Micas are difficult to separate because of their shape and because a little weathering, especially in biotite, significantly decreases the specific gravity. These differences in density in biotite may be used to concentrate weathered biotite in its various stages of alteration.

Separation of grains by heavy liquids is most effective when grains are clean. Organic matter may prevent wetting and cause grains to clump or raft together. Light coatings may cause heavy grains to float, and iron-oxide coatings may increase specific gravity. In some kinds of materials, an additional technique is to separate and weigh the magnetic fraction, either before or after the heavy-liquid separation.

Concentrate the "heavy" minerals, i.e., those with specific gravity >2.8 or 2.9, by specific-gravity separations in a heavy liquid. The reagent of choice is sodium polytungstate (density 2.8 g^{-1} mL). Dilute the sodium polytungstate with distilled water to obtain required densities <2.8 g^{-1} mL. Use a specific gravity $\approx$ 2.5 to concentrate volcanic glass, plant opal, or sponge spicules. When this liquid is used, avoid contact with skin and work in a well ventilated area.

Separation by specific gravity alone in separatory funnels, cylinders, or various kinds of tubes is usually adequate for grains >0.10 mm. Separation by centrifuging is required for grains <0.10 mm. Pointed, 15-ml centrifuge tubes should be used for these separations.

Decant the light minerals after inserting a smooth bulb glass rod to stop off the tapered end of the centrifuge tube. Alternatively, remove the heavy minerals by gravity flow using a lower stopcock or maintain the heavy minerals in place by freezing the lower part of the tube.

7.4   Optical Analysis
7.4.1   Grain Studies
7.4.1.1   Analysis and Interpretation
7.4.1.1.2   Grain Mounts, Epoxy

Grain mounts, epoxy, application: Grain counts are used to identify and quantify minerals in the coarse silt and sand fractions of soils. Results are used to classify soil pedons in mineralogy families of *Soil Taxonomy* (Soil Survey Staff, 1999), to help determine substrate provenance of source materials, and to support or identify lithologic discontinuities.

Grain mounts, epoxy, measurement: In particle-size analysis, soils are dispersed so that material <20 µm in diameter is separated by settling and decanting, and the sand and coarse silt fractions are separated by sieving. For the separation by heavy liquids of the less abundant minerals with a specific gravity >2.8 or 2.9, refer to the method of separation by heavy liquids.

Following sample selection, permanent mounts are prepared for the two most abundant particle-size fractions among the fine sand, very fine sand, and coarse silt. The grains are mounted in a thermo-setting epoxy cement with a refractive index of 1.54.

The grains are then identified and counted under a petrographic microscope (Lynn et al., 2008).

A mineralogical analysis of a sand or silt fraction may be entirely qualitative, or it may be quantitative to different degrees (Cady, 1965). The SSL performs a quantitative analysis. Data are reported as a list of minerals and an estimated quantity of each mineral as a percentage of the grains counted in the designated fraction. The percentages of minerals are obtained by identifying and counting a minimum of 300 grains on regularly spaced line traverses that are 2 mm apart.

The identification procedures and reference data on minerals are described in references on sedimentary petrography (Krumbein and Pettijohn, 1938; Durell, 1948; Milner, 1962; Kerr, 1977; Deer et al., 1992) and optical crystallography (Bloss, 1961; Stoiber and Morse, 1972; Shelley, 1978; Klein and Hurlbut, 1985; Drees and Ransom, 1994).

Mineral contents are reported as percentages of grains counted. These data are accurate number percentages for the size-fraction analyzed, but they may need to be recomputed to convert to weight percentages (Harris and Zelazny, 1985). The taxonomic criteria must be considered to determine the particle-size fraction or fractions needed for the analysis. Quantifying multiple fractions may be necessary to accurately determine the weighted average mineral content for a range of fractions (Wilson et al., 1999). Grain counts can deviate significantly from weight percentage due to platy grains and density variations. These data are reported on the mineralogy data page of the primary characterization dataset. For each grain size counted, the mineral type and amount are recorded; i.e., quartz, 87 percent of fraction, is recorded as QZ87. The percentage of resistant minerals in each fraction is reported on the SSL datasheet.

7.4   Optical Analysis
7.4.2   Platy Grains
7.4.2.1   Magnetic Separation

Platy grains, magnetic separation, application: A magnetic separator is used to separate magnetic or paramagnetic minerals from nonmagnetic mineral grains from the fine-earth fractions that range from 0.02 to 0.5 mm in size. In the SSL, the common application is to quantify the amount of platy grains (phyllosilicates) in micaceous soils. Ferrimagnetic grains, such as magnetite and ilmenite, typically are separated first with a hand magnet. The separator then concentrates grains of paramagnetic minerals, such as biotite and muscovite, from nonmagnetic minerals, such as quartz and feldspar. This method is often used in combination with static tube separation or froth flotation. Grains in each separate can be further analyzed by optical microscopy or x-ray diffraction.

Platy grains, magnetic separation, measurement: A magnetic separator applies a strong magnetic field along a shallow trough that slopes down from an entry point to an exit point. Grains travel the path under the force of gravity. The trough is also tilted perpendicular to the travel path. The magnetic field draws paramagnetic grains up the tilt slope. A divider at midslope along the path separates the paramagnetic from nonmagnetic grains. Grains exit the path into separate containers, and the two components are weighed to obtain a relative percentage. Percent platy minerals of specific analyzed fraction are reported.

Platy grains, magnetic separation, interferences: Some mafic minerals are paramagnetic and exit with the platy grains. The two groups may be separated by the static tube. Mafic minerals commonly are heavy minerals and may be separated from platy grains by density separation.

7.4   Optical Analysis
7.4.2   Platy Grains
7.4.2.2   Static Tube Separation

Platy grains, static tube separation, application: Static charge of mineral grains to glass and a magnetic separator are used to separate platy grains from nonplaty grains in the 0.02- to 2-mm fraction of soil. The separates are weighed to determine the quantity of platy minerals. The platy separates are examined by optical microscope and analyzed by x-ray diffraction to determine the kinds of minerals present.

Platy grains, static tube separation, measurement: A sample of <2-mm soil is prepared according to the method described for magnetic separation of platy grains. A small portion of the sample is introduced into the top of an inclined glass tube mounted on a vibrator. As the tube is rotated and vibrated, the platy grains adhere to the tube and the nonplaty grains (residue) roll or slide through. The residue is run through a magnetic separator to separate the coarser platy grains that did not adhere to the glass tube. Percent platy minerals of specific analyzed fraction are reported.

Platy grains, static tube separation, interferences: Large platy grains tend to slide through the tube into the residue, especially if the plates are stacked into a book.

7.4   Optical Analysis
7.4.2   Platy Grains
7.4.2.3   Froth Flotation

Platy grains, froth flotation, application: This method is used with coarse silt and very fine, fine, and medium sand fractions of soils with significant amounts of platy minerals (mica, vermiculite, chlorite, and their pseudomorphically altered weathering products). It provides weight percent data on each of the fractions. Combined use of froth flotation and magnetic separation improves separation of platy and nonplaty grains and provides a better estimate of weight percentages of components.

Platy grains, froth flotation, measurement: Platy minerals (muscovite, biotite, vermiculite, and kaolinite) are floated off over the top of a container in an agitated aqueous suspension by action of a complexer and frother, adapted from the procedure provided by Louis Schlesinger of the Minerals Research Laboratory, College of Engineering, North Carolina State University, in Asheville, North Carolina. Percent platy minerals of specific analyzed fraction are reported. Platy grains as percent of the specific particle-size fraction analyzed are reported.

7.5 Ratios, Estimates, and Calculations Related to Optical Analysis
7.5.1 Total Resistant Minerals

The sum of the grain-count percentages of resistant minerals is reported. For more detailed information on total resistant minerals, refer to Soil Survey Staff (2010). Also refer to Table 7.5.2 for the list of mineralogy codes for resistant and weatherable minerals (Soil Survey Staff, 2004).

Table 7.5.2 Mineralogy Codes

Resistant Minerals

AE = Anatase

AN = Andalusite

BY = Beryl

CD = Chalcedony (Chert, Flint,
    Jasper, Agate, Onyx)

CE = Cobaltite

CH = Cliachite (Bauxite)

CN = Corundum

CR = Cristobalite

CT = Cassiterite

FE = Iron Oxides (Goethite,
    Magnetite, Hematite, Limonite)

GD = Gold

GE = Goethite

GI = Gibbsite

GN = Garnet

HE = Hematite

HS = Hydroxy-Interlayered Smectite

HV = Hydroxy-Interlayered Vermiculite

KS = Interstratified Kaolinite-Smectite

KK = Kaolinite

KY = Kyanite

LE = Lepidocrocite

LM = Limonite

LU = Leucoxene

MD = Resistant Mineraloids

MG = Magnetite

MH = Maghemite

MZ = Monazite

OP = Opaques

OR = Other Resistant Minerals

PO = Plant Opal

PN = Pollen

QC = Clay-Coated Quartz

QI = Iron-Coated Quartz

QZ = Quartz

RA = Resistant Aggregates

RE = Resistant Minerals

RU = Rutile

SA = Siliceous Aggregates

SL = Sillimanite

SN = Spinel

SO = Staurolite

SP = Sphene

SS = Sponge Spicule

TD = Tridymite

TM = Tourmaline

TP = Topaz

ZR = Zircon

Weatherable Minerals

AC = Actinolite

AF = Arfvedsonite

AG = Antigorite

AH = Anthophyllite

AI = Aegerine-Augite

AL = Allophane

AM = Amphibole

AO = Aragonite[1]

AP = Apatite

AR = Weatherable Aggregates

AU = Augite

AY = Anhydrite[1]

BA = Barite

BC = Biotite-Chlorite

BE = Boehmite

BG = Basic Glass

BK = Brookite

BR = Brucite

BT = Biotite

BZ = Bronzite

CA = Calcite[1]

CB = Carbonate Aggregates[1]

CC = Coal

CL = Chlorite

CM = Chlorite-Mica

CO = Collophane

CY = Chrysotile

CZ = Clinozoisite

DL = Dolomite

DP = Diopside

DU = Dumortierite

EN = Enstatite

EP = Epidote

FA = Andesite

FB = Albite

FC = Microcline

FD = Feldspar

FF = Foraminifera

FG = Glass-Coated Feldspar

FH = Anorthoclase

FK = Potassium Feldspar

FL = Labradorite

FM = Ferromagnesium Mineral

FN = Anorthite

FO = Oligoclase

FP = Plagioclase Feldspar

FR = Orthoclase

FS = Sanidine

FT = Fluorapatite

FU = Fluorite[1]

FZ = Feldspathoids

GA = Glass Aggregates

GC = Glass-Coated Grain

GG = Galena

GL = Glauconite

GO = Glaucophane

GM = Glassy Materials

GY = Gypsum[1]

HA = Halite[1]

HB = Hydrobiotite

HG = Glass-Coated Hornblende

HN = Hornblende

HY = Hypersthene

ID = Iddingsite

IL = Illite (Hydromuscovite)

JO = Jarosite

KH = Halloysite

LA = Lamprobolite

LC = Analcime[1]

LI = Leucite

LO = Lepidomelane

LP = Lepidolite

LT = Lithiophorite

MC = Montmorillonite-Chlorite

ME = Magnesite[1]

MI = Mica

ML = Melilite

MM = Montmorillonite-Mica

MR = Marcasite

MS = Muscovite

MT = Montmorillonite

MV = Montmorillonite-Vermiculite

NA = Natron

NE = Nepheline

NJ = Natrojarosite

NX = Noncrystalline

OG = Glass-Coated Opaque

OT = Other

OV = Olivine

OW = Other Weatherable Minerals

PA = Palagonite

PD = Piemontite

PG = Palygorskite

PI = Pyrite

PJ = Plumbjarosite

PK = Perovskite

PL = Phlogopite

PM = Pumice

PR = Pyroxene

PT = Paragonite

PU = Pyrolusite

PY = Pyrophyllite

QC = Glass-Coated Quartz

RB = Riebeckite (Blue Amphibole)

RO = Rhodochrosite

SC = Scapolite

SE = Sepiolite

SG = Sphalerite

SI = Siderite

SM = Smectite

SR = Sericite

ST = Stilbite[1]

SU = Sulphur

SZ = Serpentine

TA = Talc

TE = Tremolite

TH = Thenardite[1]

VC = Vermiculite-Chlorite

VH = Vermiculite-Hydrobiotite

VI = Vivianite

VM = Vermiculite-Mica
VR = Vermiculite
WE = Weatherable Mineral

WV = Wavellite
ZE = Zeolite[1]
ZO = Zoisite

Glass Count Minerals and Mineraloids[2]

Volcanic Glass Grains[3]	Organic Origin Grains[4]	Other Grains
BG = Basic Glass	DI = Diatoms	OT = Other
FG = Glass-Coated Feldspar	PO = Plant Opal	
GA = Glass Aggregates	SS = Sponge Spicule	
GC = Glass-Coated Grain		
GM = Glassy Materials		
GS = Glass		
HG = Glass-Coated Hornblende		
OG = Glass-Coated Opaque		
PA = Palagonite		
PM = Pumice		
QG = Glass-Coated Quartz		

[1] Minerals not included as "weatherable minerals" as defined in *Keys to Soil Taxonomy* (Soil Survey Staff, 2010): "the intent is to include only those weatherable minerals that are unstable in a humid climate compared to other minerals such as quartz and 1:1 lattice clays, but are more resistant to weathering than calcite." This group of minerals is not part of the calculation for percent resistant minerals used in the siliceous family mineralogy class or percent weatherable minerals used as criteria for oxic horizons but are included in the calculation of "total resistant minerals" on the SSL mineralogy data sheet. Therefore, the value on the data sheet should be recalculated for strict use in soil taxonomy criteria if these minerals (e.g., calcite) are in the grain count of a selected horizon.

[2] Minerals on this list are identified during the "glass count" procedure by the SSL during the quantification of particle-size separates in the sand-silt fraction. Minerals in the "OT" category are other weatherable or resistant minerals that would be quantified during a "full grain count."

[3] Minerals and mineraloids in this column are all considered weatherable according to the SSL and are defined in *Keys to Soil Taxonomy* (Soil Survey Staff, 2010) as "volcanic glass." The percentages of these minerals are summed as "volcanic glass" and used in the criteria for andic soil properties, subgroups with the formative element "vitr(i)", families with "ashy" substitutes for particle-size class, and the glass mineralogy class in soil taxonomy.

[4] Mineraloids included in this list are regarded as resistant minerals according to the SSL and are included in the calculation of "total resistant minerals" on the SSL mineralogy data sheet.

III. Supplementary Characterization Data

Supplementary Characterization Data appear on the Supplementary Characterization Data Sheets. These data are considered the interpretive physical data for pedons analyzed at the SSL. They are primarily derived or calculated data, using the analytical data as a basis for calculation. Some Primary Characterization Data (e.g., bulk density at 33-kPa and oven-dryness and percent sand, silt, and clay) are repeated on the Supplementary Characterization Data Sheet for user convenience. The interpretive information may or may not be the exact same value as the analytical information from which they are calculated because of the procedure of rounding and significant digits in calculating the data.

The Supplementary Characterization Data are not discussed in SSIR No. 42 (Soil Survey Staff, 2004) and thus do not carry method codes. Discussion of the Supplementary Data is not tied to the SSL data sheet format. Users of this manual are referred to the discussion in the Primary Characterization Data section for some of the analytical data elements that are repeated on the Supplementary Characterization Data Sheet.

Important metadata are shown on the Supplementary Characterization Data Sheet as well as on the Primary Characterization Data Sheet (e.g., site and pedon identification numbers; SSL project numbers and names; "sampled as" and "revised to" soil names; sample layer number; depth (cm); genetic horizon; and laboratory preparation code). Refer to the introduction to this manual for a more detailed discussion of the significance of these metadata.

In the following section, the SSL interpretive data and their respective calculations are described. These calculations are presented as equations and are enumerated consecutively in the text. Some analytical and derived values are used repeatedly in a number of calculations. Discussion is logically and sequentially presented as broad groupings of supplementary characterization data types as follows:

- Engineering particle-size distribution analysis (PSDA)
- Cumulative curve fractions
- Atterberg limits
- Gradation
- Weight fractions
- Weight per unit volume
- Void ratios
- Volume fractions
- Pores
- C/N ratio
- Ratios to clay
- Linear extensibility
- Water retention difference (WRD)
- Weight fractions
- Texture
- PSDA summaries

- pH 0.01 M CaCl$_2$
- Electrical conductivity and resistivity
- Particle density

1   Engineering PSDA

Engineering PSDA, definition: The engineering particle-size distribution analysis (PSDA) data are derived from USDA PSDA data and are reported as cumulative weight percentages less than a given diameter passing a 3-in sieve (76.1 mm). Percent passing expresses the percentage of total sample that is finer than each sieve size and is not the percentage retained on sieve. Square-holed sieves are routinely used to obtain engineering and SSL PSDA data.

Particles passing 2- 3/2-, and 1-in sieves and No. 40 sieve: In interpolating between sieve sizes, a logarithmic distribution on the cumulative particle-size curve is usually assumed with a linear relationship between size fractions. Log factors for these relationships in Equations 2, 3, 4, 6, and 9 are calculated as follows:

Steps:
   1) Use sieve dimensions in millimeters to obtain logarithms of larger and smaller sieve sizes for which there are measured particle percentages.
   2) Determine range of these two logarithmic values by subtraction.
   3) Determine logarithm of sieve size for which there is an estimated particle percentage and subtract result from logarithm of larger sieve size determined in step 1.
   4) Divide result in step 3 by value determined in step 2. Result is Log Factor.

The sieve dimensions used to obtain Log Factors (Table 1) are from the engineering grain-size distribution graph (form NRCS-ENG-353).

Table 1   Sieve dimensions used to obtain Log Factors

Sieve size		Sieve size	
(in)	(mm)	(No.)	(mm)
3	76.1	4	4.76
2	50.8	10	2.00
3/2	38.1	40	0.420
1	25.4	200	0.074
3/4	19.05		
3/8	9.51		

Example: Calculate Log Factor 0.2906 (P$_{<2in}$) as follows:

where
P$_{<2in}$ = Cumulative weight percentage passing 2-in sieve on <3-in basis.

Steps:

1) Log of sieve size 76.1 mm (3 in) = 1.881
   Log of sieve size 19.05 mm (3/4 in) = 1.279
2) Determine log range by subtraction = 0.602
3) Log of sieve size 50.8 mm (2 in) = 1.706
   Subtract 1.706 from 1.881 = 0.175
4) Divide 0.175 (Step 3) by 0.602 (Step 2)
   Resulting Log Factor = 0.2906

These sieve dimensions can be related to nonflat rock fragments by diameter (USDA/NRCS, 2009b) as follows:

2-75 mm	Gravel
2-5 mm	Fine Gravel
5-20 mm	Medium Gravel
20-75 mm	Coarse Gravel
75-250 mm	Cobbles
250-600 mm	Stones
≥600 mm	Boulders

1.1 Particles Passing 3-in Sieve

Particles passing 3-in sieve are reported as a cumulative weight percentage on <3-in basis. These data should *always* be 100 percent.

Equation 1:

$$P_{<3in} = 100\%$$

where
$P_{<3in}$ = Cumulative weight percentage passing 3-in sieve on <3-in basis.

1.2 Particles Passing 2-in Sieve

Particles passing 2-in sieve are reported as a cumulative weight percentage on <3-in basis.

Equation 2:

$$P_{<2in} = 100 - (A \times 0.2906)$$

where

$P_{<2in}$ = Cumulative weight percentage of particles passing 2-in sieve on <3-in basis.

301

A	=	Weight percentage of particles with 20- to 75-mm diameter on <75-mm basis. To obtain this weight fraction in equation, subtract the percent passing the 3/4-in sieve from the percent passing the 3-in sieve.
0.2906	=	Cumulative Log Factor as determined by Steps 1 through 4.

1.3 Particles Passing 3/2-in Sieve

Particles passing 3/2-in sieve are reported as a cumulative weight percentage on <3-in basis.

Equation 3:

$$P_{<3/2in} = 100 - (A \times 0.4981)$$

where

$P_{<3/2in}$	=	Cumulative weight percentage of particles passing 3/2-in sieve on <3-in basis.
A	=	Weight percentage of particles with 20- to 75-mm diameter on <75-mm basis. To obtain this weight fraction in equation, subtract the percent passing the 3/4-in sieve from the percent passing the 3-in sieve.
0.4981	=	Calculate Log Factor using sieve dimensions as outlined in Steps 1 through 4.

1.4 Particles Passing 1-in Sieve

Particles passing 1-in sieve are reported as a cumulative weight percentage on <3-in basis.

Equation 4:

$$P_{<1in} = 100 - (A \times 0.7906)$$

where

$P_{<1in}$	=	Cumulative weight percentage of particles passing 1-in sieve on <3-in basis.
A	=	Weight percentage of particles with 20- to 75-mm diameter on <75-mm basis. To obtain this weight fraction in equation, subtract the percent passing the 3/4-in sieve from the percent passing the 3-in sieve.

302

0.7906 = Calculate Log Factor using sieve dimensions as outlined in Steps 1 through 4.

1.5 Particles Passing 3/4-in Sieve

Particles passing 3/4-in sieve are reported as a cumulative weight percentage on <3-in basis.

Equation 5:

$$P_{<3/4in} = 100 - A$$

where

$P_{<3/4in}$ = Cumulative weight percentage of particles passing 3/4-in sieve on <3-in basis.

A = Weight percentage of particles with 20- to 75-mm diameter on <75-mm basis. To obtain this weight fraction in equation, subtract the percent passing the 3/4-in sieve from the percent passing the 3-in sieve.

1.6 Particles Passing 3/8-in Sieve

Particles passing 3/8-in sieve are reported as a cumulative weight percentage on <3-in basis.

Equation 6:

$$P_{<3/8in} = A - (B \times 0.4962)$$

where

$P_{<3/8in}$ = Cumulative weight percentage of particles passing 3/8-in sieve on <3-in basis.

A = Cumulative weight percentage of particles passing 3/4-in sieve.

B = Weight percentage of particles with 5- to 20-mm diameter on <75-mm basis. To obtain this weight fraction in equation, subtract the percent passing the No. 4 sieve from the percent passing the 3/4-in sieve.

303

0.4962 = Calculate Log Factor using sieve dimensions as outlined in Steps 1 through 4.

1.7 Particles Passing No. 4 Sieve

Particles passing No. 4 sieve are reported as a cumulative weight percentage on <3-in basis.

Equation 7.1:

$$P_{<4} = 100 - (A + B)$$

OR (alternatively)

Equation 7.2:

$$P_{<4} = P_{<3/8} - B$$

where

$P_{<4}$ = Cumulative weight percentage of particles passing No. 4 sieve on <3-in basis.

A = Weight percentage of particles with 20- to 75-mm diameter on <75-mm basis. To obtain this weight fraction in equation, subtract the percent passing the 3/4-in sieve from the percent passing the 3-in sieve.

B = Weight percentage of particles with 5- to 20-mm diameter on <75-mm basis. To obtain this weight fraction in equation, subtract the percent passing the No. 4 sieve from the percent passing the 3/4-in sieve.

$P_{<3/8in}$ = Cumulative weight percentage of particles passing 3/8-in sieve on <3-in basis.

1.8 Particles Passing No. 10 Sieve

Particles passing No. 10 sieve are reported as a cumulative weight percentage on <3-in basis.

Equation 8.1:

$$P_{<10} = 100 - (A + B + C)$$

304

OR (alternatively)

Equation 8.2:

$$P_{<10} = P_{<4} - C$$

where

$P_{<10}$ = Cumulative weight percentage of particles passing No. 10 sieve on <3-in basis.

A = Weight percentage of particles with 20- to 75-mm diameter on <75-mm basis. To obtain this weight fraction in equation, subtract the percent passing the 3/4-in sieve from the percent passing the 3-in sieve.

B = Weight percentage of particles with 5- to 20-mm diameter on <75-mm basis. To obtain this weight fraction in equation, subtract the percent passing the No. 4 sieve from the percent passing the 3/4-in sieve.

C = Weight percentage of particles with 2- to 5-mm diameter on <75-mm basis. To obtain this weight fraction on <75-mm basis in equation, subtract the percent passing the No. 10 sieve from the percent passing the No. 4 sieve.

$P_{<4}$ = Cumulative weight percentage of particles passing No. 4 sieve on <3-in basis.

1.9   Particles Passing No. 40 Sieve

Particles passing No. 40 sieve are reported as a cumulative weight percentage on <3-in basis.

Equation 9:

$$P_{<40} = A - [(B + C + (D \times 0.2515)) \times (A / 100)]$$

where

$P_{<40}$ = Cumulative weight percentage of particles passing No. 40 sieve on <3-in basis.

A = Cumulative weight percentage of particles passing No. 10 sieve on <3-in basis. This cumulative weight percentage is the decimal fraction passing the No.

10 sieve and is used in converting from 2-mm to 75-mm basis percentage.

B	=	Weight percentage of particles with 1- to 2-mm diameter (very coarse sand) on <2-mm basis.
C	=	Weight percentage of particles with 0.5- to 1-mm diameter (coarse sand) on <2-mm basis.
D	=	Weight percentage of particles with 0.25- to 0.5-mm diameter (medium sand) on <2-mm basis.
0.2515	=	Calculate Log Factor using sieve dimensions as outlined in Steps 1 through 4.

1.10   Particles Passing No. 200 Sieve

Particles passing No. 200 sieve are reported as a cumulative weight percentage on <3-in basis. Calculations are based on percent very fine sand (VFS) as follows:

If very fine sand <15 percent: For the 0.074-mm size fraction, a linear relationship (non-logarithmic distribution) is assumed if the very fine sand fraction (VFS) (0.05 to 0.10 mm) is <15 percent. If the VFS fraction is <15 percent, use Equation 10 to calculate the particles passing a No. 200 sieve. To determine equation factor for Equation 10, use sieve dimensions (mm) as outlined in Steps 1 through 4 but do not convert to logarithmic values.

Equation 10.1:

$$P_{<200} = [(0.56559 \times A) + B + C] \times (D / 100)$$

where

$P_{<200}$	=	Cumulative weight percentage of particles passing No. 200 sieve on <3-in basis.
A	=	Weight percentage of particles with 0.05- to 0.10-mm diameter (very fine sand) on <2-mm basis.
B	=	Weight percentage of particles with 0.002- to 0.05-mm diameter (total silt).
C	=	Weight percentage of particles with <0.002-mm diameter (total clay).
D	=	Cumulative weight percentage of particles passing No. 10 sieve on <3-in basis. This cumulative weight percentage is the decimal fraction passing the No.

306

10 sieve and is used in converting from 2-mm to
75-mm basis percentage.

0.56559 = Calculate Equation factor as outlined in Steps 1
through 4 but with no logarithmic conversion.

If very fine sand ≥15 percent: A relationship between variables may be
approximately linear when studied over a limited range but markedly curvilinear when a
broader range is considered (Steel and Torrie, 1980). If the VFS is ≥15 percent, a short
curve is fitted to three points just above and below 0.074 mm. These points or
cumulative points are as follows:

1) Medium sand (0.50- to 0.25-mm diameter)
2) Medium sand (0.50- to 0.25-mm diameter) + Fine sand (0.25- to 0.10-mm diameter)
3) Medium sand (0.5- to 0.25-mm diameter) + Fine sand (0.25- to 0.10-mm diameter) + Very fine sand (0.10- to 0.05-mm diameter)

In those instances in which VFS ≥15 percent, the cumulative particle percentage
passing the No. 200 sieve is estimated using an adjusting parameter based on a *least
squares* method. Previous studies at the former regional SCS SSL at Riverside,
California, have shown that this procedure gives a slightly better prediction than the
linear assumption. When VFS ≥15 percent, the cumulative particle percentage passing
the No. 200 sieve is estimated as follows:

Equation 10.2:

$$P_{<200} = (A - \text{fitted parameter} + B + C) \times D$$

where
$P_{<200}$ = Cumulative weight percentage of particles passing
No. 200 sieve on <3-in basis.

A = Weight percentage of particles with 0.05- to 0.10-mm diameter (very fine sand) on <2-mm basis.

B = Weight percentage of particles with 0.002- to 0.05-mm diameter (total silt).

C = Weight percentage of particles with <0.002-mm diameter (total clay).

D = Cumulative weight percentage of particles passing
No. 10 sieve on <3-in basis. This cumulative weight
percentage is the decimal fraction passing the No.

307

10 sieve and is used in converting from 2-mm to 75-mm basis percentage.

1.11 Particles Passing 20-μm Sieve (Particles Finer Than 20 μm)
1.12 Particles Passing 5-μm Sieve (Particles Finer Than 5 μm)
1.13 Particles Passing 2-μm Sieve (Particles Finer Than 2 μm)

In Sections 1.11–1.13, the SSL PSDA are recalculated and reported as cumulative weight percentages of particles less than a specified diameter (20-, 5-, and 2-μm sieves) on a <3-in basis. These PSDA add detail in the silt and clay particle-size range. The headings for these data are misnomers, as these analytical separations are made by sedimentation based on Stokes' Law. Headings may be more correctly stated as *Particles Finer Than 20, 5, and 2μm*.

1.11 Particles Passing 20-μm Sieve (Particles Finer Than 20 μm)

Particles finer than 20 μm are reported as a cumulative weight percentage on <3-in basis.

Equation 11:

$$P_{<20\mu m} = (A + B) \times (C / 100)$$

where

$P_{<20\mu m}$	= Cumulative weight percentage of particles <20 μm on <3-in basis.
A	= Weight percentage of particles with <0.002-mm diameter (clay) on <2-mm basis.
B	= Weight percentage of particles with 0.002- to 0.02-mm diameter (fine silt) on <2-mm basis.
C	= Cumulative weight percentage of particles passing No. 10 sieve on <3-in basis. This cumulative weight percentage is the decimal fraction passing the No. 10 sieve and is used in converting from 2-mm to 75-mm basis percentage.

1.12 Particles Passing 5-μm Sieve (Particles Finer Than 5 μm)

Particles finer than 5 μm are reported as a cumulative weight percentage on <3-in basis.

Equation 12:

$$P_{<5\mu m} = [A + (B \times 0.39794)] \times (C / 100)$$

where

P_{<5μm} = Cumulative weight percentage of particles <5 μm on <3-in basis.

A = Weight percentage of particles with <0.002-mm diameter (total clay) on <2-mm basis.

B = Weight percentage of particles with 0.002- to 0.02-mm diameter (fine silt) on <2-mm basis.

C = Cumulative weight percentage of particles passing No. 10 (2.00 mm) sieve on <3-in basis. This cumulative weight percentage is the decimal fraction passing the No. 10 sieve and is used in converting from 2-mm to 75-mm basis percentage.

0.39794 = Calculate Log Factor using sieve dimensions as outlined in Steps 1 through 4.

1.13   Particles Passing 2-μm Sieve (Particles Finer Than 2 μm)

Particles finer than 2 μm are reported as a cumulative weight percentage on <3-in basis.

Equation 13:

$$P_{<2\mu m} = A \times (B / 100)$$

where

P_{<2μm} = Cumulative weight percentage of particles <2 μm on <3-in basis.

A = Weight percentage of particles with <2-μm diameter (total clay) on <2-mm basis.

B = Cumulative weight percentage of particles passing No. 10 sieve on <3-in basis. This cumulative weight percentage is the decimal fraction passing the No. 10 sieve and is used in converting from 2-mm to 75-mm basis percentage.

309

1.14 Particles <1 mm
1.15 Particles <0.5 mm
1.16 Particles <0.25 mm
1.17 Particles <0.10 mm
1.18 Particles <0.05 mm

In Sections 1.14–1.18, the SSL PSDA are recalculated and reported as cumulative weight percentages of particles less than a specified diameter (1, 0.5, 0.25, 0.10, and 0.05 mm) on a 75-mm basis. These PSDA add detail to the sand particle-size range.

1.14 Particles <1 mm

Particles with <1-mm diameter are reported as a cumulative weight percentage on <75-mm basis.

Equation 14.1:

$$P_{<1.0mm} = (A + B + C + D + E + F) \times (G / 100)$$

OR (alternatively)

Equation 14.2:

$$P_{<1.0mm} = (100 - H) \times G / 100$$

where

$P_{<1.0mm}$ =   Cumulative weight percentage of particles <1.0 mm on <75-mm basis.

A   =   Weight percentage of particles with <2-μm diameter (total clay) on <2-mm basis.

B   =   Weight percentage of particles with 0.002- to 0.05-mm diameter (total silt).

C   =   Weight percentage of particles with 0.05- to 0.10-mm diameter (very fine sand) on <2-mm basis.

D   =   Weight percentage of particles with 0.10- to 0.25-mm diameter (fine sand) on <2-mm basis.

E   =   Weight percentage of particles with 0.25- to 0.50-mm diameter (medium sand) on <2-mm basis.

F	=	Weight percentage of particles with 0.50- to 1.0-mm diameter (coarse sand) on <2-mm basis.
G	=	Cumulative weight percentage of particles passing No. 10 sieve on <3-in basis. This cumulative weight percentage is the decimal fraction passing the No. 10 sieve and is used in converting from 2-mm to 75-mm basis percentage.
H	=	Weight percentage of particles with 1.0- to 2.0-mm diameter (very coarse sand) on <2-mm basis.

1.15   Particles <0.5 mm

Particles with <0.5-mm diameter are reported as a cumulative weight percentage on <75-mm basis. Data may be determined as follows:

Equation 15.1:

$$P_{<0.5mm} = (A + B + C + D + E) \times (F / 100)$$

OR (alternatively)

Equation 15.2:

$$P_{<0.5mm} = P_{<1.0mm} - G \times (F / 100)$$

where

$P_{<0.5mm}$	=	Cumulative weight percentage of particles <0.5 mm on <75-mm basis.
A	=	Weight percentage of particles with <2-μm (<0.002-mm) diameter (total clay) on <2-mm basis.
B	=	Weight percentage of particles with 0.002- to 0.05-mm diameter (total silt).
C	=	Weight percentage of particles with 0.05- to 0.10-mm diameter (very fine sand) on <2-mm basis.
D	=	Weight percentage of particles with 0.10- to 0.25-mm diameter (fine sand) on <2-mm basis.
E	=	Weight percentage of particles with 0.25- to 0.50-mm diameter (medium sand) on <2-mm basis.

F	=	Cumulative weight percentage of particles passing No. 10 sieve on <3-in basis. This cumulative weight percentage is the decimal fraction passing the No. 10 sieve and is used in converting from 2-mm to 75-mm basis percentage.
G	=	Weight percentage of particles with 0.50- to 1.0-mm diameter (coarse sand) on <2-mm basis.
$P_{<1.0mm}$	=	Cumulative weight percentage of particles <1.0 mm on <75-mm basis.

1.16 Particles <0.25 mm

Particles with <0.25-mm diameter are reported as a cumulative weight percentage on <75-mm basis. Data may be determined as follows:

Equation 16.1:

$$P_{<0.25mm} = (A + B + C + D) \times (E / 100)$$

OR (alternatively)

Equation 16.2:

$$P_{<0.25mm} = P_{<0.5mm} - F \times (E / 100)$$

where

$P_{<0.25mm}$	=	Cumulative weight percentage of particles <0.25 mm on <75-mm basis.
A	=	Weight percentage of particles with <2-μm diameter (total clay) on <2-mm basis.
B	=	Weight percentage of particles with 0.002- to 0.05-mm diameter (total silt).
C	=	Weight percentage of particles with 0.05- to 0.10-mm diameter (very fine sand) on <2-mm basis.
D	=	Weight percentage of particles with 0.10- to 0.25-mm diameter (fine sand) on <2-mm basis.
E	=	Cumulative weight percentage of particles passing No. 10 sieve on <3-in basis. This cumulative weight percentage is the decimal fraction passing the No. 10 sieve and is

312

used in converting from 2-mm to 75-mm basis percentage.

F   =   Weight percentage of particles with 0.25- to 0.50-mm diameter (medium sand) on <2-mm basis.

$P_{<0.5mm}$ =   Cumulative weight percentage of particles <0.5 mm on <75-mm basis.

1.17   Particles <0.10 mm

Particles with <0.10-mm diameter are reported as a cumulative weight percentage on <75-mm basis. Data may be determined as follows:

Equation 17.1:

$$P_{<0.10mm} = (A + B + C) \times (D / 100)$$

OR (alternatively)

Equation 17.2:

$$P_{<0.10mm} = P_{<0.25mm} - E \times (D / 100)$$

where

$P_{<0.10mm}$ =   Cumulative weight percentage of particles <0.10 mm on <75-mm basis.

A   =   Weight percentage of particles with <2-μm diameter (total clay) on <2-mm basis.

B   =   Weight percentage of particles with 0.002- to 0.05-mm diameter (total silt).

C   =   Weight percentage of particles with 0.05- to 0.10-mm diameter (very fine sand) on <2-mm basis.

D   =   Cumulative weight percentage of particles passing No. 10 sieve on <3-in basis. This cumulative weight percentage is the decimal fraction passing the No. 10 sieve and is used in converting from 2-mm to 75-mm basis percentage.

E   =   Weight percentage of particles with 0.10- to 0.25-mm diameter (fine sand) on <2-mm basis.

$$P_{<0.25mm} \quad = \quad \text{Cumulative weight percentage of particles} <0.25 \text{ mm on} <75\text{-mm basis.}$$

1.18   Particles <0.05 mm

Particles with <0.05 diameter are reported as a cumulative weight percentage on <75-mm basis. Data may be determined as follows:

Equation 18.1:

$$P_{<0.05mm} = (A + B) \times (C / 100)$$

OR (alternatively)

Equation 18.2:

$$P_{<0.05mm} = P_{<0.10mm} - D \times (C / 100)$$

where

$P_{<0.05mm}$	=	Cumulative weight percentage of particles <0.05 mm on <75-mm basis.
A	=	Weight percentage of particles with <2-μm (<0.002 mm) diameter (total clay) on <2-mm basis.
B	=	Weight percentage of particles with 0.002- to 0.05-mm diameter (total silt).
C	=	Cumulative weight percentage of particles passing No. 10 sieve on <3-in basis. This cumulative weight percentage is the decimal fraction passing the No. 10 sieve and is used in converting from 2-mm to 75-mm basis percentage.
D	=	Weight percentage of particles with 0.05- to 0.10-mm diameter (very fine sand) on <2-mm basis.
$P_{<0.10mm}$	=	Cumulative weight percentage of particles <0.10 mm on <75-mm basis.

2   Cumulative Curve Fractions

Particle diameters in millimeters (mm) at specified percentile points on the cumulative particle-size distribution curve are reported on <3-in basis. Particles finer than the reported diameters are 60, 50, and 10 percent of <3-in material. Calculations of these diameters are embedded in a computer protocol too complicated to include in this discussion. Refer to the American Society for Testing and Materials (ASTM) Method D 2487-06 (ASTM, 2008b) for a more detailed discussion of these criteria.

2.1 Particle Diameter, 60 Percentile

Particle diameter corresponding to 60 percent finer on the cumulative particle-size distribution curve is reported. Data (D_{60}) are used in the classification of soils for engineering purposes in the Unified Soil Classification System (USCS).

2.2 Particle Diameter, 50 Percentile

Particle diameter corresponding to 50 percent finer on the cumulative particle-size distribution curve is reported. Data (D_{50}) are the particle-size geometric mean for the cumulative distribution curve.

2.3 Particle Diameter, 10 Percentile

Particle diameter corresponding to 10 percent finer on the cumulative particle-size distribution curve is reported. Data (D_{10}) are used in the classification of soils for engineering purposes in the Unified Soil Classification System (USCS).

3 Atterberg Limits

Atterberg limits: *Atterberg limits* is a general term that encompasses liquid limit (LL), plastic limit (PL), and, in some references, shrinkage limit (SL). The test method for these limits by ASTM has the designation of ASTM D 4318-05 (ASTM, 2008k). This test method is used as an integral part of several engineering classification systems, e.g., Unified Soil Classification System (USCS) and American Association of State Highway and Transportation Officials (AASHTO), to characterize the fine-grained fractions of soils (ASTM D 2487-06 and ASTM D 3282-93) and to specify the fine-grained fraction of construction materials (ASTM D 1241-07) (ASTM, 2008b, 2008a, 2008d, respectively). The LL and plasticity index (PI) of soils also are used extensively, either individually or together with other soil properties, to correlate with engineering behavior, e.g., compressibility, permeability, compactability, shrink-swell, and shear strength (ASTM, 2008b). The LL and PI are closely related to amount and kind of clay, CEC, 1500-kPa water, and engineering properties, e.g., load-carrying capacity of the soil.

In general, the AASHTO engineering system is a classification system for soils and soil-aggregate mixtures for highway construction purposes, e.g., earthwork structures, particularly embankments, subgrades, subbases, and bases. The Unified Soil Classification System (USCS) is used for general soils engineering work by many organizations, including USDA/NRCS.

3.1   Liquid Limit (LL)

Liquid limit: The *liquid limit* is the percent water content of a soil at the arbitrarily defined boundary between the liquid and plastic states. This water content is defined as the water content at which a pat of soil placed in a standard cup and cut by a groove of standard dimensions will flow together at the base of the groove for a distance of 13 mm (1/2 in) when subjected to 25 shocks from the cup being dropped 10 mm in a standard LL apparatus operated at a rate of 2 shocks s^{-3}. Refer to ASTM Method D 4318-05 (ASTM, 2008k). The LL is reported as percent water on <0.4-mm basis in this data column. If the LL is not measured, it can be estimated for use in engineering classification through the use of algorithms.

3.2   Plasticity Index (PI)

Plasticity index: The *plasticity index* is the range of water content over which a soil behaves plastically. Numerically, the PI is the difference in the water content between the LL and the PL. The PL is the percent water content of a soil at the boundary between the plastic and brittle states. The water content at this boundary is the water content at which a soil can no longer be deformed by rolling into 3.2-mm (1/8-in) threads without crumbling. Refer to ASTM Method D 4318-05 (ASTM, 2008k). If either the LL or PL cannot be determined, or if PL is $\geq$ LL, the soil is reported as nonplastic (NP). The PI is reported as percent water on <0.4-mm basis in this data column. If the PI is not measured, it can be estimated for use in engineering classification through the use of algorithms.

4   Gradation Curve

Gradation curve is the cumulative grain-size distribution curve. The coefficients of uniformity (Cu) and curvature (Cm) help to define the shape and position of the grain-size distribution curve. Characteristics of the gradation curve are used as classification criteria in the Unified Soil Classification System (USCS). Refer to ASTM Method D 2487-06 (ASTM, 2008b) for a detailed discussion of these criteria.

4.1   Uniformity

The coefficient of uniformity (Cu) is used to evaluate the grading characteristics of coarse-grained soils. The Cu is calculated as follows:

Equation 19:

$$Cu = A / B$$

where
Cu = Coefficient of uniformity.
A = Particle diameter corresponding to 60 percent finer on gradation curve.
B = Particle diameter corresponding to 10 percent finer on gradation curve.

4.2 Curvature

The coefficient of curvature (Cm) is used to evaluate the grading characteristics of coarse-grained soils. The Cm is calculated as follows:

Equation 20:

$$Cc = (A)^2 / (B \times C)$$

where
Cc = Coefficient of curvature.
A = Particle diameter corresponding to 30 percent finer on gradation curve. Value
        calculated but not reported on the SSL data sheets.
B = Particle diameter corresponding to 10 percent finer on gradation curve.
C = Particle diameter corresponding to 60 percent finer on gradation curve.

5 Weight Fractions

Weight fractions are reported as percentages on a whole-soil basis and <75-mm basis. These data add detail to the distribution of the >2-mm particle sizes. Weight percentages reported in these columns are derived from field and laboratory weights of coarse fragments, if available. If these data are not available, estimated field volume percentages and an estimated particle density are used for calculations.

5.1 Particles >2 mm, Whole-Soil Basis

Particles with >2-mm diameter are reported as a weight percentage on a whole-soil basis. As used herein, the fine-earth fraction refers to particles with <2-mm diameter, and the whole soil is all particle-size fractions, including boulders with maximum horizontal dimensions less than those of the pedon. In addition, the term *rock fragments* means particles ≥2 mm in diameter and includes all particles with horizontal dimensions smaller than the size of the pedon and is not the same as the term *coarse fragments*, which excludes stones and boulders with diameter ≥250 mm (Soil Survey Staff, 1975). The >250-mm division corresponds to the size opening in the 10-in screen (254 mm) used in engineering. Refer to Soil Survey Division Staff (1993) for additional discussion of rock fragments.

5.2 Particles 250 mm-UP, Whole-Soil Basis

Particles with ≥250-mm diameter are reported as a weight percentage on a whole-soil basis. Fragments with 250- to 600-mm particle diameter correspond to the nonflat rock fragment class *stones* (USDA/NRCS, 2009b).

5.3 Particles 75-250 mm, Whole-Soil Basis

Particles with 75- to 250-mm diameter are reported as a weight percentage on a whole-soil basis. Coarse fractions with 75- to 250-mm particle diameter correspond to the nonflat rock fragment class *cobbles* (USDA/NRCS, 2009b).

5.4 Particles 2-75 mm, Whole-Soil Basis

Particles with 2- to 75-mm diameter are reported as a weight percentage on a whole-soil basis.

5.5 Particles 20-75 mm, Whole-Soil Basis

Particles with 20- to 75-mm diameter are reported as a weight percentage on a whole-soil basis.

5.6 Particles 5-20 mm, Whole-Soil Basis

Particles with 5- to 20-mm diameter are reported as a weight percentage on a whole-soil basis.

5.7 Particles 2-5 mm, Whole-Soil Basis

Particles with 2- to 5-mm diameter are reported as a weight percentage on a whole-soil basis.

5.8 Particles <2 mm, Whole-Soil Basis

Particles with <2-mm diameter are reported as a weight percentage on a whole-soil basis.

5.9 Particles 2-75 mm, <75-mm Basis

Particles with 2- to 75-mm diameter are reported as a weight percentage on <75-mm basis. Coarse fractions with 2- to 75-mm particle diameter correspond to the nonflat rock fragment class *gravel* (USDA/NRCS, 2009b).

5.10 Particles 20-75 mm, <75-mm Basis

Particles with 20- to 75-mm diameter are reported as a weight percentage on <75-mm basis. Coarse fractions with 20- to 75-mm particle diameter correspond to the nonflat rock fragment class *coarse gravel* (USDA/NRCS, 2009b).

5.11 Particles 5-20 mm, <75-mm Basis

Particles with 5- to 20-mm diameter are reported as a weight percentage on <75-mm basis. Coarse fractions with 2- to 5-mm particle diameter correspond to the nonflat rock fragment class *medium gravel* (USDA/NRCS, 2009b).

5.12 Particles 2-5 mm, <75-mm Basis

Particles with 2- to 5-mm diameter are reported as a weight percentage on <75-mm basis. Coarse fractions with 2- to 5-mm particle diameter correspond to the nonflat rock fragment class *fine gravel* (USDA/NRCS, 2009b).

5.13 Particles <2 mm, <75-mm Basis

Particles with <2-mm diameter are reported as a weight percentage on <75-mm basis.

6 Weight per Unit Volume

The SSL reports weight per unit volume on a whole-soil basis and <2-mm basis. These data are calculated for soil survey and engineering purposes.

6.1 Weight per Unit Volume (33-kPa), Whole-Soil Basis, Soil Survey

Weight per unit volume at 33-kPa water content (ρ_{B33}) is reported in g cm^{-3} on a whole-soil basis.

Equation 21:

$$\rho_{B33ws} = 100 \, / \, \{(A \, / \, B) + [(100 - A) \, / \, C]\}$$

where

ρ_{B33ws}	=	Bulk density at 33-kPa water content on a whole-soil basis (g cm^{-3}).
A	=	Weight percentage of >2-mm fraction on a whole-soil basis.
B	=	Specific gravity of the whole soil, default value of 2.65 g cm^{-3}. For most >2-mm fractions, the specific gravity is ~ 2.65 g cm^{-3} (National Soil Survey Laboratory Staff, 1975). If these fractions are weathered, the specific gravity may be less, or if the fractions contain heavy minerals, the specific gravity may be greater than 2.65 g cm^{-3} (National Soil Survey

Laboratory Staff, 1975). Specific gravity or particle density is herein defined as the density of solid particles collectively and is expressed as the ratio of the total mass of solid particles to their total volume excluding pore spaces between particles (Blake and Hartge, 1986a).

C   =   Bulk density at 33-kPa water content on <2-mm basis (g cm^{-3}). If there is no measured value, then a ρ_{B33} estimate, ρ_B rewet (ρ_{Br}), or a default value of 1.45 g cm^{-3} may be used for mineral soils.

6.2   Weight per Unit Volume (Oven-Dry), Whole-Soil Basis, Soil Survey

Weight per unit volume at oven-dryness (ρ_{Bod}) is reported in g cm^{-3} on a whole-soil basis.

Equation 22:

$$\rho_{Bodws} = 100 / \{(A / B) + [(100 - A) / C]\}$$

where

ρ_{Bodws} =   Bulk density at oven-dryness on a whole-soil basis (g cm^{-3}).

A   =   Weight percentage of the >2-mm fraction.

B   =   Specific gravity of the whole soil, default value of 2.65 g cm^{-3}.

C   =   Bulk density at oven-dryness on <2-mm basis (g cm^{-3}).

6.3   Moist Weight per Unit Volume, Whole-Soil Basis, Engineering

Moist weight per unit volume is reported in g cm^{-3} on a whole-soil basis. These data are calculated from the bulk density and 33-kPa water content of the whole soil. For example, if the calculated density of moist soil is 1.85 g cm^{-3}, then 1 m^3 weighs 1.85 x 10^6 g or 1,850 kg. Since 1 yd^3 = 0.765 m^3 and 1 lb = 0.4536 kg, then the soil weighs $\approx$ 1.6T yd^{-3} or $\approx$ 116 lb ft^{-3}.

Equation 23:

$$MW/ V_{ws} =   A + \{[B \text{ x } (C / 100) \text{ x } A] / 100\}$$

where

MW/V$_{ws}$	=	Moist weight per unit volume on whole-soil basis (g cm^{-3}).
A	=	Refer to Equation 21.
B	=	Weight percentage of the <2-mm fraction on a whole-soil basis. To obtain Wt$_{<2mm}$ fraction in equation, subtract analytical data for Wt$_{>2mm}$ fraction from 100%.
C	=	Weight percentage of water retained at 33-kPa suction on <2-mm basis.

6.4 Saturated Weight per Unit Volume, Whole-Soil Basis, Engineering

Saturated weight per unit volume is reported in g cm^{-3} on a whole-soil basis. These data are calculated from the bulk density and 33-kPa water content of the whole soil plus the amount of water necessary to saturate the whole soil. This calculation ignores the possibility for swelling with change in water content between 33kPa and saturation. For example, if the calculated wet density of a soil is 1.95 g cm^{-3}, then 1 m^3 weighs 1,950 kg. Since 1 yd^3 = 0.765 m^3 and 1 lb = 0.4536 kg, then the soil weighs ≈ 1.6T yd^{-3} or ≈ 122 lb ft^{-3}.

Equation 24:

$$SW/V_{ws} = A + \{1 - [A / B]\}$$

where

SW/V$_{ws}$	=	Saturated weight per unit volume on a whole-soil basis (g cm^{-3}).
A	=	Refer to Equation 21.
B	=	Specific gravity of the whole soil, default value of 2.65 g cm^{-3}.

6.5 Weight per Unit Volume (33-kPa), <2-mm Basis, Soil Survey

Weight per unit volume at 33-kPa water content ($\rho_{B33)}$ is reported in g cm^{-3} on <2-mm basis.

6.6 Weight per Unit Volume (1500-kPa), <2-mm Basis, Soil Survey

Weight per unit volume at 1500-kPa water content (ρ_{B1500}) is reported in g cm^{-3} on <2-mm basis. This calculation assumes that the change in bulk density with change in water content is a straight line, but in actuality it is more sigmoidal in nature.

Equation 25:

$$\rho_{B1500<2mm} = A + \{(B - A) \times [(C - D) / (C - ((E - 1) \times 100))]\}$$

where

$\rho_{B1500<2mm}$ = Weight per unit volume at 1500-kPa water content on <2-mm basis (g cm^{-3}).

A = Bulk density at 33-kPa water content on <2-mm basis (g cm^{-3}). If there is no measured value, then a ρ_{B33} estimate, ρ_B rewet (ρ_{Br}), or a default value of 1.45 g cm^{-3} may be used for mineral soils.

B = Bulk density at oven-dryness on <2-mm basis (g cm^{-3}).

C = Weight percentage of water retained at 33-kPa suction on <2-mm basis.

D = Weight percentage of water retained at 1500-kPa suction on <2-mm basis.

E = Air-dry/Oven-dry ratio.

Bulk density, intermediate and weight per unit volume, 1500-kPa, <2-mm: The bulk density of clods at water contents intermediate between 33- or 10-kPa water retention and oven-dryness is termed *bulk density intermediate* (ρ_{Bi}). The calculation for $\rho_{B1500<2mm}$ (Equation 25) is essentially equivalent to the calculation for ρ_{Bi} (Equation 26) when $\rho_{Bi} = \rho_{B1500<2mm}$. When $\rho_{Bi} = \rho_{B1500<2mm}$, the difference in the equations is that Equation 25 uses the laboratory data for AD/OD, whereas in Equation 26 the decimal fraction for 1500-kPa water content (Factor x W_{1500}) is used to allow an approximation in the field. The ρ_{Bi} is calculated (Grossman et al., 1990) as follows:

Equation 26:

$$\rho_{Bi} = A + \{[A + (B - A) \times (C - D)] / [C - (Factor \times E)]\}$$

where

ρ_{Bi} = Bulk density intermediate between 33- or 10-kPa water content and oven-dryness on <2-mm basis (g cm^{-3}).

A = Bulk density at 33- or 10-kPa water content on <2-mm basis (g cm^{-3}).

322

B   = Bulk density at oven-dryness on <2-mm basis (g cm^{-3}).

C   = Weight percentage of water retained at 33- or 10-kPa suction on <2-mm basis.

D   = Weight percentage at intermediate water content. This weight percentage is not reported on the SSL data sheets.

Factor = Factors (F) related to linear extensibility (LE) are as follows:

F = 0.6 when LE = <0.06
F = 0.5 when LE = 0.06–0.09
F = 0.4 when LE = 0.09–0.12
F = 0.3 when LE = $\geq$0.12

E   = Weight percentage of water retained at 1500-kPa suction on <2-mm basis.

From ρ_{Bi} an intermediate coefficient of linear extensibility (COLE$_i$) may be calculated (assume no >2-mm fraction) as follows:

Equation 27:

$$\text{COLE}_i = [\rho_{Bi}/\rho_{Bf}]^{1/3} - 1$$

where

COLE$_i$ = Coefficient of linear extensibility intermediate.

ρ_{Bi} = Bulk density intermediate between 33- or 10-kPa water content and oven-dryness on <2-mm basis (g cm^{-3}). The ρ_{Bi} is not reported on the SSL data sheets.

ρ_{Bf} = Bulk density at 33- or 10-kPa water content on <2-mm basis (g cm^{-3}).

If field water content is estimated or measured, the associated ρ_{Bi} may be computed and used for prediction of root restriction. The related intermediate extensibility may be helpful in the prediction of surface-connected cracks (Grossman et al., 1990). The ρ_{Bi} and COLE$_i$ are not reported on the SSL data sheets.

6.7 Weight per Unit Volume (Oven-Dry), <2-mm Basis, Soil Survey

Weight per unit volume at oven-dryness (ρ_{Bod}) is reported in g cm^{-3} on <2-mm basis.

6.8 Moist Weight per Unit Volume, <2-mm Basis, Engineering

Moist weight per unit volume is reported in g cm^{-3} on <2-mm basis.

Equation 28:

$$MW/V_{<2mm} = A + [A \times (B / 100)]$$

where

$MW/V_{<2mm}$ = Moist weight per unit volume on <2-mm basis (g cm^{-3}).

A = Bulk density at 33-kPa water content on <2-mm basis (g cm^{-3}). If there is no measured value, then a ρ_{B33} estimate, ρ_B rewet (ρ_{Br}), or a default value of 1.45 g cm^{-3} may be used for mineral soils.

B = Weight percentage of water retained at 33-kPa suction on <2-mm basis.

6.9 Saturated Weight per Unit Volume, <2-mm Basis, Engineering

Saturated weight per unit volume is reported in g cm^{-3} on <2-mm basis.

Equation 29:

$$SW/V_{<2mm} = A + [1 - (A / B)]$$

where

$SW/V_{<2mm}$ = Saturated weight per unit volume on <2-mm basis (g cm^{-3}).

A = Bulk density at 33-kPa water content on <2-mm basis (g cm^{-3}). If there is no measured value, then a ρ_{B33} estimate, ρ_B rewet (ρ_{Br}), or a default value of 1.45 g cm^{-3} may be used for mineral soils.

B = Specific gravity of the <2-mm fraction, default value of 2.65 g cm^{-3}.

7 Void Ratios

Void ratio is defined as the ratio of volume of void space (space occupied by air and water) to volume of solids. This ratio is reported at 33-kPa water content on a whole-soil basis and on <2-mm basis.

7.1 Void Ratio, 33-kPa, Whole-Soil Basis

Void ratio at 33-kPa water content is reported on a whole-soil basis.

Equation 30:

$$Void_{33ws} = (A / B) - 1$$

where

$Void_{33ws}$	=	Void ratio at 33-kPa water content on a whole-soil basis.
A	=	Specific gravity of the whole soil, default value of 2.65 g cm^{-3}.
B	=	Refer to Equation 21.

7.2 Void Ratio, 33-kPa, <2-mm Basis

Void ratio at 33-kPa water content is reported on <2-mm basis.

Equation 31:

$$Void_{33<2mm} = (A / B) - 1$$

where

$Void_{33<2mm}$	=	Void ratio at 33-kPa water content of <2-mm fraction.
A	=	Specific gravity of the <2-mm fraction, default value of 2.65 g cm^{-3}.
B	=	Bulk density at 33-kPa water content on <2-mm basis (g cm^{-3}). If there is no measured value, then a ρ_{B33} estimate, ρ_B rewet (ρ_{Br}), or a default value of 1.45 g cm^{-3} may be used for mineral soils.

8 Volume Fractions

Volume fractions refer to the volume (percent) of components of a soil horizon at 33-kPa water content on a whole-soil basis. The 33-kPa bulk density is used to convert mass (weight) to volume percents.

8.1 Particles >2 mm

Particles with >2-mm diameter are reported as a volume percentage on a whole-soil basis.

Equation 32:

$$P_{>2mm} = \{(A / B) / [(A / B) + [(100 - A) / C]]\} \times 100$$

where

    $P_{>2mm}$ = Volume percentage of particles with >2-mm diameter on a whole-soil basis.

    A   = Weight percentage of >2-mm fraction on a whole-soil basis.

    B   = Specific gravity of >2-mm fraction, default value of 2.65 g cm^{-3}.

    C   = Bulk density at 33-kPa water content on <2-mm basis (g cm^{-3}). If there is no measured value, then a ρ_{B33} estimate, ρ_B rewet (ρ_{Br}), or a default value of 1.45 g cm^{-3} may be used for mineral soils.

8.2   Particles 250 mm-UP

Particles with ≥250-mm diameter are reported as a volume percentage on a whole-soil basis.

Equation 33:

$$P_{\geq 250mm} = \{(A / B) / [(C / B) + [(100 - C) / D]]\} \times 100$$

where

    $P_{\geq 250mm}$ = Volume percentage of particles with ≥250-mm diameter on a whole-soil basis.

    A   = Weight percentage of ≥250-mm fraction on a whole-soil basis.

    B   = Specific gravity of >2-mm fraction, default value of 2.65 g cm^{-3}. Specific gravity of ≥250 mm assumed to equal $SG_{>2mm}$.

    C   = Weight percentage of >2-mm fraction on a whole-soil basis.

    D   = Bulk density at 33-kPa water content on <2-mm basis (g cm^{-3}). If there is no measured value, then a ρ_{B33} estimate, ρ_B rewet (ρ_{Br}), or a default value of 1.45 g cm^{-3} may be used for mineral soils.

8.3   Particles 75-250 mm

Particles with 75- to 250-mm diameter are reported as a volume percentage on a whole-soil basis.

Equation 34:

$$P_{75\text{-}250mm} = \{(A / B) / [(C / B) + [(100 - C) / D]]\} \times 100$$

where

$P_{75\text{-}250mm}$	=	Volume percentage of particles with 75- to 250-mm diameter on a whole-soil basis.
A	=	Weight percentage of 75- to 250-mm fraction on a whole-soil basis.
B	=	Specific gravity of >2-mm fraction, default value of 2.65 g cm^{-3}. Specific gravity of 75- to 250-mm assumed to equal $SG_{>2mm}$.
C	=	Weight percentage of >2-mm fraction on a whole-soil basis.
D	=	Bulk density at 33-kPa water content on <2-mm basis (g cm^{-3}). If there is no measured value, then a ρ_{B33} estimate, ρ_B rewet (ρ_{Br}), or a default value of 1.45 g cm^{-3} may be used for mineral soils.

8.4   Particles 2-75 mm

Particles with 2- to 75-mm diameter are reported as a volume percentage on a whole-soil basis.

Equation 35:

$$P_{2\text{-}75mm} = \{(A / B) / [(C / B) + [(100 - C) / D]]\} \times 100$$

where

$P_{2\text{-}75mm}$	=	Volume percentage of particles with 2- to 75-mm diameter on a whole-soil basis.
A	=	Weight percentage of 2- to 75-mm fraction on a whole-soil basis.
B	=	Specific gravity of >2-mm fraction, default value of 2.65 g cm^{-3}. Specific gravity of 2- to 75-mm assumed to equal $SG_{>2mm}$.

C	= Weight percentage of >2-mm fraction on a whole-soil basis.
D	= Bulk density at 33-kPa water content on <2-mm basis (g cm^{-3}). If there is no measured value, then a ρ_{B33} estimate, ρ_B rewet (ρ_{Br}), or a default value of 1.45 g cm^{-3} may be used for mineral soils.

8.5   Particles 20-75 mm

Particles with 20- to 75-mm diameter are reported as a volume percentage on a whole-soil basis.

Equation 36:

$$P_{20\text{-}75mm} = \{(A / B) / [(C / B) + [(100 - C) / D]]\} \times 100$$

where

$P_{20\text{-}75mm}$	= Volume percentage of particles with 20- to 75-mm diameter on a whole-soil basis.
A	= Weight percentage of 20- to 75-mm fraction on a whole-soil basis.
B	= Specific gravity of >2-mm fraction, default value of 2.65 g cm^{-3}. Specific gravity of 20- to 75-mm assumed to equal SG$_{>2mm}$.
C	= Weight percentage of >2-mm fraction on a whole-soil basis.
D	= Bulk density at 33-kPa water content on <2-mm basis (g cm^{-3}). If there is no measured value, then a ρ_{B33} estimate, ρ_B rewet (ρ_{Br}), or a default value of 1.45 g cm^{-3} may be used for mineral soils.

8.6   Particles 5-20 mm

Particles with 5- to 20-mm diameter are reported as a volume percentage on a whole-soil basis.

Equation 37:

$$P_{5\text{-}20mm} = \{(A / B) / [(C / B) + [(100 - C) / D]]\} \times 100$$

where

$P_{5\text{-}20mm}$	=	Volume percentage of particles with 5- to 20-mm diameter on a whole-soil basis.
A	=	Weight percentage of 5- to 20-mm fraction on a whole-soil basis.
B	=	Specific gravity of >2-mm fraction, default value of 2.65 g cm^{-3}. Specific gravity of 5- to 20-mm assumed to equal SG$_{>2mm}$.
C	=	Weight percentage of >2-mm fraction on a whole-soil basis.
D	=	Bulk density at 33-kPa water content on <2-mm basis (g cm^{-3}). If there is no measured value, then a ρ_{B33} estimate, ρ_B rewet (ρ_{Br}), or a default value of 1.45 g cm^{-3} may be used for mineral soils.

8.7   Particles 2-5 mm

Particles with 2- to 5-mm diameter are reported as a volume percentage on a whole-soil basis.

Equation 38:

$$P_{2\text{-}5mm} = \{(A / B) / [(C / B) + [(100 - C) / D]]\} \times 100$$

where

$P_{2\text{-}5mm}$	=	Volume percentage of particles with 2- to 5-mm diameter on a whole-soil basis.
A	=	Weight percentage of 2- to 5-mm fraction on a whole-soil basis.
B	=	Specific gravity of >2-mm fraction, default value of 2.65 g cm^{-3}. Specific gravity of 2- to 5-mm assumed to equal SG$_{>2mm}$.
C	=	Weight percentage of >2-mm fraction on a whole-soil basis.
D	=	Bulk density at 33-kPa water content on <2-mm basis (g cm^{-3}). If there is no measured value, then a ρ_{B33} estimate, ρ_B rewet (ρ_{Br}), or a default value of 1.45 g cm^{-3} may be used for mineral soils.

8.8 Particles <2 mm

Particles with <2-mm diameter are reported as a volume percentage on a whole-soil basis.

Equation 39:

$$P_{2-5mm} = \{(A / B) / [(A / B) + [(100 - A) / C]]\} \times 100$$

where

$P_{<2mm}$ = Volume percentage of particles with <2-mm diameter on a whole-soil basis.

A = Weight percentage of <2-mm fraction on a whole-soil basis.

B = Bulk density at 33-kPa water content on <2-mm basis (g cm^{-3}). If there is no measured value, then a ρ_{B33} estimate, ρ_B rewet (ρ_{Br}), or a default value of 1.45 g cm^{-3} may be used for mineral soils.

C = Specific gravity of >2-mm fraction, default value of 2.65 g cm^{-3}.

8.9 Particles 0.05-2.0 mm

Particles with 0.05- to 2-mm diameter are reported as a volume percentage on a whole-soil basis.

Equation 40:

$$P_{0.05-2.0mm} = \{[(100 - A) \times (B / 100)] / C\} / \{(A / D) + [(100 - A) / E]\}$$

where

$P_{0.05-2.0mm}$ = Volume percentage of particles with 0.05- to 2-mm diameter (total sand) on a whole-soil basis.

A = Weight percentage of >2-mm fraction on a whole-soil basis.

B = Weight percentage of 0.05- to 2-mm fraction on <2-mm basis.

C = Specific gravity of <2-mm fraction, default value of 2.65 g cm^{-3}. Specific gravity of 0.05-

to 2.0-mm assumed to equal specific gravity of >2-mm fraction.

D   = Specific gravity of >2-mm fraction, default value of 2.65 g cm^{-3}.

E   = Bulk density at 33-kPa water content on <2-mm basis (g cm^{-3}). If there is no measured value, then a ρ_{B33} estimate, ρ_B rewet (ρ_{Br}), or a default value of 1.45 g cm^{-3} may be used for mineral soils.

8.10   Particles 0.002-0.05 mm

Particles with 0.002- to 0.05-mm diameter are reported as a volume percentage on a whole-soil basis.

Equation 41:

$$P_{0.002\text{-}0.5mm} = \{[(100 - A) \times (B / 100)] / C\} / \{(A / D) + [(100 - A) / E]\}$$

where

$P_{0.002\text{-}0.05mm}$ = Volume percentage of particles with 0.002- to 0.05-mm particle diameter (total silt) on a whole-soil basis.

A   = Weight percentage of >2-mm fraction on a whole-soil basis.

B   = Weight percentage of 0.002- to 0.05-mm fraction on <2-mm basis.

C   = Specific gravity of <2-mm fraction, default value of 2.65 g cm^{-3}. Specific gravity of 0.002- to 0.05-mm assumed to equal specific gravity of >2-mm fraction.

D   = Specific gravity of >2-mm fraction, default value of 2.65 g cm^{-3}.

E   = Bulk density at 33-kPa water content on <2-mm basis (g cm^{-3}). If there is no measured value, then a ρ_{B33} estimate, ρ_B rewet (ρ_{Br}), or a default value of 1.45 g cm^{-3} may be used for mineral soils.

8.11   Particles <0.002 mm (<2 μm)

Particles with <0.002-mm diameter are reported as a volume percentage on a whole-soil basis.

Equation 42:

$$P_{<0.002} = \{[(100 - A) \times (B/100)]/C\}/\{(A/D) + [(100 - A)/E]\}$$

where

$P_{<0.002mm}$	=	Volume percentage of particles with <0.002-mm diameter (total clay) on a whole-soil basis.
A	=	Weight percentage of >2-mm fraction on a whole-soil basis.
B	=	Weight percentage of <0.002-mm fraction on <2-mm basis.
C	=	Specific gravity of <2-mm fraction, default value of 2.65 g cm^{-3}. Specific gravity of <0.002 mm assumed to equal specific gravity of >2-mm fraction.
D	=	Specific gravity of >2-mm fraction, default value of 2.65 g cm^{-3}.
E	=	Bulk density at 33-kPa water content on <2-mm basis (g cm^{-3}). If there is no measured value, then a ρ_{B33} estimate, ρ_B rewet (ρ_{Br}), or a default value of 1.45 g cm^{-3} may be used for mineral soils.

9   Pore Volume, 33-kPa Water Content

Drained and filled pores at 33-kPa water content are reported as percentages on a whole-soil basis. Total porosity (drained + filled pores) at 33-kPa water content is calculated as follows:

Equation 43:

$$Pores_T = 100 - [(A/B) \times 100]$$

where

$Pores_T$	=	Percent total porosity (drained + filled pores) at 33-kPa water content on a whole-soil basis.
A	=	Refer to Equation 21.

B = Specific gravity of the whole soil, default value of 2.65 g cm^{-3}.

9.1 Pores, Drained (D), 33-kPa Water Content

Pores drained at 33-kPa water content are reported as percent on a whole-soil basis. These have been defined as noncapillary pores.

Equation 44:

$$Pores_D = Pores_T - [A \times (B / 100) \times C]$$

where

Pores$_D$ = Percent pores drained at 33-kPa water content on a whole-soil basis.

Pores$_T$ = Percent total porosity (drained + filled pores) at 33-kPa water content on a whole-soil basis.

A = Refer to Equation 21.

B = Weight percentage of <2-mm fraction on a whole-soil basis.

C = Weight percentage of water retained at 33-kPa suction on <2-mm basis (g H$_2$O 100 g^{-3} soil).

9.2 Pores, Filled (F), 33-kPa Water Content

Pores filled at 33-kPa water content are reported as percent on a whole-soil basis. Capillary pores.

Equation 45:

$$Pores_F = Pores_T - Pores_D$$

where

Pores$_F$ = Percent filled pores at 33-kPa water content on a whole-soil basis.

Pores$_T$ = Percent total porosity at 33-kPa water content on a whole-soil basis.

Pores$_D$ = Refer to Equation 44.

10   Carbon/Nitrogen (C/N) Ratio

C:N ratio, application. The C:N ratio provides information on soils relating to fertility and organic matter decomposition. Decomposed humified organic materials as found in a mollic epipedon typically have a C:N ratio of 10 or 12 to 1. Higher ratios in soils may suggest low decomposition levels or low N levels in plant residues and soils. Lower ratios may be a result of the accumulation of inorganic N in the soil horizon. Care is required in interpreting this ratio, especially when the data are numerically small. Extremely low N levels can significantly inflate this ratio. Refer to additional discussion about the C:N ratio in the Primary Characterization Data section in this manual.

C:N ratio, calculation: In the past, the SSL determined organic C by the wet combustion method, Walkley-Black modified acid-dichromate digestion, $FeSO_4$ titration, automatic titrator (6A1c, method obsolete; Soil Survey Staff, 2004). Currently, the SSL measures total C and N by the dry combustion method. This method measures both organic and inorganic C and N forms. For those soils with organic C data by the obsolete Walkley-Black method, the C:N ratios are calculated based on organic C:total N. Currently, the C:N ratios are calculated based on total C:total N values. If carbonates are present, this value is subtracted from the total C value and the C:N ratio is calculated as follows:

Equation 46:

$$\text{Total C (\%)} \times 0.12 = \text{Inorganic C (\%)}$$

where:
Atomic weights, $CaCO_3$
Ca = 40
O = 16
C = 12

C = [40 + (3 x 16) + 12] = 12% of $CaCO_3$

Equation 47:

$$\text{Total C (\%)} - \text{Inorganic C (\%)} = \text{Organic C (\%)}$$

Equation 48:

$$\text{C:N} = \text{Organic C(\%)} / \text{Total N (\%)}$$

For example calculations, refer to the Primary Characterization Data Sheet section under Ratios, Estimates, and Calculations Related to Organic Matter.

11 Total Clay, Ratios

Ratios of some properties to clay are provided. All of these ratios are expressed on <2-mm basis.

11.1 Fine Clay/Total Clay

An increase in the ratio of fine clay to total clay with depth commonly occurs in soils that have argillic horizons. Refer to Soil Survey Staff (2010) for a more detailed discussion of this ratio and its use as a diagnostic criterion for argillic horizons. Both total and fine clay are measured values and are reported. The ratio of fine clay to total clay is reported on <2-mm basis.

11.2 CEC-8.2/Total Clay

Total clay and CEC by sum of cations (CEC-8.2) are measured values. The ratio of CEC-8.2 to total clay is reported on <2-mm basis.

11.3 CEC-7/Total Clay

In general, soils contain negatively charged colloids, and their CEC increases with increasing pH. This increase in CEC is due to pH-dependent charge prevalent in oxide minerals and organic matter. As a general rule, the CEC-8.2 > CEC-7 > ECEC. The ratio of CEC-7 to total clay is reported on <2-mm basis.

11.4 1500-kPa Water Content/Total Clay

The 1500-kPa and total clay percentage on air-dry samples are measured values. The ratio of 1500-kPa water to total clay is reported on <2-mm basis.

11.5 LEP/Total Clay, <2-mm Basis

The *coefficient of linear extensibility* (COLE) can be expressed as percent, i.e., *linear extensibility percent* (LEP). LEP = COLE x 100. The LEP is not the same as LE. In *Soil Taxonomy*, the LE of a soil layer is the product of the thickness, in centimeters, multiplied by the COLE of the layer in question, whereas the LE of a soil is the sum of these products for all soil horizons (Soil Survey Staff, 1975). The SSL reports the ratio of LEP at 33-kPa water to total clay on <2-mm basis.

Linear extensibility percent (LEP or COLE x 100) is a function of several factors, including the kind and amount of clay. A ratio of LEP to clay takes out the factor of clay content and permits a more direct comparison of LEP and clay mineralogy (National Soil Survey Laboratory Staff, 1983). Cemented aggregates and $CaCO_3$ decrease the linear extensibility, i.e., decrease the potential to swell with increase in water content. Smectites typically have high LEP to clay ratios; kaolinites have low ratios; and micas or mixed-layer clays have intermediate ratios. In a study of soils of the

Western United States (National Soil Survey Laboratory Staff, 1990), some LEP/clay relationships were determined as follows: smectite 0.18 ($r^2 = 0.76$, n = 291); clay mica 0.13 ($r^2 = 0.77$, n = 118); and kaolinite 0.08 ($r^2 = 0.44$, n = 71). Some general rules for LEP/clay ratios as related to clay mineralogy (National Soil Survey Laboratory Staff, 1983) are as follows:

LEP/clay	Mineralogy
>0.15	Smectites
0.05–0.15	Micas (illites) chlorites
<0.05	Kaolinites

Explanations should be sought if data deviate widely from the ranges indicated. The LEP/clay ratio as well as the CEC/clay ratio are useful as internal checks of the data and as estimators of mineralogy when mineralogy data are not available (National Soil Survey Laboratory Staff, 1983). The ratio of linear extensibility to total clay is reported on <2-mm basis.

Coefficient of linear extensibility and linear extensibility percent, fine-earth fraction: The COLE and LEP values for the fine-earth fraction alone are referred to as $COLE_f$ and LE_f, respectively. The LE_f is a function of clay content, clay mineralogy, organic matter, and carbonate clay. The LE_f correlates with clay content (r = 0.90 to 0.95) and to a lesser extent with 1500-kPa water and cation-exchange capacity (National Soil Survey Laboratory Staff, 1975).

With each absolute 10 percent increase in clay content, some general rules for $COLE_f$ increase in soils with 0 to 60 percent clay (National Soil Survey Laboratory Staff, 1975) are as follows:

Mineralogy	$COLE_f$ increase
nonexpanding 1:1 layer clays	0.005
nonexpanding 2:1 layer clays	0.01
nonexpanding 1:1 and 2:1 clays	0.01
mixed expanding and nonexpanding clays	0.06
expanding 2:1 layer clays	0.015–0.02

OR (alternatively)

LEP_f = 0 to 3% nonexpanding 1:1 layer clays
LEP_f = 0 to 6% nonexpanding 2:1 layer clays
LEP_f = 0 to 6% mixed nonexpanding 1:1 and 2:1 clays
LEP_f = 0 to 10% mixed expanding and nonexpanding clays
LEP_f = 0 to 15–20% expanding 2:1 layer clays

The shrink-swell classes based on LEP and COLE values, as defined in the National Soil Survey Handbook (USDA/NRCS, 2009b), are as follows:

Class	LEP	COLE
Low	<3	<0.03
Moderate	3–6	0.03–0.06
High	6–9	0.06–0.09
Very high	>9	>0.09

12   Linear Extensibility Percent

The SSL reports the Linear Extensibility Percent (LEP) between 33- and 1500-kPa water content on a whole-soil basis and between 33-kPa and oven-dryness on a whole-soil basis. The SSL also reports the LEP between 33- and 1500-kPa water content on <2-mm basis and between 33-kPa and oven-dryness on <2-mm basis.

12.1   LEP, 33- to 1500-kPa Water Content, Whole-Soil Basis

Linear extensibility between 33- and 1500-kPa water content is reported as percent on a whole-soil basis. The calculation of bulk density at 1500-kPa water content on a whole-soil basis ($\rho_{B1500ws}$) is required in order to calculate the LEP between 33- and 1500-kPa water content. The $\rho_{B1500ws}$ is calculated as follows:

Equation 49:

$$\rho_{B\,1500ws} = \{[(A - B) \times (C - D) \times (E / 100)] / [E \times (C / 100)] - [((F - 1) \times 100) \times (E / 100)]\} + B$$

where

$\rho_{B1500ws}$	=	Bulk density at 1500-kPa water content on a whole-soil basis (g cm^{-3}).
A	=	Bulk density at oven-dryness on a whole-soil basis (g cm^{-3}).
B	=	Bulk density at 33-kPa water content on a whole-soil basis (g cm^{-3}).
C	=	Weight percentage of water retained at 33-kPa suction on <2-mm basis.
D	=	Weight percentage of water retained at 1500-kPa suction on <2-mm basis.
E	=	Weight percentage of <2-mm fraction on a whole-soil basis.
F	=	Air-dry/Oven-dry ratio (AD/OD).

Equation 50:

$$LEP_{33 \text{ to } 1500ws} = [[\rho_{B1500ws} / \rho_{B33ws}]^{1/3} - 1] \times 100$$

where

$LEP_{33 \text{ to } 1500ws}$ =   Linear extensibility percent between 33- and 1500-kPa water content on a whole-soil basis.

$\rho_{B1500ws}$ =   Bulk density at 1500-kPa water content on a whole-soil basis (g cm^{-3}).

ρ_{B33ws} =   Bulk density at 33-kPa water content on a whole-soil basis (g cm^{-3}).

12.2   LEP, 33 kPa to Oven-Dryness, Whole-Soil Basis

Linear extensibility between 33-kPa water content and oven-dryness is reported as percent on a whole-soil basis.

Equation 51:

$$LEP_{33 \text{ to } odws} = [[\rho_{Bodws} / \rho_{B33ws}]^{1/3} - 1] \times 100$$

where

$LEP_{33 \text{ to } odws}$ =   Linear extensibility percent between 33-kPa and oven-dryness on a whole-soil basis.

ρ_{Bodws} =   Bulk density at oven-dryness on a whole-soil basis (g cm^{-3}).

ρ_{B33ws} =   Bulk density at 33-kPa water content on a whole-soil basis (g cm^{-3}).

12.3   LEP, 33 to 1500 kPa, <2-mm Basis

Linear extensibility between 33- and 1500-kPa water content is reported as percent on <2-mm basis.

Equation 52:

$$LEP_{33 \text{ to } 1500<2mm} = [[\rho_{B1500<2mm} / \rho_{B33<2mm}]^{1/3} - 1] \times 100$$

where

$LEP_{33 \text{ to } 1500<2mm}$ =   Linear extensibility percent between 33- and 1500-kPa water content on <2-mm basis.

$\rho_{B1500<2mm}$ = Bulk density at 1500-kPa water content on <2-mm basis (g cm^{-3}).

$\rho_{B33<2mm}$ = Bulk density at 33-kPa water content on <2-mm basis (g cm^{-3}). If there is no measured value, then a ρ_{B33} estimate, ρ_B rewet (ρ_{Br}), or a default value of 1.45 g cm^{-3} may be used for mineral soils.

12.4   LEP, 33 kPa to Oven-Dryness, <2-mm Basis

Linear extensibility between 33-kPa water content and oven-dryness is reported as percent on <2-mm basis.

Equation 53:

$$LEP_{33 \text{ to } od<2mm} = [[\rho_{Bod<2mm} / \rho_{B33<2mm}]^{1/3} - 1] \times 100$$

where

$LEP_{33 \text{ to } od<2mm}$ = Linear extensibility percent 33- to 1500-kPa water content on <2-mm basis.

$\rho_{Bod<2mm}$ = Bulk density at oven-dryness on <2-mm basis (g cm^{-3}).

$\rho_{B33<2mm}$ = Bulk density at 33-kPa water content on <2-mm basis (g cm^{-3}). If there is no measured value, then a ρ_{B33} estimate, ρ_B rewet (ρ_{Br}), or a default value of 1.45 g cm^{-3} may be used for mineral soils.

13   Water Retention Difference (WRD)

The WRD between 1500- and 33-kPa suctions is reported on a whole-soil basis and <2-mm basis. The units are expressed as in in^{-3}, but the numbers do not change when other units, e.g., cm cm^{-3} or ft ft^{-3}, are needed.

13.1   Water Retention Difference (WRD), Whole-Soil Basis

The WRD is reported between 1500- and 33-kPa suctions on a whole-soil basis (in in^{-3}).

Equation 54:

$$WRD_{ws} = [(A - B) \times (C / 100) \times D] / 100$$

339

where

WRD$_{ws}$	=	Volume fraction (in^3 in^{-3}) of water retained in the whole soil between 33-kPa and 1500-kPa suction reported in inches of water per inch of soil (in in^{-3}). This is numerically equivalent to cm cm^{-3}.
A	=	Weight percentage of water retained at 33-kPa suction on <2-mm basis.
B	=	Weight percentage of water retained at 1500-kPa suction on <2-mm basis. If available, moist 1500-kPa is the first option in the WRD calculation; otherwise, dry 1500-kPa is used.
C	=	Weight percentage of <2-mm fraction on a whole-soil basis.
D	=	Refer to Equation 21.

13.2   Water Retention Difference (WRD), <2-mm Basis

The WRD is reported between 33- and 1500-kPa water content on <2-mm basis (in in^{-3}).

Equation 55:

$$WRD_{<2mm} = [(A - B) \times C] / 100$$

where

WRD$_{<2mm}$	=	Volume fraction (in^3 in^{-3}) of water retained in the <2-mm fraction between 33-kPa and 1500-kPa suction reported in inches of water per inch of soil (in in^{-3}). This is numerically equivalent to cm cm^{-3}.
A	=	Weight percentage of water retained at 33-kPa suction on <2-mm basis.
B	=	Weight percentage of water retained at 1500-kPa suction on <2-mm basis. If available, moist 1500-kPa is the first option in the WRD calculation; otherwise, dry 1500-kPa is used.
C	=	Bulk density at 33-kPa water content on <2-mm basis (g cm^{-3}). If there is no measured value, then a ρ_{B33} estimate, ρ_B rewet (ρ_{Br}), or a default value of 1.45 g cm^{-3} may be used for mineral soils.

340

14 Weight Fractions, Clay-Free Basis

Weight fractions, data assessment: Laboratory data may be used to assess whether a field-designated discontinuity is corroborated and to see if any data show evidence of discontinuity not observed in the field, and if so, to help sort the lithological changes from the pedogenic changes (National Soil Survey Laboratory Staff, 1983). In some cases, the quantities of sand and coarser fractions are not altered significantly by soil-forming processes. As particle size increases, the change in particle size by weathering decreases; i.e., an abrupt change in sand content is a clue to lithological change (National Soil Survey Laboratory Staff, 1983). The gross soil mineralogy and the resistant mineral suite are other clues.

Weight fractions, sand ratios: Another aid used to assess lithological changes is ratios of one sand separate to another (National Soil Survey Laboratory Staff, 1983). The ratios can be computed and examined as a numerical array, or they can be plotted in graphical form. The ratios work well if sufficient quantities of the two fractions are present. Low quantities magnify changes in ratios, especially if the denominator is low.

Weight fractions, clay-free basis: A common manipulation in assessing lithological change is to compute sand and silt separates on a clay-free basis. Clay distribution is subject to pedogenic change and may mask inherited lithological differences (National Soil Survey Laboratory Staff, 1983). The numerical array on a clay-free basis can be inspected visually or plotted in graphical form.

14.1 Particles >2 mm, Whole Soil, Clay-Free Basis

Particles with >2-mm diameter are reported as a weight percentage on a whole-soil clay-free basis. Whole-soil clay-free basis = >2-mm fraction + total sand + total silt.

Equation 56:

$$P_{>2mm} = \{A / [100 - ((B \times C) / 100)]\} \times 100$$

where

$P_{>2mm}$	=	Weight percentage of particles with >2-mm diameter on a whole-soil clay-free basis.
A	=	Weight percentage of >2-mm fraction.
B	=	Weight percentage of <2-mm fraction on a whole-soil basis.
C	=	Weight percentage of particles with <0.002-mm diameter (total clay).

14.2 Particles 2-75 mm, Whole Soil, Clay-Free Basis

Particles with 2- to 75-mm diameter are reported as a weight percentage on a whole-soil clay-free basis.

Equation 57:

$$P_{>2\text{-}75mm} = \{A / [100 - ((B \times C) / 100)]\} \times 100$$

where

$P_{2\text{-}75mm}$	=	Weight percentage of particles with 2- to 75-mm diameter on a whole-soil clay-free basis.
A	=	Weight percentage of particles with 2- to 75-mm diameter on a whole-soil basis.
B	=	Weight percentage of <2-mm fraction on a whole-soil basis.
C	=	Weight percentage of particles with <0.002-mm diameter (total clay).

14.3 Particles 2-20 mm, Whole Soil, Clay-Free Basis

Particles with 2- to 20-mm diameter are reported as a weight percentage on a whole-soil clay-free basis.

Equation 58:

$$P_{>2\text{-}20mm} = \{A / [100 - ((B \times C) / 100)]\} \times 100$$

where

$P_{2\text{-}20mm}$	=	Weight percentage of particles with 2- to 20-mm diameter on a whole-soil clay-free basis.
A	=	Weight percentage of particles with 2- to 20-mm diameter on a whole-soil basis.
B	=	Weight percentage of <2-mm fraction on a whole-soil basis.
C	=	Weight percentage of particles with <0.002-mm diameter (total clay).

14.4 Particles 0.05-2 mm, Whole Soil, Clay-Free Basis

Particles with 0.05- to 2-mm diameter are reported as a weight percentage on a whole-soil clay-free basis.

Equation 59:

$$P_{0.05\text{-}2mm} = \{[[A \ \times \ (B \ / \ 100)]] \ / \ \{100 \ - \ [(A \ \times \ C) \ / \ 100]]\} \ \times \ 100$$

where

$P_{0.05\text{-}2mm}$	=	Weight percentage of particles with 0.05- to 2-mm diameter (total sand) on a whole-soil clay-free basis.
A	=	Weight percentage of <2-mm fraction on a whole-soil basis.
B	=	Weight percentage of particles with 0.05- to 2-mm diameter (total sand) on <2-mm basis.
C	=	Weight percentage of particles with <0.002-mm diameter on <2-mm basis.

14.5 Particles 0.002-0.05 mm, Whole Soil, Clay-Free Basis

Particles with 0.002- to 0.05-mm diameter are reported as a weight percentage on a whole-soil clay-free basis.

Equation 60:

$$P_{0.002\text{-}0.05mm} = \{[[A \ \times \ (B \ / \ 100)]] \ / \ \{100 \ - \ [(A \ \times \ C) \ / \ 100]]\} \ \times \ 100$$

where

$P_{0.002\text{-}0.05mm}$	=	Weight percentage of particles with 0.002- to 0.05-mm diameter (total silt) on a whole-soil clay-free basis.
A	=	Weight percentage of <2-mm fraction on a whole-soil basis.
B	=	Weight percentage of particles with 0.002- to 0.05-mm diameter (total silt) on <2-mm basis.
C	=	Weight percentage of particles with <0.002-mm diameter (total clay).

343

14.6 Particles <0.002 mm, Whole Soil, Clay-Free Basis

Particles with <0.002-mm diameter are reported as a weight percentage on a whole-soil clay-free basis. These data have application in assessing discontinuities only if all the clay is inherited in the sediment; i.e., there is no pedogenic clay.

Equation 61:

$$P_{<0.002mm} = \{[[A \times (B / 100)]] / \{100 - [(A \times B) / 100]]\} \times 100$$

where

$P_{<0.002mm}$	=	Weight percentage of particles with <0.002-mm diameter (total clay) on a whole-soil clay-free basis.
A	=	Weight percentage of <2-mm fraction on a whole-soil basis.
B	=	Weight percentage of particles with <0.002-mm diameter on <2-mm basis.

14.7 Very Coarse Sand (1.0-2.0 mm), <2-mm Fraction, Clay-Free Basis

Very coarse sand (1.0- to 2.0-mm particle diameter) is reported as a weight percentage on <2-mm clay-free basis. The <2-mm clay-free basis = total sand + total silt.

Equation 62:

$$P_{1.0-2.0mm} = [A / (100 - B)] / 100$$

where

$P_{1.0-2.0mm}$	=	Weight percentage of particles with 1.0- to 2.0-mm diameter (very coarse sand) on <2-mm clay-free basis.
A	=	Weight percentage of particles with 1.0- to 2.0-mm diameter (very coarse sand) on <2-mm basis.
B	=	Weight percentage of particles with <0.002-mm diameter (total clay).

14.8   Coarse Sand (0.5-1.0 mm), <2-mm Fraction, Clay-Free Basis

Coarse sand (0.5- to 1.0-mm particle diameter) is reported as a weight percentage on <2-mm clay-free basis.

Equation 63:

$$P_{0.50\text{-}1.0mm} = [A / (100 - B)] \times 100$$

where

$P_{0.50\text{-}1.0mm}$	= Weight percentage of particles with 0.5- to 1.0-mm diameter (coarse sand) on <2-mm clay-free basis.
A	= Weight percentage of particles with 0.50- to 1.0-mm diameter (coarse sand) on <2-mm basis.
B	= Weight percentage of particles with <0.002-mm diameter (total clay).

14.9   Medium Sand (0.25-0.50 mm), <2-mm Fraction, Clay-Free Basis

Medium sand (0.25- to 0.50-mm particle diameter) is reported as a weight percentage on <2-mm clay-free basis.

Equation 64:

$$P_{0.25\text{-}0.50mm} = [A / (100 - B)] \times 100$$

where

$P_{0.25\text{-}0.50mm}$	= Weight percentage of particles with 0.25- to 0.50-mm diameter (medium sand) on <2-mm clay-free basis.
A	= Weight percentage of particles with 0.25- to 0.50-mm diameter (medium sand) on <2-mm basis.
B	= Weight percentage of particles with <0.002-mm diameter (total clay).

14.10   Fine Sand (0.10-0.25 mm), <2-mm Fraction, Clay-Free Basis

Fine sand (0.10- to 0.25-mm particle diameter) is reported as a weight percentage on <2-mm clay-free basis.

Equation 65:

$$P_{0.10\text{-}0.25mm} = [A / (100 - B)] \times 100$$

where

$P_{0.10\text{-}0.25mm}$	=	Weight percentage of particles with 0.10- to 0.25-mm diameter (fine sand) on <2-mm clay-free basis.
A	=	Weight percentage of particles with 0.10- to 0.25-mm diameter (fine sand) on <2-mm basis.
B	=	Weight percentage of particles with <0.002-mm diameter (total clay).

14.11 Very Fine Sand (0.05-0.10 mm), <2-mm Fraction, Clay-Free Basis

Very fine sand (0.05- to 0.10-mm particle diameter) is reported as a weight percentage on <2-mm clay-free basis.

Equation 66:

$$P_{0.05\text{-}0.10mm} = [A / (100 - B)] \times 100$$

where

$P_{0.05\text{-}0.10mm}$	=	Weight percentage of particles with 1.0- to 2.0-mm diameter (very fine sand) on <2-mm clay-free basis.
A	=	Weight percentage of particles with 0.05- to 0.10-mm diameter (very fine sand) on <2-mm basis.
B	=	Weight percentage of particles with <0.002-mm diameter (total clay).

14.12 Coarse Silt (0.02-0.05 mm), <2-mm Fraction, Clay-Free Basis

Coarse silt (0.02- to 0.05-mm particle diameter) is reported as a weight percentage on <2-mm clay-free basis.

Equation 67:

$$P_{0.02\text{-}0.05mm} = [A / (100 - B)] \times 100$$

346

where

$P_{0.02-0.05mm}$	=	Weight percentage of particles with 0.02- to 0.05-mm diameter (coarse silt) on <2-mm clay-free basis.
A	=	Weight percentage of particles with 0.02- to 0.05-mm diameter (coarse silt) on <2-mm basis.
B	=	Weight percentage of particles with <0.002-mm diameter (total clay).

14.13   Fine Silt (0.002-0.02 mm), <2-mm Fraction, Clay-Free Basis

Fine silt (0.002- to 0.02-mm particle diameter) is reported as a weight percentage on <2-mm clay-free basis.

Equation 68:

$$P_{0.002-0.02mm} = [A / (100 - B)] \times 100$$

where

$P_{0.002-0.02mm}$	=	Weight percentage of particles with 0.002- to 0.02-mm diameter (fine silt) on <2-mm clay-free basis.
A	=	Weight percentage of particles with 0.002- to 0.02-mm diameter (fine silt) on <2-mm basis.
B	=	Weight percentage of particles with <0.002-mm diameter (total clay).

14.14   Clay (<0.002 mm), <2-mm Fraction, Clay-Free Basis

Total clay (<0.002-mm particle diameter) is reported as a weight percentage on <2-mm clay-free basis.

Equation 69:

$$P_{<0.002mm} = [A / (100 - A)] \times 100$$

where

$P_{<0.002mm}$	=	Weight percentage of particles with 1.0- to 2.0-mm diameter (total clay) on <2-mm clay-free basis.

A	= Weight percentage of particles with <0.002-mm diameter (total clay).

15   Texture Determined, PSDA

Texture, definition: The term *texture* is defined in Webster's Dictionary as "something composed of closely interwoven elements; the visual or tactile surface characteristics and appearance of something; a basic scheme or overall structure." Although soil texture is a seemingly simple basic concept in soil science, its consistent application has not been easy (Soil Survey Staff, 1951). Historically, the textural terms were related not only to the qualities of texture but also had some connotations of both consistence and structure, as these soil properties are related in part to texture (Soil Survey Staff, 1951). As investigations of all soils continued, however, scientists realized that structure, consistence, and texture had to be measured or observed separately. In addition, the early laboratory methods for soil dispersion were so inadequate that fine granules of clay were reported as silt or sand. Structure and consistence are related not only to the amount of clay but also to the kind and condition of the clay as well as other constituents and living tissue in the soil (Soil Survey Staff, 1951). Textural class names are not used to express differences in consistence or structure; else the names lose their fundamental significance (Soil Survey Staff, 1951).

Soil texture, data reporting: Soil texture class names are reported as codes (abbreviations). Texture class names are based first on the distribution of sand, silt, and clay and then, for some classes, on the distribution of several size fractions of sand. Names are based on PSDA data to the nearest 1 percent applied to definitions of the texture classes (Soil Survey Staff, 1951). The texture class codes are as follows:

COS	- coarse sand	VFSL	- very fine sandy loam
S	- sand	L	- loam
FS	- fine sand	SIL	- silt loam
VFS	- very fine sand	SI	- silt
LCOS	- loamy coarse sand	SCL	- sandy clay loam
LS	- loamy sand	CL	- clay loam
LFS	- loamy fine sand	SICL	- silty clay loam
LVFS	- loamy very fine sand	SC	- sandy clay
COSL	- coarse sandy loam	SIC	- silty clay
SL	- sandy loam	C	- clay
FSL	- fine sandy loam		

Soil texture, laboratory-determined: The SSL-determined PSDA soil texture is reported. The laboratory-determined texture may or may not agree with the field-determined texture. In the past, the field texture was reported on the Supplementary Characterization Data Sheet; it is currently reported as metadata on the Primary Characterization Data Sheet.

16 PSDA, Particles <2 mm

In the following sections, SSL PSDA summaries are reported for the soil horizons. These summaries (sand, silt, and clay) are expressed on <2-mm basis as follows:

16.1 Total Sand, 0.05-2.0 mm

Total sand (0.05- to 2.0-mm diameter) is reported as a weight percentage on <2-mm basis. Refer to additional discussion on total sand in the Primary Characterization Data section of this manual.

16.2 Total Silt, 0.002-0.05 mm

Total silt (0.002- to 0.05-mm diameter) is reported as a weight percentage on <2-mm basis. Refer to additional discussion on total silt in the Primary Characterization Data section of this manual.

16.3 Total Clay, <0.002 mm (<2 μm)

Total clay (<0.002-mm diameter) is reported as a weight percentage on <2-mm basis. Refer to additional discussion on total clay in the Primary Characterization Data section of this manual.

17 pH, CaCl$_2$ 0.01 M

Refer to the discussion on 0.01 M CaCl$_2$ pH in the Primary Characterization Data section of this manual.

18 Electrical

18.1 Electrical Resistivity, Saturated Paste

Refer to the discussion on electrical resistivity of the saturated paste (R_s) in the Primary Characterization Data section of this manual.

18.2 Electrical Conductivity, Saturation Extract

Refer to the discussion on electrical conductivity of the saturated extract (EC_s) in the Primary Characterization Data section of this manual.

19   Particle Density

Refer to the discussion on particle density in the Primary Characterization Data section of this manual.

IV. References

Anonymous. 1987. Petrographic sample preparation for micro-structural analysis. Buehler Digest, Vol. 24, No.1.

Aase, J.K., and J.L. Pikul. 1995. Crop and soil response to long term tillage practices in the northern great plains. Agron. J. 87:652-656.

Abrams, M.M., and W.M. Jarrell. 1995. Soil phosphorus as a potential nonpoint source for elevated stream phosphorus levels. J. Environ. Qual. 24:132-138.

Adachi, T. 1973. Studies of the humus of volcanic ash soils—regional differences of humus composition in volcanic ash soils in Japan. Bull. Natl. Inst. Agric. Sci. Series B, 24:127-264 (in Japanese, with English abstract).

Adams, F. 1974. Soil solution. pp. 441-481. *In* E.W. Carson (ed.), The plant root and its environment. University Press of Virginia, Charlottesville, VA.

Adams, F. 1984. Crop response to lime in the Southern United States. pp. 211-265. *In* F. Adams (ed.), Soil acidity and liming. 2nd ed. Agron. Monogr. 12. ASA and SSSA, Madison, WI.

Adams, F., and B.L. Moore. 1983. Chemical factors affecting root growth in subsoil horizons of Coastal Plain soils. Soil Sci. Soc. Am. J. 47:99-102.

Aguilera, N.H., and M.L. Jackson. 1953. Iron oxide removal from soils and clays. Soil Sci. Soc. Am. Proc. 17:359-364.

Alam, S.M., and W.A. Adams. 1979. Effects of aluminum on nutrient composition and yield of oats. J. Plant Nutr. 1:365-375.

Alexander, E.B., and R. Burt. 1996. Soil development on moraines of Mendenhall Glacier, southeast Alaska. 1. The moraines and soil morphology. Geoderma 72:1-17.

Alexander, E.B., W.E. Wildman, and W.C. Lynn. 1985. Ultramafic (serpentinitic) mineralogy class. pp. 135-146. *In* J.A. Kittrick (ed.), Mineral classification of soils. Am. Soc. Agron. and Soil Sci. Soc. Am. Spec. Publ. 16. Madison, WI.

Alexander, M. 1977. Introduction to soil microbiology. 2nd ed. John Wiley & Sons Inc., New York, NY.

Alexander, M. 1980. Effects of acidity on microorganisms and microbial processes in soils. pp. 363-364. *In* T. Hutchinson and M. Havas (eds.), Effects of acid precipitation on terrestrial ecosystems. Plenum Publishing Corporation, NY.

Alexiades, C.A., and M.L. Jackson. 1966. Quantitative clay mineralogical analysis of soils and sediments. Clays and Clay Minerals 14:35-52.

Allen, B.L., 1985. Micromorphology of Aridisols. *In* L.A. Douglas and M.L. Thompson (eds.), Soil micromorphology and soil classification. Soil Sci. Soc. Am. Spec. Publ. 15. Madison, WI.

Allen, B.L., and B.F. Hajek. 1989. Mineral occurrence in soil environments. pp. 199-278. *In* J.B. Dixon and S.B. Weed (eds.), Minerals in soil environments. 2nd ed. Soil Sci. Soc. Am. Book Series 1.

Allison, L.E. 1960. Wet-combustion apparatus and procedure for organic and inorganic carbon in soil. Soil Sci. Soc. Am. Proc. 24:36-40.

Allison, L.E., and C.D. Moodie. 1965. Carbonate. pp. 1379-1400. *In* C.A. Black, D.D. Evans, J.L. White, L.E. Ensminger, and F.E. Clark (eds.), Methods of soil analysis. Part 2. 2nd ed. Agron. Monogr. 9. ASA, CSSA, and SSSA, Madison, WI.

Amer, F.A., A. Mahmoud, and V. Sabel. 1985. Zeta potential and surface area of calcium carbonate as related to phosphate sorption. Soil Sci. Soc. Am. J. 49:1137-1142.

American Association of Analytical Communities (AOAC), International. 2000. Phosphorus (citrate-soluble) in fertilizers and phosphorus (citrate-insoluble) in fertilizers, methods 960.01 and 963.03, respectively. Official Methods of Analysis.

American Society for Testing and Materials (ASTM). 1958. Book of ASTM Standards. Construction. Section 4. ASTM, Philadelphia, PA.

American Society for Testing and Materials (ASTM). 1963. Book of ASTM Standards. Construction. Section 4. ASTM, Philadelphia, PA.

American Society for Testing and Materials (ASTM). 2008a. Standard practice for classification of soils and soil-aggregate mixtures for highway construction purposes. D 3282-93. Annual Book of ASTM Standards 04.08:363-368. ASTM, Philadelphia, PA.

American Society for Testing and Materials (ASTM). 2008b. Standard practice for classification of soils for engineering purposes (Unified Soil Classification System). D 2487-06. Annual Book of ASTM Standards 04.08:260-271. ASTM, Philadelphia, PA.

American Society for Testing and Materials (ASTM). 2008c. Standard practice for description and identification of soils (visual-manual procedure). D 2488-06. Annual Book of ASTM Standards 04.08:272-282. ASTM, Philadelphia, PA.

American Society for Testing and Materials (ASTM). 2008d. Standard specification for materials for soil-aggregate subbase, base, and surface courses. D 1241-07. Annual Book of ASTM Standards 04.08:127-129. ASTM, Philadelphia, PA.

American Society for Testing and Materials (ASTM). 2008e. Standard test method for CBR (California Bearing Ratio) of laboratory-compacted soils. D 1883-99. Annual Book of ASTM Standards 04.08:186-194. ASTM, Philadelphia, PA.

American Society for Testing and Materials (ASTM). 2008f. Standard test method for CBR (California Bearing Ratio) of soils in place. D 4429-04. Annual Book of ASTM Standards 04.08:664-667. ASTM, Philadelphia, PA.

American Society for Testing and Materials (ASTM). 2008g. Standard test method for determining deformability and strength of weak rock by an in situ uniaxial compressive test. D 4555-01. Annual Book of ASTM Standards 04.08:753-756. ASTM, Philadelphia, PA.

American Society for Testing and Materials (ASTM). 2008h. Standard test method for dispersive characteristics of clay soil by double hydrometer. D 4221-99. Annual Book of ASTM Standards 04.08:524-526. ASTM, Philadelphia, PA.

American Society for Testing and Materials (ASTM). 2008i. Standard test method for particle-size analysis of soils. D 422-63. Annual Book of ASTM Standards 04.08:10-17. ASTM, Philadelphia, PA.

American Society for Testing and Materials (ASTM). 2008j. Standard test methods for laboratory compaction characteristics of soil using modified effort (56,000 ft-lbf/ft^3 (2,700 kN-m/m^3)). D 1557-07. Annual Book of ASTM Standards 04.08:145-157. ASTM, Philadelphia, PA.

American Society for Testing and Materials (ASTM). 2008k. Standard test methods for liquid limit, plastic limit, and plasticity index of soils. D 4318-05. Annual Book of ASTM Standards 04.08:581-596. ASTM, Philadelphia, PA.

Amonette, J.E., and R.W. Sanders. 1994. Nondestructive techniques for bulk elemental analysis. pp. 1-48. *In* J.E. Amonette and L.W. Zelazny (eds.), Quantitative methods in soil mineralogy. Soil Sci. Soc. Am. Misc. Publ. SSSA, Madison, WI.

Amonette, J.E., and L.W. Zelazny. 1994. Quantitative methods in soil mineralogy. Soil Sci. Soc. Am. Misc. Publ. SSSA, Madison, WI.

Amos, D.F., and E.P. Whiteside. 1975. Mapping accuracy of a contemporary soil survey in an urbanizing area. Soil Sci. Soc. Am. Proc. 39:937-942.

Andersen, M.K., K. Raulund-Rasmussen, H.C.B. Hansen, and B.W. Strobel. 2002. Distribution and fractionation of heavy metals in pairs of arable and afforested soils in Denmark. Eur. J. Soil Sci. 53:491-502.

Anderson, H.A., M.L. Berrow, V.C. Farmer, A. Hepburn, J.D. Russell, and A.D. Walker. 1982. A reassessment of podzols formation processes. J. Soil. Sci. 33:125-136.

Anderson, J.P.E., and K.H. Domsch. 1978. Mineralization of bacteria and fungi in chloroform fumigated soils. Soil. Biol. Biochem. 10:207-213.

Andrew, L.E., and F.W. Stearns. 1963. Physical characteristics of four Mississippi soils. Soil Sci. Soc. Am. Proc. 27:693-697.

Arndt, W. 1965. The nature of mechanical impedance to seedlings by soil surface seals. Aust. J. Soil Res. 3:45-54.

Arnold, J.G., K.W. King, and J.R. Williams. 2002. Erosion by water, hybrid models. pp. 463-467. *In* R. Lal (ed.), Encyclopedia of soil science. Marcel Dekker, Inc.

Arnon, D.L., and P.R. Stout. 1939. The essentiality of certain elements in minute quantity for plants with special reference to copper. Plant Physiol. 14:371-375.

Artieda, O., J. Herrero, and P.J. Drohan. 2006. A refinement of the differential water loss method for gypsum determination in soils. Soil Sci. Soc. Am. J. 70:1932-1935.

Ashaye, T.L. 1969. Sesquioxide status and particle size distribution in twelve Nigerian soils derived from sandstones. African Soils 14:85-96.

Atterberg, A. 1912. Die mechanische bodenanalyse und die klassification der mineralboden schwedens. Intern. Mitt. Bodenk. 2:312-342.

Ayers, R.S., and D.W. Westcot. 1985. Water quality for agriculture. FAO Irrigation and Drainage Paper 29. FAO, Rome.

Azzam, R.A.I. 1980. Agricultural polymers, polyacrylamide preparation, application, and prospects. Commun. Soil Sci. Plant Anal. 11:767-834.

Bache, B.W. 2002. Ion exchange. pp. 726-729. *In* R. Lal (ed.), Encyclopedia of soil science. Marcel Dekker, Inc.

Baeumer, K., and W.A. Bakermans. 1970. Zero tillage. Adv. Agron. 25:77-123.

Bailey, R.G. 1976. Ecoregions of the United States. Map (scale 1:1,750,000). USDA, Forest Service, Intermountain Region, Ogden, UT.

Bailey, R.G. 1996. Ecosystem geography. Springer-Verlag, New York, NY.

Bailey, S.W. 1980. Summary of recommendations of AIPEA nomenclature committee on clay minerals. Am. Mineral 65:1-7.

Bain, D.C., and B.F.L. Smith. 1994. Chemical analysis. pp. 300-332. *In* M.J. Wilson (ed.), Clay mineralogy: Spectroscopic and chemical determinative methods. Chapman and Hall, Inc., London, England.

Bandick, A.K., and R.P. Dick. 1999. Field management effects on soil enzyme activities. Soil Biol. Biochem. 31:1471-1479.

Banks, H.O. 1982. Six-state High Plains-Ogallala Aquifer area regional study. Report to U.S. Congress.

Baril, R., and G. Bitton. 1969. Teneurs elevees de fer libre et identification taxonomique de certain sols due Quebec contenant de la magnetite. Can. J. Soil Sci. 49:1-9.

Barrow, C.J. 1994. Land degradation. Cambridge University Press.

Barrow, N.J. 1974. Effect of previous additions of phosphate on phosphate adsorption by soils. Soil Sci. 118:82-89.

Barrow, N.J., and V.C. Cox. 1992. The effects of pH and chloride concentration on mercury sorption: I. By goethite. J. Soil Sci. 43:295-304.

Bartelli, L.J., A.A. Klingebiel, J.V. Baird, and M.R. Heddleson. 1966. Soil surveys and land use planning. SSSA and ASA, Madison, WI.

Bartlett, R.J. 1972. Field test for spodic character based on pH-dependent phosphorus adsorption. Soil Sci. Soc. Am. Proc. 36:642-644.

Bartlett, R.J., and B.R. James. 1996. Chromium. pp. 683-701. *In* D.L. Sparks (ed.), Methods of soil analysis. Part 3. Chemical methods. Soil Sci. Soc. Am. Book Series 5. ASA and SSSA, Madison, WI.

Barton, H., W.G. McCully, H.M.Taylor, and J.H. Box. 1966. Influence of soil compaction on emergence and first-year growth of seeded grasses. J. Range Manage. 19:118-12.

Bascomb, C.L., and M.G. Jarvis. 1976. Variability in three areas of the Denchworth soil map unit. I: Purity of the map unit and property variability within it. J. Soil Sci. 27:420-437.

Bates, R.L., and J.A. Jackson (eds.). 1987. Glossary of geology. 3rd ed. American Geological Institute, Alexandria, VA.

Bates, T.E. 1993. Soil handling and preparation. pp. 19-24. *In* M.R. Carter (ed.), Soil sampling and methods of analysis. Canadian Society of Soil Science. Lewis Publishers, CRC Press, Boca Raton, FL.

Baumer, O.W. 1992. Diagnosis and treatment of saline and sodic soils. Lecture notes.

Baver, L.D. 1956. Soil physics. 3rd ed. John Wiley & Sons Inc., New York, NY.

Beaton, James D., G.R. Burns, and J. Platou. 1968. Determination of sulphur in soils and plant material. Tech. Bull. 14. The Sulfur Institute, Washington, DC.

Beauchemin, S., and R.R. Simard. 1999. Soil phosphorus saturation degree: Review of some indices and their suitability for P management in Quebec, Canada. Can. J. Soil Sci. 79:615-625.

Beckett, P.H.T., and R. Webster. 1971. Soil variability: A review. Soils Fert. 34:1-15.

Bedunah, D.J., and R.E. Sosebee (eds.). 1995. Wildland plants: Physiological ecology and developmental morphology. Society for Range Management, Denver, CO.

Bell, M.A., and J. van Keulen. 1995. Soil pedotransfer functions of four Mexican soils. Soil Sci. Soc. Am. J. 59:865-871.

Benham, E.C. 2003. Soil resistivity temperature correction. USDA/NRCS, Lincoln, NE.

Benites, V.M., P.L.O.A. Machado, E.C.C. Fidalgo, M.R. Coelho, and B.E. Madari. 2007. Pedotransfer functions for estimating soil bulk density from existing soil survey reports in Brazil. Geoderma 139:90-97.

Bennett, O.L., D.A. Ashley, and B.D. Doss. 1964. Methods of reducing soil crusting to increase cotton seedling emergence. Agron. J. 56:162-165.

Berdanier, C.R., W.C. Lynn, and G.W. Threlkeld. 1978. Illitic mineralogy in Soil Taxonomy: X-ray vs. total potassium. Agronomy Abstracts, 1978 Annual Meetings.

Bernas, B. 1968. A new method for decomposition and comprehensive analysis of silicates by atomic absorption spectrometry. Anal. Chem. 40:1682-1686.

Bernstein, Leon. 1964. Salt tolerance of plants. USDA Info. Bull. 283.

Bertsch, P.M., and P.R. Bloom. 1996. Aluminum. pp. 517-530. *In* D.L. Sparks (ed.), Methods of soil analysis. Part 3. Chemical methods. Soil Sci. Soc. Am. Book Series 5. ASA and SSSA, Madison, WI.

Bhiyan, L.R., and J.E. Sedberry, Jr. 1995. Apparent phosphorus fixation by selected soils of Louisiana. Commun. Soil Sci. Plant Anal. 26:21-34.

Bidwell, R.G.S. 1979. Plant physiology. Macmillan Publishing Company, Inc., NY.

Birkeland, P.W., R.M. Burke, and J.B. Benedict. 1989. Pedogenic gradients of iron and aluminum accumulation and phosphorus depletion in arctic and alpine soils as a function of time and climate. Quat. Res. 32:193-204.

Birrell, K.S. 1961. Ion fixation by allophane. N.Z. J. Sci. 4:393-414.

Blackburn, W.H., and C.M. Skau. 1974. Infiltration rates and sediment production of selected plant communities in Nevada. J. Range Manage. 27:476-480.

Blair, G.J., R. Lefroy, and L. Lise. 1995. Soil carbon fractions based on their degree of oxidation, and the development of a carbon management index for agricultural systems. Austr. J. Agric. Res. 46:1459-1466.

Blair, G.J., R. Lefroy, A.Whitbread, N. Blair, and A. Conteh. 2001. The development of the $KMnO_4$ oxidation technique to determine labile carbon in soil and its use in a carbon management index. pp. 323-337. *In* R. Lal, J. Kimble, R. Follet, and B. Stewart (eds.), Assessment methods for soil carbon. Lewis Publishers, Boca Raton, FL.

Blake, G.R., and K.H. Hartge. 1986a. Bulk density. pp. 363-375. *In* A. Klute (ed.), Methods of soil analysis. Part 1. Physical and mineralogical methods. 2nd ed. Agron. Monogr. 9. ASA and SSSA, Madison, WI.

Blake, G.R., and K.H. Hartge. 1986b. Particle density. pp. 377-382. *In* A. Klute (ed.), Methods of soil analysis. Part 1. Physical and mineralogical methods. 2nd ed. Agron. Monogr. 9. ASA and SSSA, Madison, WI.

Blakemore, L.C., P.L. Searle, and B.K. Daly. 1987. Methods for chemical analysis of soils. N.Z. Soil Bur. Sci. Rep. 80. Lower Hutt, New Zealand.

Blevins, R.L., N. Holowaychuk, and L.P. Wilding. 1970. Micromorphology of soil fabric at tree root-soil interface. Soil Sci. Soc. Am. Proc. 34:460-465

Bloom, R.P., U.L. Skyllberg, and M.E. Sumner. 2005. Soil acidity. pp. 411-459. *In* Chemical processes in soils. Soil Sci. Soc. Am. Book Series 8. Madison, WI.

Bloss, D.F. 1961. An introduction to the methods of optical crystallography. Holt, Rinehart, and Winston, NY.

Boardman, J. 2002. Erosion assessment. pp. 399-401. *In* R. Lal (ed.), Encyclopedia of soil science. Marcel Dekker, Inc.

Bockheim, J.G. 1980. Solution and use of chronosfunctions in studying soil development. Geoderma 24:71-85.

Bohn, H.L., B.L. McNeal, and G.A. O'Connor. 1979. Soil chemistry. John Wiley & Sons Inc., New York, NY.

Bohn, H.L., B.L. McNeal, and G.A. O'Connor. 1985. Soil chemistry. 2nd ed. John Wiley & Sons Inc., New York, NY.

Boischot, P., M. Coppenet, and J. Hebert. 1950. Fixation de l'acide phosphorique sur le calcaire des sols. Plant Soil 2:311-322.

Bollard, E.G., and G.W. Butler. 1966. Mineral nutrition of plants. Ann. Rev. Plant Physiol. 17:77-112.

Bonnifield, P. 1979. The Dust Bowl: Men, dirt, and depression. University of New Mexico Press, Albuquerque.

Bouyoucos, G.J. 1927. The hydrometer as a new method for the mechanical analysis of soils. Soil Sci. 23:343-353.

Bouyoucos, G.J. 1929. The ultimate natural structure of soil. Soil Sci. 28:27-37.

Bouyoucos, G.J. 1951. A recalibration of the hydrometer method for making mechanical analysis of soils. Agron. J. 43:435-438.

Bouyoucos, G.J. 1962. Hydrometer method improved for making particle size analysis of soils. Agron. J. 54:464-465.

Bower, C.A. 1950. Fixation of ammonium in difficultly exchangeable forms under moist conditions by some soils of semi-arid regions. Soil Sci. 70:375-383.

Bradford, J.M., and R.W. Blanchar. 1977. Profile modification of a Fragiudalf to increase crop production. Soil Sci. Soc. Am. J. 41:127-131.

Bradford, J.M., and C. Huang. 1992. Mechanisms of crust formation: Physical components. pp. 55-72. *In* M.E. Sumner and B.A. Stewart (eds.), Soil crusting, chemical and physical processes. Advances in Soil Science. Lewis Publishers.

Brady, N.C. 1974. The nature and properties of soils. 9th ed. Macmillan Publishing Company, Inc., New York, NY.

Brasher, B.R., D.P. Franzmeier, V.T. Volassis, and S.E. Davidson. 1966. Use of saran resin to coat natural soil clods for bulk density and water retention measurements. Soil Sci. 101:108.

Bray, R.H. 1934. A chemical study of soil development in the Peorian loess region of Illinois. Am. Soil Surv. Assoc. Bull. 15:58-65.

Bray, R.H., and L.T. Kurtz. 1945. Determination of total, organic, and available forms of phosphorus in soils. Soil Sci. 59:39-45.

Breeuwsma, A., and S. Silva. 1992. Phosphorus fertilization and environmental effects in the Netherlands and Po Region (Italy). Report 57. Agricultural Research Department. Winand Staring Centre for Integrated Land, Soil and Water Research. Wageningen, The Netherlands.

Brejda, J.J., T.B. Moorman, D.L. Karlen, and T.H. Dao. 2000. Identification of regional soil quality factors and indicators: I. Central and Southern High Plains. Soil Sci. Soc. Am. J. 64:2115-2124.

Bremner, J.M. 1996. Nitrogen—Total. pp. 1085-1121. *In* D.L. Sparks (ed.), Methods of soil analysis. Part 3. Chemical methods. Soil Sci. Soc. Am. Book Series 5. ASA and SSSA, Madison, WI.

Bremner, J.M., G.A. Breitenback, and A.M. Blackmer. 1981. Effect of anhydrous ammonia fertilizer on emission of nitrous oxide from soils. J. Environ. Qual. 10:77-80.

Bremner, J.M., and C.S. Mulvaney. 1982. Nitrogen—Total. pp. 595-624. *In* A.L. Page, R.H. Miller, and D.R. Keeney (eds.), Methods of soil analysis. Part 2. Chemical and microbiological properties. 2nd ed. Agron. Monogr. 9. ASA and SSSA, Madison, WI.

Brennan, R.F., M.D.A. Bolland, R.C. Jeffery, and D.G. Allen. 1994. Phosphorus adsorption by a range of Western Australian soils related to soil properties. Commun. Soil Sci. Plant Anal. 25:2785-2795.

Bresler, E., B.L. McNeal, and D.L. Carter. 1982. Saline and sodic soils. Principles-dynamics-modeling. Adv. Series in Agric. Sci. No. 10. Springer-Verlag, New York, NY.

Brewer, R. 1964. Fabric and mineral analysis of soils. John Wiley & Sons Inc., New York, NY.

Brewer, R. 1976. Fabric and mineral analysis of soils. Reprint of 1964 ed., with supplemental material. Robert E. Cringer Publishing Co., Huntington, NY.

Brewer, R., and A.V. Blackmore. 1976. Relationships between subplasticity rating, optically oriented clay, cementation and aggregate stability. Austr. J. Soil Res. 14:237-248.

Brewer, R., and S. Pawluk. 1975. Investigations of some soils developed in hummocks of the Canadian sub-arctic and southern-arctic regions. I. Morphology and micromorphology. Can. J. Soil Sci. 55:301-319.

Brewer, R., and J.R. Sleeman. 1963. Pedotubules: Their definition, classification and interpretation. J. Soil Sci. 15:66-78.

Brewer, R., J.R. Sleeman, and R.C. Foster. 1983. The fabric of Australian soils. pp. 439-476. *In* Division of soils, CSIRO (ed.), Soils: An Australian viewpoint. CSIRO: Melbourne/Academic Press, London.

Briggs, L.J., and H.L. Shantz. 1911. A wax seal method for determining the lower limit of available soil moisture. Bot. Gaz. 51:210-219.

Briggs, L.J., and H.L. Shantz. 1912. The relative wilting coefficients for different plants. Bot. Gaz. 53:229-235.

Brimhall, G.H., C.J. Lewis, C. Ford, J. Bratt, G. Taylor, and O. Warin. 1991. Quantitative geochemical approach to pedogenesis: Importance of parent material reduction, volumetric expansion, and eolian influx in laterization. Geoderma 51:51-91.

Broadbent, F.E. 1953. The soil organic fraction. Adv. Agron. 5:153-183.

Brooks, R.R. 1987. Serpentine and its vegetation. A multidisciplinary approach. *In* T.R. Dudley (ed.), Ecology, phytogeography, and physiology series, Vol. 1. Dioscorides Press, Portland, OR.

Brown, G. 1980. Appendix I (Tables for the determination of d in Å from 2θ for the Kα and Kβ radiations of copper, cobalt, and iron). pp. 439-475. *In* G.W. Brindley and G. Brown (eds.), Crystal structures of clay minerals and their x-ray identification. Mineralogical Soc. Monogr. 5. Mineralogical Society of Great Britain.

Brown, G.R., and J.F. Thilenius. 1976. A low-cost machine for separation of roots from soil material. J. Range Manage. 29:506-507.

Brown, I.C., and H.G. Byers. 1932. The fractionation, composition and hypothetical constitution of certain colloids derived from the great soil groups. USDA Tech. Bull. 319.

Brown, L.R. 1984. The global loss of topsoil. J. Soil Water Conserv. 36(5):255-260.

Brubaker, S.C., C.S. Holzhey, and B.R. Brasher. 1992. Estimating the water-dispersible clay content of soils. Soil Sci. Soc. Am. J. 56:1226-1232.

Bruce, R.R., G.W. Langdale, and A.L. Dillard. 1990. Tillage and crop rotation effect on characteristics of a sandy surface soil. Soil Sci. Soc. Am. J. 54:1744-1747.

Bruce, R.R., and R.J. Luxmoore. 1986. Water retention: Field methods. pp. 663-686. *In* A. Klute (ed.), Methods of soil analysis. Part 1. Physical and mineralogical methods. 2nd ed. Agron. Monogr. 9. ASA and SSSA, Madison, WI.

Brumester, C.H., M.G. Patterson, and D.W. Reeves. 1995. Challenges of no-till cotton production on silty clay soils in Alabama. Conservation Tillage for Special Report 169. University of Arkansas, Fayetteville, AR.

Buck, B.J., K. Wolff, D. Merkler, and N. McMillan. 2006. Soil mineralogy of Las Vegas Wash, Nevada: Morphology and subsurface evaporation. Soil Sci. Soc. Am. J. 70:1639-1651.

Buck, R., M.D. Mays, E.C. Benham, and M.A. Wilson. 2002. Phosphorus fertilization and correlation with properties of selected benchmark soils of the United States. Commun. Soil Sci. Plant Anal. 33:117-141.

Bullock, P., N. Fedoroff, A. Jongerius, G. Stoops, T. Tursina, and others. 1985. Handbook for soil thin section description. Waine Research Publications, Walverhampton, England.

Bundy, L.G., and J.M. Bremner. 1972. A simple titrimetric method for determination of inorganic carbon in soils. Soil Sci. Soc. Am. Proc. 36:273-275.

Buol, S.W., F.D. Hole, and R.J. McCracken. 1980. Soil genesis and classification. 2nd ed. Iowa State University Press, Ames, IA.

Burt, R. 1996. Sample collection procedures for laboratory analysis in the United States soil survey program. Commun. Soil. Sci. Plant Anal. 27:1293-1298.

Burt, R., and E.B. Alexander. 1996. Soil development on moraines of Mendenhall Glacier, southeast Alaska. 2. Chemical transformations and soil micromorphology. Geoderma 72:19-36.

Burt, R., M. Filmore, M.A. Wilson, E.R. Gross, R.W. Langridge, and D.A. Lammers. 2001a. Soil properties of selected pedons on ultramafic rocks in Klamath Mountains, Oregon. Commun. Soil Sci. Plant Anal. 32:2145-2175.

Burt, R., M.D. Mays, E.C. Benham, and M.A. Wilson. 2002a. Phosphorus characterization and correlation with properties of selected benchmark soils of the United States. Commun. Soil Sci. Plant Anal. 33:117-141.

Burt, R., and M.A. Wilson. 2006. Serpentinitic soils. *In* R. Lal (ed.), Encyclopedia of soil science. 2nd ed. Marcel Dekker, Inc.

Burt, R., M.A.Wilson, C.W. Kanyanda, J.K.R.Spurway, and J.D. Metzler. 2001b. Properties and effects of management on selected granitic soils in Zimbabwe. Geoderma 101:119-141.

Burt, R., M.A. Wilson, T.J. Keck, B.D. Dougherty, D.E. Strom, and J.A. Lindhal. 2002b. Trace element speciation in selected smelter-contaminated soils in Anaconda and Deer Lodge Valley, Montana, USA. Adv. Environ. Res. 8:51-67.

Burt, R., M.A. Wilson, M.D. Mays, T.J. Keck, M. Filmore, A.W. Rodman, E.B. Alexander, and L. Hernandez. 2000. Trace and major elemental analysis application in the U.S. Cooperative Soil Survey program. Commun. Soil Sci. Plant Anal. 31:1757-1771.

Burt, R., M.A. Wilson, M.D. Mays, and C.W. Lee. 2003. Major and trace elements of selected pedons in the USA. J. Environ. Qual. 32:2109-2121.

Busacca, A., and D. Chandler. 2002. Erosion by wind: Climate change. pp. 484-488. *In* R. Lal (ed.), Encyclopedia of soil science. Marcel Dekker, Inc.

Butler, B.E. 1974. A contribution towards the better specification of parna and some other aeolian clays in Australia. Z. Geomorph. N.F. Suppl. Bd. 20:106-116. Berlin, Germany.

Buurman, P., B. van Lagen, and E.J. Velthorst. 1996. Manual for soil and water analysis. Backhuys Publishing, Leiden, The Netherlands.

Cady, J.G. 1965. Petrographic microscope techniques. pp. 604-631. *In* D.D. Evans, L.E. Ensminger, J.L. White, and F.E. Clark (eds.), Methods of soil analysis. Part 1. Physical and mineralogical properties, including statistics of measurement and sampling. 1st ed. Agron. Monogr. 9. ASA and SSSA, Madison, WI.

Cady, J.G., L.P. Wilding, and L.R. Drees. 1986. Petrographic microscope techniques. pp. 198-204. *In* A. Klute (ed.), Methods of soil analysis. Part 1. Physical and mineralogical methods. 2nd ed. Agron. Monogr. 9. ASA and SSSA, Madison, WI.

Cain, Z., and S. Lovejoy. 2004. History and outlook for Farm Bill conservation programs. Choices Magazine. 4th quarter 19(4).

Cambardella, C.A., and E.T. Elliott. 1992. Particulate soil organic-matter changes across a grassland cultivation sequence. Soil Sci. Soc. Am. J. 56:777-783.

Campbell, C.A., B.H. Ellert, and Y.W. Jame. 1993. Nitrogen mineralization potential in soils. pp. 341-357. *In* Martin R. Carter (ed.), Soil sampling and methods of analysis. Canadian Society of Soil Science. Lewis Publishers, Boca Raton, FL.

Canadian National Committee on Irrigation and Drainage (CANCID). 1999. Annual report.

Cannell, R.Q., and J.D. Hawes. 1994. Trends in tillage practices in relation to sustainable crop production with special reference to temperate climates. Soil and Tillage Res. 30:245-282.

Carlson, S.J., and W.W. Donald. 1986. A washer for removing thickened roots from soils. Weed Sci. 34:794-799.

Carter, D.L. 1986. Effects of erosion on soil productivity. pp. 1131-1138. *In* Proceedings, Water Forum '86. Long Beach, CA. Aug. 4-6, 1986. Vol. 2. American Society of Civil Engineers, New York, NY.

Carter, D.L., R.B. Berg, and B.J. Sanders. 1985. The effect of furrow irrigation erosion on crop productivity. Soil Sci. Soc. Am. J. 49:207-211.

Carter, D.L., M.M. Mortland, and W.D. Kemper. 1986. Specific surface. *In* A. Klute (ed.), Methods of soil analysis. Part 1. Physical and mineralogical methods. 2nd ed. Agron. Monogr. 9. SSSA, Madison, WI.

Carter, L.M., and J.R. Tavernetti. 1968. Influence of precision tillage and soil compaction on cotton yields. Trans. of ASAE 11(1):66-68, 73.

Carter, M., and S.P. Bentley. 1991. Correlations of soil properties. Pentech Press, London.

Casagrande, A. 1934. Die Aräometer-Methode zur Bestimmung de Kornverteilung von Böden und anderen Materialen. Julius Springer, Berlin.

Cassel, D.K., and D.R. Nielsen. 1986. Field capacity and available water capacity. *In* A. Klute (ed.), Methods of soil analysis. Part 1. Physical and mineralogical properties. 2nd ed. Agron. Monogr. 9:901-926. SSSA, Madison, WI.

Chadwick, O.A., and W.D. Nettleton. 1990. Micromorphologic evidence of adhesive and cohesive forces in soil cementation. pp. 207-212. *In* L.A. Douglas (ed.), Developments in Soil Science. Vol. 19. Soil micro-morphology: A basic and applied science. Elsevier, New York, NY.

Chancellor, W.J. 1976. Compaction of soil by agricultural equipment. Bull. 1881, Agricultural Sciences Division, University of California, Davis.

Chang, J.H., and M.L. Jackson. 1957. Fractionation of soil phosphorus. Soil Sci. 84:133-144.

Chapman, S.L., and M.E. Horn. 1968. Parent material uniformity and origin of silty soils in northwest Arkansas based on zirconium-titanium contents. Soil Sci. Soc. Am. Proc. 32:265-271.

Chen, M., L.Q. Ma, and W.G. Harris. 1999. Baseline concentrations of 15 trace elements in Florida surface soils. J. Environ. Qual. 28:1173-1181.

Chen, M., L.Q. Ma, and W.G. Harris. 2002. Arsenic concentrations in Florida surface soils: Influence of soil type and properties. Soil Sci. Soc. Am. J. 66:632-640.

Chen, T., Z. Huang, Y. Huang, and M. Lei. 2004. Distribution of arsenic and essential elements in pinna of arsenic hyperaccumulator *Pteris vittata* L. Science in China Ser. C. Life Sci. 48(1):13-19.

Chen, Y., J. Tarchitzky, J. Brouwer, J. Morin, and A. Basin. 1980. Scanning electron microscope observations on soil crusts and their formation. Soil Sci. 130:49-55.

Childs, C.W., R.L. Parfitt, and R. Lee. 1983. Movement of aluminum as an inorganic complex in some podsolized soils. New Zealand. Geoderma 29:139-155.

Childs, S., and A.L. Flint. 1990. Physical properties of forest soils containing rock fragments. pp. 95-121. *In* S.P. Gessel, D.S. Lacate, G.F. Weetman, and R.F. Power (eds.), Sustained productivity of forest soils. Proceedings of the 7th North American Forest Soils Conference, University of British Columbia, Vancouver, BC.

Chlopecka, A., J.R. Bacon, M.J. Wilson, and J. Kay. 1996. Forms of cadmium, lead, and zinc in contaminated soils from southwest Poland. J. Environ. Qual. 25:69-79.

Chlou, C.T., and D.W. Rutherford. 1993. Sorption of N2 and EGME vapors on some soils, clays, and mineral oxides and determination of sample surface areas by use of sorption data. Environ. Sci. Technol. 27:1587-1594.

Churchman, G.J., J.S. Whitton, G.G.C. Claridge, and B.K.G. Theng. 1984. Intercalation method using formamide for differentiating halloysite from kaolinite. Clays and Clay Minerals 32:241-248.

Claff, S.R., L.A. Sullivan, E.D. Burton, and R.T. Bush. 2010. A sequential extraction procedure for acid sulfate soils: Partitioning of iron. Geoderma 155:224-230.

Clark, B.R. 1970. Mechanical formation of preferred orientation in clays. Am. J. Sci. 269:250-266.

Cleland, D.T., J.B. Hart, K.S. Pregitzer, and C.W. Ramm. 1985. Classifying oak ecosystems for management. *In* J.E. Johnson (ed.), Proceedings, Challenges for oak management and utilization, March 28-29, 1985. University of Wisconsin, Madison, WI.

Codling, E.E., and J.C. Ritchie. 2005. Eastern gamagrass uptake of lead and arsenic from lead arsenate contaminated soil amended with lime and phosphorus. Soil Sci. 170(6):413-424.

Coleman, T.R., E.J. Kamprath, and S.B. Weed. 1958. Liming. Adv. Agron. 10:475.

Cong, Tu, and L.Q. Ma. 2002. Arsenic accumulation in the hyperaccumulator Chinese brake and its utilization potential for phytoremediation. J. Environ. Qual. 31:1671-1675.

Cook, R.L., and B.G. Ellis. 1987. Soil management: A world view of conservation and production. John Wiley & Sons Inc., New York, NY.

Cooke, M. 1936. Report of Great Plains Drought Area Committee to President Roosevelt, August 27, 1936. Franklin D. Roosevelt Library, Hopkins Papers.

Corey, R.B. 1987. Soil test procedures: Correlation. pp. 15-22. *In* J.R. Brown (ed.), Soil testing: Sampling, correlation, calibration, and interpretation. Soil Sci. Soc. Am. Spec. Publ. 21, SSSA, Madison, WI.

Corwin, D.L. 2007. Salinity measurement, monitoring, and mapping. *In* USDA Proceedings of International Workshop to Improve Agricultural Water Management in Iraq. Amman, Jordan, Aug. 5-9, 2007.

Council for Agricultural Science and Technology (CAST). 1996. Future of irrigated agriculture. Task Force Rep. 127. Ames, IA.

Coutinet, S. 1965. Methodes d'analyse utilisables pour les sols sales, calcaires et gypseux. Agronomie Tropicales, Paris 12:1242-1253.

Cradwick, P.D.G., V.C. Farmer, J.D. Russell, C.R. Masson, K. Wada, and N. Yoshinaga. 1972. Imogolite, a hydrated aluminum silicate of tubular structure. Nature Phys. Sci. 240:187-189.

Cresser, M., K. Killham, and T. Edwards. 1993. Soil chemistry and its applications. Cambridge University Press, Great Britain.

Crock, M.J. 1996. Mercury. pp. 769-791. *In* D.L. Sparks (ed.), Methods of soil analysis. Part 3. Chemical methods. Soil Sci. Soc. Am. Book Series 5. ASA and SSSA, Madison, WI.

Cross, A.F., and W.H. Schlesinger. 1995. A literature review and evaluation of Hedley fractionation: Applications to the biogeochemical cycle of soil phosphorus in natural ecosystems. Geoderma 64:197-214.

Crosson, L.S., and R. Protz. 1974. Quantitative comparison of two closely related soil mapping units. Can. J. Soil Sci. 54:7-14.

Crosson, P. 1985. Impact of erosion on land productivity and water quality in the United States. pp. 217-236. *In* S.A. El-Swaify, W.C. Moldenhauer, and A. Lo (eds.), Soil erosion and conservation. Soil Conservation Society of America, Ankeny, IA.

Dahlgren, R., S. Shoji, and M. Nanzyo. 1993. Mineralogical characteristics of volcanic ash soils. pp. 101-143. *In* Developments in Soil Science. Vol. 21. Volcanic ash soils. Genesis, properties, and utilization. Elsevier, Amsterdam, The Netherlands.

Dan, J., D.H. Yaalon, R. Moshe, and S. Nissim. 1982. Evolution of Reg soils in southern Israel and Sinai. Geoderma 28:173-202.

Dane, J.H., and J. W. Hopmans. 2002. Water retention and storage: Laboratory. pp. 675-720. *In* J.H. Dane and G.C. Topp (eds.), Methods of soil analysis. Part 4. Physical methods. Soil Sci. Soc. Am. Book Series 5. ASA and SSSA, Madison, WI.

Daniels, R.B., E.E. Gamble, and J.G. Cady. 1970. Some relationships among Coastal Plains soils and geomorphic surfaces in North Carolina. Soil. Sci. Soc. Am. Proc. 34:648-653.

Darmondy, R.G., J.C. Marlin, J. Talbott, R.A. Green, E.F. Brewer, and C. Stohr. 2004. Dredged Illinois River sediments: Plant growth and metal uptake. J. Environ. Qual. 33:458-464.

Davies, B.E. 1980. Trace element pollution. pp. 287-351. *In* B.E. Davies (ed.), Applied soil trace elements. John Wiley & Sons Inc., New York, NY.

Davis, F.L. 1936. A study of the uniformity of soil types and of the fundamental differences between the different series. Ala. Agric. Exp. Bull. 244.

Davis, R.O., and H. Bryan. 1910. The electrical bridge for the determination of soluble salts in soils. USDA, Bur. Soils Bull. 61.

Day, P.R. 1956. Report of the committee on physical analysis, 1954-1955. Soil Sci. Soc. Am. Proc. 20:167-169.

Day, P.R. 1965. Particle fractionation and particle-size analysis. *In* C.A. Black, D.D. Evans, J.L. White, L.E. Ensminger, and F.E. Clark (eds.), Methods of soil analysis. Part 1. Physical and mineralogical properties, including statistics of measurement and sampling. 1st ed. Agron. 9:545-567.

D'Amore, J.J., S.R. Al-Abed, K.G. Scheckel, and J.A. Ryan. 2005. Methods for speciation of metals in soils. J. Environ. Qual. 34:1707-1745.

De Bivort, L.H. 1975. World agricultural development strategy and the environment. Agric. Environ. 2:1-14.

De Connick, F. 1980. Major mechanisms in formation of spodic horizons. Geoderma 24:101-128.

De Gryze, S., J. Six, C. Brits, and R. Merckx. 2005. A quantification of short-term macroaggregate dynamics: Influences of wheat residue input and texture. Soil Biology & Biochemistry 37:55-66. Elsevier.

de Jong, E. 1999. Comparison of three methods of measuring surface area of soils. Can. J. Soil Sci. 79:345-351.

Deer, W.A., R.A. Howie, and J. Zussman. 1992. An introduction to the rock-forming minerals. 2nd ed. Longman Scientific and Technical, Essex, England.

Del Val, C., J.M. Barea, and C. Azon-Aguilar. 1999. Diversity of arbuscular mycorrhizal fungus populations in heavy-metal contaminated soils. Appl. Environ. Microbiol. 65:718-723.

Delvigne, J.E. 1998. Atlas of micromorphology of mineral alteration and weathering. Canadian Mineralogist Spec. Publ. 3. Mineralogical Association of Canada, Ontario.

Derici, M.R. 2002. Degradation, chemical. pp. 268-271. *In* R. Lal (ed.), Encyclopedia of soil science. Marcel Dekker, Inc.

DeYoung, W. 1925. The relation of specific surface to moisture equivalent of soils. M.S. thesis, University of Missouri.

Diaz-Zorita, M., E. Perfect, and J.H. Grove. 2002. Disruptive methods for assessing soil structure. Soil and Tillage Res. 64:3-22. Elsevier.

Doering, O.C. 1997. An overview of conservation and agricultural policy: Questions from the past and observations about the present. Agriculture and Conservation Policies, A Workshop in Honor of A. Berg.

Doner, H.E., and W.C. Lynn. 1989. Carbonate, halide, sulfate, and sulfide minerals. pp. 279-330. *In* J.B. Dixon and S.B. Weed (eds.), Minerals in soil environments. 2nd ed. Soil Sci. Soc. Am. Book Series 1. SSSA, Madison, WI.

Doran, J.W., and T.B. Parkin. 1994. Defining and assessing soil quality. pp. 3-21. *In* J.W. Doran, D.C. Coleman, D.F. Bezdicek, and B.A. Stewart (eds.), Defining soil quality for a sustainable environment. SSSA, Madison, WI.

Drake, E.H., and H.L. Motto. 1982. An analysis of the effect of clay and organic matter content on the cation exchange capacity of New Jersey soils. Soil Sci. 133:282-288.

Drees, L.R., and M.D. Ransom. 1994. Light microscope techniques in quantitative mineralogy. pp.137-176. *In* J.E. Amonette and L.W. Zelazny (eds.), Quantitative methods in soil mineralogy. Soil Sci. Soc. Am. Misc. Publ. SSSA, Madison, WI.

Dregne, H.E. 1998. Desertification assessment. *In* R. Lal, W.H. Blum, C. Valentine, and B.A. Stewart (eds.), Methods for assessment of soil degradation. Advances in Soil Science. CRC Press.

Dreimanis, A. 1962. Quantitative gasometric determination of calcite and dolomite by using Chittick apparatus. J. Sediment. Petrol. 32:520-529.

Driscoll, R.S. 1984. An ecological land classification framework for the United States. USDA, Forest Service Misc. Publ. 1439.

Drosdoff, M., and E.F. Miles. 1938. Action of hydrogen peroxide on weathered mica. Soil Sci. 46:391-395.

Dubbin, W.E., A.R. Mermut, and H.P.W. Rostad. 1993. Clay mineralogy of parent material derived from uppermost Cretaceous and Tertiary sedimentary rocks in southern Saskatchewan. Can. J. Soil Sci. 73:447-457.

Duggar, J.F., and M.J. Funchess. 1911. Lime for Alabama soils. Alabama Agric. Exp. Stn. Bull. 161.

Dumas, J.B.A. 1831. Procédés de l'analyse organique. Ann. Chim. Phys. 247:198-213.

Durrell, Cordell. 1948. A key to common rock-forming minerals in thin section. W.H. Freeman and Co., San Francisco, CA.

Dyer, B. 1894. On the analytical determination of probable mineral plant food in soils. Trans. Chem. Soc. 65:115-167.

Edil, T.B., and R.J. Krizek. 1976. Influence of fabric and soil suction on the mechanical behavior of a kaolinitic clay. Geoderma 15:323-341.

Edwards, F.E. 1966. Cotton seedling emergence. Mississippi Farm Res. 29:4-5.

Egawa, T., A. Sato, and T. Nishimura. 1960. Release of OH ions from clay minerals treated with various anions, with special reference to the structure and chemistry of allophane. Adv. Clay Sci. (Tokyo) 2:252-262.

Elliott, E.T., and C.A. Cambardella. 1991. Physical separation of soil organic matter. Agric. Ecosyst. Environ. 34:407-419.

Ellis, B.G. 1986. Highlights of research in Division S-2—Soil chemistry. SSSA Golden Anniversary Papers. Soil Sci. Soc. Am. J. 50:555-556.

Elrashidi, M.A., D. Hammer, C.A. Seybold, R.J. Engel, R. Burt, and P. Jones. 2007. Application of equivalent gypsum content to estimate potential subsidence of gypsiferous soils. Soil Sci. 172:209-224.

Elrashidi, M.A., M.D. Mays, and C.W. Lee. 2003. Assessment of Mehlich3 and ammonium bicarbonate–DTPA extraction for simultaneous measurement of fifteen elements in soils. Commun. Soil Sci. Plant Anal. 34:2817-2838.

El-Swaify, S.A. 1980. Physical and mechanical properties of Oxisols. pp. 303-324. *In* B.K.G. Theng (ed.), Soils with variable charge. New Zealand Society of Soil Science, Lower Hutt, New Zealand.

El-Swaify, S.A., T.S. Walker, and S.M. Virmani. 1984. Dryland management alternatives and research needs for Alfisols in the semi-arid tropics. International Crop Research Institute for the Semi-Arid Tropics. Patancheru P.O., Andhra Pradesh, India.

Espinoza, W., R.H. Rust, and R.S. Adams, Jr. 1975. Characterization of mineral forms in Andepts from Chile. Soil Sci. Soc. Am. Proc. 39:556-561.

Eswaran, H. 1983. Characterization of domains with the scanning electron microscope. Pedology 33:41-54.

Eswaran, H., R. Lal, and P.F. Reich. 2001. Land degradation: An overview. *In* E.M. Bridges, I.D. Hannam, L.E. Olderman, F.W.T. Pening de Vries, S.J. Scherr, and S. Sompatpanit (eds.), Responses to land degradation. Proceedings, 2nd International Conference on Land Degradation and Desertification, Khon Kaen, Thailand. Oxford Press, New Delhi, India.

Eswaran, H., and G. Zi-Tong. 1991. Properties, genesis, classification, and distribution of soils with gypsum. pp. 89-120. *In* W.D. Nettleton (ed.), Occurrence, characteristics, and genesis of carbonate, gypsum, and silica accumulations in soils. Soil Sci. Soc. Am. Spec. Publ. 26. SSSA, Madison, WI.

Evangelou, V.P., L.D. Whittig, and K.K. Tanji. 1984. An automated manometric method for quantitative determination of calcite and dolomite. Soil Sci. Soc. Am. J. 48:1236-1239.

Evans, C.E., and E.J. Kamprath. 1970. Lime response as related to percent Al saturation, solution Al, and organic matter content. Soil Sci. Soc. Am. Proc. 34:893-955.

Fanning, D.S., V.Z. Keramidas, and M.A. El-Desoky. 1989. Micas. pp. 551-634. *In* J.B. Dixon and S.B. Weed (eds.), Minerals in soil environments. 2nd ed. Soil Sci. Soc. Am. Book Series 1. SSSA, Madison, WI.

Farmer, V.C., J.D. Russell, and M.L. Berrow. 1980. Imogolite and proto-imogolite in spodic horizons: Evidence for a mobile aluminum silicate complex in podzol formation. J. Soil Sci. 31:673-684.

Federal Register. 1996. Pesticides and ground water state management plan regulation: Proposed rule. Fed. Regist. 61(124):33259–33301.

Fendorf, Scott, M.J. La Force, and L. Guangchao. 2004. Temporal changes in soil partitioning and bioaccessibility of arsenic, chromium, and lead. J. Environ. Qual. 33:2049–2055.

Fenneman, N.M. 1931. Physiography of the western United States. McGraw-Hill Co., New York, NY.

Fenneman, N.M. 1938. Physiography of the eastern United States. McGraw-Hill Co., New York, NY.

Fenneman, N.M. 1946 (reprinted 1957). Physical divisions of the United States. U.S. Geological Survey, map (1 sheet), 1:7,000,000. GPO, Washington, DC.

Fey, M.V., and J. LeRoux. 1976. Quantitative determinations of allophane in soil clays. Proc. Int. Clay Conf. 1975 (Mexico City). 5:451-463. Applied Publishing Ltd., Wilmette, IL.

Fields, M., and K.W. Perrott. 1966. The nature of allophane in soils. Part 3. Rapid field and laboratory test for allophane. N.Z. J. Sci. 9:623-629.

Fio, J.L., R. Fujii, and S.J. Deverel. 1991. Selenium mobility and distribution in irrigated and nonirrigated alluvial soils. Soil. Sci. Soc. Am. J. 55:1313-1320.

FitzPatrick, E.A. 1984. Micromorphology of soils. Chapman and Hall, Ltd., London.

FitzPatrick, E.A. 1993. Soil microscopy and micromorphology. John Wiley & Sons Inc., New York, NY.

Flach, K.W. 1960. Sols bruns acides in the northeastern United States. Genesis, morphology, and relationships to associated soils. Ph.D. thesis, Cornell University, Ithaca, NY.

Flanagan, D.C. 2002. Erosion. pp.395-398. *In* R. Lal (ed.), Encyclopedia of soil science. Marcel Dekker, Inc.

Flanagan, D.C., and M.A. Nearing (eds.). 1995. USDA water erosion prediction project (WEPP) hillslope profile and watershed model documentation, NSERL Report 10. National Erosion Research Laboratory, USDA/ARS, West Lafayette, IN.

Flint, A.L., and S. Childs. 1984. Physical properties of rock fragments and their effect on available water in skeletal soils. pp. 91-103. *In* J.E. Box, Jr. (ed.), Erosion and productivity of soils containing rock fragments. Soil Sci. Soc. Am. Spec. Publ. 13. SSSA, Madison, WI.

Flint, A.L., and L.E. Flint. 2002. Particle density. pp. 229-240. *In* J.H. Dane and G.C. Topp (eds.), Methods of soil analysis. Part 4. Physical methods. Soil Sci. Soc. Am. Book Series 5. ASA and SSSA, Madison, WI.

Follet, E.A.C., W.J. McHardy, B.D. Mitchell, and B.F.L. Smith. 1965. Chemical dissolution techniques in the study of soil clays. I and II. Clay Miner. 6:23-43.

Follett, R.F., and E. Pruessner. 1997. POM procedure. USDA/ARS, Ft. Collins, CO.

Forstner, U. 1995. Land contamination by metals: Global scale and magnitude of problem. pp. 1-33. *In* H.E. Allen, C.P. Huang, G.W. Bailey, and A.R. Bowers (eds.), Metal speciation and contamination in soil. Lewis Publishers, Boca Raton, FL.

Foth, H.D. 1984. Fundamentals of soil science. 7th ed. John Wiley & Sons Inc., New York, NY.

Foth, H.D., and B.G. Ellis. 1988. Soil fertility. 1st ed. John Wiley & Sons Inc., New York, NY.

Foth, H.D., and B.G. Ellis. 1997. Soil fertility. 2nd ed. John Wiley & Sons Inc., New York, NY.

Fox, C.A., R.K. Guertin, E. Dickson, S. Sweeney, R. Protz, and A.R. Mermut. 1993. Micromorphology methodology for inorganic soils. pp. 683-709. *In* M.R. Carter (ed.), Soil sampling and methods of analysis. Lewis Publishers, Boca Raton, FL.

Fox, C.A., and L.E. Parent. 1993. Micromorphology methodology for organic soils. pp. 473-485. *In* M.R. Carter (ed.), Soil sampling and methods of analysis. Lewis Publishers, Boca Raton, FL.

Foy, C.D. 1973. Manganese and plants. *In* Manganese. National Academy of Science, National Research Council, Washington, DC.

Foy, C.D. 1974. Effects of aluminum on plant growth. pp. 601-642. *In* E.W. Carson (ed.), The plant root and its environment. University Press of Virginia, Charlottesville.

Foy, C.D. 1984. Physiological effects of hydrogen, aluminum, and manganese toxicities in acid soil. *In* F. Adams (ed.), Soil acidity and liming. 2nd ed. Agron. 12:57-97.

Foy, C.D., R.L. Chaney, and M.C. White. 1978. The physiology of metal toxicity in plants. Ann. Rev. Plant Physiol. 29:511-566.

Foy, C.D., and A.L. Fleming. 1978. The physiology of plant tolerance to excess available aluminum and manganese in acid soils. pp. 301-328. *In* G.A. Jung (ed.), Crop tolerance to suboptimal land conditions. ASA Spec. Publ. 32, ASA, Madison, WI.

Franks, C.D. Separating roots from the soil by hand-sieving. 1998. USDA/NRCS, National Soil Survey Center, Lincoln, Nebraska. Available online at ftp://ftp-fc.sc.egov.usda.gov/NSSC/Analytical_Soils/sep_root.pdf (verified January 14, 2011).

Franks, C.D., J.M. Kimble, S.E. Samson-Liebig, and T.M. Sobecki. 2001. Organic carbon methods, microbial biomass, root biomass, and sampling design under development by NRCS. pp.105-113. *In* R. Lal, J.M. Kimble, R.F. Follett, and B.A. Stewart (eds.), Assessment methods for soil carbon. CRC Press, Boca Raton, FL.

Franzluebbers, A.J., F.M. Hons, and D.A. Zuberer. 1994. Long-term changes in soil carbon and nitrogen pools in wheat management systems. Soil Sci. Soc. Am. J. 58:1639-1645.

Franzmeier, D.P., and S.J. Ross, Jr. 1968. Soil swelling: Laboratory measurement and relation to other soil properties. Soil Sci. Soc. Am. Proc. 32:573-577.

Frazier, B.E., D.K. McCool, and C.F. Engle. 1983. Soil erosion in the Palouse: An aerial perspective. J. Soil Water Conserv. 38(2):70-74.

Freundlich, H. 1926. Colloid and capillary chemistry. Methuen & Co., London.

Fribourg, H.A. 1953. A rapid method for washing roots. Agron. J. 45:334-335.

Friedel, B. 1978. Zur Bestimmung von Gips in Böden. Zeitschrift für Pflanzenernährung und Bodenkunde 141:231-239.

Friedman, G.M. 1959. Identification of carbonate minerals by staining methods. J. Sedimentary Petrology 29:87-97.

Frossard, E., C. Feller, H. Tiessen, J.W.B. Stewart, J.C. Fardeau, and J.L. Morel. 1993. Can an isotopic method allow for the determination of the phosphate-fixing capacity of soils? Commun. Soil Sci. Plant Anal. 24:367-377.

Furr, J.A., and J.O. Reeve. 1945. The range of soil moisture percentages through which plants undergo permanent wilting in some soils from semi-arid, irrigated areas. J. Agric. Res. 71:149-170.

Gabriels, D., R. Horn, M.M. Villagra, and R. Hartmann. 1998. Assessment, prevention, and rehabilitation of soil structure caused by soil surface sealing, crusting, and compaction. pp. 129-165. *In* R. Lal, W.H. Blum, C. Valentine, and B.A. Stewart (eds.), Methods for assessment of soil degradation. Advances in Soil Science. CRC Press.

Gallivan, G.J., G.A. Surgeoner, and J. Kovach. 2001. Pesticide risk reduction on crops in the province of Ontario. J. Environ. Qual. 30:798-813.

Gamble, E.E., R.B. Daniels, and W.D. Nettleton. 1970. Geomorphic surfaces and soils in the Black Creek Valley, Johnston County, North Carolina. Soil Sci. Soc. Am. Proc. 34:276-281.

Gambrell, R.P. 1994. Trace and toxic metals in wetlands–A review. J. Environ. Qual. 23:883-891.

Gambrell, R.P. 1996. Manganese. pp. 665-682. *In* D.L. Sparks (ed.), Methods of soil analysis. Part 3. Chemical methods. Soil Sci. Soc. Am. Book Series 5. ASA and SSSA, Madison, WI.

Gardner, W.H. 1986. Water content. pp. 493-544. *In* A. Klute (ed.), Methods of soil analysis. Part 1. Physical and mineralogical methods. 2nd ed. Agron. Monogr. 9. ASA and SSSA, Madison, WI.

Gardner, W.R., and R.H. Nieman. 1964. Lower limit of water availability to plants. Science 143:1460-1462.

Gartley, K.L., and J.T. Sims. 1994. Phosphorus soil testing: Environmental uses and implications. Commun. Soil Sci. Plant Anal. 25:1565-1582.

Gebhardt, H., and N.T. Coleman. 1984. Anion adsorption of allophanic tropical soils: III. Phosphate adsorption. pp. 237-248. *In* K.H. Tan (ed.), Anodosols. Benchmark Papers in Soil Science Series. Van Nostrand Reinhold Co., Melbourne, Australia.

Gee, G.W., and J.W. Bauder. 1979. Particle size analysis by hydrometer: A simplified method for routine textural analysis and a sensitivity test of measurement parameters. Soil Sci. Soc. Am. J. 43:1004-1007.

Gee, G.W., and J.W. Bauder. 1986. Particle-size analysis. pp. 383-411. *In* A. Klute (ed.), Methods of soil analysis. Part 1. Physical and mineralogical methods. 2nd ed. Agron. Monogr. 9. ASA and SSSA, Madison, WI.

Gee, G.W., and D. Or. 2002. Particle-size analysis. pp. 255-293. *In* J.H. Dane and G.C. Topp (eds.), Methods of soil analysis. Part 4. Physical methods. Soil Sci. Soc. Am. Book Series 5. ASA and SSSA, Madison, WI.

Geiger, A.F. 1957. Sediment yields from small watersheds in the United States. Wallingford, UK.

Germida, J.J. 1993. Cultural methods for soil microorganisms. pp. 263-275. *In* Martin R. Carter (ed.), Soil sampling and methods of analysis. Canadian Society of Soil Science. Lewis Publishers, Boca Raton, FL.

Gile, L.H., J.W. Hawley, and R.B. Grossman. 1981. Soils and geomorphology in the basin and range area of southern New Mexico—Guidebook to the desert project. N.M. Bur. Mines and Miner. Resour. Mem. 39.

Gilkes, R.J., and J.C. Hughes. 1994. Sodium fluoride pH of South-Western Australian soils as an indicator of P-sorption.

Gill, W.R. 1971. Economic assessment of soil compaction. ASAE Monograph. St. Joseph, MI.

Gillman, G.P. 1973. Influence of net charge on water dispersible clay and sorbed sulfate. Aust. J. Soil Res. 1:173-176.

Gilroy, G. 1928. The compressibility of sand-mica mixtures. Proc. Am. Soc. Civ. Eng. 54:555-568.

Glantz, Michael (ed.). 1989. Forecasting by analogy: Societal responses to regional climatic change. Summary Report, Environmental and Societal Impacts Group NCAR. Summary based on a study by Donald A. Wilhite, Center for Agricultural Meteorology and Climatology, University of Nebraska, Lincoln, NE.

Goldberg, S., D.A. Martens, H.S. Forster, and M.J. Herbel. 2006. Speciation of selenium (IV) and selenium (VI) using coupled ion chromatography-hydride generation atomic absorption spectrometry. Soil Sci. Soc. Am. J. 70:41-47.

Goldberg, S., and G. Sposito. 1984. A chemical model of phosphate adsorption by soils. II. Noncalcareous soils. Soil Sci. Soc. Am. J. 48:779-783.

Goodwin, W.A. 1965. Bearing capacity. *In* C.A. Black (ed.), Methods of soil analysis. Part 1. Agron. 9:485-498. ASA, Madison, WI.

Grattan, S.R., and C.M. Grieve. 1999. Salinity-mineral nutrient relations in horticultural crops. Sci. Hort. 78:127-157.

Greene-Kelly, R. 1974. Shrinkage of clay soils: A statistical correlation with other soil properties. Geoderma 11:243-257.

Gregorich, E.G. 2002. Quality. pp. 1058-1061. *In* R. Lal (ed.), Encyclopedia of soil science. Marcel Dekker, Inc.

Gregorich, E.G., M.R. Carter, D.A. Angers, C.M. Monreal, and B.H. Ellert. 1994. Towards a minimum data set to assess soil organic matter quality in agricultural soil. Can. J. Soil Sci. 74:367-385.

Gregorich, E.G., M.R. Carter, J.W. Doran, C.E. Pankhurst, and L.M. Dwyer. 1997. Biological attributes of soil quality. pp. 81-113. *In* E.G. Gregorich and M.R. Carter (eds.), Soil quality for crop production and ecosystem health. Elsevier, New York, NY.

Gregorich, E.G., R.G. Kachanoski, and R.P. Voroney. 1988. Ultrasonic dispersion of aggregates: Distribution of organic matter in size fractions. Can. J. Soil. Sci. 68:395-403.

Grossman, R.B., B.R. Brasher, D.P. Franzmeier, and J.L. Walker. 1968. Linear extensibility as calculated from natural-clod bulk density measurements. Soil Sci. Soc. Am. Proc. 32:570-573.

Grossman, R.B., and J.L. Millet. 1961. Carbonate removal from soils by a modification of the acetate buffer method. Soil Sci. Soc. Am. Proc. 25:325-326.

Grossman, R.B., F.B. Pringle, R.J. Bigler, W.D. Broderson, and T.M. Sobecki. 1990. Systematics and field morphology for a use-dependent, temporal soil data base for edaphological and environmental applications. pp. 452-460. *In* International Symposium on Advanced Technology in Natural Resource Management, 2nd ed. Georgetown University, Washington, DC.

Grossman, R.B., and T.G. Reinsch. 2002. Bulk density and linear extensibility. pp. 201-228. *In* J.H. Dane and G.C. Topp (eds.), Methods of soil analysis. Part 4. Physical methods. Soil Sci. Soc. Am. Book Series 5. ASA and SSSA, Madison, WI.

Grossman, R.B., T. Sobecki, and P. Schoeneberger. 1994. Soil interpretations for resource soil scientists. Soil water (technical workbook). USDA/NRCS, National Soil Survey Center, Lincoln, NE.

Gunnison Basin & Grand Valley Selenium Task Forces. Web page. http://www.seleniumtaskforce.org/ (verified January 14, 2011).

Guzel, N., and H. Ibrikci. 1994. Distribution and fractionation of soil phosphorus in particle-size separates by soils of western Turkey. Commun. Soil Sci. Plant Anal. 25:2945-2958.

Hagan, L.J.A. 1991. Wind erosion prediction system to meet user needs J. Soil Water Conserv. 46:106-111.

Hakansson, I., and W.B. Voorhees. 1998. Soil compaction. pp.167-179. *In* R. Lal, W.H. Blum, C. Valentine, and B.A. Stewart (eds.), Methods for assessment of soil degradation. Advances in Soil Science. CRC Press.

Hall, A., A.C. Duarte, M.T. Caldeeira, and M.F. Lucas. 1987. Sources and sinks of mercury in the coastal lagoon of Aveiro, Portugal. Sci. Total Environ. 64:75-87.

Hall, D.G.M., M.J. Reeve, A.J. Thomasson, and V.F. Wright. 1977. Water retention, porosity, and density of field soils. Soil Survey Tech. Monogr. 9:1-67. Rothamsted Experimental Station, Harpenden, England.

Hamad, M.E., D.L. Rimmer, and J.K. Syers. 1992. Effect of iron oxide on phosphate sorption by calcite and calcareous soils. J. Soil Sci. 43:273-281.

Hamel, S., J. Heckman, and S. Murphy. 2010. Lead contaminated soil: Minimizing health risks. Rutgers NJAES Coop. Exten. Publ. FS336. Rutgers University.

Harradine, F.F. 1949. The variability of soil properties in relation to stage of profile development. Soil Sci. Soc. Am. Proc. 14:302-311.

Harris, W.G., J.C. Parker, and L.W. Zelazny. 1984a. Effects of mica content on engineering properties of sand. Soil Sci. Soc. Am. J. 48:501-505.

Harris, W.G., and G.N. White. 2008. X-ray diffraction techniques for soil mineral identification. pp. 81-115. *In* A.L. Ulery and R. Drees (eds.), Methods of soil analysis. Part 5. Mineralogical methods. SSSA, Madison, WI.

Harris, W.G., and L.W. Zelazny. 1985. Criteria assessment for micaceous and illitic classes in soil taxonomy. pp. 147-160. *In* J.A. Kittrick (ed.), Mineral classification of soils. SSSA Spec. Publ. 16. Madison, WI.

Harris, W.G., L.W. Zelazny, J.C. Parker, J.C. Baker, R.S. Weber, and J.H. Elder. 1984b. Engineering properties of soils as related to mineralogy and particle-size distribution. Soil Sci. Soc. Am. J. 48:978-982.

Harter, R.D. 1969. Phosphorus adsorption sites in soils. Soil Sci. Soc. Am. J. 33:630-632.

Hartgrove, N.T., D.F. Clendenon, and J.T. Ammons. 2006. Influence of long-term agriculture on properties of some Tennessee soils. Soil Surv. Hor. 47:51-55.

Harwood, R.R., M.A. Cavigelli, S.R. Deming, L.A. Frost, and L.K. Probyn (eds.). 1998. Michigan field crop ecology: Managing biological processes for productivity and environmental quality. Michigan State Univ. Ext. Bull. E-2646.

Hashimoto, I., and M.L. Jackson. 1960. Rapid dissolution of allophane and kaolinite-halloysite after dehydration. Clays and Clay Minerals 7:102-113.

Hassink, J. 1995. Decomposition rate constants of size and density fractions of soil organic matter. Soil Sci. Soc. Am. J. 59:1631-1635.

Havlin, J.L., D.E. Kissel, L.D. Maddux, M.M. Claasen, and J.H. Long. 1990. Crop rotation and tillage effects on soil organic carbon and nitrogen. Soil Sci. Soc. Am. J. 54:448-452.

Haynes, H.R. 1982. Effects of liming on phosphate availability in acid soils. Plant and Soil 63:289-308.

Haynes, R.J., and G.S. Francis. 1993. Changes in microbial biomass C, soil carbohydrate composition and aggregate stability induced by growth of selected crop and forage species under field conditions. J. Soil Sci. 44:665-675.

Heady, H.F., and D. Child. 1994. Rangeland ecology and management. Westview Press. Boulder, CO.

Heckrath, G., P.C. Brookes, P.R. Poulton, and K.W.T. Goulding. 1995. Phosphorus leaching from soils containing different phosphorus concentrations in the Broadbalk Experiment. J. Environ. Qual. 24:904-910.

Hedley, M.J., J.B. Stewart, and B.S. Chauhan. 1982. Changes in inorganic and organic soil phosphorus fractions induced by cultivation practices and by laboratory incubation. Soil Sci. Soc. Am. J. 46:970-976.

Helling, C.S., G. Chesters, and R.B. Corey. 1964. Contribution of organic matter and clay to soil cation-exchange capacity as affected by the pH of the saturating solution. Soil Sci. Soc. Am. J. 28:517-520.

Helms, G. 1990. Conserving the plains: The Soil Conservation Service in the Great Plains. Agricultural History 64:58-73.

Herreo, J., and J. Porta. 2000. The terminology and the concepts of gypsum-rich soils. Geoderma 96:47-61.

Hesse, P.R. 1974. Methods of soil analysis—Texture analysis of gypsic soils. The Euphrates pilot irrigation project. FAO No. AGON/SF/SYR/67/522.

Heuscher, S.A., C.C. Brandt, and P.M. Jardine. 2005. Using soil physical and chemical properties to estimate bulk density. Soil Sci. Soc. Am. J. 69:1-7.

Higashi, T., and H. Ikeda. 1974. Dissolution of allophane by acid oxalate solution. Clay Sci. 4:205-212.

Higashi, T., and K. Wada. 1977. Size fractionation, dissolution analysis and infrared spectroscopy of humus complexes in Ando soils. J. Soil Sci. 28:653-663.

Hills, G.A. 1952. The classification and evaluation of site for forestry. Ontario Department of Lands and Forests. Resource Div. Report 24.

Hodges, S.C., and L.W. Zelazny. 1980. Determination of noncrystalline soil components by weight difference after selective dissolution. Clays and Clay Minerals 28:35-42.

Hoffman, G.J., T.A. Howell, and K.H. Solomon. 1990. Management of farm irrigation systems. pp. 667-715. *In* G.J. Hoffman, J.D. Rhoades, J. Letey, and F. Sheng (eds.), Salinity management. ASCE. St. Joseph, MI.

Holford, I.C.R., and G.E.G. Mattingly. 1975a. Phosphate sorption by Jurassic oolitic limestones. Geoderma 13:257-264.

Holford, I.C.R., and G.E.G. Mattingly. 1975b. Surface areas of calcium carbonate in soils. Geoderma 13:247-255.

Holmgren, G.G.S. 1968. Nomographic calculation of linear extensibility in soils containing coarse fragments. Soil Sci. Soc. Am. Proc. 32:568-570.

Holmgren, G.G.S. 1973. Quantitative calcium carbonate equivalent determination in the field. Soil Sci. Soc. Am. Proc. 37:304-307.

Holmgren, G.G.S., R.L. Juve, and R.C. Geschwender. 1977. A mechanically controlled variable rate leaching device. Soil Sci. Soc. Am. J. 41:1207-1208.

Holmgren, G.G.S., M.W. Meyer, R.L. Chaney, and R.B. Daniels. 1993. Cadmium, lead, zinc, copper, and nickel in agricultural soils of the United States of America. J. Environ. Qual. 22:335-348.

Holzhey, C.S., W.D. Nettleton, and R.D. Yeck. 1974. Microfabric of some argillic horizons in udic, xeric, and torric soil environments of the United States. pp. 747-760. *In* G.K. Rutherford (ed.), Soil microscopy. Proceedings of 4th International Working-Meeting on Soil Micromorphology. Limestone Press, Kingston, Ontario, Canada.

Honna, T., S. Yamamoto, and K. Matsui. 1988. A simple procedure to determine melanic index that is useful for differentiating melanic from fulvic Andisols. Pedologist 32:69-78.

Horwath, W.R., and E.A. Paul. 1994. Microbial mass. pp. 753-771. *In* R.W. Weaver, J.S. Angle, and P.S. Bottomley (eds.), Methods of soil analysis. Part 2. Microbiological and biochemical properties. ASA and SSSA, Madison, WI.

Houba, V.J.G., Th. M. Lexmond, I. Novozamsky, and J.J. van der Lee. 1996. State of the art and future developments in soil analysis for bioavailability assessment. Sci. Total Environ. 178:21-28.

Howard, D.D., and F. Adams. 1965. Calcium requirement for penetration of subsoils by primary cotton roots. Soil Sci. Soc. Am. Proc. 29:558-562.

Howard, P.M. 1981. The water erosion problem in the United States. pp. 25-34. *In* Proceedings of the American Society of Agricultural Engineers Conference on Crop Production with Conservation in the 1980's.

Howell, T.A. 2001. Enhancing water use efficiency in irrigated agriculture. Agron. J. 93:281-289.

Huang, P.M., and R. Fujii. 1996. Selenium and arsenic. pp. 793-831. *In* D.L. Sparks (ed.), Methods of soil analysis. Part 3. Chemical analysis. Soil Sci. Soc. Am. Book Series 5. ASA and SSSA, Madison, WI.

Hughes, R.E., D.M. Moore, and H.D. Glass. 1994. Qualitative and quantitative analysis of clay minerals in soils. pp. 330-359. *In* J.E. Amonette and L.W. Zelazny (eds.), Quantitative methods in soil mineralogy. Soil Sci. Soc. Am. Misc. Publ. SSSA, Madison, WI.

Hugi, T.H. 1945. Gesteinsbildend wichtige Karbonate und deren Nachweis mittels Farbmethoden:Schweizerische Mineralog. und Petogr. Mitt. 25:114-140.

Hunt, C.B. 1967. Physiography of the United States. W.H. Freeman & Co., London, England.

Hurt, R.D. 1981. The Dust Bowl: An agricultural and social history. Nelson-Hall Inc., Chicago, IL.

Hutson, S.S., N.L. Barber, J.F. Kenny, K.S. Linsey, S.L. Lumia, and M.A. Maupin. 2004. Estimated use of water in the United States in 2000: U.S. Geological Survey Circular 1268. Available online at http://pubs.usgs.gov/circ/2004/circ1268 (verified January 14, 2011).

Ike, A.F., and J.D. Cotter. 1968. The variability of forest soils of the Georgia Blue Ridge Mountains. Soil Sci. Soc. Am. Proc. 32:284-288.

Inacio, M.M., V. Pereira, and M.S. Pinto. 1998. Mercury contamination in sandy soils surrounding an industrial emission source (Estarreja, Portugal). Geoderma 85:325-339.

Indiati, R. 2000. Addition of phosphorus to soils with low to medium phosphorus retention capacities. I. Effect of soil phosphorus sorption properties. Commun. Soil Sci. Plant Anal. 31:1179-1194.

Indorante, S.J. 2002. Sodium-affected soils in humid areas. pp. 1229-1232. *In* R. Lal (ed.), Encyclopedia of soil science. Marcel Dekker, Inc.

Innes, R.P., and D.J. Pluth. 1970. Thin section preparation using an epoxy impregnation for petrographic and electron microprobe analysis. Soil Sci. Soc. Am. Proc. 34:483-485.

Isbell, R.F. 1996. The Australian soil classification. CSIRO Publishing, Collingwood, Australia.

Islam, K.R., and R.R. Weil. 2000. Soil quality indicator properties in mid-Atlantic soils as influenced by conservation management. J. Soil Water Conserv. 55:69-78.

Jackson, M.L. 1956. Soil chemical analysis. Advanced course. University of Wisconsin, Madison.

Jackson, M.L. 1958. Soil chemical analysis. Prentice-Hall, Inc., Englewood Cliffs, NJ.

Jackson, M.L. 1969. Soil chemical analysis. Advanced course. 2nd ed. University of Wisconsin, Madison.

Jackson, M.L. 1979. Soil chemical analysis. Advanced course. 2nd ed. 11th printing. M.L. Jackson, Madison, WI.

Jackson, M.L., C.H. Lim, and L.W. Zelazny. 1986. Oxides, hydroxides, and aluminosilicates. pp. 101-150. *In* A. Klute (ed.), Methods of soil analysis. Part 1. Physical and mineralogical methods. 2nd ed. Agron. Monogr. 9. ASA and SSSA, Madison, WI.

Jackson, T.L., and G.E. Carter. 1976. Nutrient uptake by Russet Burbank potatoes as influenced by fertilization. Agron. J. 68:9-12.

Jackson, T.L., and H.M. Reisenauer. 1984. Crop response to lime in the Western United States. pp. 333-347. *In* F. Adams (ed.), Soil acidity and liming. 2nd ed. Agron. 12. ASA, CSSA, and SSSA, Madison, WI.

Janzen, H.H. 1987. Effect of fertilizer on soil productivity in long-term spring wheat rotations. Can. J. Soil Sci. 67:165-174.

Janzen, H.H., C.A. Campbell, E.G. Gregorich, and B.H. Ellert. 1998. Soil carbon dynamics in Canadian agroecosystems. pp. 57-80. *In* R. Lal, J.M. Kimble, R.F. Follett, and B.A. Stewart (eds.), Soil processes and the carbon cycle. Advances in Soil Science. CRC Press.

JCPDS. 1980. Mineral powder diffraction file—Data book. International Centre for Diffraction Data, Swarthmore, PA.

Jeanroy, E., and B. Guillet. 1981. The occurrence of suspended ferruginous particles in pyrophosphate extracts of some soil horizons. Geoderma 26:95-105.

Jenkinson, D.S., and D.S. Powlson. 1976. The effects of biocidal treatments on metabolism in soil—V.A. method for measuring soil biomass. Soil Biol. Biochem. 8:209-213.

Jenkinson, D.S., and J.H. Rayner. 1977. The turnover of soil organic matter in some of the Rothamsted classical experiments. Soil Sci. 123:298-305.

Jenny, H. 1941. Factors of soil formation. McGraw-Hill Co., New York, NY.

Jenny, H. 1980. The soil resource. Springer-Verlag, New York, NY.

Jensen, M.E., W.R. Rangeley, and P.J. Dielman. 1990. Irrigation trends in world agriculture. pp. 31-67. *In* Irrigation of agricultural crops. Am. Soc. Agron. Monogr. 30. ASA, Madison, WI.

Jermyn, M.A. 1958. Fungal cellulases. Aust. J. Biol. Sci. 11:114-126.

Jersak, J., R. Amundson, and G. Brimhall, Jr. 1997. Trace metal geochemistry in Spodosols of the Northeastern United States. J. Environ. Qual. 26:551-521.

Jim, C.Y. 1986. Experimental study of soil microfabrics induced by anistropic stresses of confined swelling and shrinking. Geoderma 37:91-112.

Joergensen, R.G., T. Mueller, and V. Wolters.1996. Total carbohydrates of the soil microbial biomass in 0.5 M K_2SO_4 soil extracts. Soil Biol. Biochem. 28:9:1147-1153.

Johnson, G.V. 1987. Sulfate: Sampling, testing, and calibration. pp. 89-96. *In* J.R. Brown (ed.), Soil testing: Sampling, correlation, calibration, and interpretation. Soil Sci. Soc. Am. Spec. Publ. 21. SSSA, Madison, WI.

Johnston, W.R., and J. Proctor. 1979. Ecological studies on the Lime Hill Serpentine, Scotland. Trans. Proc. Bot. Soc. Edinb. 43:145-150.

Jones, C.A., W.A. Inskeep, and D.R. Neuman. 1997. Arsenic transport in contaminated mine tailings following liming. J. Environ. Qual. 26:433-439.

Jones, C.A., A.N. Sharpley, and J.R. Williams. 1984. A simplified soil and plant phosphorus model. III. Testing. Soil Sci. Soc. Am. J. 48:810-813.

Jones, R.C., C.J. Babcock, and W.B. Knowlton. 2000. Estimation of the total amorphous content of Hawaii soils by the Rietveld method. Soil Sci. Soc. Am. J. 64:1100-1108.

Jones, R.C., and G. Uehara. 1973. Amorphous coatings on mineral surfaces. Soil Sci. Soc. Am. Proc. 37:792-798.

Jones, R.K. 1983. Field guide to forest ecosystem classification for the clay belt, site region 3e. Ministry of Natural Resources. Ontario, Canada.

Jordan, J.K. 1982. Application of an integrated land classification. pp. 65-82. *In* Proceedings, artificial regeneration of conifers in the Upper Lakes Region. October 26-28, 1982. Green Bay, WI.

Joseph, A.F. 1925. Clay as soil colloids. Soil Sci. 20:89-94.

Kabala, C., and B.R. Singh. 2001. Fractionation and mobility of copper, lead, and zinc in soil profiles in the vicinity of a copper smelter. J. Environ. Qual. 30:485-492.

Kabata-Pendias, A., and H. Pendias. 1992. Trace elements in soils and plants. 2nd ed. CRC Press, Boca Raton, FL.

Kabata-Pendias, A., and H. Pendias. 2001. Trace elements in soils and plants. 3rd ed. CRC Press, Boca Raton, FL.

Kahle, M., M. Kleber, and R. Jahn. 2002. Review of XRD-based quantitative analyses of clay minerals in soils: The suitability of mineral intensity factors. Geoderma 109:191-205.

Kamprath, E.J., and C.D. Foy. 1972. Lime-fertilizer-plant interactions in acid soils. pp. 105-151. *In* R.W. Olsen et al. (eds.), Fertilizer technology and use. 2nd ed. SSSA, Madison, WI.

Kansas State University. 2006. The Western Kansas irrigation research project. Kansas State University Agricultural Experiment Station and Cooperative Extension Service.

Kapur, S., and E. Akca. 2002. Degradation, global assessment of. pp. 296-306. *In* R. Lal (ed.), Encyclopedia of soil science. Marcel Dekker, Inc.

Karathanasis, A.D. 2008. Thermal analysis of soil minerals. pp. 81-115. *In* A.L. Ulery and R. Drees (eds.), Methods of soil analysis. Part 5. Mineralogical methods. SSSA, Madison, WI.

Karathanasis, A.D., and B.F. Hajek. 1982a. Quantitative evaluation of water adsorption on soil clays. Soil Sci. Soc. Am. J. 46:1321-1325.

Karathanasis, A.D., and B.F. Hajek. 1982b. Revised methods for rapid quantitative determination of minerals in soil clays. Soil Sci. Soc. Am. J. 46:419-425.

Karathanasis, A.D., and W.G. Harris. 1994. Quantitative thermal analysis of soil minerals. pp. 360-411. *In* J.E. Amonette and L.W. Zelazny (eds.), Quantitative methods in soil mineralogy. Soil Sci. Soc. Am. Misc. Publ. SSSA, Madison, WI.

Kassim, J.K., S.N. Gafoor, and W.A. Adams. 1984. Ferrihydrite in pyrophosphate extracts of podzol B horizons. Clay Miner. 19:99-106.

Kaup, B.S. and B.J. Carter. 1987. Determining Ti source and distribution within a Paleustalf by micromorphology, submicroscopy, and elemental analysis. Geoderma 40:141-156.

Kay, B.D., and D.A. Angers. 2002. Soil structure. pp. 249-295. *In* A.W. Warrick (ed.), Soil physics companion. CRC Press.

Keeney, D.R., and J.M. Bremner. 1966. Comparison and evaluation of laboratory methods of obtaining an index of soil nitrogen availability. Agron. J. 58:498-503.

Keeney, D.R. and T.H. DeLuca. 1993. Des Moines River nitrate in relation to watershed agricultural practices: 1945 versus 1980's. J. Environ. Qual. 22:267-272.

Keeney, D.R., and D.W. Nelson. 1982. Nitrogen—Inorganic forms. pp. 643-698. *In* A.L. Page, R.H. Miller, and D.R. Keeney (eds.), Methods of soil analysis. Part 2. Chemical and microbiological properties. 2nd ed. Agron. Monogr. 9. ASA and SSSA, Madison, WI.

Keller, C., and J.C. Vedy. 1994. Distribution of copper and cadmium fractions in two forest soils. J. Environ. Qual. 23:987-999.

Kelley, J.A. 2006. Report of the mica research project. Southern Regional Cooperative Soil Survey Conference, Oklahoma City, OK, June 2006. Available online at ftp://ftp-fc.sc.egov.usda.gov/NSSC/NCSS/Conferences/regional/2006/south/Papers/Mica%20Research%20Project%20Report-SRCSSC.pdf (verified January 14, 2011).

Kemper, W.D., and R.C. Rosenau. 1986. Aggregate stability and size distribution. pp. 425-442. *In* A. Klute (ed.), Methods of soil analysis. Part 1. Physical and mineralogical methods. 2nd ed. Agron. Monogr. 9. ASA and SSSA, Madison, WI.

Kenney, T.C. 1967. The influence of mineral composition on the residual strength of natural soils. pp. 123-129. *In* Proceedings, Geotechnical Conference, Vol. 1. June 1967. Norwegian Geotechnical Society, Oslo.

Keren, R., and P. Kauschansky. 1981. Coating of calcium carbonate on gypsum particle surfaces. Soil Sci. Soc. Am. J. 45:1242-1244.

Kern, J.S. 1994. Spatial patterns of soil organic carbon in the contiguous United States. Soil Sci. Soc. Am. J. 58:439-455.

Kerr, P.F. 1977. Optical mineralogy. McGraw-Hill Co., New York, NY.

Khym, J.X. 1974. Analytical ion-exchange procedures in chemistry and biology: Theory, equipment, techniques. Prentice-Hall, Inc., Englewood Cliffs, NJ.

Kilmer, V.J., and L.T. Alexander. 1949. Methods of making mechanical analyses of soils. Soil Sci. 68:15-24.

Kimble, J.M., C.S. Holzhey, and G.G.S. Holmgren. 1984. An evaluation of potassium extractable aluminum in Andepts (Andisols). Soil Sci. Soc. Am. J. 48:1366-1369.

Kingery, W.L., C.W. Wood, D.P. Delaney, J.C. Williams, and G.L. Mullins. 1994. Impact of long-term application of broiler litter on environmentally related soil properties. J. Environ. Qual. 23:139-147.

Kjeldahl, J. 1883. Neue Methode zur Bestimmung des Stickstoffs in organischen Körpern. Z. Anal. Chem. 22:366-382.

Kladivko, E.J. 2002. Structure. pp.123-125. *In* R. Lal (ed.), Encyclopedia of soil science. Marcel Dekker, Inc.

Klein, C., and C.S. Hurlbut, Jr. 1985. Manual of mineralogy. John Wiley & Sons Inc., New York, NY.

Klingebiel, A.A., and P.H. Montgomery. 1961. Land capability classification. Soil Conservation Service, U.S. Department of Agriculture Handbook 210.

Klute, A. 1986. Water retention: Laboratory methods. pp. 635-662. *In* A. Klute (ed.), Methods of soil analysis. Part 1. Physical and mineralogical methods. 2nd ed. Agron. Monogr. 9. ASA and SSSA, Madison, WI.

Kononova, M.M. 1966. Soil organic matter. 2nd ed. Pergamon Press, Inc., Elmsford, NY.

Kovalenko, T.A. 1972. Determination of gypsum in soils. Sov. Soil Sci. 3:373-376.

Kraimer, R.A., H.C. Monger, and R.L. Steiner. 2005. Mineralogical distinctions of carbonates in desert soils. Soil Sci. Soc. Am. J. 69:1773-1781.

Kramer, P.J. 1969. Water in the soil. pp. 46-72. *In* Plant and soil water relationships. McGraw-Hill Co., New York, NY.

Krumbein, W.C., and F.J. Pettijohn. 1938. Manual of sedimentary petrography. Appleton-Century-Crofts, New York, NY.

Krupkin, P.I. 1963. Movement of salt solutions and soil materials. Sov. Soil Sci. (Engl. transl.) 6:567-574.

Kubiena, W.L. 1938. Micropedology. Collegiate Press, Ames, IA.

Kubota, T. 1972. Aggregate-formation of allophanic soils: Effects of drying on the dispersion of the soils. Soil Sci. and Plant Nutr. 18:79-87.

Kukier, U., and R.L. Chaney. 2001. Amelioration of nickel phytotoxicity in muck and mineral soils. J Environ. Qual. 30:1949-1960.

Kumada, K. 1987. Chemistry of soil organic matter. Japan Scientific Societies Press. Elsevier, Tokyo.

Kumada, K., and M.H. Hurst. 1967. Green humic acid and its possible origin as a fungal metabolite. Nature 214:631-633.

Kuo, S. 1996. Phosphorus. pp. 869-919. *In* D.L. Sparks (ed.), Methods of soil analysis. Part 3. Chemical methods. Soil Sci. Soc. Am. Book Series 5. ASA and SSSA, Madison, WI.

LACHAT Instruments. 1989. QuikChem method 12-115-01-1-A, phosphorus as orthophosphate, 0.4 to 20 mg P L^{-1}. LACHAT Instruments, 6645 West Mill Rd., Milwaukee, WI.

LACHAT Instruments. 1992. QuikChem method 12-107-04-1-B, nitrate in 2 M (1 M) KCl soil extracts, 0.02 to 20.0 mg N L^{-1}. LACHAT Instruments, 6645 West Mill Rd., Milwaukee, WI.

LACHAT Instruments. 1993. QuikChem method 10-115-01-1-A, orthophosphate in waters, 0.02 to 2.0 mg P L^{-1}. LACHAT Instruments, 6645 West Mill Rd., Milwaukee, WI.

Laflen, J.M. 2002. Erosion by water, empirical models. pp. 457-462. *In* R. Lal (ed.), Encyclopedia of soil science. Marcel Dekker, Inc.

Lagerwerff, J.V., G.W. Akin, and S.W. Moses. 1965. Detection and determination of gypsum in soils. Proc. Soil Sci. Soc. Am. 29:535-540.

Lal, R. 1979. Physical characteristics of soils of the tropics: Determination and management. pp. 3-44. *In* R. Lal and D.J. Greenland (eds.), Soil physical properties of crop production in the tropics. John Wiley & Sons Inc., New York, NY.

Lal, R. 1981. Physical properties. pp. 144-148. *In* D.J. Greenland (ed.), Characterization of soils in relation to their classification and management for crop production: Examples from some areas of the humid tropics. Clarendon Press, Oxford.

Lal, R. 1990. Soil erosion and land degradation: Their global risks. pp. 129-172. *In* J. Lal and B.A. Stewart (eds.), Soil degradation. Advances in Soil Science. Springer-Verlag.

Lal, R. 1993. Tillage effects on soil degradation, soil resilience, soil quality, and sustainability. Soil and Tillage Res. 27:1-8.

Lal, R. 1998. Soil erosion impact on agronomic productivity and environmental quality. Critical Reviews in Plant Sci. 19:319-464.

Lal, R., and M. Ahamdi. 2000. Axle load and tillage effects on crop yield for two soils in central Ohio. Soil and Tillage Res. 54:111-119.

Lal, R., and J.M. Kimble. 2006. Soil C sink in U.S. cropland. Nature. Berkeley, CA.

Lal, R., J.M. Kimble, R.F. Follett, and C.V. Cole. 1998a. The potential of U.S. cropland to sequester C and mitigate the greenhouse effect. Sleeping Bear Press, Ann Arbor, MI.

Lal, R., J.M. Kimble, R.F. Follett, and B.A. Stewart. 1998b. Management of carbon sequestration in soil. Advances in Soil Science. CRC Press.

Lal, R., J.M. Kimble, R.F. Follett, and B.A. Stewart. 1998c. Soil processes and the carbon cycle. Advances in Soil Science. CRC Press.

Lane, L.J., and M.A. Nearing. 1989. USDA water erosion prediction project (WEPP) hillslope profile and watershed model documentation, NSERL Report 2. National Soil Erosion Laboratory. USDA/ARS, West Lafayette, IN.

Lapham, M.H. 1932. Genesis and morphology of desert soils. Am. Soil Surv. Assoc. Bull. 13:34-52.

Larson, W.E. 1981. Protecting the soil resource base. J. Soil Water Conserv. 36:13-16.

Larson, W.E., and F.J. Pierce. 1991. Conservation and enhancement of soil quality. *In* Evaluation for sustainable land management in the developing world. Vol. 2. IBSRAM Proc. 12(2). International Board for Soil Research and Management. Bangkok, Thailand.

Lathwell, D.J., and W.S. Reid. 1984. Crop response to lime in the northeastern United States. pp. 305-332. *In* F. Adams (ed.), Soil acidity and liming. 2nd ed. Agron. 12. ASA, CSSA, and SSSA, Madison, WI.

Lauenroth, W.K., and W.C. Whitman. 1971. A rapid method for washing roots. J. Range Mgt. 24:308-309.

Law, J.P., G.V. Skogerboe, and J.D. Denit. 1972. The need for implementing irrigation return flow control. pp. 1-17. *In* Managing irrigated agriculture to improve water quality. Proc. Math. Conf. Manag. Irrig. Agric. Improve Water. Avail. Denver, CO, May, 1972. Graphics Management Corp., Washington, DC.

Lebron, I., J. Herrero, and D.A. Robinson. 2009. Determination of gypsum content in dryland soils exploiting the gypsum-bassanite phase change. Soil Sci. Soc. Am. J. 73:403-411.

Lee, B.D., R.C. Graham, T.E. Laurent, C. Amrhein, and R.M. Creasy. 2001. Spatial distributions of soil chemical conditions in a serpentinitic wetland and surrounding landscape. Soil Sci. Soc. Am. J. 65:1183-1196.

Lee, B.D., R.C. Graham, T.E. Laurent, and C. Amrhein. 2003a. Pedogenesis in a wetland meadow and surrounding serpentinitic landslide terrain, northern California, USA. Geoderma 118:303-320.

Lee, B.D., S.K. Sears, R.C. Graham, C. Amrhein, and H. Vali. 2003b. Secondary mineral genesis from chlorite and serpentine in an ultramafic soil toposequence. Soil Sci. Soc. Am. J. 67:1309-1317.

Lee, R., B.W. Bache, M.J. Wilson, and G.S. Sharp. 1985. Aluminum release in relation to the determination of cation exchange capacity of some podzolized New Zealand soils. J. Soil Sci. 36:239-253.

Lee, R., J.M. Bailey, R.D. Northey, P.R. Barker, and E.J. Gibson. 1975. Variations in some chemical and physical properties of three related soil types: Dannevirke silt loam, Kirvitea silt loam, and Morton silt loan. N.Z. J. Agric. Res. 18:29-36.

Lehmann, J. 2007. Bio-energy in the black. Front. Ecol. Environ. 5:381-387.

LeRoy, L.W. 1950. Stain analysis. *In* Subsurface geologic methods. Colorado School of Mines, Golden, CO.

Letey, J., Jr., C. Roberts, M. Penberth, and C. Vasek. 1986. An agricultural dilemma: Drainage water and toxics disposal in the San Joaquin Valley. Univ. Calif. Agric. Exp. Stn. Spec. Publ. 3319.

Liang, B., J. Lehmann, D. Solomon, J. Kinyangi, J. Grossman, B. O'Neill, J.O. Skjemstad, J. Thies, F.J. Luizao, J. Petersen, and E.G. Neves. 2006. Black carbon increases cation exchange capacity in soils. Soil Sci. Soc. Am. J. 70:1719-1730.

Liao, X.Y., T.B. Chen, H. Xie, and Y.R. Liu. 2005. Soil as contamination and its risk assessment in areas near the industrial districts of Chenzhou City, Southern China. Environ. Intl. 31:791-798.

Lim, C.H., and M.L. Jackson. 1982. Dissolution of total elemental analysis. *In* A.L Page, R.H. Miller, and D.R. Keeney (eds.), Methods of soil analysis. Part 2. Chemical and microbiological properties. 2nd ed. Agron. 9:1-12.

Lima, H.N., C.E.R. Schaefer, J.W.V. Mello, R.J. Gilkes, and J.C. Ker. 2002. Pedogenesis and pre-Colombian land use of "Terra Preta Anthrosols" ("Indian black earth") of Western Amazonia. Geoderma 110:1-17.

Lindsay, W.L. 1979. Chemical equilibria in soils. John Wiley & Sons Inc., New York, NY.

Lindsay, W.L., and L.G. Vlek. 1977. Phosphate minerals. *In* J.B. Dixon and S.B. Weed (eds.), Minerals in soil environments. 1st ed. Soil Sci. Soc. Am. Book Series 1:639-672.

Lindstrom, M.J., W.B. Voorhees, and G.W. Randall. 1981. Long-term tillage effects on interrow runoff and infiltration. Soil Sci. Soc. Am. J. 45:945-948.

Loeppert, R.H., and W.P. Inskeep. 1996. Iron. pp. 639-664. *In* D.L. Sparks (ed.), Methods of soil analysis. Part 3. Chemical methods. Soil Sci. Soc. Am. Book Series 5. ASA and SSSA, Madison, WI.

Loeppert, R.H., and D.L. Suarez. 1996. Carbonate and gypsum. pp. 437-474. *In* D.L. Sparks (ed.), Methods of soil analysis. Part 3. Chemical methods. Soil Sci. Soc. Am. Book Series 5. ASA and SSSA, Madison, WI.

Loganathan, P., N.O. Isirimah, and D.A. Nwachuku. 1987. Phosphorus sorption by Ultisols and Inceptisols of the Niger Delta in Southern Nigeria. Soil Sci. 144:330-338.

Lombi, E., F.J. Zhao, S.J. Dunham, and S.P. McGrath. 2001. Phytoremediation of heavy metal-contaminated soils: Natural hyperaccumulation versus chemically enhanced phytoremediation. J. Environ. Qual. 30:1919-1926.

Long, F.L. 1979. Runoff water quality as affected by surface-applied dairy cattle manure. J. Environ. Qual. 8:215-218.

Lorenz, O.A., and C.J. Johnson. 1953. Nitrogen fertilization as related to the availability of phosphorus in certain California soils. Soil Sci. 75:119-129.

Loveday, J. (ed.). 1974. Methods of analysis of irrigated soils. II: Particle-size analysis. Commonwealth Bureau of Soils Technical Communication 54.

Loveland, P.J., and P. Digby. 1984. The extraction of Fe and Al by 0.1M pyrophosphate solutions: A comparison of some techniques. J. Soil Sci. 35:243-250.

Lund, Z., and B. Doss. 1980. Coastal bermudagrass yield and soil properties as affected by surface-applied dairy manure and its residue. J. Environ. Qual. 9:157-162.

Lundstrom, U.S., N. Van Breemen, and D. Bain. 2000. The podzolization process: A review. Geoderma 94:91-107.

Lundstrom, U.S., N. Van Breemen, and A.G. Jongmans. 1995. Evidence for microbial decomposition of organic acids during podzolization. European J. Soil Sci. 46:489-496.

Lyle, E.S., and F. Adams. 1971. Effect of available soil calcium on taproot elongation of loblolly pine (*Pinus taeda* L.) seedlings. Soil Sci. Soc. Am. Proc. 35:800-805.

Lynn, W.C., R.J. Ahrens, and A.L. Smith. 2002. Soil minerals, their geographic distribution, and soil taxonomy. pp. 691-701. *In* J.B. Dixon and D.G. Schulze (eds.), Soil mineralogy with environmental applications. Soil Sci. Soc. Am. Book Series 7. SSSA, Madison, WI.

Lynn, W.C., W.E. McKenzie, and R.B. Grossman. 1974. Field laboratory tests for characterization of Histosols. pp. 11-20. *In* Histosols: Their characteristics, use, and classification. Soil Sci. Soc. Am. Spec. Publ. 6. SSSA, Madison, WI.

Lynn, W., J.E. Thomas, and L.E. Moody. 2008. Petrographic microscope techniques for identifying soil minerals in grain mounts. pp. 161-190. *In* A.L. Ulery and R. Drees (eds.), Methods of soil analysis. Part 5. Mineralogical methods. SSSA, Madison, WI.

Ma, L.Q., and G.N. Rao. 1997. Chemical fractionation of cadmium, copper, nickel, and zinc in contaminated soils. J. Environ. Qual. 26:259-264.

Ma, L.Q., F. Tan, and W.G. Harris. 1997. Concentrations and distributions of eleven metals in Florida soils. J. Environ. Qual. 26:769-775.

Mackenzie, R.C. 1970. Simple phyllosilicates based on gibbsite- and brucite-like sheets. pp. 498-537. *In* R.C. Mackenzie (ed.), Differential thermal analysis. Vol. 1. Academic Press, London.

Mackenzie, R.C., and G. Berggen. 1970. Oxides and hydroxides of higher-valency elements. pp. 272-302. *In* R.C. Mackenzie (ed.), Differential thermal analysis. Academic Press, London.

Mackenzie, R.C., and S. Caillere. 1975. The thermal characteristics of soil minerals and the use of these characteristics in the qualitative and quantitative determination of clay minerals in soils. pp. 529-571. *In* J.E. Gieseking (ed.), Soil components. Vol. 2. Inorganic components. Springer-Verlag, New York, NY.

Mackenzie, R.C., and B.D. Mitchell. 1972. Soils. pp. 267-297. *In* R.C. Mackenzie (ed.), Differential thermal analysis. Vol. 2. Academic Press, London.

MacNaughton, M.G., and R.O. James. 1974. Adsorption of aqueous mercury (II) complexes at the oxide/water interface. J. Colloid Interface Sci. 47:431-440.

Maeda, T., H. Takenaka, and B.P. Warkentin. 1977. Physical properties of allophane soils. pp. 229-263. *In* N.C. Brady (ed.), Advances in Agronomy. Academic Press, New York, NY.

Major, J. 1969. Historical development of the ecosystem concept. pp. 9-22. *In* G.M. Van Dyne (ed.), The ecosystem concept in natural resource management. Academic Press, New York, NY.

Mannering, J.V. 1967. The relationship of some physical and chemical properties of soil to surface sealing. Ph.D. thesis. Purdue University, West Lafayette, IN.

Marion, G.M., D.M. Hendricks, G.R. Dutt, and W.H. Fuller. 1976. Aluminum and silica solubility in soils. Soil Sci. 121:76-85.

Marquez, C.O., V.J. Garcia, C.A. Cambardella, R.C. Schultz, and T.M. Isenhart. 2004. Aggregate-size stability distribution and soil stability. Soil Sci. Soc. Am. J. 68:725-735.

Marshall, C.E. 1935. Mineralogical methods for the study of silts and clays. Z. Krist. (A) 90:8-34.

Martel, Y.A., C.R. De Kimpe, and M.R. Laverdiere. 1978. Cation-exchange capacity of clay-rich soils in relation to organic matter, mineral composition, and surface area. Soil Sci. Soc. Am. J. 42:764-767.

Martel, Y.A., and A.F. MacKenzie. 1980. Long-term effects of cultivation and land use on soil quality in Quebec. Can. J. Soil Sci. 60:411-420.

Martens, D.A., and D.L. Suarez. 1997. Selenium speciation of marine shales, alluvial soils, and evaporation basin soils of California. J. Environ. Qual. 26:424-432.

Martens, R. 1987. Estimation of microbial biomass in soil by the respiration method: Importance of soil pH and flushing methods of the respired CO_2. Soil Biol. Biochem. 19:77-81.

Martin, S.E., and R. Reeve. 1955. A rapid manometric method for determining soil carbonate. Soil Sci. 79:187-197.

Matar, A.E., and T. Douleimy. 1978. Note on proposed method for the mechanical analysis of gypsiferous soils. ACSAD Publ. The Arab Center for the Studies of Arid Zones and Dry Lands. Damascus, Syria.

Mausbach, M.J., B.R. Brasher, R.D. Yeck, and W.D. Nettleton. 1980. Variability of measured properties in morphologically matched pedons. Soil Sci. Soc. Am. J. 44:358-363.

Mausbach, M.J., and J.L. Richardson. 1994. Biogeochemical processes in hydric soil formation. pp. 68-127. *In* Current topics in wetland biogeochemistry. Vol. 1. Wetland Biogeochemical Institute, Louisiana State University, Baton Rouge.

Maynard, D.G., and Y.P. Kalra. 1993. Nitrate and exchangeable ammonium nitrogen. pp. 25-38. *In* Martin R. Carter (ed.), Soil sampling and methods of analysis. Canadian Society of Soil Science. Lewis Publishers, Boca Raton, FL.

McBride, M.B. 1989. Surface chemistry of soil minerals. pp. 35-88. *In* J.B. Dixon and S.B. Weed (eds.), Minerals in soil environments. 2nd ed. Soil Sci. Soc. Am. Book Series 1. SSSA, Madison, WI.

McBride, M.B., E.A. Nibarger, B.K. Richards, and T. Steenhuis. 2003. Trace metal accumulation by red clover grown on sewage sludge-amended soils and correlations to Mehlich3 and calcium chloride-extractable metals. Soil Sci. 168(1):29-38.

McCollum, R.E. 1991. Buildup and decline in soil phosphorus: 30-year trends on a Typic Umprabuult. Agron. J. 83:77-85.

McCormack, D.E., and L.P. Wilding. 1969. Variation of soil properties within mapping units of soils with contrasting substrata in northwestern Ohio. Soil Sci. Soc. Am. Proc. 33:587-593.

McDaniel, P.A., and M.A. Wilson. 2007. Physical and chemical characteristics of ash-influenced soils of inland Northwest forest. pp. 31-45. *In* D. Page-Dumroese, R. Miller, J. Mital, P. McDaniel, and D. Miller (eds.), Properties and implications for management and restoration. November 9-10, 2005. Coeur d'Alene, ID. Proceedings RMRS-P-44, USDA, Forest Service, Rocky Mountain Research Station. Fort Collins, CO.

McGill, W.B., C.A. Campbell, J.F. Dormaar, E.A. Paul, and D.W. Anderson. 1981. Soil organic matter losses. pp.72-133. *In* Agricultural land. Our disappearing heritage. Proceedings, 18th Annual Alberta Soil Science Workshop. Alberta Soil and Feed Testing Laboratory, Edmonton, Alberta.

McKeague, J.A., J.E. Brydon, and N.M. Miles. 1971. Differentiation of forms of extractable iron and aluminum in soils. Soil Sci. Soc. Am. Proc. 35:33-38.

McKeague, J.A., and J.H. Day. 1966. Dithionite- and oxalate-extractable Fe and Al as aids in differentiating various classes of soils. Can. J. Soil Sci. 46:13-22.

McKeague, J.A., F. De Connick, and D.P. Franzmeier. 1983. Spodosols. pp. 217-248. *In* L.P. Wilding, N.E. Smeck, and G.F. Hall (eds.), Pedogenesis and soil taxonomy. II. The soil orders. Elsevier, New York, NY.

McKeague, J.A., and P.A. Schuppli. 1982. Changes in concentration of iron and aluminum in pyrophosphate extracts of soil and composition of sediment resulting from ultracentrifugation in relation to spodic horizon criteria. Soil Sci. 134:265-270.

McKeague, J.A., C. Wang, and G.C. Topp. 1982. Estimating saturated hydraulic conductivity from soil morphology. Soil Sci. Soc. Am. J. 46:1239-1244.

McKenzie, R.M. 1989. Manganese oxides and hydroxides. pp. 439-466. *In* J.B. Dixon and S.B. Weed (eds.), Minerals in soil environments. 2nd ed. Soil Sci. Soc. Am. Book Series 1. SSSA, Madison, WI.

McLean, E.O. 1982. Soil pH and lime requirement. pp. 199-224. *In* A.L Page, R.H. Miller, and D.R. Keeney (eds.), Methods of soil analysis. Part 2. Chemical and microbiological properties. 2nd ed. Agron. Monogr. 9. ASA and SSSA, Madison, WI.

McLean, E.O., and J.R. Brown. 1984. Crop response to lime in the Midwestern USA. pp. 267-303. *In* F. Adams (ed.), Soil acidity and liming. 2nd ed. Agronomy 12. ASA, CSSA, and SSSA, Madison, WI.

McNab, W.H. 1987. Rationale for a multifactor forest site classification system for the southern Appalachians. pp. 283-294. *In* Proceedings of 6th Central Hardwood Forest Conference. February 24-26. Knoxville, TN.

Mehlich, A. 1939. Use of triethanolamine acetate-barium hydroxide buffer for the determination of some base exchange properties and lime requirement of soil. Soil Sci. Soc. Am. J. 3:162-166.

Mehlich, A. 1943. The significance of percentage of base saturation and pH in relation to soil differences. Soil Sci. Soc. Am. Proc. 7:167-174.

Mehlich, A. 1948. Determination of cation- and anion-exchange properties of soils. Soil Sci. 66:429-445.

Mehlich, A. 1984. Mehlich 3 soil text extractant: A modification of Mehlich 2 extractant. Commun. Soil Sci. Plant Anal. 15:1409-1416.

Mehra, O.P., and M.L. Jackson. 1960. Iron oxide removal from soils and clays by a dithionite-citrate system buffered with sodium bicarbonate. Clays and Clay Minerals 7:317-327.

Mehuys, G.R. 1984. Soil degradation of agricultural land in Quebec. A review and impact assessment. Science Council of Canada, Ottawa.

Meisinger, J.J., and G.B. Randall. 1991. Estimating nitrogen budgets for soil-crop systems. *In* R.F. Follett, D.R. Keeney, and R.M. Cruse (eds.), Managing nitrogen for groundwater quality and farm profitability. SSSA, Madison, WI.

Melsted, S.W., and T.R. Peck. 1977. The Mitscherlich-Bray growth function. pp. 1-18. *In* T.R. Peck, J.T. Cope, Jr., and D.A. Whitney (eds.), Soil testing: Correlating and interpreting the analytical results. Am. Soc. Agron. Spec. Publ. 29. ASA, Madison, WI.

Mendoza, R.E., and N.J. Barrow. 1987. Ability of three soil extractants to reflect the factors that determine the availability of soil phosphate. Soil Sci. 144:319-329.

Mermut, A.R., J.C. Jain, L. Song, R. Kerrich, L. Kozak, and S. Jana. 1996. Trace element concentrations of selected soils and fertilizers in Saskatchewan, Canada. J. Environ. Qual. 25:845-853.

Mesuere, K., R.E. Martin, and W. Fish. 1991. Identification of copper contamination in sediments by a microscale partial extraction technique. J. Environ. Qual. 20:114-118.

Metherell, A.K., L.A. Harding, C.V. Cove, and W.J. Parton. 1993. CENTURY soil organic matter model environment, technical documentation agroecosystem, Version 4.0. Great Plains System Res. Unit Tech. Rep. 4, USDA/ARS, Ft. Collins, CO.

Metson, A.J. 1956. Methods of chemical analysis for soil survey samples. Soil Bur. Bull. 12. N.Z. Department of Scientific and Industrial Research.

Middleton, H.E. 1930. Properties of soils which influence soil erosion. USDA Tech. Bull. 178.

Mikhail, E.H., and G.P. Briner. 1978. Routine particle size analysis of soils using sodium hypochlorite and ultrasonic dispersion. Aust. J. Soil Res. 16:241-244.

Miles, D.L. 1977. Salinity in the Arkansas Valley of Colorado. Interagency Agreement Report EPA-AIC-D4-0544. USEPA, Denver, CO.

Miller, D.E. 1971. Formation of vesicular structure in soil. Soil Sci. Soc. Am. J. 35:635-637.

Miller, M.H. 2000. Arbuscular mycorrhizae and the phosphorus nutrition of maize: A review of Guelph studies. Can. J. Plant Sci. 80:47-52.

Miller, M.H., T.P. McGonigle, and H.D Addy. 1995. Functional ecology of vesicular arbuscular mycorrhizas as influenced by phosphate fertilization and tillage in an agricultural ecosystem. Critical Review Biotechnology 15:241-255.

Miller, W.F. 1970. Inter-regional predictability of cation-exchange capacity by multiple regression. Plant Soil 33:721-725.

Miller, W.P., and M.K. Baharuddin. 1987. Particle size of interrill-eroded sediments from highly weathered soils. Soil Sci. Soc. Am. J. 51:1610-1615.

Miller, W.P., and D.E. Radcliffe. 1992. Soil crusting in the Southeastern United States. pp. 233-266. *In* M.E. Sumner and B.A. Stewart (eds.), Soil crusting, chemical and physical processes. Advances in Soil Science. Lewis Publishers.

Miller, W.P., C.C. Truman, and G.W. Langsdale. 1988. The influence of degree of previous erosion on crusting behavior of Cecil soils. J. Soil Water Conserv. 42:338-341.

Mills, H.A., and J. Benton Jones. 1996. Plant analysis handbook II. A practical sampling, preparation, analysis, and interpretation guide. MicroMacro Publishing, Inc., Athens, GA.

Milner, H.B. 1962. Sedimentary petrography. 4th ed. The Macmillan Co., New York, NY.

Mitchell, B.D., V.C. Farmer, and W.J. McHardy. 1964. Amorphous inorganic materials in soils. Adv. Agron. 16:327-383.

Monger, H.C., L.A. Daugherty, and L.H. Gile. 1991. A microscopic examination of pedogenic calcite in an Aridisol of southern New Mexico. pp. 37-60. *In* W.D. Nettleton (ed.), Occurrence, characteristics, and genesis of carbonates, gypsum, and silica accumulations in soils. Soil Sci. Soc. Am. Spec. Publ. 26. SSSA, Madison, WI.

Monreal, C.M., M. Schnitzer, H.R. Schulten, C.A. Campbell, and D.W. Anderson. 1995a. Soil organic structures in macro and microaggregates of a cultivated Brown Chernozem. Soil Biol. Biochem. 27:845-853.

Monreal, C.M., R.P. Zentner, and J.A. Robertson. 1995b. The influence of management on soil loss and yield of wheat in Chernozemic and Luvisolic soils. Can. J. Soil Sci. 75:567-574.

Montavalli, P.P., C.Q. Palm, W.J. Parton, E.T. Elliott, and S.D. Frey. 1994. Comparison of laboratory and modeling simulation methods for estimating soil carbon pools in tropical forest soils. Soil Biol. Biochem. 26:935-944.

Moore, C.A. 1971. Effect of mica on K_o compressibility of two soils. Proc. Am. Soc. Civil Eng., Soil Mech. and Found. Div. 97:1275-1291.

Moore, D.M., and R.C. Reynolds, Jr. 1997. X-ray diffraction and the identification and analysis of clay minerals. Oxford University Press. New York, NY.

Moore, J., and J.V. Hefner. 1977. Irrigation with saline water in the Pecos Valley of West Texas. pp. 339-394. *In* Proceedings of the International Salinity Conference on Managing Saline Water for Irrigation. Texas Tech University, Lubbock, TX.

Mubiru, D.N., and A.D. Karathanasis. 1994. Phosphorus-sorption characteristics of intensely weathered soils in south-central Kentucky. Commun. Soil Sci. Plant Anal. 25:2745-2759.

Muhs, D.R., E.A. Bettis III, J. Been, and J.P. McGeehin. 2001. Impact of climate and parent material on chemical weathering in loess-derived soils of the Mississippi River Valley. Soil Sci. Soc. Am. J. 65:1761-1777.

Mulvaney, R.L. 1996. Nitrogen—Inorganic forms. pp. 1123-1184. *In* D.L. Sparks (ed.), Methods of soil analysis. Part 3. Chemical methods. Soil Sci. Soc. Am. Book Series 5. ASA and SSSA, Madison, WI.

Murphy, C.P. 1986. Thin section preparation of soils and sediments. Academic Publishers, Great Britain.

Murphy, J., and J.R. Riley. 1962. A modified single solution method for the determination of phosphate in natural waters. Anal. Chem. Acta 27:31-36.

Murray-Darling Basin Authority. 2010. Detailed assessment of acid sulfate soils in the Murray-Darling Basin: Protocols for sampling, field characterization, laboratory analysis and data presentation. MDBA Publ. 57/10.

National Association of Conservation Districts (NACD). 1988. NACD Farm Bill issue paper. Washington, DC.

National Research Council. 1993. Soil and water quality. An agenda for agriculture. Committee on long-range soil and water conservation. The National Academies Press, Washington, DC.

National Research Council. 1996. A new era for irrigation. The National Academies Press, Washington, DC.

National Soil Survey Laboratory Staff. 1975. Proposed tables for soil survey reports. RSSIU. USDA/SCS, Lincoln, NE.

National Soil Survey Laboratory Staff. 1983. Principles and procedures for using soil survey laboratory data. USDA/SCS, Lincoln, NE.

National Soil Survey Laboratory Staff. 1990. Examples and descriptions of laboratory information. USDA/SCS, Lincoln, NE.

Nelson, D.G., and W.R. Johnston. 1984. San Joaquin drainage—Development and impact. pp. 424-432. *In* Proceedings, Specialty Conference, July 24-26, Flagstaff, AZ. American Society of Civil Engineers, New York.

Nelson, D.W., and L.E. Sommers. 1982. Total carbon, organic carbon, and organic matter. *In* A.L. Page, R.H. Miller, and D.R. Kenney (eds.), Methods of soil analysis. Part 2. Chemical and microbiological properties. 2nd ed. Agron. 9:539-579.

Nelson, D.W., and L.E. Sommers. 1996. Total carbon, organic carbon, and organic matter. pp. 961-1010. *In* D.L. Sparks (ed.), Methods of soil analysis. Part 3. Chemical methods. Soil Sci. Soc. Am. Book Series 5. ASA and SSSA, Madison, WI.

Nelson, L.A., and R.J. McCracken. 1962. Properties of Norfolk and Portsmouth soils: Statistical summarization and influence on corn yields. Soil Sci. Soc. Am. Proc. 26:497-502.

Nelson, R.E. 1982. Carbonate and gypsum. pp. 181-197. *In* A.L. Page, R.H. Miller, and D.R. Keeney (eds.), Methods of soil analysis. Part 2. Chemical and microbiological properties. 2nd ed. Agron. Monogr. 9. ASA and SSSA, Madison, WI.

Nelson, R.E., L.C. Klameth, and W.D. Nettleton. 1978. Determining soil gypsum content and expressing properties of gypsiferous soils. Soil Sci. Soc. Am. J. 42:659-661.

Nelson, R.E., and W.D. Nettleton. 1975. Some properties of smectite and mica in clays of soils in dry regions. Geoderma 14:247-253.

Nettleton, W.D. 2004. Micromorphology. pp. 143-153. *In* R. Burt (ed.), Soil survey laboratory methods manual. Ver. 4.0. USDA/NRCS, Soil Survey Investigations Report No. 42. U.S. Government Printing Office, Washington, DC.

Nettleton, W.D., B.R. Brasher, and S.L. Baird. 1991. Carbonate clay characterization by statistical methods. pp. 75-88. *In* W.D. Nettleton (ed.), Occurrence, characteristics, and genesis of carbonate, gypsum, and silica accumulations in soils. Soil Sci. Soc. Am. Spec. Publ. 26. SSSA, Madison, WI.

Nettleton, W.D., S.H. Brownfield, R. Burt, E.C. Benham, S.L. Baird, K. Hipple, C.L. McGrath, and H.R. Sinclair. 1999. Reliability of Andisol field texture clay estimates. Soil Surv. Hor. 40:36-49.

Nettleton, W.D., K.W. Flach, and B.R. Brasher. 1969. Argillic horizons without clay skins. Soil Sci. Soc. Am. Proc. 33:121-125.

Nettleton, W.D., K.W. Flach, and R.E. Nelson. 1970. Pedogenic weathering of tonalite in southern California. Geoderma 4:387-402.

Nettleton, W.D., R.B. Grossman, and B.R. Brasher. 1990. Concept of argillic horizons in Aridisols—Taxonomic implications. pp. 167-176. *In* J.M. Kimble and W.D. Nettleton (eds.), Proceedings of the 4th International Soil Correlation Meeting (ISCOM). Characterization, classification, and utilization of Aridisols in Texas, New Mexico, Arizona, and California, October 3-17, 1987. Part A: Papers.

Nettleton, W.D., R.J. McCracken, and R.B. Daniels. 1968. Two North Carolina Coastal Plain catenas. II. Micromorphology, composition, and fragipan genesis. Soil Sci. Soc. Am. Proc. 32:582-587.

Nettleton, W.D., R.E. Nelson, B.R. Brasher, and P.S. Derr. 1982. Gypsiferous soils in the western United States. pp. 147-168. *In* J.A. Kittrick, D.S. Fanning, and L.R. Hossner (eds.), Acid sulfate weathering. Soil Sci. Soc. Am. Spec. Publ. 10. SSSA, Madison, WI.

Nettleton, W.D., and F.F. Peterson. 1983. Aridisols. pp. 165-215. *In* L.P. Wilding, N.E. Smeck, and G.F. Hall (eds.), Pedogenesis and soil taxonomy. II. The soil orders. Elsevier, New York, NY.

Nicholas, D.J.D. 1969. Minor mineral nutrients. Ann. Rev. Plant Physiol. 12:63-90.

Nielsen, D.R., R.D. Jackson, J.W. Cary, and D.D. Evans (eds). 1972. Soil water. ASA, Madison, WI.

Nimmo, J.R., and K.S. Perkins. 2002. Aggregate stability and size distribution. pp. 317-328. *In* J.H. Dane and G.C. Topp (eds.), Methods of soil analysis. Part 4. Physical methods. Soil Sci. Soc. Am. Book Series 5. ASA and SSSA, Madison, WI.

Noggle, C.R., and G.J. Fritz. 1976. Introductory plant physiology. 1st ed. Prentice-Hall, Inc., Englewood Cliffs, NJ.

Nommik, H., and K. Vahtras. 1982. Retention and fixation of ammonium and ammonia in soils. *In* F.J. Stevenson (ed.), Nitrogen in agricultural soils. Agron. 22:123-171.

North Central Regional Research Publication (NCRRP). 1979. Water infiltration into representative soils in the North Central Region (Appendixes). North Central Regional Research Publication No. 259, Agricultural Experiment Station, University of Illinois Bulletin No. 760. University of Illinois, Urbana, IL.

North Central Regional Research Publication (NCRRP). 1988. Recommended chemical soil test procedures for the North Central region. North Central Regional Research Publication No. 221. Agricultural Experiment Station (IL, IN, IA, KS, MI, MN, MS, NE, ND, OH, SD, and WI).

Northwest Hydraulic Consultants, Ltd. 1980. Erosional features and processes in the upper Oldman River Basin. Alberta Environment Planning Division, Canada.

Oden, Sven. 1921-1922. On clays as disperse systems. Trans. Faraday Soc. 17:327-348.

O'Geen, A.T., R. Elkins, and D. Lewis. 2006. Erodibility of agricultural soils, with examples in Lake and Mendocino Counties. Publ. 8194. University of California, Division of Agriculture and Natural Resources. Oakland, CA.

Oh, Y.M., D.L. Hesterberg, and P.V. Nelson. 1999. Comparison of phosphate adsorption on clay minerals for soilless root media. Commun. Soil Sci. Plant Anal. 30:747-756.

Okamura, Y., and K. Wada. 1983. Electric charge characteristics of horizons of Ando (B) and Red-Yellow B soils and weathered pumices. J. Soil Sci. 34:287-295.

Olsen, R.E. 1974. Shearing strengths of kaolinite, illite, and montmorillonite. Proc. Am. Soc. Civ. Eng., Geotech. Eng. Div. 100:1215-1229.

Olsen, S.R., and C.V. Cole, F.S. Watanabe, and L.A. Dean. 1954. Estimation of available phosphorus in soils by extraction with sodium bicarbonate. USDA Circ. 939. U.S. Government Printing Office, Washington, DC.

Olsen, S.R., and L.E. Sommers. 1982. Phosphorus. pp. 403-430. *In* A.L. Page, R.H. Miller, and D.R. Keeney (eds.), Methods of soil analysis. Part 2. Chemical and microbiological properties. 2nd ed. Agron. Monogr. 9. ASA and SSSA, Madison, WI.

Olsen, S.R., and F.S. Watanabe. 1957. A method of determining phosphorus adsorption maximum of soils as measured by the Langmuir Isotherm. Soil Sci. Soc. Am. Proc. 21:144-149.

Olson, C.G., M.L. Thompson, and M.A. Wilson. 1999. Section F. Soil mineralogy. Chapter 2. Phyllosilicates. pp. F-77 to F-13. *In* M.E. Sumner (ed.), Handbook of soil science. CRC Press, Boca Raton, FL.

O'Neill, R.V., D.L. DeAngelis, J.B. Waide, and T.F.H. Allen. 1986. A hierarchical concept of ecosystems. Monographs in Population Biology 23:1-272.

Orlov, D.S. 1968. Absorption spectra and distribution of P type humic acids in USSR soils. Soviet Soil Sci. 1384-1393.

Orlov, D.S. 1985. Humus acids of soils. (In English; translated from Russian.) Moscow University Publishing, Moscow, Russia.

Orr, B. 2006. Defining rangeland management. A comparison of three textbooks. Rangelands West.

Osborne, T.B. 1887. The methods of mechanical soil analysis. pp. 143-162. *In* Connecticut Agricultural Experiment Station Annual Report.

Owusu-Bennoah, E., and D.K. Acguaye. 1989. Phosphate sorption characteristics of selected major Ghanaian soils. Soil Sci. 148:114-123.

Pais, Istvan, and J.B. Jones, Jr. 1997. The handbook of trace elements. St. Lucie Press, Boca Raton, FL.

Pankhurst, C.E., B.M. Doube, and V.V.S.R. Gupta (eds.). 1997. Biological indicators of soil health. CAB International, Wallingford, UK.

Pankhurst, C.E., B.G. Hawke, H.J. McDonald, C.A. Kirkby, J.C. Buckerfield, P. Michelsen, K.A. O'Brien, V.V.S.R. Gupta, and B.M. Doube. 1995. Evaluation of soil biological properties as potential bioindicators of soil health. Aust. J. Exp. Agric. 35:1015-1028.

Papendick, R.I., and J.F. Parr. 1992. Soil quality—The key to a sustainable agriculture. Am. J. Altern. Agric. 7:2-3.

Pardue, J.H., R.D. DeLaune, and W.H. Patrick, Jr. 1992. Metal to aluminum correlation in Louisiana wetlands: Identification of elevated metal concentrations. J. Environ. Qual. 21:539-545.

Parfitt, R.L. 1990. Allophane in New Zealand—A review. Aust. J. Soil Res. 28:343-360. Available online at http://www.publish.csiro.au/nid/84/paper/SR9900343.htm (verified January 20, 2011).

Parfitt, R.L., and C.W. Childs. 1988. Estimation of forms of Fe and Al: A review of and analysis of contrasting soils by dissolution and Moessbauer methods. Aust. J. Soil Res. 26:121-144.

Parfitt, R.L., and T. Henmi. 1982. Comparison of an oxalate-extraction method and an infrared spectroscopic method for determining allophane in soil clays. Soil Sci. Nutr. 28:183-190.

Parfitt, R.L., and J.M. Kimble. 1989. Conditions for formation of allophane in soils. Soil Sci. Soc. Am. J. 53:971-977.

Parfitt, R.L., and A.D. Wilson. 1985. Estimation of allophane and halloysite in three sequences of volcanic soils, New Zealand. Catena Suppl. 7:1-9.

Parkinson, D., and E.A. Paul. 1982. Microbial biomass. pp. 815-820. *In* A.L. Page, R.H. Miller, and D.R. Kenney (eds.), Methods of soil analysis. Part 2. Chemical and microbiological properties. 2nd ed. SSSA, Madison, WI.

Parsons, R.B., C.A. Balster, and A.O. Ness. 1970. Soil development and geomorphic surfaces, Willamette Valley, Oregon. Soil Sci. Soc. Am. Proc. 34:485-491.

Parton, W.J., D.S. Ojima, C.V. Cole, and D.S. Schimel. 1994. A general model for soil organic matter dynamics: Sensitivity to litter chemistry, texture and management. pp. 147-167. *In* Quantitative modeling of soil forming processes, Soil Sci. Soc. Am. Spec. Publ. 39. SSSA, Madison, WI.

Parton, W.J., D.S. Schimel, C.V. Cole, and D.S. Ojima. 1987. Analysis of factors controlling soil organic matter levels in Great Plains grasslands. Soil Sci. Soc. Am. J. 51:1173-1179.

Patterson, G.T. 1993. Collection and preparation of soil samples: Site description. pp. 1-4. *In* Martin R. Carter (ed.), Soil sampling and methods of analysis. Canadian Society of Soil Science. Lewis Publishers, Boca Raton, FL.

Paul, E.A., and F.E. Clark. 1989. Soil microbiology and biochemistry. Academic Press, Inc., Harcourt Brace Jovanovich, San Diego, CA.

Peech, M., L.T. Alexander, L.A. Dean, and J.F. Reed. 1947. Methods of soils analysis for soil fertility investigations. USDA Circ. 757.

Peech, M., R.L. Cowan, and J.H. Baker. 1962. A critical study of the $BaCl_2$-triethanolamine and the ammonium acetate methods for determining the exchangeable hydrogen content of soils. Soil Sci. Soc. Am. J. 26:37-40.

Peele, T.C., E.E. Latham, and O.W. Beale. 1945. The effect of raindrop impact on the dynamics of soil surface crusting and water movement in the profile. J. Hydrology 52:321-335.

Pennell, K.D. 2002. Specific surface area. *In* J.H. Dane and G.C. Topp (eds.), Methods of soil analysis. Part 4. Physical methods. Soil. Sci. Soc. Am. Book Series 5. SSSA, Madison, WI.

Pennell, K.D., S.A. Boyd, and L.M. Abriola. 1995. Surface area of soil organic matter reexamined. Soil Sci. Soc. Am. J. 59:1012-1018.

Petersen, R.G., and L.D. Calvin. 1986. Sampling. pp. 33-51. *In* A. Klute (ed.), Methods of soil analysis. Part 1. Physical and mineralogical methods. 2nd ed. Agron. Monogr 9. ASA and SSSA, Madison, WI.

Phillips, R.E., and D. Kirkham. 1962. Soil compaction in the field and corn growth. Agron. J. 54:29-34.

Pierce, F.J., W.E. Larson, R.H. Dowdy, and W.A.P. Graham. 1983. Productivity of soils: Assessing long-term changes and erosion. J. Soil Water Cons. 138:39-44.

Pierre, W.H., and W.L. Banwart. 1973. Excess-base and excess-base nitrogen ratio of various crops species and parts of plants. Agron. J. 65:91-96.

Pierzynski, G.M. 2000. Methods of phosphorus analysis for soils, sediments, residuals, and waters. SERA-IEG 17, USDA-CSREES Regional Committee Minimizing Agricultural Phosphorus Losses for Protection of the Water Resource. Southern Coop. Ser. Bull. 396. Kansas State University, Manhattan, KS.

Pierzynski, G.M., and A.P. Schwab. 1993. Bioavailability of zinc, cadmium, and lead in a metal-contaminated alluvial soil. J. Environ. Qual. 22:247-254.

Pierzynski, G.M., J.T. Sims, and G.F. Vance. 1994. Soils and environmental quality. 1st ed. Lewis Publishers, Boca Raton, FL.

Pierzynski, G.M., J.T. Sims, and G.F. Vance. 2000. Soils and environmental quality. 2nd ed. Lewis Publishers, Boca Raton, FL.

Pimentel, D., E.C. Terhune, R. Dyson-Hudson, S. Rochereau, R. Samis, E.Q. Smith, D. Denman, D. Reifschneider, and M. Shepard. 1976. Land degradation: Effects on food and energy resources. Sci. 194:149-155.

Pinsker, L.M. 2001. Health hazards: Arsenic. Geotimes, Nov., pp. 32-33.

Polyzopoulos, N.A., V.Z. Keramidas, and H. Koisse. 1985. Phosphate sorption by some Alfisols as described by commonly used isotherms. Soil Sci. Soc. Am. J. 49:81-84.

Pons, L.J., and I.S. Zonneveld. 1965. Soil ripening and soil classification. Initial soil formation in alluvial deposits and a classification of the resulting soils. Int. Inst. Land Reclam. and Impr. Publ. 13. Wageningen, The Netherlands.

Porta, J., and J. Herrero. 1990. Micromorphology and genesis of soils enriched with gypsum. *In* L.A. Douglas (ed.), Soil micromorphology: A basic and applied science. Proceedings of the 8th International Working Meeting, Soil Micromorphology, San Antonio, TX, July, 1988.

Pote, D.H., T.C. Daniel, A.N. Sharpley, P.A. Moore, Jr., D.R. Edwards, and D.J. Nichols. 1996. Relating extractable soil phosphorus to phosphorus losses in runoff. Soil Sci. Soc. Am. J. 60:855-859.

Powell, J.C., and M.E. Springer. 1965. Composition and precision of classification of several mapping units of the Appling, Cecil, and Lloyd series in Walton County, Georgia. Soil Sci. Soc. Am. Proc. 29:454-458.

Pratt, P.F., and N., Holowaychuk. 1954. A comparison of ammonium acetate, barium acetate, and buffered barium chloride methods of determining cation exchange capacity. Soil Sci. Soc. Am. J. 18:365-368.

Pregitzer, K.S., and B.V. Barnes. 1984. Classification and comparison of upland hardwood and conifer ecosystems of the Cyrus H. McCormick Experimental Forest, Upper Michigan. Can. J. For. Res. 14:362-375.

Presley, B.J. 1975. A simple method for determining calcium carbonate in sediment samples. J. Sediment. Petrol. 45:745-746.

Proctor, J., and S.R.J. Woodell. 1975. The ecology of serpentine soils. Adv. Ecol. Res. 9:255-366.

Prunty, L. 2004. Soil science and pedology disciplines in the North Dakota 1862 land grant college of agriculture: A review. North Dakota State University White Paper.

Quantachrome Instruments. 2003. Penta-pycnometer instruction manual. Quantachrome Corporation, Boynton Beach, FL

Quirk, J.P., and R.S. Murray. 1999. Appraisal of the ethylene glycol monoethyl ether method for measuring hydratable surface area of clays and soils. Soil Sci. Soc. Am. J. 63:839-849.

Rabenhorst, M.C., L.T. West, and L.P. Wilding. 1991. Genesis of calcic and petrocalcic horizons in soils over carbonate rocks. pp. 61-74. *In* W.D. Nettleton (ed.), Occurrence, characteristics, and genesis of carbonate, gypsum, and silica accumulations in soils. Soil Sci. Soc. Am. Spec. Publ. 26. ASA and SSSA, Madison, WI.

Raghavan, G.S.V., E. McKyes, G. Gendron, B. Borghum, and H.H. Lee. 1978. Effects of soil compaction in development of yield of corn (maize). Can. J. Plant Sci. 58:435-443.

Rao, S.C., and J. Ryan. 2004. Challenges and strategies for dryland agriculture. Crop Sci. Soc. Am. Spec. Publ. 32. CSSA, Madison, WI.

Rebertus, R.A., and S.W. Buol. 1989. Influence of micaceous minerals on mineralogy class placement of loamy and sandy soils. Soil Sci. Soc. Am. J. 53:196-201.

Reed, S.T., and D.C. Martens. 1996. Copper and zinc. pp. 703-722. *In* D.L. Sparks (ed.), Methods of soil analysis. Part 3. Chemical methods. Soil Sci. Soc. Am. Book Series 5. ASA and SSSA, Madison, WI.

Reeder, J.D., C.D. Franks, and D.G. Milchunas. 2001. Root biomass and microbial biomass. pp. 139-166. *In* R.F. Follett, J.M. Kimble, and R. Lal (eds.), The potential of U.S. grazing lands to sequester carbon and mitigate the greenhouse effect. Lewis Publishers, Boca Raton, FL.

Reeder, R.C. 2002. Controlled traffic. pp. 233-235. *In* R. Lal (ed.), Encyclopedia of soil science. Marcel Dekker, Inc.

Reeve, N.G., and M.E. Sumner. 1972. Amelioration of subsoil acidity in Natal Oxisols by leaching of surface-applied amendments. Agrochemophysica 4:1-6.

Reeves, D.W. 1997. The role of organic matter in maintaining soil quality in continuous cropping systems. Soil and Tillage Res. 43:131-167.

Reeves, D.W., and C.W. Wood. 1994. A sustainable winter-legume conservation tillage system for maize: Effects on soil quality. pp. 1011-1016. *In* Proceedings of the 13th International Conference, International Soil Tillage Research Organization (ISTRO). Vol. II. The Royal Veterinary and Agricultural University and the Danish Institute of Plant and Soil Science, Aalborg, Denmark. July 24-29, 1994.

Reinsch, T.G., and R.B. Grossman. 1995. A method to predict bulk density of tilled Ap horizons. Soil and Tillage Res. 34:95-104.

Remley, P.A., and J.M. Bradford. 1989. Relationship of soil crust morphology to interrill erosion parameters. Soil Sci. Soc. Am. J. 53:1215-1221.

Renard, K.G., G.R. Foster, G.A. Weesies, D.K. McCool, and D.C. Yoder. 1997. Predicting soil erosion by water: A guide to conservation planning with the revised Universal Soil Loss Equation (RUSLE). USDA/ARS, Agric. Handb. 703. U.S. Government Printing Office, Washington, DC.

Renella, G., P. Adamo, M.R. Bianco, L. Landi, P. Violante, and P. Nannipieri. 2004. Availability and speciation of cadmium added to a calcareous soil under various managements. European J. Soil Sci. 55:123-133.

Rengasamy, P. 1997. Sodic soils. pp. 265-277. *In* R. Lal, W.H. Blum, C. Valentine, and B.A. Stewart (eds.), Methods of assessment of soil degradation. CRC Press, New York.

Rengasamy, P. 2002. Sodic soils. pp. 1210-1212. *In* R. Lal (ed.), Encyclopedia of soil science. Marcel Dekker, Inc.

Renshaw, C.E., B.C. Bostick, X. Feng, C.K. Wong, E.S. Winston, R. Karimi, C.L. Folt, and C.Y. Chen. 2006. Impact of land disturbance on the fate of arsenical pesticides. J. Environ. Qual. 35:61-67.

Rhoades, J.D. 1982a. Cation exchange capacity. pp. 149-157. *In* A.L. Page, R.H. Miller, and D.R. Keeney (eds.), Methods of soil analysis. Part 2. Chemical and microbiological properties. 2nd ed. Agron. Monogr. 9. ASA and SSSA, Madison, WI.

Rhoades, J.D. 1982b. Soluble salts. pp. 167-179. *In* A.L Page, R.H. Miller, and D.R. Keeney (eds.), Methods of soil analysis. Part 2. Chemical and microbiological properties. 2nd ed. Agron. Monogr. 9. ASA and SSSA, Madison, WI.

Rhoades, J.D. 1990a. Principal effects of salts on soils and plants. *In* A. Kandiah (ed.), Water, soil, and crop management relating to the use of saline water. FAO (AGL) Misc. Ser. Publ. 16/90. FAO, Rome, Italy.

Rhoades, J.D. 1990b. Soil salinity—Causes and controls. pp. 109-134. *In* A.S. Goude (ed.), Techniques for desert reclamation. Wiley, New York.

Rhoades, J.D. 1998. Use of saline and brackish waters for irrigation: Implications and role in increasing food production, conserving water, sustaining irrigation, and controlling soil and water degradation. pp. 261-304. *In* R. Ragab and G. Pearce (eds.), Proceedings of the International Workshop on "The Use of Saline and Brackish Waters for Irrigation: Implications for the Management of Irrigation, Drainage, and Crops" at the 10th Afro-Asian Conference of the International Committee on Irrigation and Drainage, Bali, Indonesia. July 23-24.

Rhoades, J.D. 1999. Use of saline drainage water for irrigation. pp. 619-657. *In* R.W. Skaggs and J. van Schilfgaarde (eds.), Agricultural drainage. Am. Soc. Agron. Monogr. 38. ASA, Madison, WI.

Rhoades, J.D. 2002. Irrigation and soil salinity. pp.750-753. *In* R. Lal (ed.), Encyclopedia of soil science. Marcel Dekker, Inc.

Rhoades, J.D., A. Kandiah, and A.M. Mashali. 1992. The use of saline waters for crop production. FAO Irrigation and Drainage Paper 48. FAO, Rome, Italy.

Richards, L.A. 1960. Advances in soil physics. Trans. 7th Int. Congr. Soil Sci. 1:67-69.

Richards, L.A., and L.R. Weaver. 1943. Fifteen-atmosphere percentages as related to the permanent wilting percentage. Soil Sci. 56:331-339.

Richards, L.A., and L.R. Weaver. 1944. Moisture retention by some irrigated soils as related to soil moisture tension. J. Agric. Res. 69:215-235.

Richards, S.J., and C.H. Wadleigh. 1952. Soil water and plant growth. pp. 73-251. *In* B.T. Shaw (ed.), Soil physical conditions and plant growth. Academic Press Inc., New York, NY.

Rigler, F.W. 1968. Further observations inconsistent with the hypothesis that the molybdenum blue method measures orthophosphate in lake water. Limnology and Oceanography 13:7-13.

Robertson, E.G., D.C. Coleman, C.S. Bledsoe, and P. Sollins (eds.). 1999. Standard soil methods for long-term ecological research. Oxford University Press.

Robinson, C.A., R.M. Cruse, and M. Ghaffarzadeh. 1996. Cropping system and nitrogen effects on Mollisol organic carbon. Soil Sci. Soc. Am. J. 60:264-269.

Robinson, G.W. 1936. Soils, their origin, constitution and classification. 2nd ed. Thomas Murby & Co., London.

Robinson, G.W., and E.W. Lloyd. 1915. Probable error of sampling in soil surveys. J. Agric. Sci. 7:144-153.

Rode, A.A. 1969. Theory of soil moisture. Vol. 1. Moisture properties of soils and movement of soil moisture. (In English, translated from Russian.) Israel Program for Scientific Translations, Jerusalem. p. 38.

Rodgers, J. 1940. Distinction between calcite and dolomite on polished surfaces. Am. J. Sci. 238:788-798.

Rogers, H.T, and D.L. Thurlow. 1973. Soybeans restricted by soil compaction. Highlights Agric. Res. 20:10. Auburn University Agricultural Experiment Station, Auburn, AL.

Romkens, P.F.A.M., and W. Salomons. 1998. Cd, Cu, and Zn solubility in arable and forest soils: Consequences of land use changes for metal mobility and risk assessment. Soil Sci. 163:859-871.

Rose, C.W. 1998. Modeling erosion by water and wind. pp. 57-88. *In* R. Lal, W.H. Blum, C. Valentine, and B.A. Stewart (eds.), Methods for assessment of soil degradation. Advances in Soil Science. CRC Press.

Rose, C.W. 2002. Erosion by water, modeling. pp. 468-472. *In* R. Lal (ed.), Encyclopedia of soil science. Marcel Dekker, Inc.

Ross, C.S., and P.F. Kerr. 1934. Halloysite and allophane. U.S. Geol. Surv. Prof. Paper 185 G:135-148.

Rostagno, C.M. 1989. Infiltration and sediment production as affected by soil surface conditions in a shrubland of Patagonia, Argentina. J. Range Manage. 42:383-385.

Rouston, R.C., J.A. Kittrick, and E.H. Hope. 1972. Interlayer hydration and the broadening of the 10A x-ray peak in illite. Soil Sci. 113:167-174.

Rowe, J.S. 1980. The common denominator in land classification in Canada: An ecological approach to mapping. Forest Chronicle 56:19-20.

Rowe, J.S. 1984. Forest land classification: Limitations of the use of vegetation. pp. 132-147. Proceedings of the symposium on forest land classification. March 18-20. Madison, WI.

Ruhe, R.V. 1975. Geomorphology. Geomorphic processes and surficial geology. Houghton Mifflin Co., Boston.

Ruhe, R.V., R.B. Daniels, and J.G. Cady. 1967. Landscape evolution and soil formation in southwestern Iowa. USDA Tech. Bull. 1349.

Ryan, J., D. Curtin, and M.A. Cheema. 1984. Significance of iron oxides and calcium carbonate particle-size in phosphate sorption by calcareous soils. Soil Sci. Soc. Am. J. 48:74-76.

Ryker, S.J. 2001. Mapping arsenic in groundwater: A real need, but a hard problem. Geotimes, Nov., pp. 34-36.

Sabbe, E., and D.B. Marx. 1987. Soil sampling: Spatial and temporal variability. pp. 1-14. *In* J.R. Brown (ed.), Soil testing: Sampling, correlation, calibration, and interpretation. Soil Sci. Soc. Am. Spec. Publ. 21. SSSA, Madison, WI.

Sahrawat, K.L. 1983. An analysis of the contribution of organic matter and clay to cation exchange capacity of some Philippine soils. Commun. Soil Sci. Plant Anal. 14:803-809.

Said, M.B., and A. Dakermanji. 1993. Phosphate adsorption and desorption by calcareous soils of Syria. Commun. Soil Sci. Plant Anal. 24:197-210.

Salter, P.J., and J.B. Williams. 1965. The influence of texture on the moisture characteristics of soil determining the available water capacity and moisture characteristic curve of a soil. Soil Sci. 16:1-15.

Saly, R. 1967. Use of ultrasonic vibration for dispersing of soil samples. Sov. Soil Sci. 11:1547-1559.

Sampson, A.W. 1923. Range and pasture management. John Wiley & Sons Inc., New York, NY.

Sanford, W.E., and W.W. Wood. 2001. Hydrology of the coastal sabkhas of Abu Dhabi, United Arab Emirates. Hydrogeology J. 9:358-366.

Sawhney, B.L., and D.E. Stilwell. 1994. Dissolution and elemental analysis of minerals, soils, and environmental samples. pp. 49-82. *In* J.E. Amonette and L.W. Zelazny (eds.), Quantitative methods in soil mineralogy. Soil Sci. Soc. Am. Misc. Publ. SSSA, Madison, WI.

Sayegh, A.H., N.A. Khan, P. Khan, and J. Ryan. 1978. Factors affecting gypsum and cation exchange capacity determinations in gypsiferous soils. Soil Sci. 125:294-300.

Schnitzer, M. 1982. Organic matter characterization. *In* A.L. Page, R.H. Miller, and D.R. Keeney (eds.), Methods of soil analysis. Part 2. Chemical and microbiological properties. 2nd ed. Agron. 9:581-594.

Schnitzer, M., and S.U. Khan. 1978. Soil organic matter. Developments in Soil Sci. 8. Elsevier, New York, NY.

Schnitzer, M., and H. Kodama. 1977. Reactions of minerals with soil humic substances. pp. 741-770. *In* J.B. Dixon and S.B. Weed (eds.), Minerals in soil environments. 2nd ed. Soil Sci. Soc. Am. Book Series 1. SSSA, Madison, WI.

Schoeneberger, P.J., and D.A. Wysocki. 2004. Geomorphology. pp. 1-3. *In* R. Burt (ed.), Soil survey laboratory methods manual. Ver. 4.0. USDA/NRCS, Soil Survey Investigations Report No. 42. U.S. Government Printing Office, Washington, DC.

Schoeneberger, P.J., D.A. Wysocki, E.C. Benham, and W.D. Broderson. 2002. Field book for describing and sampling soils. Ver. 2.0. USDA/NRCS, National Soil Survey Center, Lincoln, NE.

Schraer, S.M., D.R. Shaw, M. Boyette, R.H. Coupe, and E.M. Thurman. 2000. Comparison of enzyme-linked immunosorbent assay and gas chromatography procedures for the detection of cyanazine and metolachlor in surface water samples. J. Agric. and Food Chem. 48:5881-5886.

Schreier, H.E., J.A. Omueti, and L.M. Lavkulich. 1987. Weathering processes of asbestos-rich serpentinitic sediments. Soil Sci. Soc. Am. J. 51:993-999.

Schulte, E.E., and B.G. Hopkins. 1996. Estimation of organic matter by weight loss-on-ignition. pp. 21-31. *In* F.R. Magdoff et al. (eds.), Soil organic matter: Analysis and interpretation. Soil Sci. Soc. Am. Spec. Publ. 46. SSSA, Madison, WI.

Schulze, D.G. 1989. An introduction to soil mineralogy. pp. 1-34. *In* J.B. Dixon and S.B. Weed (eds.), Minerals in soil environments. Soil Sci. Soc. Am. Book Series 1. SSSA, Madison, WI.

Schumacher, B.A. 2002. Methods for the determination of total organic carbon (TOC) in soils and sediments. NCEA-C 1282, EMASC, April 2002. U.S. Environmental Protection Agency, Environmental Science Division, National Exposure Research Lab, Las Vegas, NV.

Schuppli, P.A., G.J. Ross, and J.A. McKeague. 1983. The effective removal of suspended materials from pyrophosphate extracts of soils from tropical and temperate regions. Soil Sci. Soc. Am. J. 47:1026-1032.

Schwertmann, U. 1959. Die fraktionierte Extraktion der freien Eisenoxide in Böden, ihre mineralogischen Formen und ihre Entstehungsweisen. Z. Pflanzenernähr. Düng. Bodenk. 84:194-204.

Schwertmann, U. 1964. The differentiation of iron oxides in soil by extraction with ammonium oxalate solution. Z. Pflanzenernähr. Düng. Bodenk. 105:194-202.

Schwertmann, U. 1973. Use of oxalate for iron extraction from soils. Can. J. Soil Sci. 53:244-246.

Schwertmann, U. 1985. The effect of pedogenic environments on iron oxide minerals. pp. 172-200. In B.A. Stewart (ed.), Advances in soil science. Vol. 1.

Schwertmann, U. 1992. Relations between iron oxides, soil color, and soil formation. In J.M. Bigham and E.J. Ciolkosz (eds.), Soil color. Soil Sci. Soc. Am. Spec. Publ. 31. SSSA, Madison, WI.

Schwertmann, U., R.W. Fitzpatrick, and J. LeRoux. 1977. Al substitution and differential disorder in soil hematites. Clays and Clay Minerals 25:373-374.

Schwertmann, U., and R.M. Taylor. 1977. Iron oxides. pp. 145-180. In J.B. Dixon and S.B. Weed (eds.), Minerals in soil environments. Soil Sci. Soc. Am. Book Series 1. SSSA, Madison, WI.

Schwertmann, U., and R.M. Taylor. 1989. Iron oxides. pp. 370-438. In J.B. Dixon and S.B. Weed (eds.), Minerals in soil environments. 2nd ed. Soil Sci. Soc. Am. Book Series 1. SSSA, Madison, WI.

Scofield, R.K. 1942. Pecos River joint investigation: Reports of participating agencies. pp. 263-334. U.S. National Resources Planning Board, Washington, DC.

Seifferlein, E.R., P. Jones, R. Ferguson, R. Burt, and M.D. Mays. 2005. Extractable acidity by a centrifuge method. Commun. Soil Sci. Plant Anal. 36:2067-2083.

Seybold, C.A., R.B. Grossman, and T.G. Reinsch. 2005. Predicting cation exchange capacity for soil survey using linear models. Soil Sci. Soc. Am. J. 69:856-863.

Shacklette, H.T., and J.G. Boerngen. 1984. Elemental concentrations in soils and surficial materials of the conterminous United States. U.S. Geol. Surv. Prof. Paper 1270. U.S. Government Printing Office, Washington, DC.

Shahid, S.A., M.A. Abdelfattah, and M.A. Wilson. 2007. A unique anhydrite soil in the Coastal Sabkha of Abu Dhabi Emirate. Soil Surv. Horiz. 48:75-79.

Shainberg, I. 1992. Chemical and mineralogical components of crusting. In M.E. Sumner and B.A. Stewart (eds.), Soil crusting, chemical and physical processes. Advances in Soil Science. Lewis Publishers.

Sharpley, A.N. 1985. Depth of surface soil-runoff interaction as affected by rainfall, soil, slope, and management. Soil Sci. Soc. Am. J. 49:1010-1015.

Sharpley, A.N. 1995. Dependence of runoff phosphorus on extractable soil phosphorus. J. Environ. Qual. 24:920-926.

Sharpley, A.N. 1996. Availability of residual phosphorus in manured soils. Soil Sci. Soc. Am. J. 60:1459-1466.

Sharpley, A.N., B.J. Carter, B.J.Wagner, S.J.Smith, E.L. Cole, and G.A. Sample. 1991. Impact of long-term swine and poultry manure applications on soil and water resources in eastern Oklahoma. Oklahoma State Univ. Tech. Bull. T169. Oklahoma State University, Stillwater.

Sharpley, A.N., C.A.Jones, C. Gray, and C.V. Cole. 1984. A simplified soil and plant phosphorus model. II. Prediction of labile, organic, and sorbed phosphorus. Soil Sci. Soc. Am. J. 48:805-809.

Sharpley, A.N., C.A. Jones, C. Gray, C.V. Cole, H. Tiessen, and C.S. Holzhey. 1985. A detailed phosphorus characterization of seventy-eight soils. USDA/ARS, ARS-31. National Technical Information Service, Springfield, VA.

Sharpley, A.N., and R.G. Menzel. 1987. The impact of soil and fertilizer phosphorus on the environment. Adv. Agron. 41:297-324.

Sharpley, A.N., U. Singh, G. Uehara, and J. Kimble. 1989. Modeling soil and plant phosphorus dynamics in calcareous and highly weathered soils. Soil Sci. Soc. Am. J. 53:153-158.

Sharpley, A.N., and S.J. Smith. 1985. Fractionation of inorganic and organic phosphorus in virgin and cultivated soils. Soil Sci. Soc. Am. J. 49:127-130.

Sharpley, A.N., and J.R. Williams. 1990. EPIC—erosion/productivity impact calculator: I. Model documentation. USDA Tech. Bull. 1768.

Shelley, David. 1978. Manual of optical mineralogy. Elsevier. North-Holland, Inc., NY.

Shoji, S., M. Nanzyo, and R.A. Dahlgren. 1993. Volcanic ash soils. Genesis, properties, and utilization. Developments in soil science. 21. Elsevier, Amsterdam, The Netherlands.

Shuman, L.M. 1985. Fractionation method for soil microelements. Soil Sci. 140:11-22.

Simpson, J.R., A. Pinkerton, and J. Lazdovskis. 1977. Effects of subsoil calcium on the root growth of some lucerne genotypes (*Medicago sativa* L.) in acidic soil profiles. Aust. J. Agric. Res. 28:629-638.

Sims, G.K. 1990. Biological degradation of soil. pp. 289-330. *In* R. Lal and B.A. Stewart (eds.), Soil degradation. Advances in Soil Science. Springer-Verlag.

Sims, J.T. 1989. Comparison of Mehlich 1 and Mehlich 3 extractants for P, K, Ca, Mg, Mn, Cu, Zn in Atlantic Coastal Plain soils. Commun. Soil Sci. Plant Anal. 20:1707-1726.

Sims, J.T., R.R. Simard, and B.C. Joern. 1998. Phosphorus loss in agricultural drainage: Historical perspective and research. J. Environ. Qual. 27:277-293.

Sinclair, T.R. 1994. Limits to crop yield? pp. 509-532. *In* K.J. Boone (ed.), Physiology and determination of crop yield. ASA, Madison, WI.

Singer, M.J., and D.N. Munns. 1987. Soils: An introduction. Macmillan Publishing Company, New York, NY.

Singer, M.J., and D.N. Warrington. 1992. Crusting in the Western United States. pp.179-204. *In* M.E. Sumner and B.A. Stewart (eds.), Soil crusting, chemical and physical processes. Advances in Soil Science. Lewis Publishers.

Singleton, G.A., and L.M. Lavkulich. 1987. Phosphorus transformations in soil chronosequence, Vancouver Island, British Columbia. Can. J. Soil Sci. 67:787-793.

Skinner, H.C.W., M. Ross, and C. Frondel. 1988. Asbestos and other fibrous materials. Oxford University Press, United Kingdom.

Skoog, D.A. 1985. Principles of instrumental analysis. 3rd ed. Saunders College Publishing, New York, NY.

Skopp, J. 1992. Concepts of soil physics. Course notes for Agronomy 461/861. University of Nebraska, Lincoln, NE.

Skujins, J. 1967. Enzymes in soil. pp. 371-414. In A.D. McLaren and G.H. Peterson (eds.), Soil Biochemistry. Vol. 1. Marcel Dekker, Inc.

Skujins, J. 1976. Extracellular enzymes in soil. Critical Reviews in Microbiology 4:383-421.

Slatyer, R.O. 1957. The significance of the permanent wilting percentage in studies of plant and soil water relations. Bot. Rev. 23:585-636.

Small, R.J. 1975. The study of landforms. A textbook of geomorphology. 2nd ed. Cambridge University Press.

Smalley, G.W. 1986. Site classification and evaluation for the Interior Uplands. USDA, Forest Service, Southern Region. Tech. Publ. R8-TP9. Atlanta, GA.

Smith, F.W., B.G. Ellis, and J. Grava. 1957. Use of acid-fluoride solutions for the extraction of available phosphorus in calcareous soils and in soils to which rock phosphate has been added. Soil Sci. Soc. Am. Proc. 21:400-404.

Smith, J.L., and L.F. Elliott. 1990. Tillage and residue management effects on soil organic matter dynamics in semiarid regions. pp. 69-88. *In* R.P. Singh, J.F. Parr, and B.A. Stewart (eds.), Dryland agriculture. Strategies for sustainability. Advances in Soil Science. Springer-Verlag.

Smucker, A.J.M., S.L. McBurney, and A.K. Srivastava. 1982. Quantitative separation of roots from compacted soil profiles by hydropneumatic elutriation system. Agron. J. 74:500-503.

Snell, F.D., and C.T. Snell. 1949. Colorimetric methods of analysis. Phosphorus. Vol 2. 3rd ed. pp. 630-681. D. Van Nostrand Co., Inc.

Soil and Plant Analysis Council. 1999. Handbook on reference methods for soil analysis. Council on Soil Testing and Plant Analysis. CRC Press, Boca Raton, FL.

Soil Quality Institute. 1999. Soil quality test kit guide. USDA/ARS and USDA/NRCS. Available online at http://soils.usda.gov/sqi/assessment/test_kit.html (verified January 24, 2011).

Soil Science Society of America (SSSA). 1937. New size limits for silt and clay. Proceedings 2.

Soil Science Society of America (SSSA). 1993. A reference collection for soil micromorphology. CD Rom. ASA and SSSA, Madison, WI.

Soil Science Society of America (SSSA). 2010. Glossary of soil science terms. SSSA, Madison, WI. Available online at https://www.soils.org/publications/soils-glossary/ (verified January 24, 2011).

Soil Survey Division Staff. 1993. Soil survey manual. USDA Handb. 18. U.S. Government Printing Office, Washington, DC.

Soil Survey Staff. 1951. Soil survey manual. USDA/SCS. U.S. Government Printing Office, Washington, DC.

Soil Survey Staff. 1975. Soil taxonomy: A basic system of soil classification for making and interpreting soil surveys. USDA/SCS. Agric. Handb. 436. U.S. Government Printing Office, Washington, DC.

Soil Survey Staff. 1989. Soil survey laboratory methods manual. Ver. 1.0. USDA/NRCS, Soil Survey Investigations Report No. 42. U.S. Government Printing Office, Washington, DC.

Soil Survey Staff. 1992. Soil survey laboratory methods manual. Ver. 2.0. USDA/NRCS, Soil Survey Investigations Report No. 42. U.S. Government Printing Office, Washington, DC.

Soil Survey Staff. 1995. Soil survey laboratory information manual. Ver. 1.0. USDA/NRCS, Soil Survey Investigations Report No. 45. U.S. Government Printing Office, Washington, DC. Available online at http://soils.usda.gov/survey/nscd/lim/index.html (verified January 24, 2011).

Soil Survey Staff. 1996. Soil survey laboratory methods manual. Ver. 3.0. USDA/NRCS, Soil Survey Investigations Report No. 42. U.S. Government Printing Office, Washington, DC.

Soil Survey Staff. 1998. Glossary of landforms and geologic materials. Part 629, National soil survey handbook. USDA/NRCS, National Soil Survey Center, Lincoln, NE.

Soil Survey Staff. 1999. Soil taxonomy: A basic system of soil classification for making and interpreting soil surveys. 2nd ed. USDA/NRCS. Agric. Handb. 436. U.S. Government Printing Office, Washington, DC. Available online at ftp://ftp-fc.sc.egov.usda.gov/NSSC/Soil_Taxonomy/tax.pdf (verified January 24, 2011).

Soil Survey Staff. 2004. R. Burt (ed.). Soil survey laboratory methods manual. Ver. 4.0. USDA/NRCS. Soil Survey Investigations Report No. 42. U.S. Government Printing Office, Washington, DC. Available online at http://soils.usda.gov/technical/lmm/ (verified January 24, 2011).

Soil Survey Staff. 2009. R. Burt (ed.). Soil survey field and laboratory methods manual. Ver. 1.0. USDA/NRCS, Soil Survey Investigations Report No. 51. Available online at http://www.soils.usda.gov/technical/ (verified January 24, 2011).

Soil Survey Staff. 2010. Keys to soil taxonomy. 11th ed. USDA/NRCS. U.S. Government Printing Office, Washington, DC. Available online at ftp://ftp-fc.sc.egov.usda.gov/NSSC/Soil_Taxonomy/keys/2010_Keys_to_Soil_Taxonomy.pdf (verified January 24, 2011).

Solis, P., and J. Torrent. 1989. Phosphate sorption by calcareous Vertisols and Inceptisols of Spain. Soil Sci. Soc. Am. J. 53:456-459.

Sombroek, W., M.L. Ruivo, P.M. Fearnside, B. Glaser, and J. Lehmann. 2003. Amazonian dark earths as carbon stores and sinks. In J. Lehmann, D.C. Kern, B. Glaser, and W.I. Woods (eds.), Amazonian dark earths: Origin, properties, management. Kluwer Academic Publishers, Dordrecht, The Netherlands.

Sosebee, R.E. (ed.) 1977. Rangeland plant physiology. Range Sci. Series 4. Society for Range Management, Denver, CO.

Spalding, R.F., D.G. Watts, D.D. Snow, D.A. Cassada, M.E. Exner, and J.S. Schepers. 2003. Herbicide loading to shallow ground water beneath Nebraska's management systems evaluation area. J. Environ. Qual. 32:84-91.

Sparrow, H.O. 1984. Soil at risk. Canada's eroding future. Standing Senate Committee on Agriculture, Fisheries, and Forestry. Ottawa, Ontario, Canada.

Spies, T.A., and B.V. Barnes. 1985. A multifactor ecological classification of the northern hardwood and conifer ecosystems of Sylvania Recreation Area, Upper Peninsula, Michigan. Can. J. For. Res. 15:949-960.

Sposito, G. 1989. The chemistry of soils. Oxford University Press, New York, NY.

Springer, M.E. 1958. Desert pavement and vesicular layer of some soils of the desert of the Lahontan Basin, Nevada. Soil Sci. Soc. Am. Proc. 22:63-66.

Stahlberg, S., S. Sombatpanit, and J. Stahlberg. 1976. Manganese relationships in soil and plant: II. Studies on manganese fixation. Acta Agric. Scand. 26:65-81.

Steel, R.G.D., and J.H. Torrie. 1980. Principles and procedures of statistics: A biometrical approach. McGraw-Hill Co., New York, NY.

Steinnes, E., T.E. Sjobakk, C. Donisa, and M.L. Brannvall. 2005. Quantification of pollutant lead in forest soils. Soil Sci. Soc. Am. J. 69:1399-1404.

Stevenson, F.J. 1982. Origin and distribution of nitrogen in soil. *In* F.J. Stevenson (ed.), Nitrogen in agricultural soils. Agron. 22:1-39.

Stewart, B.A., R. Lal, and S.A. El-Swaify. 1991. Sustaining the resource base of an expanding world argriculture. pp. 125-144. *In* R. Lal and F.J. Pierce (eds.), Soil management for sustainability. Soil and Water Conservation Society, Ankeny, IA.

Stewart, J.W.B., and J.R. Bettany. 1982a. Dynamics of organic sulphur. Proc. Alberta Soil Sci. Workshop, p. 184. Edmonton, Alberta, February 23-24, 1982.

Stewart, J.W.B., and J.R. Bettany. 1982b. Dynamics of soil organic phosphorus and sulfur. Publ. R290, Saskatchewan Institute of Pedology, Saskatoon.

Stewart, J.W.B., and H. Tiessen. 1987. Dynamics of soil organic phosphorus. Biogeochemistry 4:41-60.

Stewart, M.A., P.M. Jardine, M.O. Barnett, T.L. Mehlhorn, L.K. Hyder, and L.D. McKay. 2003. Influence of soil geochemical and physical properties on the sorption and bioaccessibility of chromium (III). J. Environ. Qual. 32:129-137.

Stichling, W. 1973. Sediment loads in Canadian rivers. Proceedings of the Ninth Hydrology Symposium, Edmonton, May 8-9, 1973. Fluvial processes and sedimentation.

Stilwell, D.E., and K.D. Gorny. 1997. Contamination of soil with copper, chromium, and arsenic under decks built from pressure treated wood. Bull. Environ. Contam. Toxicol. 58:22-29.

Stoddart, L.A., A.D. Smith, and T.W. Box. 1975. Range management. 3rd ed. McGraw-Hill Co., New York, NY.

Stoiber, R.E., and S.A. Morse. 1972. Microscopic identification of crystals. The Ronald Press Company, New York, NY.

Stoops, G. 2003. Guidelines for analysis and description of soil and regolith thin sections. SSSA, Madison, WI.

Stoops, G., and A. Jongerius. 1975. Proposal for micromorphological classification in soil materials. I. A classification of the related distribution of coarse and fine particles. Geoderma 13:189-200.

Storm, G.L., G.J. Fosmire, and E.D. Bellis. 1994. Persistence of metals in soil and selected vertebrates in the vicinity of the Palmerton zinc smelters. J. Environ. Qual. 23:508-514.

Strickland, T.C., and P. Sollins. 1987. Improved method for separating light- and heavy-fraction organic material from soils. Soil Sci. Soc. Am. J. 51:1390-1393.

Suarez, D.L., J.D. Wood, and S.M. Lesch. 2008. Infiltration into cropped soils: Effects of rain and sodium adsorption ratio-impacted irrigation water. J. Environ. Qual. 37:S-169-S-179.

Suave, S., and D.R. Parker. 2005. Chemical speciation of trace elements in soil solution. pp. 655-688. *In* M.A. Tabatabai and D.L. Sparks (eds.), Chemical processes in soils. Soil Sci. Soc. Am. Book Series 8. SSSA, Madison, WI.

Sudom, M.D., and R.J. St. Arnaud. 1971. Use of quartz, zirconium, and titanium as indices in pedological studies. Can. J. Soil Sci. 51:385-396.

Sumner, M.E. 1970. Aluminum toxicity—A growth limiting factor in some Natal sands. Proc. S. Afr. Sug. Tech. Assoc. 1-6.

Sumner, M.E. 1992. The electrical double layer and clay dispersion. pp. 1-32. *In* M.E. Sumner and B.A. Stewart (eds.), Soil crusting, chemical and physical processes. CRC Press Inc., Boca Raton, FL.

Sumner, M.E. 1995. Amelioration of subsoil acidity with minimum disturbance. pp. 147-185. *In* N.S. Jayawardane and B.A. Stewart (ed.), Subsoil management techniques. Lewis Publishers, Boca Raton, FL.

Sumner, M.E. 1998. Acidification. pp.213-228. *In* R. Lal, W.H. Blum, C. Valentine, and B.A. Stewart (eds.), Methods for assessment of soil degradation. Advances in Soil Science. CRC Press.

Sumner, M.E., and W.P. Miller. 1996. Cation exchange capacity and exchange coefficients. pp. 1201-1229. *In* D.L. Sparks (ed.), Methods of soil analysis. Part 3. Chemical methods. Soil Sci. Soc. Am. Book Series 5. ASA and SSSA, Madison, WI.

Sumner, M.E., H. Shahandeh, J. Bouton, and J.E. Hammel. 1986. Amelioration of an acid soil profile through deep liming and surface application of gypsum. Soil Sci. Soc. Am. J. 50:1254-1258.

Sumner, M.E., and B.A. Stewart (eds.). 1992. Soil crusting, chemical and physical processes. Advances in Soil Science. Lewis Publishers.

Swanson, C.L.W., and J.B. Peterson. 1942. The use of micrometric and other methods for the evaluation of soil structure. Soil Sci. 53:173-183.

Sweeten, J.M., M.L. Wolfe, E.S. Chasteen, M. Sanderson, B.A. Auvermann, and G.D. Alston. 1995. Dairy lagoon effluent irrigation: Effects on runoff quality, soil chemistry, and forage yield. pp. 99-106. *In* K. Steele (ed.), Animal wastes and land-water interface. Lewis Publishers, Boca Raton, FL.

Switzer, G., J.M. Axelrod, M.L. Lindberg, and E.S. Larsen III. 1948. Tables of d spacings for angle 2Θ. U.S. Department of the Interior, Geological Survey Circular 29. Washington, DC.

Syers, J.K., M.G. Brownman, G.W. Smillie, and R.B. Corey. 1973. Phosphate sorption by soils evaluated by the Langmuir adsorption equation. Soil Sci. Soc. Am. J. 37:358-363.

Syers, J.K., T.D. Evans, J.D.H. Williams, and J.T. Murdocks. 1971. Phosphorus sorption parameters of representative soils from Rio Grande do Sul, Brazil. Soil Sci. 112:267-275.

Syers, J.K., J.D.H. Williams, and T.W. Walker. 1968. The determination of total phosphorus in soils and parent materials. New Zealand J. Agric. Res. 11:757-762.

Sykes, D.J., and W.E. Loomis. 1967. Plant and soil factors in permanent wilting percentages and field capacity storage. Soil Sci. 104:162-173.

Tabatabai, M.A. 1982. Sulfur. *In* A.L. Page, R.H. Miller, and D.R. Keeney (eds.), Methods of soil analysis. Part 2. Chemical and microbiological properties. 2nd ed. Agron. 9:501-538.

Tabatabai, M.A. 1996. Sulfur. pp. 921-960. *In* D.L. Sparks (ed.), Methods of soil analysis. Part 3. Chemical methods. Soil Sci. Soc. Am. Book Series 5. ASA and SSSA, Madison, WI.

Talibudeen, O., and P. Arambarri. 1964. The influence of the amount and the origin of calcium carbonates on the isotopically-exchangeable phosphate in calcareous soils. J. Agric. Sci. 62:93-97.

Tan, K.H., and B.F. Hajek. 1977. Thermal analysis of soils. pp. 865-884. *In* J.B. Dixon and S.B. Weed (eds.), Minerals in soil environments. 2nd ed. Soil Sci. Soc. Am. Book Series 1. ASA and SSSA, Madison, WI.

Tan, K.H., B.F. Hajek, and I. Barshad. 1986. Thermal analysis techniques. pp. 151-183. *In* A. Klute (ed.), Methods of soil analysis. Part 1. Physical and mineralogical properties. 2nd ed. Agron. Monogr. 9. ASA and SSSA, Madison, WI.

Tanji, K.K. 1990. Nature and extent of agricultural salinity. pp. 1-17. *In* K.K. Tanji (ed.), Agricultural salinity assessment and management. American Society of Civil Engineers, New York.

Taylor, H.M. 1971. Soil conditions as they affect plant establishment, root development and yield. pp. 292-305. *In* Compaction of agricultural soils. American Society of Agricultural Engineers. St. Joseph, MO.

Taylor, H.M., and R.R. Bruce. 1968. Effects of soil strength on root growth and crop yield in the southern United States. pp. 803-811. *In* Trans. 9th Int. Congr. Soil Sci., Adelaide, Australia. International Society of Soil Science and Angus and Robertson Ltd., Sydney, NSW, Australia.

Taylor, H.M., and I.E. Burnett. 1964. Influence of soil strength on the root growth habits of plants. Soil Sci. 98:174-180.

Taylor, J.K. 1988. Quality assurance of chemical measurements. Lewis Publishers, Inc., Chelsea, MI.

Tedrow, J.C.F. 1968. Pedogenic gradients of the polar region. J. Soil Sci. 19:197-204.

Tessier, A., P.G.C. Campbell, and M. Bisson. 1979. Sequential extraction procedure for the speciation of particulate trace metals. Anal. Chem. 51:844-851.

Thomas, G.W. 1982. Exchangeable cations. *In* A.L. Page, R.H. Miller, and D.R. Keeney (eds.), Methods of soil analysis. Part 2. Chemical and microbiological properties. 2nd ed. Agron. 9:159-165.

Thomas, G.W. 1996. Soil pH and soil acidity. pp. 475-490. *In* D.L. Sparks (ed.), Methods of soil analysis. Part 3. Chemical methods. Soil Sci. Soc. Am. Book Series 5. ASA and SSSA, Madison, WI.

Thomas, G.W., and W. L. Hargrove. 1984. The chemistry of soil acidity. *In* F. Adams (ed.), Soil acidity and liming. 2nd ed. Agron. 12:4-49.

Thomasson, A.J. 1967. The moisture regimes and morphology of some fine-textured soils. M.Sc. thesis, University of Nottingham.

Thornley, J.H.M. 1995. Shoot:root allocation with respect to C, N and P: An investigation and comparison of resistance and teleonomic models. Annals of Botany 75:391-405.

Thorp, J., and G.D. Smith. 1949. Higher categories of soil classification: Order, suborder, and great soil group. Soil Sci. 67:117-126.

Thurman, E.M., and D.S. Aga. 2001. Detection of pesticides and pesticide metabolites using the cross reactivity of enzyme immunoassays. J. AOAC Intl. 84:162-167.

Thurman, E.M., D.A. Goolsby, M.T. Meyer, M.S. Mills, M.L. Pomes, and D.W. Koplin. 1992. A reconnaissance study of herbicides and their metabolites in surface-water of the Midwestern United States using immunoassay and gas chromatography/mass spectrometry. Environ. Sci. Technol. 26:2440–2447.

Tibke, G.L. 2002. Erosion by wind, control measures. pp.489-494. *In* R. Lal (ed.), Encyclopedia of soil science. Marcel Dekker, Inc.

Tiessen, H., J.W.B. Stewart, and C.V. Cole. 1984. Pathways of phosphorus transformations in soils of differing pedogenesis. Soil Sci. Soc. Am. J. 48:853-858.

Tiller, K.G. 1989. Heavy metals in soils and their environmental significance. pp. 113-142. *In* B.A. Stewart (ed.), Advances in soil science. Vol. 9.

Tisdale, S.L, and W.L. Nelson. 1975. Soil fertility and fertilizers. 3rd ed. Macmillan Publishing Co., New York, NY.

Tisdale, S.L., W.L. Nelson, and J.D. Beaton. 1985. Soil fertility and fertilizers. 4th ed. Macmillan Publishing Co., New York, NY.

Tisdall, J.M., and J.M. Oades. 1982. Organic matter and water stable aggregates in soils. J. Soil Sci. 33:141-163.

Tokashiki, Y., and K. Wada. 1975. Weathering implications of the mineralogy of clay fractions of two Ando soils, Kyushu. Geoderma 14:47-62.

Topp, G.C., Y.T. Galganov, B.C. Ball, and M.R. Carter. 1993. Soil water desorption curves. pp. 569-579. *In* M.R. Carter (ed.), Soil sampling and methods of analysis. Canadian Society of Soil Science. Lewis Publishers, Boca Raton, FL.

Tran, T.S., M. Giroux, J. Guilbeault, and P. Audesse. 1990. Evaluation of Mehlich 3 extractant to estimate available P in Quebec soils. Commun. Soil Sci. Plant Anal. 21:1-28.

Tran, T.S., and R.R. Simard. 1993. Mehlich III-extractable elements. pp. 43-50. *In* M.R. Carter (ed.), Soil sampling and methods of analysis. Canadian Society of Soil Science. Lewis Publishers, Boca Raton, FL.

Trimble, S.W. 1974. Man-induced soil erosion on the Southern Piedmont 1700-1970. Soil and Water Conservation Society, Ankeny, IA.

Troeh, F.R., J.A. Hobbs, and R.L. Donahue. 1980. Soil and water conservation for productivity and environmental protection. Prentice-Hall, Inc., Englewood Cliffs, NJ.

Truog, E., J.R. Taylor, R.W. Pearson, M.W. Weeks, and R.W. Simonson. 1936. Procedure for special type of mechanical and mineralogical soil analysis. Soil Sci. Soc. Am. Proc. 1:101-112.

Tugel, A.J., and A.M. Lewandowski (eds.). 2001. Soil biology primer. Soil quality— Soil Biology Technical Note No. 1. Available online at http://soils.usda.gov/sqi/concepts/soil_biology/biology.html (verified January 24, 2011).

Uehara, G., and G. Gillman. 1981. The mineralogy, chemistry, and physics of tropical soils with variable charge clays. Westview Tropical Agricultural Series 4. Westview Press, Boulder, CO.

Uehara, G., and H. Ikawa. 1985. Family criteria for soils dominated by amorphous and poorly crystalline materials. pp. 95-102. *In* J.A. Kittrick (ed.), Mineral classification of soils. Soil Sci. Soc. Am. Spec. Publ. 16. SSSA and ASA, Madison, WI.

Uehara, G., and R.C. Jones. 1974. Particle surfaces and cementing agents. *In* J.W. Cary and D.D. Evans (eds.), Soil crusts. Agric. Exp. Stn., Univ. of Arizona Tech. Bull. 214.

Ugolini, R.C., and R.A. Dahlgren. 1987. The mechanism of podzolization as revealed through soil solution studies. pp. 195-203. *In* D. Righi and A. Chauvel (eds.), Podzols et podzolisation. AFES-INRA, Plaisir-Grignon, Paris.

Umali, D.L. 1993. Irrigation-induced salinity: A growing problem for development and environment. Technical Paper World Bank. Washington, DC.

United States Bureau of the Census. 1943. Sixteenth census of the United States, 1940. Abandoned or idle land. Special Agricultural Study.

United States Department of Agriculture. 1957. The yearbook of agriculture. 85th Congress, 1st Session, House Document No. 30. USDA. U.S. Government Printing Office, Washington, DC.

United States Department of Agriculture. 1989. Wind erosion worst in over 30 years. pp. 788-789. USDA news release.

United States Department of Agriculture. 1995. 1994 farm and ranch irrigation survey. USDA, National Agricultural Statistics Service. Vol. 3, Special Studies Part 1.

United States Department of Agriculture. 2004. 2002 census of agriculture. USDA, National Agricultural Statistics Service. Vol. 1, Geographic Area Series Part 51. Available online at http://www.nass.usda.gov/Census_of_Agriculture/ (verified January 24, 2011).

United States Department of Agriculture. 2005. 2003 farm and ranch irrigation survey. USDA, National Agricultural Statistics Service. Vol. 3, Special Studies Part 1.

United States Department of Agriculture. 2007. Proceedings of International Workshop to Improve Agricultural Water Management in Iraq. Amman, Jordan, Aug. 5-9, 2007.

United States Department of Agriculture, Natural Resources Conservation Service (USDA/NRCS). 1997. Revision 1 December 2003. Updated 06/02/08. Inventorying and monitoring grazing land resources. pp. 4i-4ex21. *In* National range and pasture handbook. USDA/NRCS. Government Printing Office, Washington, DC. Available online at http://www.glti.nrcs.usda.gov/technical/publications/nrph.html (verified January 24, 2011).

United States Department of Agriculture, Natural Resources Conservation Service (USDA/NRCS). 1999. Phosphorus management: Bridging the interface between agriculture and environment. Phosphorus management for environmental risk assessment. Washington, DC.

United States Department of Agriculture, Natural Resources Conservation Service (USDA/NRCS). 2001. Annual national resources inventory (NRI). Urbanization and development of rural land. Available online at http://www.nrcs.usda.gov/technical/NRI/2001/nri01dev.html (verified January 24, 2011).

United States Department of Agriculture, Natural Resources Conservation Service (USDA/NRCS). 2004. Soil biology and land management. Soil quality—Soil Biology Technical Note No. 4. Available online at http://soils.usda.gov/sqi/publications/files/soilbiolandmgt.pdf (verified January 24, 2011).

United States Department of Agriculture, Natural Resources Conservation Service (USDA/NRCS). 2005. Interpreting indicators of rangeland health. Technical Reference 1734-6. Version 4. USDA/NRCS, USGS, USDA/ARS, and BLM. Available online at http://www.glti.nrcs.usda.gov/ (verified January 24, 2011).

United States Department of Agriculture, Natural Resources Conservation Service (USDA/NRCS). 2006. Land resource regions and major land resource areas of the United States, the Caribbean, and the Pacific Ocean. USDA/NRCS Agric. Handb. 296. U.S. Government Printing Office, Washington, DC. Available online at http://soils.usda.gov/survey/geography/mlra/index.html (verified January 24, 2011).

United States Department of Agriculture, Natural Resources Conservation Service (USDA/NRCS). 2009a. National range and pasture handbook. Revision 1. Updated 03/02/09. Available online at http://www.glti.nrcs.usda.gov/technical/publications/nrph.html (verified January 24, 2011).

United States Department of Agriculture, Natural Resources Conservation Service (USDA/NRCS). 2009b. National soil survey handbook, title 430-VI. (Online): Available at http://soils.usda.gov/technical/handbook/ (verified January 24, 2011).

United States Department of Agriculture, Natural Resources Conservation Service (USDA/NRCS), and U.S. Environmental Protection Agency (USEPA). 1999. Phosphorus management: Bridging the interface between agriculture and environment. Phosphorus management for environmental risk assessment. Washington, DC.

United States Department of Agriculture, Soil Conservation Service (USDA/SCS). 1955. Recommendations of the Soil Conservation Service to the Departmental Committee on Land Use Problems in the Great Plains, May 12, 1955. Historical SCS Reports File, Great Plains Conservation Program Files, Soil Conservation Service, Washington, DC.

United States Department of Agriculture, Soil Conservation Service (USDA/SCS). 1965. Land resource regions and major land resource areas of the United States. USDA/SCS Agric. Handb. 296. U.S. Government Printing Office, Washington, DC.

United States Department of Agriculture, Soil Conservation Service (USDA/SCS). 1966a. Soil survey and descriptions for some soils of Kansas. Soil Survey Investigations Report No. 4. USDA/SCS, in cooperation with the Kansas Agricultural Experiment Station.

United States Department of Agriculture, Soil Conservation Service (USDA/SCS). 1966b. Soil survey and descriptions for some soils of Montana. Soil Survey Investigations Report No. 7. USDA/SCS, in cooperation with the Montana Agricultural Experiment Station.

United States Department of Agriculture, Soil Conservation Service (USDA/SCS). 1966c. Soil survey and descriptions for some soils of North Dakota. Soil Survey Investigations Report No. 2. USDA/SCS, in cooperation with the North Dakota Agricultural Experiment Station.

United States Department of Agriculture, Soil Conservation Service (USDA/SCS). 1966d. Soil survey and descriptions for some soils of Wyoming. Soil Survey Investigations Report No. 8. USDA/SCS, in cooperation with the Wyoming Agricultural Experiment Station.

United States Department of Agriculture, Soil Conservation Service (USDA/SCS). 1967a. Soil survey and descriptions for some soils of Colorado. Soil Survey Investigations Report No. 10. USDA/SCS, in cooperation with the Colorado Agricultural Experiment Station.

United States Department of Agriculture, Soil Conservation Service (USDA/SCS). 1967b. Soil survey and descriptions for some soils of Oklahoma. Soil Survey Investigations Report No. 11. USDA/SCS, in cooperation with the Oklahoma Agricultural Experiment Station.

United States Department of Agriculture, Soil Conservation Service (USDA/SCS). 1970. Soil survey and descriptions for some soils of Nevada. Soil Survey Investigations Report No. 23. USDA/SCS, in cooperation with the Nevada Agricultural Experiment Station.

United States Department of Agriculture, Soil Conservation Service (USDA/SCS). 1973. Soil survey and descriptions for some soils of California. Soil Survey Investigations Report No. 24. USDA/SCS, in cooperation with the California Agricultural Experiment Station.

United States Department of Agriculture, Soil Conservation Service (USDA/SCS). 1971. Guide for interpreting engineering uses of soils. U.S. Government Printing Office, Washington, DC.

United States Department of Agriculture, Soil Conservation Service (USDA/SCS). 1972. Soil survey laboratory methods and procedures for collecting soil samples. Soil Survey Investigations Report No. 1. U.S. Government Printing Office, Washington, DC.

United States Department of Agriculture, Soil Conservation Service (USDA/SCS). 1982. Soil survey laboratory methods and procedures for collecting soil samples. Soil Survey Investigations Report No. 1. U.S. Government Printing Office, Washington, DC.

United States Department of Agriculture, Soil Conservation Service (USDA/SCS). 1984. Soil survey laboratory methods and procedures for collecting soil samples. Soil Survey Investigations Report No. 1. U.S. Government Printing Office, Washington, DC.

United States Department of Agriculture, Soil Conservation Service (USDA/SCS). 1989. The second RCA appraisal: Soil, water, and related resources on non-federal land in the United States. USDA/SCS, Washington, DC.

United States Department of Agriculture, Soil Conservation Service (USDA/SCS), and U.S. Salinity Laboratory Staff. 1993. Management of saline and sodic soils. U.S. Government Printing Office, Washington, DC.

United States Department of the Interior, U.S. Geological Survey. 1993. Methods for determination of inorganic substances in water and fluvial sediments. Book 5. Chapter Al. Method I-2601-78.

United States Environmental Protection Agency. 1983. Methods for chemical analysis of water and wastes. EPA-600/4-79-020. Revised March, 1983, Method 353.2.

United States Environmental Protection Agency. 1992. Handbook of laboratory methods for forest health monitoring. G.E. Byers, R.D. Van Remortel, T.E. Lewis, and M. Baldwin (eds.). Part III. Soil analytical laboratory. Section 10. Mineralizable N. U.S. Environmental Protection Agency, Office of Research and Development, Environmental Monitoring Systems Laboratory, Las Vegas, NV.

United States Salinity Laboratory Staff. 1954. L.A. Richards (ed.). Diagnosis and improvement of saline and alkali soils. USDA Agric. Handb. 60. U.S. Government Printing Office, Washington, DC.

Valentin, C., and L.M. Bresson. 1992. Morphology, genesis and classification of surface crusts in loamy and sandy soils. Geoderma 55:225-245.

Valentin, C., and L.M. Bresson. 1998. Soil crusting. pp. 89-107. *In* R. Lal, W.H. Blum, C. Valentine, and B.A. Stewart (eds.), Methods for assessment of soil degradation. Advances in Soil Science. CRC Press.

Van Breemen, N. 1982. Genesis, morphology, and classification of acid sulfate soils in Coastal Plains. pp. 95-108. *In* J.A. Kittrick, D.S. Fanning, and L.R. Hossner (eds.), Acid sulfate weathering. Soil Sci. Soc. Am. Spec. Publ. 10. ASA and SSSA, Madison, WI.

van der Zee, S.E.A.T.M., L.G.J. Fokkink, and W.H. van Riemsdijk. 1987. A new technique for assessment of reversibly adsorbed phosphate. Soil Sci. Soc. Am. J. 51:599-604.

Van Doren, D.M., Jr. 1986. Highlights of research in Division S-6—Soil and water management and conservation. SSSA Golden Anniversary Papers. Soil Sci. Soc. Am. J. 50:271-272.

van Olphen, H. 1971. Amorphous clay materials. Science 171:91-92.

van Olphen, H. 1977. An introduction to clay colloid chemistry. 2nd ed. John Wiley & Sons Inc., New York, NY.

van Riemsdijk, W.H., A.M.A. van der Linden, and L.J.M. Boumans. 1984. Phosphate sorption by soils. III. The diffusion-precipitation model tested for three acid sandy soils. Soil Sci. Soc. Am. J. 48:545-548.

Van Veen, J.A., R. Merckx, and S.C. Van De Geijn. 1989. Plant- and soil related controls of the flow of carbon from roots through the soil microbial mass. Plant and Soil. Kluwer Academic Publishers, The Netherlands.

Van Wambeke, A.R. 1962. Criteria for classifying tropical soils by age. J. Soil Sci. 13:124-132.

Van Wambeke, A. 1992. Soils of the tropics. Properties and appraisal. McGraw-Hill Co., New York, NY.

Vander Pluym, H.S.A. 1978. Extent, causes and control of dryland seepage in the northern Great Plains region of North America. pp.1-48. *In* H.S. Vander Pluym (ed.), Dryland-saline-seep control. Alberta Agr., Lethbridge.

Veihmeyer, F.J., and A.H. Hendrickson. 1931. The moisture equivalent as a measure of the field capacity of soils. Soil Sci. 32:181-194.

Veihmeyer, F.J., and A.H. Hendrickson. 1948. Soil density and root penetration. Soil Sci. 65:487-493.

Velthorst, E.J. 1996. Water analysis. pp. 121-242. *In* P. Buurman, B. van Lagen, and E.J. Velthorst (eds.), Manual for soil and water analysis. Backhuys Publishers, Leiden.

Viebel, S. 1950. p-Glucosidase. *In* The Enzymes, Vol. I, Part 1, pp. 583-620. Academic Press Inc., New York, NY.

Vieillefon, J. 1979. Contribution to the improvement of analysis of gypsiferous soils. Cahiers ORSTOM, Série Pédologie 17:195-223.

Vrindts, E., A.M. Mouazen, M. Reyniers, K. Maertens, M.R. Maleki, H. Ramon, and J. De Baerdemaeker. 2005. Management zones based on correlation between soil compaction, yield and crop data. Biosystems Engineering 92:419-428.

Wada, K. 1977. Allophane and imogolite. pp. 603-638. *In* J.B. Dixon and S.B. Weed (eds.), Minerals in soil environments. 1st ed. Soil Sci. Soc. Am. Book Series 1.

Wada, K. 1980. Mineralogical characteristics of Andisols. pp. 87-107. *In* B.K.G. Theng (ed.), Soils with variable charge. New Zealand Society of Soil Science, Lower Hutt, New Zealand.

Wada, K. 1985. The distinctive properties of Andosols. pp. 173-229. *In* B.A. Stewart (ed.), Advances in soil science. Vol. 2. Springer-Verlag, New York, NY.

Wada, K. 1989. Allophane and imogolite. pp. 1051-1087. *In* J.B. Dixon and S.B. Weed (eds.), Minerals in soil environments. 2nd ed. Soil Sci. Soc. Am. Book Series 1.

Wada, K., and K. Kakuto. 1985a. A spot test with toluidine blue for allophane and imogolite. Soil Sci. Soc. Am. J. 49:276-278.

Wada, K., and K. Kakuto. 1985b. Embyronic halloysites in Ecuador soils derived from volcanic ash. Soil Sci. Soc. Am. J. 49:1309-1318.

Wada, K., and Y. Tange. 1984. Interaction of methyl- and ethyl-ammonium ions and piperidinium ions with soils. Soil Sci. 137:315-323.

Wada, K., and S. Wada. 1976. Clay mineralogy of the B horizons of two Hydrandepts, a Torrox and a Humitropept in Hawaii. Geoderma 16:139-157.

Wahrhaftig, C. 1965. Physiographic divisions of Alaska. U.S. Geol. Surv. Prof. Paper 482.

Walker, A.L. 1983. The effects of magnetite on oxalate- and dithionite-extractable iron. Soil Sci. Soc. Am. J. 47:1022-1025.

Walker, T.W. 1974. Phosphorus as an index of soil development. Trans. Int. Congr. Soil Sci. 10:451-457.

Walker, T.W., and A.F.R. Adams. 1958. Studies on soil organic matter. 1. Influence of phosphorus content of parent materials on accumulation of carbon, nitrogen, sulfur, and organic phosphorus in grassland soils. Soil Sci. 85:307-318.

Walker, T.W., and J.K. Syers. 1976. The fate of phosphorus during pedogenesis. Geoderma 15:1-19.

Walkley, A., and I.A. Black. 1934. An examination of the Degtjareff method for determining soil organic matter and a proposed modification of the chromic acid titration method. Soil Sci. 37:29-38.

Waltman, S.W., and N.B. Bliss. 1997. Estimates of SOC content. USDA/NRCS, Lincoln, NE.

Warne, S. St. J. 1962. A quick field or laboratory staining scheme for the differentiation of the major carbonate minerals. J. Sedimentary Petrology 32:29-38.

Warren, A. 2002. Global hot spots of erosion by wind. pp. 508-511. *In* R. Lal (ed.), Encyclopedia of soil science. Marcel Dekker, Inc.

Washington, H.S. 1930. The chemical analysis of rocks. 4th ed. John Wiley & Sons Inc., New York, NY.

Watanabe, M., H. Tanaka, K. Sakagami, K. Aoki, and S. Sugiyama. 1996. Evaluation of Pg absorption strength of humic acids as a paleoenvironmental indicator in buried paleosols of tephra beds, Japan. Quaternary Intl. 34-36:197-203.

Watson, J.R. 1971. Ultrasonic vibration as a method of soil dispersion. Soil Fertil. 34:127-134.

Weaver, C.E., and L.D. Pollard. 1973. The chemistry of clay minerals. Elsevier, Amsterdam, The Netherlands.

Weaver, R.M., J.K. Syers, and M.L. Jackson. 1968. Determination of silica in citrate-bicarbonate-dithionite extracts of soils. Soil Sci. Soc. Am. Proc. 32:497-501.

Webster, R., and P.H.T. Beckett. 1972. Matric suctions to which soils in South Central England drain. J. Agric. Sci. Camb. 78:379-387.

Weesies, G.A., D.L. Schertz, and W.F. Kuenstler. 2002. Erosion control: Agronomic practices. pp. 402-406. *In* R. Lal (ed.), Encyclopedia of soil science. Marcel Dekker, Inc.

Weil, R.R., R.I. Kandikar, M.A. Stine, J.B. Gruver, and S.E. Samson-Liebig. 2003. Estimating active carbon for soil quality assessment: A simplified method for laboratory and field use. Am. J. Alternative Agric. 18(1):3-17.

Wendlandt, W.W. 1986. Thermal analysis. 3rd ed. John Wiley & Sons Inc., New York, NY.

Wertz, W.A., and J.A. Arnold. 1972. Land systems inventory. Ogden, UT. USDA, Forest Service, Intermountain Region.

West, L.T., L.R. Drees, L.P. Wilding, and M.C. Rabenhorst. 1988. Differentiation of pedogenic and lithogenic carbonate forms in Texas. Geoderma 43:271-287.

West, L.T., W.P. Miller, G.W. Langdale, R.R. Bruce, and A.W. Thomas. 1991.Cropping system effects on interrill soil loss in the Georgia Piedmont. Soil Sci. Soc. Am. J. 55:460-466.

West, T.O., and W.M. Post. 2002. Soil organic carbon sequestration rates by tillage and crop rotation: A global data analysis. Soil Sci. Soc. Am. J. 66:1930-1946.

Whatmuff, M.S. 2002. Applying biosolids to acid soils in New South Wales: Are guideline soil metal limits from other countries appropriate? Aust. J. Soil Res. 40:1041-1056.

Whitney, M. 1896. Methods of the mechanical analysis of soils. U.S. Bur. Soils Bull. 4.

Whitney, M., and T.H. Means. 1897. An electrical method of determining the soluble salt content of soils. USDA, Div. Soils Bull. 8.

Whittig, L.D., and W.R. Allardice. 1986. X-ray diffraction techniques. pp. 331-362. *In* A. Klute (ed.), Methods of soil analysis. Part 1. Physical and mineralogical methods. 2nd ed. Agron. Monogr. 9. ASA and SSSA, Madison, WI.

Wilcke, W., and W. Amelung. 1996. Small-scale heterogeneity of aluminum and heavy metals in aggregates along a climatic transect. Soil Sci. Soc. Am. Proc. 60:1490-1495.

Wilcke, W., S. Muller, N. Kanchanakool, and W. Zech. 1998. Urban soil contamination in Bangkok: Heavy metal and aluminum partitioning in topsoils. Geoderma 86:211-228.

Wilding, L.P., R.B. Jones, and G.M. Shafer. 1965. Variation of soil morphological properties within Miami, Celina, and Crosby mapping units in west-central Ohio. Soil Sci. Soc. Am. Proc. 29:711-717.

Wilding, L.P., and E.M. Rutledge. 1966. Cation-exchange capacity as a function of organic matter, total clay, and various clay fractions in a toposequence. Soil Sci. Soc. Am. Proc. 30:782-785.

Wilding, L.P., G.M. Shafer, and R.B. Jones. 1964. Morely and Blount soils: A statistical summary of certain physical and chemical properties of some selected profiles from Ohio. Soil Sci. Soc. Am. Proc. 28:674-679.

Williams, J.R., C.A. Jones, and P.T. Dyke. 1984. A modeling approach to determining the relationship between erosion and soil productivity. Trans. ASAE 27:129-144.

Wilson, M.A., S.P. Anderson, K.D. Arroues, S.B. Southard, R.L. D'Agostino, and S.L. Baird. November 15, 1994. No. 6 publication of laboratory data in soil surveys. National Soil Survey Center Soil Technical Note Handbook 430-VI Amendment 5.

Wilson, M.A., R. Burt, S.J. Indorante, A.B. Jenkins, J.V. Chiaretti, M.G. Ulmer, and J.M. Scheyer. 2008. Geochemistry in the modern soil survey program. Environ. Monit. Assess. 139:151-171.

Wilson, M.A., R. Burt, and C.W. Lee. 2006. Improved elemental recoveries in soils with heating boric acid following microwave total digestion. Commun. Soil Sci. Plant Anal. 37:513-524.

Wilson, M.A., R. Burt, W.C. Lynn, and L.C. Klameth. 1997. Total elemental analysis digestion method evaluation on soils and clays. Commun. Soil Sci. Plant Anal. 28:407-426.

Wilson, M.A., R. Burt, and M.D. Mays. 2001. Application of trace elements in the U.S. cooperative soil survey program. p. 324. In Proc. 6th Int. Conf. on the Biogeochemistry of Trace Elements (ICOBTE), Guelph, Ontario, Canada. July 29 to August 2, 2001.

Wilson, M.A., R. Burt, R.J. Ottersberg, D.A. Lammers, T.D. Thorson, R.W. Langridge, and A.E. Kreger. 2002. Isotic mineralogy: Criteria review and application in Blue Mountains, Oregon. Soil Sci. 167:465-477.

Wilson, M.A., R. Burt, T.M. Sobecki, R.J. Engel, and K. Hipple. 1996. Soil properties and genesis of pans in till-derived Andisols, Olympic Peninsula, Washington. Soil Sci. Soc. Am. J. 60:206-218.

Wilson, M.A., R. Burt, T.D. Thorson, and J.E. Thomas. 1999. Volcanic glass analyses in multiple fine-earth fractions. Soil Surv. Hor. 40:29-35.

Wilson, M.A., and D. Righi. 2010. Spodic materials. In G. Stoops, V. Marcelino, and F. Mees (eds.), Interpretations of micromorphological features of soils and regoliths. Elsevier, Amsterdam, The Netherlands. In Press.

Wilson, M.J. 1994. Clay mineralogy: Spectroscopic and chemical determinative methods. Chapman and Hall, Inc., New York, NY.

Wischmeier, W.H., and W.D. Smith. 1978. Predicting rainfall erosion losses—A guide to conservation planning. USDA Agric. Handb. 537. Washington, DC.

Wolf, A.M., and D.E. Baker. 1985. Comparison of soil test phosphorus by the Olsen, Bray P-1, Mehlich 1 and Mehlich 3 methods. Commun. Soil Sci. Plant Anal. 16:457-484.

Wolf, K.H., and S. St. J. Warne. 1960. Remarks on application of Friedman's staining methods. J. Sedimentary Petrology 30:496-497.

Wood, C.W., J.H. Edwards, and C.G. Cummins. 1991. Tillage and crop rotation effects on soil organic matter in a Typic Hapludult of northern Alabama. J. Sustainable Agric. 2:31-41.

Woodell, S.R.J., H.A. Mooney, and H. Lewis. 1975. The adaptation to serpentine soils in California of the annual species *Linanthus Androsaceus (Polemoniaceae)*. Bull. Torrey Bot. Clu. 102:232-238.

Woodruff, N.P., and F.H. Siddoway. 1965. A wind erosion equation. Soil Sci. Soc. Am. J. 29:602-608.

Woolhouse, H.W. 1983. Toxicity and tolerance in the responses of plant to metals. *In* O.L. Lange (ed.), Encyclopedia of plant physiology. Vol. 12C. Springer, New York, NY.

Worster, D. 1979. Dust Bowl: The southern plains in the 1930s. Oxford Press, New York, NY.

Wright, R.J., J.L. Hern, V.C. Baligar, and O.L. Bennett. 1985. The effect of surface applied soil amendments on barley root growth in an acid subsoil. Comm. Soil Sci. Plant Anal. 16:179-192.

Wysocki, D.A., P.J. Schoeneberger, and H.E. LaGarry. 2000. Geomorphology of soil landscapes. *In* M.E. Sumner (ed.), Handbook of soil science. CRC Press LLC, Boca Raton, FL.

Yang, C.Y., H.F. Chiu, S.S. Tsai, C.C. Chang, and F.C. Sung. 2002. Magnesium in drinking water and the risk of delivering a child of very low birth weight. Magnes. Res. 15:207-213.

Yerima, B.P.K., F.G. Calhoun, A.L. Senkayi, and J.B. Dixon. 1985. Occurrence of interstratified kaolinite-smectite in El Salvador Vertisols. Soil Sci. Soc. Am. J. 49:462-466.

Yerima, B.P.K., L.P. Wilding, F.G. Calhoun, and C.T. Hallmark. 1987. Volcanic ash-influenced Vertisols and associated Mollisols of El Salvador: Physical, chemical, and morphological properties. Soil Sci. Soc. Am. J. 51:699-708.

Yin, Y., H.E. Allen, Y. Li, C.P. Huang, and P.F. Sanders. 1996. Adsorption of mercury(II) by soil: Effects of pH, chloride, and organic matter. J. Environ. Qual. 25:837-844.

Yost, R.S., E.J. Kamprath, G.C. Naderman, and E. Lobato. 1981. Residual effects of phosphorus applications on a high phosphorus adsorbing Oxisol of central Brazil. Soil Sci. Soc. Am. J. 45:540-543.

Young, J.L., and R.W. Aldag. 1982. Inorganic forms of nitrogen in soil. pp. 43-66. *In* F.J. Stevenson (ed.), Nitrogen in agricultural soils. Agron. 22. ASA and SSSA, Madison, WI.

Young, K.K., and J.D. Dixon. 1966. Overestimation of water content at field capacity from sieved-sample data. Soil Sci. 101:104-107.

Yuan, G., L.M. Lavkulich, and C. Wang. 1993. A method for estimating organic-bound iron and aluminum contents in soils. Commun. Soil Sci. Plant Anal. 24(11&12), 1333-1343.

Yuan, T.L., N. Gammon, Jr., and R.G. Leighty. 1967. Relative contribution of organic and clay fractions to cation-exchange capacity of sandy soils from several soil groups. Soil Sci. 104:123-128.

Zobeck, T.M., N.A. Rolong, D.W. Fryear, J.D. Bilbro, and B.L. Allen. 1995. Properties and productivity of recently tilled grass sod and 70-year cultivated soil. J. Soil Water Conserv. 50:210-215.

Zoebisch, M.A., and A.R. Dexter. 2002. Degradation, physical. pp. 311-316. *In* R. Lal (ed.), Encyclopedia of soil science. Marcel Dekker, Inc.

V. Appendices

This section provides example pedon data sheets, including the primary, supplementary, and taxonomy sheets and grain-size distribution curves and water retention curves for selected pedons. These data sheets are used in a number of example pedon calculations presented throughout this manual, such as weight to volume conversions, weighted averages, and other estimates.

1 Chowchow

Pedon ID: S04WA-027-009

*** Primary Characterization Data ***
(Grays Harbor, Washington)

Sampled as on Sep 30, 2004 : Chowchow ; Loamy, isotic, dysic, isomesic Terric Haplosaprist
Revised to correlated on Aug 15, 2007 : Chowchow ; Loamy, isotic, dysic, isomesic Terric Haplosaprist

SSL   - Project   C2005USWA011 Grays Harbor
      - Site ID   S04WA-027-009 Lat: 47° 18' 20.80" north Long: 124° 5' 31.40" west NAD83 MLRA: 4A
      - Pedon No. 05N0175
      - General Methods 1B1A, 2A1, 2B

United States Department of Agriculture
Natural Resources Conservation Service
National Soil Survey Center
Soil Survey Laboratory
Lincoln, Nebraska 68508-3866

Layer	Horizon	Orig Hzn	Depth (cm)	Field Label 1	Field Label 2	Field Label 3	Field Texture	Lab Texture
05N00981	Oi	Oi	0-9	S04WA-027-009-1	CHOW CHOW		PEAT	
05N00982	Oa	Oa	9-20	S04WA-027-009-2	CHOW CHOW		MUCK	
05N00983	2Oe	2Oe	20-33	S04WA-027-009-3	CHOW CHOW		MPT	
05N00984	2Oa	2Oa	33-58	S04WA-027-009-4	CHOW CHOW		MUCK	
05N00985	3Cg1	3Cg1	58-91	S04WA-027-009-5	CHOW CHOW		SIL	SIL
05N00986	3Cg2	3Cg2	91-132	S04WA-027-009-6	CHOW CHOW		SIL	SIL

Calculation Name   Pedon Calculations   Result   Units of Measure

LE, Whole Soil, Summed to 1m   14   cm/m

PSDA & Rock Fragments

		-1-	-2-	-3-	-4-	-5-	-6-	-7-	-8-	-9-	-10-	-11-	-12-	-13-	-14-	-15-	-16-	-17-
		(----- Total -----)			(- - Clay - - -)		(- - - Silt - - - -)		(------------- Sand -------------)				(Rock Fragments (mm))				
		Clay	Silt	Sand	Fine	CO₃	Fine	Coarse	VF	F	M	C	VC	(------ Weight ------)				>2 mm
		<	.002	.05	<	<	.002	.02	.05	.10	.25	.5	1	2	5	20	.1-	wt %
		.002	-.05	-2	.0002	.002	-.02	-.05	-.10	-.25	-.50	-1	-2	-5	-20	-75	75	whole
Layer		(------------------------ % of <2mm Mineral Soil -------------------------)												(------ % of <75mm ------)				soil
	Depth (cm)	Horz	Prep																	
05N00981	0-9	Oi	S	3A1a1a	3A1a1a	3A1a1a	3A1a1a		3A1a1a	3A1a1a	3A1a1a	3A1a1a	3A1a1a	3A1a1a	--	--	--	--	--
05N00982	9-20	Oa	S												--	--	--	--	--
05N00983	20-33	2Oe	S												7	--	--	--	7
05N00984	33-58	2Oa	S												3	--	--	--	3
05N00985	58-91	3Cg1	S	9.3	71.6	19.1	0.3		50.2	21.4	10.3	6.7	1.9	0.2	tr	--	--	--	9	--
05N00986	91-132	3Cg2	S	21.3	63.9	14.8	2.6		43.8	20.1	9.3	4.5	0.9	0.1	tr	--	--	--	6	--

*** Primary Characterization Data ***

Pedon ID: S04WA-027-009

Sampled As : Chowchow

(Grays Harbor, Washington)

Loamy, isotic, dysic, isomesic Terric Haplosaprist

USDA-NRCS-NSSC-Soil Survey Laboratory

; Pedon No. 05N0175

Bulk Density & Moisture

			-1-	-2-	-3-				-9-	-10-	-11-	-12-	-13-
			(Bulk Density)		Cole	(-------- Water Content --------)		1500 kPa	Ratio	WRD	Aggst	(-- Ratio/Clay --)	
			33 kPa	Oven Dry	Whole Soil	6 kPa	10 kPa	33 kPa	1500 kPa	Moist	AD/OD	Whole Soil	Stabl 2-0.5mm	CEC7	1500 kPa
	Depth		(- - g cm⁻³ - -)			(-------- % of <2mm --------)					cm³ cm⁻³ %			
Layer	(cm)	Horz	Prep	DbWR1	DbWR1		DbWR1		3C2a1a	3C2a1a	3D1			
05N00981	0-9	Oi	S	0.35	0.71	0.266	138.9		81.4		1.168	0.20		
05N00981	0-9	Oi	M						250.0		7.560			
05N00982	9-20	Oa	S	0.36	0.78	0.294	202.3		66.8		1.147	0.49		
05N00982	9-20	Oa	M						154.5		4.434			
05N00983	20-33	2Oe	S	0.36	0.88	0.341	157.5		60.7		1.148	0.34		
05N00984	33-58	2Oa	S	0.40	0.52	0.091	124.5		46.5		1.161	0.31		
05N00985	58-91	3Cg1	S	0.99	1.11	0.039	48.2		16.8		1.051	0.31	2.27	1.81
05N00986	91-132	3Cg2	S	1.09	1.17	0.024	41.2		18.2		1.038	0.25	0.79	0.85

Water Content

			-1-	-2-	-3-	-4-	-5-	-6-	-7-	-8-	-9-	-10-	-11-	-12-	-13-
			(-- Atterberg --)		(---- Bulk Density ----)			(------------- Water Content -------------)						
			(--- Limits ---)		Field	Recon	Recon	Field	(------ Sieved Samples ------)						
			LL	PI		33 kPa	Oven Dry		33 kPa	6 kPa	10 kPa	33 kPa	100 kPa	200 kPa	500 kPa
	Depth		pct <0.4mm		(------------ g cm⁻³ ------------)			(---------------- % of <2mm ----------------)						
Layer	(cm)	Horz	Prep	3H						Recon 33 kPa	6 kPa	10 kPa	33 kPa	100 kPa	200 kPa	500 kPa
05N00985	58-91	3Cg1	S	NP											

Carbon & Extractions

			-1-	-2-	-3-	-4-	-5-	-6-	-7-	-8-	-9-	-10-	-11-	-12-	-13-	-14-	-15-	-16-	-17-	-18-
			(---- Total ----)			Org	C/N	(-- Dith-Cit Ext ---)			(---- Ammonium Oxalate Extraction ----)						(--- Na Pyro-Phosphate --)		
			C	N	S	C	Ratio	Fe	Al	Mn	Al+½Fe	ODOE	Fe	Al	Si	Mn	C	Fe	Al	Mn
	Depth		(------ % of <2 mm ------)					(-- % of <2mm --)			(------ % of <2mm ------)					mg kg⁻¹	(----- % of <2mm ----)		
Layer	(cm)	Horz	Prep	4H2a	4H2a	4H2a	4H2a		4G1	4G1	4G1	4G2a	4G2a	4G2a	4G2a	4G2a	4G2a	4G2a		
05N00981	0-9	Oi	S	46.73	3.07	0.24		15												
05N00982	9-20	Oa	S	41.06	2.89	0.19		14												
05N00983	20-33	2Oe	S	46.72	1.99	0.26		23												
05N00984	33-58	2Oa	S	27.35	1.26	0.15		22												
05N00985	58-91	3Cg1	S	3.71	0.27	0.07		14	0.2	1.7	--	2.12	0.38	0.03	2.11	0.68	--			
05N00986	91-132	3Cg2	S	3.58	0.18	0.06		19	0.1	1.1	--	1.20	0.22	0.02	1.19	0.33	--			

*** Primary Characterization Data ***

(Grays Harbor, Washington)

Pedon ID: S04WA-027-009

Sampled As : Chowchow   Loamy, isotic, dysic, isomesic Terric Haplosaprist

USDA-NRCS-NSSC-Soil Survey Laboratory   ;   Pedon No. 05N0175

CEC & Bases

Layer	Depth (cm)	Horz	Prep	-1- Ca 4B1a1a	-2- Mg 4B1a1a	-3- Na 4B1a1a	-4- K 4B1a1a	-5- Sum Bases	-6- Acid- ity 4B2b1a1	-7- Extr Al 4B3a1a	-8- KCl Mn 4B3a1a	-9- CEC8 Sum Cats	-10- CEC7 NH4 OAC 4B1a1a	-11- ECEC Bases +Al	-12- Al Sat	-13- Sum	-14- NH4OAC
				(----- NH4OAC Extractable Bases -----) (-------------- cmol(+) kg^-1 --------------)							mg kg^-1	(----- cmol(+) kg^-1 ---)			(----- Base ----) (- Saturation -) (----- % -----)	
05N00981	0-9	Oi	S	0.8	1.1	--	0.8		153.9	4.9	0.4		74.1			2	
05N00982	9-20	Oa	S	0.5	0.4	--	0.5		189.0	8.8	0.3		87.5			1	
05N00983	20-33	2Oe	S	0.1	0.2	0.1	0.2	0.6	219.6	9.0	0.4	220.2	92.8	9.6	94	0	1
05N00984	33-58	2Oa	S	0.2	0.1	--	0.1	0.4	235.9	7.2	0.2	236.3	99.5	7.6	95	0	0
05N00985	58-91	3Cg1	S	0.2	--	--	0.1	0.3	40.6	2.9	0.2	40.9	21.1	3.2	91	0	1
05N00986	91-132	3Cg2	S	0.2	0.1	--	tr	0.3	30.2	3.9	0.2		16.8			1	2

Salt

Layer	Depth (cm)	Horz	Prep	-1- Ca	-2- Mg	-3- Na	-4- K	-5- CO3	-6- HCO3	-7- F	-8- Cl	-9- PO4	-10- Br	-11- OAC	-12- SO4	-13- NO2	-14- NO3	-15- H2O	-16- Total Salts	-17- Elec Cond	-18- Pred Elec Cond	-19- Exch Na	-20- SAR
				(---------------------- Water Extracted From Saturated Paste ----------------------)														(------ % ------)	(-- dS m^-1 --)		%	
				(----- mmol(+) L^-1 -----) (----------------------- mmol(-) L^-1 -----------------------)																		
05N00983	20-33	2Oe	S																			tr	
05N00984	33-58	2Oa	S																			--	--
05N00985	58-91	3Cg1	S																			--	--
05N00986	91-132	3Cg2	S																			--	--

*** Primary Characterization Data ***

(Grays Harbor, Washington)

Loamy, isotic, dysic, isomesic Terric Haplosaprist

;   Pedon No. 05N0175

pH & Carbonates

				-1-	-2-	-3-	-4-	-5-	-6-	-7-	-8-	-9-	-10-	-11-
					(------------- pH -------------)					(-- Carbonate --)		(-- Gypsum --)		
					CaCl₂		Sat			As CaCO₃		As CaSO₄*2H₂O	Resist	
	Depth				0.01M	H₂O	Paste			<2mm	<20mm	<2mm	<20mm	ohms
Layer	(cm)	Horz	Prep	KCl	1:2	1:1	Paste	Sulf	NaF	%				cm⁻¹
					4C1a2a	4C1a2a			4C1a1a14E1a1a1a1					
05N00981	0-9	Oi	S		3.9	4.7								
05N00981	0-9	Oi	M		4.0	5.1								
05N00982	9-20	Oa	S		3.9	4.2								
05N00982	9-20	Oa	M		4.0	4.5								
05N00984	20-33	2Oe	S		4.1	4.2								
05N00985	33-58	2Oa	S		4.4	4.6								
05N00985	58-91	3Cg1	S		4.7	5.1			11.3	tr				
05N00986	91-132	3Cg2	S		4.5	5.3			10.9					

Organic

				-1-	-2-	-3-	-4-	-5-	-6-	-7-	-8-	-9-	-10-	-11-	-12-	-13-	-14-	-15-	-16-	-17-
				Mineral	OM	OM+	(- Total -)			Fiber Content		NaPyro	Decomp	Limnic	(-- pH --)		(-- Bulk Density --)			Proj
				Content		TC*1.724	C	N	C/N	Unrub	Rub	Color	State	Matter	CaCl₂	H₂O	33 kPa	33 kPa	OD	Subs
	Depth					Min				% (by vol)							rewet			
Layer	(cm)	Horz	Prep	5A	%		4H2a	4H2a	ratio	5?	5?	5?			5?	4C1a2a	g cm⁻³			cm cm⁻¹
05N00981	0-9	Oi	S		81	89	46.73	3.07	15	5?	5?	5?	hemic		5?	4.7				
05N00981	0-9	Oi	M													5.1				
05N00981	0-9	Oi	MW	8																--
05N00982	9-20	Oa	S		71	90	41.06	2.89	14	52	28	10YR 7/4	sapric		3.8	4.2				
05N00982	9-20	Oa	M													4.5				
05N00982	9-20	Oa	MW	19	81	97	46.72	1.99	23	68	10	10YR 6/4	sapric		3.5					--
05N00983	20-33	2Oe	S													4.2				
05N00983	20-33	2Oe	MW	16						16	tr	10YR 5/3.5			4.1					1
05N00984	33-58	2Oa	S		47	80	27.35	1.26	22	4	tr	10YR 4/4	sapric		4.2	4.6				tr
05N00984	33-58	2Oa	MW	33																
05N00985	58-91	3Cg1	S				3.71	0.27	14							5.1				--
05N00986	91-132	3Cg2	S				3.58	0.18	19							5.3				--

*** Primary Characterization Data ***

(Grays Harbor, Washington)

Pedon ID: S04WA-027-009

Sampled As   :   Chowchow

Loamy, isotic, dysic, isomesic Terric Haplosaprist

USDA-NRCS-NSSC-Soil Survey Laboratory

;   Pedon No. 05N0175

Phosphorous

Layer	Depth (cm)	Horz	Prep	-1- Melanic Index	-2- NZ % 4D8a1	-3- Acid Oxal 4G2a	-4- Bray 1	-5- Bray 2	-6- Olsen	-7- H₂O	-8- Citric Acid	-9- Mehlich III	-10- Mehlich Extr NO₃
						(- - - - - - - - - - - - - - - - - - Phosphorous -)							
						(- - - - - - - - - - - - - - - - - mg kg⁻¹ -) mg kg⁻¹							
05N00985	58-91	3Cg1	S		96	495.8							
05N00986	91-132	3Cg2	S		84	452.9							

Pedon ID: S04WA-027-009

Print Date: Aug 10 2010 8:29AM

Sampled As : Chowchow

(Grays Harbor, Washington)

USDA-NRCS-NSSC-Soil Survey Laboratory

Loamy, isotic, dysic, isomesic Terric Haplosaprist

;   Pedon No. 05N0175

Clay Mineralogy (<.002 mm)

	-1-	-2-	-3-	-4-	-5-	-6-	-7-	-8-	-9-	-10-	-11-	-12-	-13-	-14-	-15-	-16-	-17-	-18-
			X-Ray					Thermal					Elemental				EGME	Inter preta tion
										SiO_2	Al_2O_3	Fe_2O_3	MgO	CaO	K_2O	Na_2O	Retn	
Fraction	(- - - - - - - peak size - - - - - - -)						(- - - - - - - % - - - - - - -)		(- - - - - - - - - - - - - % - - - - - - - - - - - - -)								$mg\ g^{-1}$	
Depth			7A1a1															
Layer (cm) Horz	tcly	QZ 1	VR 2	KK 1														VERM
05N00985 58-91 3Cg1																		

FRACTION INTERPRETATION:
tcly - Total Clay, <0.002 mm

MINERAL INTERPRETATION:
KK - Kaolinite   QZ - Quartz   VR - Vermiculite

RELATIVE PEAK SIZE: 5 Very Large   4 Large   3 Medium   2 Small   1 Very Small   6 No Peaks

*** Primary Characterization Data ***

Pedon ID: S04WA-027-009

Sampled As   :   Chowchow

(Grays Harbor, Washington)

Loamy, isotic, dysic, isomesic Terric Haplosaprist

USDA-NRCS-NSSC-Soil Survey Laboratory   ;   Pedon No. 05N0175

Sand - Silt Mineralogy (2.0-0.002 mm)

	Depth	Horz	Fract ion	-1-	-2-	-3-	-4-	-5-	-6-	-7-	-8-	-9-	-10-	-11-	-12-	-13-	-14-	-15-	-16-	-17-	-18-
						X-Ray				Thermal			Tot Re			Optical Grain Count 7B1a2				EGME Retn	Inter preta tion
Layer	(cm)			(- - - - - - - peak size - - - - - - - -)						(- - - - - - - % - - - - - - - -)			(- % - - - - - - - - - - - - - - - - - - -)							mg g^{-1}	
05N00985	58-91	3Cg1	csi											OT 98	GS 2	GA tr					
05N00986	91-132	3Cg2	csi											OT 98	GS 2	GA tr	PO tr				

FRACTION INTERPRETATION:

csi - Coarse Silt, 0.02-0.05 mm

MINERAL INTERPRETATION:

GA - Glass Aggregates   GS - Glass   OT - Other   PO - Plant Opal

Pedon ID: S04WA-027-009

*** Supplementary Characterization Data ***
(Grays Harbor, Washington)

Sampled as on Sep 30, 2004
Revised to correlated on Aug 15, 2007

SSL		
- Project	C2005USWA011	Grays Harbor
- Site ID	S04WA-027-009	Lat: 47° 18' 20.80" north Long: 124° 5' 31.40" west NAD83 MLRA: 4A
- Pedon No.	05N0175	
- General Methods	1B1A, 2A1, 2B	

Chowchow ; Loamy, isotic, dysic, isomesic Terric Haplosaprist
Chowchow ; Loamy, isotic, dysic, isomesic Terric Haplosaprist

United States Department of Agriculture
Natural Resources Conservation Service
National Soil Survey Center
Soil Survey Laboratory
Lincoln, Nebraska 68508-3866

Tier 1

Engineering PSDA — Percentage Passing Sieve (% of Whole Soil)

Layer	Depth (cm)	Horz	Prep	3" (-1-)	2 (-2-)	3/2 (-3-)	1 (-4-)	3/4 (-5-)	3/8 (-6-)	4 (-7-)	10 (-8-)	40 (-9-)	200 (-10-)
05N00981	0-9	Oi	S	100	100	100	100	100	100	100	100	100	100
05N00982	9-20	Oa	S	100	100	100	100	100	100	100	100	100	100
05N00983	20-33	2Oe	S	100	100	100	100	100	100	100	100	93	93
05N00984	33-58	2Oa	S	100	100	100	100	100	100	100	100	97	97
05N00985	58-91	3Cg1	S	100	100	100	100	100	100	100	100	100	100
05N00986	91-132	3Cg2	S	100	100	100	100	100	100	100	100	100	100

Cumulative Curve Fractions (<75mm) — USDA Less Than Diameters (mm), Percentile Diameters, Atterberg, Gradation

Layer	1. (-11-)	.5 (-12-)	.25 (-13-)	.10 (-14-)	.05 (-15-)	20µm (-16-)	5µm (-17-)	2µm (-18-)	60 (-19-)	50 (-20-)	10 (-21-)	LL (-22-)	PI (-23-)	CU (-24-)	CC (-25-)
05N00981															
05N00982															
05N00983															
05N00984															
05N00985	100	100	98	91	81	60	29	9	0.02	0.013	0.002		NP	9.9	0.6
05N00986	100	100	99	95	85	65	39	21	0.02	0.009	0.001			25.9	1.1

Tier 2

Weight Fractions — Whole Soil (% of Whole Soil) and <75 mm Fraction (% of <75 mm)

Layer	Depth (cm)	Horz	Prep	>2 (-26-)	250-UP (-27-)	250-75 (-28-)	75-20 (-29-)	20-5 (-30-)	5-2 (-31-)	<2 (-32-)	75-20 (-33-)	20-5 (-34-)	5-2 (-35-)	<2 (-36-)
05N00981	0-9	Oi	S	--	--	--	--	--	--	100	--	--	--	100
05N00982	9-20	Oa	S	--	--	--	--	--	--	100	--	--	--	100
05N00983	20-33	2Oe	S	7	--	--	7	--	7	93	7	--	3	93
05N00984	33-58	2Oa	S	3	--	--	3	--	3	97	3	--	3	97
05N00985	58-91	3Cg1	S	--	--	--	--	--	--	100	--	--	--	100
05N00986	91-132	3Cg2	S	--	--	--	--	--	--	100	--	--	--	100

Weight Per Unit Volume (g cm⁻³) and Ratios At 33 kPa

Layer	Whole Soil — Soil Sur 33 kPa (DbWR1) (-40-)	Whole Soil — Oven-dry (-41-)	Whole Soil — Engineering Moist (-42-)	Whole Soil — Saturated (-43-)	<2 mm — Soil Survey 33 kPa (DbWR1) (-44-)	<2 mm — 1500 kPa (-45-)	<2 mm — Oven-dry (-46-)	<2 mm — Engineering Moist (-47-)	<2 mm — Saturated (-48-)	Ratios Whole Soil (-49-)	Ratios <2 mm Soil (-50-)
05N00981	0.35	0.71	0.84	1.22	0.35	0.52	0.71	0.84	1.22	6.57	6.57
05N00982	0.36	0.78	1.09	1.22	0.36	0.66	0.78	1.09	1.22	6.36	6.36
05N00983	0.38	0.92	0.93	1.24	0.36	0.71	0.88	0.93	1.22	5.97	6.36
05N00984	0.41	0.53	0.91	1.26	0.40	0.49	0.52	0.90	1.25	5.46	5.63
05N00985	0.99	1.11	1.47	1.62	0.99	1.08	1.11	1.47	1.62	1.68	1.68
05N00986	1.09	1.17	1.54	1.68	1.09	1.14	1.17	1.54	1.68	1.43	1.43

Sampled As : Chowchow

USDA-NRCS-NSSC-Soil Survey Laboratory

*** Supplementary Characterization Data ***

(Grays Harbor, Washington)

Loamy, isotic, dysic, isomesic Terric Haplosaprist

Pedon No. 05N0175

Tier 3

Layer	Depth (cm)	Horz	Prep	-51- >2	-52- 250 -UP	-53- 250 -75	-54- 75 -2	-55- 20 -5	-56- 5 -2	-57- 2-	-58- <2	-59- 2- .05	-60- .05 .002	-61- LT .002	-62- Pores D	-63- Pores F	-64- C/N Rat-io	-65- Fine Clay	-66- CEC Sum Cats	-67- NH4- OAC	-68- 1500 H2O kPa	-69- LEP 33 kPa	-70- 1500 kPa	-71- Oven -dry	-72- 1500 kPa	-73- Oven -dry	-74- WRD Whole Soil	-75- WRD <2 mm Soil
05N00981	0-9	Oi	S	--	--	--	--	--	--	--	100				38	49	15						14.1	14.1	26.6	26.6	0.20	0.20
05N00982	9-20	Oa	S	--	--	--	--	--	--	--	100				13	73	14						22.4	22.4	29.4	29.4	0.49	0.49
05N00983	20-33	2Oe	S	--	--	1	--	--	1		99				31	55	23						25.4	25.4	34.3	34.7	0.34	0.35
05N00984	33-58	2Oa	S	--	--	tr	--	tr	tr		100				35	50	22	0.03	4.40				6.8	7.0	8.9	9.1	0.31	0.31
05N00985	58-91	3Cg1	S	--	--	--	--	--	--	7	100	27	3		15	48	14				1.81	0.419	2.9	2.9	3.9	3.9	0.31	0.31
05N00986	91-132	3Cg2	S	--	--	--	--	--	--	6	100	26	9		14	45	19	0.12			0.85	0.113	1.5	1.5	2.4	2.4	0.25	0.25

(-- Volume Fractions -- Whole Soil (mm) At 33 kPa — % of Whole Soil) ; (-- Ratios To Clay -- <2 mm Fraction) : Fine Clay / CEC / NH4- OAC / H2O ; Cats, OAC, H2O ; (-- Linear Extensibility --) Whole Soil 33 kPa / <2 mm to % ; 1500 Oven Oven-dry / 1500 Oven ; (-- WRD --) Whole Soil <2 mm (---in³/in³---)

Tier 4

Layer	Depth (cm)	Horz	Prep	-76- >2	-77- 75 -20	-78- 20 -2	-79- 2- .05	-80- .05 .002	-81- < .002	-82- VC	-83- C	-84- M	-85- F	-86- VF	-87- C	-88- F	-89- ay	-90- Text-ure	-91- Sand 2- .05	-92- Silt .05- .002	-93- Clay < .002	-94- pH CaCl2 .01M	-95-	-96- Elect. Res. ohms	-97- Con-duct dS m⁻¹	-98- Part-icle Den-sity g cm⁻³
05N00981	0-9	Oi	S																			3.9				
05N00981	0-9	Oi	M																			4.0				
05N00982	9-20	Oa	S																			3.9				
05N00982	9-20	Oa	M																			4.0				
05N00983	20-33	2Oe	S																			4.1				
05N00984	33-58	2Oa	S																			4.4				
05N00985	58-91	3Cg1	S	--	--	21	79	10		tr	2	7	11	24	55			sil	19.1	71.6	9.3	4.7				
05N00986	91-132	3Cg2	S	--	--	19	81	27		tr	1	6	12	26	56			sil	14.8	63.9	21.3	4.5				

(-- Weight Fractions - Clay Free --) (----- Whole Soil -----) (----- <2 mm Fraction -----) (--- Sands ---) (--- Silts ---) (--- % of Sand and Silt ---) ; (- % of >2 mm Sand and Silt -) ; PSDA (mm) by 2-.05, .05-.002, <.002 ; 3A1a1a4C1a2a / 3A1a4C1a2a

*** Taxonomy Characterization Data ***
(Grays Harbor, Washington)

Sampled as on Sep 30, 2004 :
Revised to correlated on Aug 15, 2007 :

Chowchow ; Loamy, isotic, dysic, isomesic Terric Haplosaprist
Chowchow ; Loamy, isotic, dysic, isomesic Terric Haplosaprist

SSL - Project   C2005USWA011 Grays Harbor
    - Site ID   S04WA-027-009 Lat: 47° 18' 20.80" north Long: 124° 5' 31.40" west NAD83 MLRA: 4A
    - Pedon No.   05N0175
    - General Methods 1B1A, 2A1, 2B

United States Department of Agriculture
Natural Resources Conservation Service
National Soil Survey Center
Soil Survey Laboratory
Lincoln, Nebraska 68508-3866

Taxonomy Tier 1

Layer	Depth (cm)	Horz	Prep	-1- Clay <.002	-2- Fine Clay <.0002	-3- CaCO3 Clay <.002	-4- 1500 kPa /Clay	-5- Clay Est	-6- .1-75 mm Frac	-7- Bulk Den 33 kPa	-8- Cole Whole Soil	-9- Vol % of Whole	-10- Resist Min %
				(------ % of <2 mm ------)				(---- % ----)		g cm⁻³ DbWR1	cm cm⁻¹		
				3A1a1a	3A1a1a								
05N00981	0-9	Oi	S						--	0.35	0.266	--	
05N00982	9-20	Oa	S						--	0.36	0.294	--	
05N00983	20-33	2Oe	S						--	0.36	0.341	1	
05N00984	33-58	2Oa	S						--	0.40	0.091	tr	
05N00985	58-91	3Cg1	S	9.3	0.3		1.81	32.7	9	0.99	0.039	--	
05N00986	91-132	3Cg2	S	21.3	2.6		0.85	36.6	6	1.09	0.024	--	

Taxonomy Tier 2

Layer	Depth (cm)	Horz	Prep	-1- pH H₂O	-2- pH NaF	-3- Org C	-4- Tot C	-5- Al+½Fe Oxal	-6- ODOE	-7- CO3 as CaCO3	-8- NH4	-9- Bases	-10- NZ P Ret	-11- ECEC cmol(+)	-12- CEC7 /Clay	-13- ECEC /Clay	-14- Al Sat %	-15- E C dS m⁻¹	-15- ESP %
				4C1a2a	4C1a1a1	4C1a1a1	4H2a	4G2a	4G2a	4E1a1a1	4E1a1a1	(--- Base Sat ---)	4D8a1	(----- kg⁻¹ -----)					
05N00981	0-9	Oi	S	4.7															
05N00981	0-9	Oi	M	5.1			46.73				2								
05N00982	9-20	Oa	S	4.2															
05N00982	9-20	Oa	M	4.5			41.06												
05N00983	20-33	2Oe	S	4.2			46.72				1	0							
05N00984	33-58	2Oa	S	4.6			27.35				0	0		9.6			94		tr
05N00985	58-91	3Cg1	S	5.1	11.3		3.71	2.12	0.38		1	1	96	7.6	2.27		95		--
05N00986	91-132	3Cg2	S	5.3	10.9		3.58	1.20	0.22	tr	2	1	84	3.2	0.79	0.34	91		--

Pedon Calculations

Calculation Name	Result	Units of Measure
LE, Whole Soil, Summed to 1m	14	cm/m

PEDON DESCRIPTION

Print Date: 08/10/2010
Description Date: 9/23/2004
Describer: Ed Brincken
Site ID: 04-EAB-10
Site Note:
Pedon ID: 04WA027009
Pedon Note:
Lab Source ID: SSL
Lab Pedon #: 05N0175

Soil Name as Described/Sampled: Chowchow
Soil Name as Correlated: Chowchow
Classification: Loamy, isotic, dysic, isomesic Terric Haplosaprists
Pedon Type: within range of series
Pedon Purpose: full pedon description
Taxon Kind: series
Associated Soils:
Physiographic Division:
Physiographic Province:
Physiographic Section:
State Physiographic Area:
Local Physiographic Area:
Geomorphic Setting: on talf till plain on talf proglacial lake
Upslope Shape:
Cross Slope Shape:
Particle Size Control Section: 0 to 122 cm.
Description origin: NASIS
Diagnostic Features: histic epipedon 0 to 58 cm.

Country:
State: Washington
County: Grays Harbor
MLRA: 4A -- Sitka Spruce Belt
Soil Survey Area: WA728 -- Quinault Indian Reservation, Washington
Map Unit: 6 -- Chowchow peat, 0 to 2 percent slopes
Quad Name: Macafee Hill, Washington
Location Description: about 2 miles North of the Moclips Highway (F-5 road)
Legal Description: about 1240 feet East and 1145 feet North of the Southwest Corner of Section 16, Township 21N , Range 11W
Latitude: 47 degrees 18 minutes 20.80 seconds north
Longitude: 124 degrees 5 minutes 31.40 seconds west
Datum: NAD83
UTM Zone: 10
UTM Easting:
UTM Northing:

Primary Earth Cover: Grass/herbaceous cover
Secondary Earth Cover: Culturally induced barren
Existing Vegetation: clubmoss, rush, sedge
Parent Material: organic material over glaciolacustrine deposits
Bedrock Kind:
Bedrock Depth:
Bedrock Hardness:
Bedrock Fracture Interval:
Surface Fragments:
Description database: NSSL

Slope (%)	Elevation (meters)	Aspect (deg)	MAAT (C)	MSAT (C)	MWAT (C)	MAP (mm)	Frost-Free Days	Drainage Class	Slope Length (meters)	Upslope Length (meters)
0.5	77.0					2,667	200	very poorly		

Oi--0 to 9 centimeters; dark brown (7.5YR 3/2) rubbed peat, very dark gray (7.5YR 3/1) exterior, dry; 50 percent unrubbed fiber, 42 percent rubbed; many fine roots throughout and many very fine roots throughout; very strongly acid, pH 5.0, Bromcresol green; abrupt smooth boundary. Lab sample # 05N00981

Oa--9 to 20 centimeters; black (10YR 2/1) rubbed muck, black (10YR 2/1) exterior, dry; 20 percent unrubbed fiber, 5 percent rubbed; many fine roots throughout and common medium roots throughout and many very fine roots throughout; very strongly acid, pH 4.8, Bromcresol green; abrupt wavy boundary. Lab sample # 05N00982

2Oe--20 to 33 centimeters; very dark brown (10YR 2/2) rubbed mucky peat, black (10YR 2/1) exterior, dry; 30 percent unrubbed fiber, 20 percent rubbed; many fine roots throughout and common medium roots throughout; neutral, pH 6.6, Chlorophenol red; abrupt wavy boundary. Lab sample # 05N00983

2Oa--33 to 58 centimeters; black (10YR 2/1) rubbed muck, black (10YR 2/1) exterior, dry; 10 percent unrubbed fiber, 1 percent rubbed; common medium roots throughout; very strongly acid, pH 5.0, Bromcresol green; abrupt smooth boundary. Lab sample # 05N00984

3Cg1--58 to 91 centimeters; 70 percent very dark grayish brown (2.5Y 3/2) broken face silt loam, 70 percent light brownish gray (2.5Y 6/2) broken face, dry; structureless massive; firm, hard, slightly sticky, moderately plastic; common medium roots throughout; 30 percent fine prominent threadlike 2.5Y 2.5/1, moist, iron-manganese masses with clear boundaries on surfaces along root channels; strongly acid, pH 5.4, Chlorophenol red; gradual wavy boundary. Lab sample # 05N00985

3Cg2--91 to 122 centimeters; olive gray (5Y 5/2) broken face silt loam, light gray (5Y 7/1) broken face, dry; structureless massive; firm, hard, slightly sticky, moderately plastic; few medium roots throughout; very strongly acid, pH 4.8, Bromcresol green. Lab sample # 05N00986

2 Wildmesa

Pedon ID: 89CA027004

*** Primary Characterization Data ***
(Inyo, California)

Sampled as on May 01, 1989 :   Wildmesa ; Fine, montmorillonitic, mesic Xerollic Paleargid
Revised to correlated on Sep 01, 1996 :   Wildmesa ; Fine, smectitic, mesic Xeralfic Paleargid

SSL   - Project   CP89CA091 CHINA LAKE
      - Site ID   89CA027004 Lat: 35° 58' 26 00" north Long: 117° 35' 42 00" west MLRA: 29
      - Pedon No.   89P0325
      - General Methods 1B1A, 2A1, 2B

United States Department of Agriculture
Natural Resources Conservation Service
National Soil Survey Center
Soil Survey Laboratory
Lincoln, Nebraska 68508-3866

Layer	Horizon	Orig Hzn	Depth (cm)	Field Label 1	Field Label 2	Field Label 3	Field Texture	Lab Texture
89P01799	A	A	0-8				VFSL	FSL
89P01800	AB	AB	8-15				L	FSL
89P01801	2Bt	2BT	15-46				C	CL
89P01802	2B k1	2BTK	46-74				C	CL
89P01803	2B k2	2BTK	74-109				C	C
89P01804	3Bkqm	3BQKM	109-153					COSL

Calculation Name

	Pedon Calculations		
	Result	Units of Measure	
CEC Activity, CEC7/Clay, Weighted Average	0.83	(NA)	
Clay, carbonate free, Weighted Average	36	% wt	
Weighted Particles, 0.1-75mm, 75 mm Base	21	% wt	
Volume, >2mm, Weighted Average	1	% vol	
Clay, total, Weighted Average	36	% wt	
LE, Whole Soil, Summed to 1m	0	cm/m	

Weighted averages based on control section: 15-65 cm

PSDA & Rock Fragments

			-1-	-2-	-3-	-4-	-5-		-6-	-7-	-8-	-9-	-10-	-11-	-12-	-13-	-14-	-15-	-16-	-17-
			(- - - - Total - - - - - -)			(- - Clay - - -)			(- - - - Silt - - - - -)		(- - - - - - - - Sand - - - - - - - - -)					(Rock Fragments (mm))				
			Clay	Silt	Sand	Fine	CO₃		Fine	Coarse	VF	F	M	C	VC	(- - - - - - Weight - - - - - -)				>2 mm
			<	.002	.05	<	<		.002	.02	.05	.10	.25	.5	1	2	5	20	.1-	wt %
			.002	-.05	-2	.0002	.002		-.02	-.05	-.10	-.25	-.50	-1	-2	-5	-20	-75	75	whole
									% of <2mm Mineral Soil							(- - -% of <75mm - - - - -)				soil
		Prep	3A1	3A1	3A1	3A1	3A1		3A1	3A1	3A1	3A1	3A1	3A1	3A1	3B1	3B1	3B1	3B1	
Layer	Depth (cm)	Horz																		
89P01799	0-8	A	S	7.8	23.5	68.7				14.7	8.8	28.3	34.1	3.6	1.4	1.3	1	4	21	56	26
89P01800	8-15	AB	S	10.6	24.3	65.1	0.8			16.7	7.6	26.5	32.7	3.8	1.3	0.8	1	2	7	45	10
89P01801	15-46	2Bt	S	34.9	21.8	43.3	10.9	0.3		15.1	6.7	20.7	18.1	2.6	1.0	0.9	1	1	--	24	2
89P01802	46-74	2Btk1	S	38.1	26.7	35.2	10.2	0.9		16.9	9.8	18.2	13.6	2.2	0.9	0.3	tr	tr	--	17	tr
89P01803	74-109	2Btk2	S	44.0	26.5	29.5	10.3	1.4		17.1	9.4	13.9	12.2	2.1	0.9	0.4	tr	1	--	17	1
89P01804	109-153	3Bkqm	GP	7.0	19.0	74.0				12.4	6.6	15.2	16.5	12.0	14.9	15.4					

*** Primary Characterization Data ***

(Inyo, California)
Fine, montmorillonitic, mesic Xerollic Paleargid
; Pedon No. 89P0325

Bulk Density & Moisture

Layer	Depth (cm)	Horz	Prep	-1- (Bulk Density) 33 kPa 4A1d	-2- Oven Dry 4A1h	-3- Cole Whole Soil	-4- Water Content 6 kPa	-5- 10 kPa	-6- 33 kPa 4B1c	-7- 1500 kPa 4B2a	-8- 1500 kPa Moist	-9- AD/OD Ratio 4B5	-10- WRD Whole Soil 4C1	-11- Aggst Stabl 2-0.5mm 4G1	-12- CEC7 8D1	-13- 1500 kPa 8D1
				(--- g cm⁻³ ---)			(-------- % of < 2mm --------)						cm³ cm⁻³	%	(- - Ratio/Clay - -)	
89P01799	0-8	A	S	1.47	1.55	0.015			16.5	5.1		1.009	0.14	3	1.54	0.65
89P01800	8-15	AB	S	1.60	1.70	0.019			14.6	5.0		1.009	0.14		1.04	0.47
89P01801	15-46	2Bt	S	1.45	1.72	0.058			24.3	17.7		1.027	0.09		0.82	0.51
89P01802	46-74	2Btk1	S	1.38	1.63	0.057			29.6	16.3		1.031	0.18		0.85	0.43
89P01803	74-109	2Btk2	S	1.26	1.62	0.087			37.8	20.0		1.037	0.22		0.80	0.45
89P01804	109-153	3Bkqm	S													1.24
89P01804	109-153	3Bkqm	GP							8.7		1.017				

Water Content

Layer	Depth (cm)	Horz	Prep	-1- (- -Atterberg - -) LL pct <0.4mm 4F1	-2- (- - Limits - -) PI 4F	-3- Field	-4- Recon 33 kPa	-5- Recon Oven Dry	-6- Field	-7- Recon 33 kPa	-8- Recon 6 kPa	-9- 10 kPa	-10- Sieved Samples 33 kPa	-11- 100 kPa	-12- 200 kPa	-13- 500 kPa
						(------- Bulk Density ------)			(------- Water Content -------)							
						(------- g cm⁻³ -------)			(------- % of < 2mm -------)							
89P01799	0-8	A	S	42	NP											
89P01801	15-46	2Bt	S	51	27											
89P01802	46-74	2Btk1	S		33											

*** Primary Characterization Data ***

(Inyo, California)
Fine, montmorillonitic, mesic Xerollic Paleargid
;   Pedon No. 89P0325

Carbon & Extrac ions

			-1-	-2-	-3-	-4-	-5-	-6-	-7-	-8-	-9-	-10-	-11-	-12-	-13-	-14-	-15-	-16-	-17-	-18-
			(---- Total ----)			Org	C/N	(--- Dith-Cit Ext ---)			(----- Ammonium Oxalate Extraction -----)					(--- Na Pyro-Phosphate ---)		(----- % of < 2mm -----)		
			C	N	S	C	Ratio	Fe	Al	Mn	Al+½Fe ODOE Fe	Al	Si	Mn	Mn	C	C	Fe	Al	Mn
Layer	Depth (cm)	Horz	Prep	(----- % of <2 mm -----)			6A1c		(----- % of < 2mm -----)							mg kg⁻¹					
89P01799	0-8	A	S				0.45														
89P01800	8-15	AB	S				0.15														
89P01801	15-46	2Bt	S				0.19														
89P01802	46-74	2Btk1	S				0.15														
89P01803	74-109	2Btk2	S				0.12														
89P01804	109-153	3Bkqm	GP				0.09														

CEC & Bases

			-1-	-2-	-3-	-4-	-5-	-6-	-7-	-8-	-9-	-10-	-11-	-12-	-13-	-14-
			(----- NH₄OAC Extractable Bases -----)				Sum	Acid-	Extr	KCl	CEC8 Sum	CEC7 NH₄	ECEC Bases	Al	(--- Base ---) (- Saturation -)	
			Ca	Mg	Na	K	Bases	ity	Al	Mn	Cats	OAC	+Al	Sat	Sum	NH₄OAC
Layer	Depth (cm)	Horz	Prep	(----- cmol(+) kg⁻¹ -----)							mg kg⁻¹		(----- cmol(+) kg⁻¹ ---)		(----- % -----)		
			6N2e	6O2d	6P2b	6Q2b						5A8b			5C3	5C1
89P01799	0-8	A	S	10.0*	2.7	0.2	1.1						12.0			100	100
89P01800	8-15	AB	S	7.9*	2.2	0.4	0.7						11.0			100	100
89P01801	15-46	2Bt	S	20.3*	4.1	4.1	0.7						28.7			100	100
89P01802	46-74	2Btk1	S	42.6*	4.6	7.1	0.5						32.3			100	100
89P01803	74-109	2Btk2	S	54.8*	5.8	11.3	0.5						35.4			100	100

*Extractable Ca may contain Ca from calcium carbonate or gypsum., CEC7 base saturation set to 100.

Salt

			-1-	-2-	-3-	-4-	-5-	-6-	-7-	-8-	-9-	-10-	-11-	-12-	-13-	-14-	-15-	-16-	-17-	-18-	-19-	-20-
			(-------------------------------- Water Extracted From Saturated Paste --------------------------------)															Total	Elec	Pred Elec	Exch	
			Ca	Mg	Na	K	CO₃	HCO₃	F	Cl	PO₄	Br	SO₄ OAC	NO₂	NO₃	H₂O	Salts	Cond	Cond	Na	SAR
Layer	Depth (cm)	Horz	Prep	(---- mmol(+) L⁻¹ ----)				(------------------------------ mmol(-) L⁻¹ ------------------------------)									(--- % ---)		(--- dS m⁻¹ ---)		%	
			6N1b	6O1b	6P1b	6Q1b	6I1b	6J1b	6U1a	6K1c			6L1c	6W1a	6M1c	8A	8D5	8A3a	8I	5D2	5E
89P01799	0-8	A	S																	0.08	2	
89P01800	8-15	AB	S																	0.07	4	
89P01801	15-46	2Bt	S																	0.19	14	
89P01802	46-74	2Btk1	S	2.9	0.6	19.3	0.1	--	0.7	tr	20.4			4.1	--	--	63.6	0.1	2.28	0.96	18	15
89P01803	74-109	2Btk2	S	21.6	4.8	60.8	0.1	--	0.4	2.8	80.6			18.5	--	--	71.9	0.4	7.61	3.16	20	17

Pedon ID: 89CA027004
Sampled As   :   Wildmesa
USDA-NRCS-NSSC-Soil Survey Laboratory

Print Date: Aug 10 2010 8:37AM

*** Primary Characterization Data ***
(Inyo, California)
Fine, montmorillonitic, mesic Xerollic Paleargid
;   Pedon No. 89P0325

pH & Carbonates

				-1-	-2-	-3-	-4-	-5-	-6-	-7-	-8-	-9-	-10-	-11-
				(----------------- pH -----------------)						(-- Carbonate -- -)		(-- Gypsum -- -)		
					$CaCl_2$		Sat			As $CaCO_3$		As $CaSO_4*2H_2O$		Resist
					0.01M	H_2O	Paste	Sulf	NaF	<2mm	<20mm	<2mm	<20mm	ohms
	Depth				1:2	1:1	Paste			(-------------------- % --------------------)				cm^{-1}
Layer	(cm)	Horz	Prep	KCl	4C1a2a	4C1a2a	8C1b			6E1g		6F1a		8E1
89P01799	0-8	A	S		7.2	7.9				tr				
89P01800	8-15	AB	S		7.2	8.3				--				
89P01801	15-46	2Bt	S		7.4	8.5				--				
89P01802	46-74	2Btk1	S		8.1	8.7	8.0			2				
89P01803	74-109	2Btk2	S		8.0	8.5	7.7			3		--		270

*** Primary Characterization Data ***

Pedon ID: 89CA027004

Sampled As   :   Wildmesa

USDA-NRCS-NSSC-Soil Survey Laboratory

(Inyo, California)

Fine, montmorillonitic, mesic Xerollic Paleargid

;   Pedon No. 89P0325

Clay Mineralogy (<.002 mm)

Layer	Depth (cm)	Horz	Fract ion	-1-	-2-	-3-	-4-	-5-	-6-	-7-	-8-	-9-	-10-	-11-	-12-	-13-	-14-	-15-	-16-	-17-	-18-
						X-Ray				Thermal						Elemental				EGME	Inter
						7A2i										7C3				Retn	preta tion
				(------- peak size -------)				(------------ % ------------)					SiO_2	Al_2O_3	Fe_2O_3	MgO	CaO	K_2O	Na_2O		mg g^{-1}
														(------------ % ------------)								
89P01800	8-15	AB	tcly	MI 3		MT 2	MM 1	KK 1	LE 1					16	6.1			2.3			
89P01801	15-46	2Bt	tcly	MT 3		MI 2	KK 1							17	6.7			1.8			
89P01802	46-74	2Btk1	tcly	MT 3		MI 2	KK 1							16	6.9			1.3			

FRACTION INTERPRETATION:

tcly - Total Clay, <0.002 mm

MINERAL INTERPRETATION:

KK - Kaolinite   LE - Lepidocrocite   MI - Mica   MM - Montmorillonite-Mica   MT - Montmorillonite

RELATIVE PEAK SIZE:

5 Very Large   4 Large   3 Medium   2 Small   1 Very Small   6 No Peaks

*** Primary Characterization Data ***

Pedon ID: 89CA027004

Sampled As : Wildmesa

(Inyo, California)

Fine, montmorillonitic, mesic Xerollic Paleargid

USDA-NRCS-NSSC-Soil Survey Laboratory ; Pedon No. 89P0325

Sand - Silt Mineralogy (2 0-0.002 mm)

			-1-	-2-	-3-	-4-	-5-	-6-	-7-	-8-	-9-	-10-	-11-	-12-	-13-	-14-	-15-	-16-	-17-	-18-
					X-Ray				Thermal			Tot Re			Optical Grain Count 7B1a				EGME Retn	Inter preta tion
		Fract																mg g^{-1}	
Layer	Depth (cm)	Horz	ion		(- - - - - - - peak size - - - - - - -)			(- - - - - - - % - - - - - - -)			(- - - - - - - - - - - - - - - - - - - % - - - - - - - - - - - - - - - - - -)									
89P01800	8-15	AB	fs									66	QZ 57	FK 17	OP 7	OT 6	HN 5	BT 4		
													GE 1	ZR 1	GS tr	RU tr	FP tr	CL tr		

FRACTION INTERPRETATION:

fs - Fine Sand, 0.1-0.25 mm

MINERAL INTERPRETATION:

BT - Bio ite

GS - Glass

RU - Ru ile

CL - Chlorite

HN - Hornblende

ZR - Zircon

FK - Potassium Feldspar

OP - Opaques

FP - Plagioclase Feldspar

OT - O her

GE - Goethite

QZ - Quartz

*** Supplementary Characterization Data ***
(Inyo, California)

Sampled as on May 01, 1989 : Wildmesa : Fine, montmorillontic, mesic Xerollic Paleargid
Revised to correlated on Sep 01, 1996 : Wildmesa : Fine, smecttic, mesic Xeralfic Paleargid

SSL
- Project CP89CA091 CHINA LAKE
- Site ID 89CA027004 Lat: 35° 58' 26 00" north Long: 117° 35' 42 00" west MLRA: 29
- Pedon No. 89P0325
- General Methods 1B1A, 2A1, 2B

United States Department of Agriculture
Natural Resources Conservation Service
National Soil Survey Center
Soil Survey Laboratory
Lincoln, Nebraska 68508-3866

Tier 1

Engineering PSDA — Percentage Passing Sieve

Layer	Depth (cm)	Horz	Prep	3	2	3/2	1	3/4	3/8	4	10	40	200	20	5	2	1.	.5	.25	.10	.05
				(—Inches—)				(—Number—)					(—Millimeter—)						(—Microns—)		
89P01799	0-8	A	S	100	94	90	83	79	77	75	74	71	35	17	10	6	73	72	69	44	23
89P01800	8-15	AB	S	100	98	97	94	93	92	91	90	87	45	25	16	10	89	88	85	55	31
89P01801	15-46	2Bt	S	100	100	100	100	100	100	99	98	95	68	49	40	34	97	96	94	76	56
89P01802	46-74	2Btk1	S	100	100	100	100	100	100	100	100	98	76	55	45	38	100	99	97	83	65
89P01803	74-109	2B k2	S	100	100	100	100	100	100	99	99	97	78	60	50	44	99	98	96	84	70
89P01804	109-153	3Bkqm	S	100	100	100	100	100	100	100	100	67	35	19	12	7	85	70	58	41	26

Cumulative Curve Fractions (<75mm) — USDA Less Than Diameters (mm) at Percentile / Atterberg / Gradation

Layer	60	50	10	LL (4F1)	PI (4F)	Uni-fmty CU	Cur-vtur CC
89P01799	0.18	0.12	0.005	—	NP	36.4	4.5
89P01800	0.12	0.086	0.002			54.0	6.9
89P01801	0.06	0.023	—	42	27	>100	0.1
89P01802	0.03	0.010	—	51	33	87.2	0.1
89P01803	0.02	0.005	—			55.2	0.2
89P01804	0.29	0.163	0.003			81.8	3.6

Tier 2

Weight Fractions

Layer	Depth (cm)	Horz	Prep	>2	250 -UP	250 -75	75 -20	20 -5	5 -2	<2	75 -20	20 -5	5 -2	<2
				(————— % of Whole Soil —————)							(— % of <75 mm —)			
											3B1	3B1	3B1	3B1
89P01799	0-8	A	S	26	--	--	21	4	1	74	26	21	4	74
89P01800	8-15	AB	S	10	--	--	7	2	1	90	10	7	2	90
89P01801	15-46	2Bt	S	2	--	--	2	1	tr	98	2	2	1	98
89P01802	46-74	2Btk1	S	tr	--	--	--	--	tr	100	--	--	tr	100
89P01803	74-109	2Btk2	S	tr	--	--	--	--	tr	99	1	--	tr	99
89P01804	109-153	3Bkqm	S	1	--	--	--	--	1	100	1	1	--	100

Weight Per Unit Volume (g cm^{-3})

		Whole Soil				<2 mm Fraction					Ratios At 33 kPa	
Layer		Soil Sur 33 kPa (4A1d)	Eng. Moist	Satur-ated	Oven-dry	Soil Survey 33 kPa (4A1d)	1500 kPa (4A1h)	Oven-dry	Eng. Moist Satur-ated	Oven-dry	Whole Soil	<2 Soil mm
89P01799		1.66	1.74	1.86	2.03	1.47	1.53	1.55	1.71	1.92	0.60	0.80
89P01800		1.66	1.76	1.88	2.03	1.60	1.67	1.70	1.83	2.00	0.60	0.66
89P01801		1.46	1.73	1.81	1.91	1.45	1.53	1.72	1.80	1.90	0.82	0.83
89P01802		1.38	1.63	1.79	1.79	1.38	1.51	1.63	1.79	1.86	0.92	0.92
89P01803		1.27	1.63	1.74	1.79	1.26	1.45	1.62	1.74	1.78	1.09	1.10
89P01804		1.45										

*** Supplementary Characterization Data ***

(Inyo, California)

Fine, montmorillonitic, mesic Xerollic Paleargid

Pedon No. 89P0325

Tier 3

Columns: -51- through -75-

Layer	Depth (cm)	Horz	Prep	>2	250-UP	250-75	75-20	20-5	5-2	-2	<2	2-.05	.05-.002	.002 LT	Pores D	Pores F	C/N Ratio	Fine Clay	Sum Cats	OAC	H2O 1500 (8D1)	LEP 33 kPa	LE Whole Soil 33 kPa	LE 1500 Oven	WRD Whole Soil to %	WRD <2 mm Soil Oven 4C1
89P01799	0-8	A	S	16	--	--	16	13	3	1	84	32	11	4	17	20					0.65					
89P01800	8-15	AB	S	6	--	--	6	4	1	1	94	37	14	6	15	22					0.47	0.230	1.2	1.3	0.14	0.17
89P01801	15-46	2Bt	S	1	--	--	1	--	1	1	99	23	12	19	10	35		0.08			0.51	0.190	1.4	1.4	0.14	0.15
89P01802	46-74	2Btk1	S	tr	--	--	tr	--	tr	tr	100	18	14	20	7	41		0.31			0.43	0.170	1.8	1.8	0.09	0.10
89P01803	74-109	2Btk2	S	1	--	--	1	--	1	tr	100	14	13	21	5	47		0.27			0.45	0.150	2.8	3.0	0.18	0.18
89P01804	109-153	3Bkqm	S	--	--	--	--	--	--	--	100	40	10	4	45			0.23			1.24	0.200	4.8	4.8	0.22	0.22

Tier 4

Columns: -76- through -98-

Layer	Depth (cm)	Horz	Prep	>2	75-20	20-2	2-.05	.05-.002	.002	VC	C	M	F	VF	Silt C	Silt F	Clay	Texture	PSDA Sand	PSDA Silt	PSDA Clay	pH CaCl2 .01M (4C1a2a8E1)	Elect. Resist. ohms	Particle Density g cm⁻³ (8A3a)
89P01799	0-8	A	S	28	28	5	54	18	6	2	1	4	37	31	10	16	8	fsl	68.7	23.5	7.8	7.2		
89P01800	8-15	AB	S	11	11	3	65	24	11	1	1	4	37	30	9	19	12	fsl	65.1	24.3	10.6	7.2		
89P01801	15-46	2Bt	S	3	3	3	64	32	52	1	2	4	28	32	10	23	54	cl	43.3	21.8	34.9	7.4		
89P01802	46-74	2Btk1	S				57	43	62	tr	1	4	22	29	16	27	62	cl	35.2	26.7	38.1	8.1		
89P01803	74-109	2Btk2	S	2	2	2	52	47	77	1	2	4	22	25	17	31	79	c	29.5	26.5	44.0	8.0		2.28
89P01804	109-153	3Bkqm	S							17	16	13	18	16	7	13	8						270	7.61
89P01804	109-153	3Bkqm	GP	2	2	2	80	20	8									cosl	74.0	19.0	7.0			

*** Taxonomy Characterization Data ***
(Inyo, California)

Pedon ID: 89CA027004

Sampled as on May 01, 1989 :
Revised to correlated on Sep 01, 1996 :

Wildmesa ; Fine, montmorillonitic, mesic Xerollic Paleargid
Wildmesa ; Fine, smectitic, mesic Xeralfic Paleargid

SSL - Project   CP89CA091 CHINA LAKE
    - Site ID   89CA027004 Lat: 35° 58' 26.00" north Long: 117° 35' 42.00" west MLRA: 29
    - Pedon No. 89P0325
    - General Methods 1B1A, 2A1, 2B

United States Department of Agriculture
Natural Resources Conservation Service
National Soil Survey Center
Soil Survey Laboratory
Lincoln, Nebraska 68508-3866

Taxonomy Tier 1

Layer	Depth (cm)	Horz	Prep	-1- Clay <.002	-2- Fine Clay <.0002	-3- CaCO₃ Clay <.002	-4- 1500 kPa /Clay	-5- Clay Est	-6- .1-75 mm Frac	-7- Bulk Den 33 kPa	-8- Cole Whole Soil	-9- Vol % of Whole	-10- Resist Min %
				3A1	3A1	3A1	8D1			g cm⁻³ 4A1d	cm cm⁻¹		7B1a
				(—— % of <2 mm ——)				(—— % ——)					
89P01799	0-8	A	S	7.8			0.65		56	1.47	0.015	16	
89P01800	8-15	AB	S	10.6	0.8		0.47		45	1.60	0.019	6	66
89P01801	15-46	2Bt	S	34.9	10.9		0.51		24	1.45	0.058	1	
89P01802	46-74	2Btk1	S	38.1	10.2	0.3	0.43		17	1.38	0.057	tr	
89P01803	74-109	2Btk2	S	44.0	10.3	0.9	0.45		17	1.26	0.087	1	
89P01804	109-153	3Bkqm	S	7.0		1.4	1.24					--	
89P01804	109-153	3Bkqm	GP										

Taxonomy Tier 2

Layer	Depth (cm)	Horz	Prep	-1- pH H₂O	-2- pH NaF	-3- Org C	-4- Tot C	-5- Al+½ Fe Oxal	-5- ODOE	-6- CO₃ as CaCO₃	-7-	-8- NH₄	-9- (— Base Sat —) Bases	-10- NZ P Ret
				4C1a2a		6A1c				6E1g		5C1	5C3	
						(——			——)	%				
89P01799	0-8	A	S	7.9		0.45				tr		100*	100	
89P01800	8-15	AB	S	8.3		0.15				--		100*	100	
89P01801	15-46	2Bt	S	8.5		0.19				--		100*	100	
89P01802	46-74	2Btk1	S	8.7		0.15				2		100*	100	
89P01803	74-109	2Btk2	S	8.5		0.12				3		100*	100	
89P01804	109-153	3Bkqm	GP			0.09								

	-11- ECEC cmol(+)	-12- CEC7 /Clay	-13- ECEC /Clay	-14- Al Sat %	-15- E C dS m⁻¹	-15- ESP %
	kg⁻¹	8D1			8A3a	5D2
89P01799		1.54				2
89P01800		1.04				4
89P01801		0.82				14
89P01802		0.85			2.28	18
89P01803		0.80			7.61	20

*Extractable Ca may contain Ca from calcium carbonate or gypsum.

Pedon Calculations

Calculation Name	Result	Units of Measure
CEC Activity, CEC7/Clay, Weighted Average	0.83	(NA)
Clay, carbonate free, Weighted Average	36	% wt
Weighted Particles, 0.1-75mm, 75 mm Base	21	% wt
Volume, >2mm, Weighted Average	1	% vol
Clay, total, Weighted Average	36	% wt
LE, Whole Soil, Summed to 1m	0	cm/m

Weighted averages based on control section: 15-65 cm

PEDON DESCRIPTION

Print Date: 08/10/2010

Description Date: 5/1/1989

Describer:

Site ID: 89CA027004

Site Note:

Pedon ID: 89CA027004

Pedon Note: Physiography: basalt flow basins/alluvial fill basins. Site was disturbed from recent fire, very little vegetation left, may have had CORA on site before fire. Landuse: military, wildlife, and grazing. This soil occurs in associ

Lab Source ID: SSL

Lab Pedon #: 89P0325

Soil Name as Described/Sampled: Wildmesa Series

Soil Name as Correlated: Wildmesa

Classification: Fine, smectitic, mesic Xeralfic Paleargids

Pedon Type:

Pedon Purpose: full pedon description

Taxon Kind:

Associated Soils:

Physiographic Division:

Physiographic Province:

Physiographic Section:

State Physiographic Area:

Local Physiographic Area:

Geomorphic Setting: None Assigned

Upslope Shape: concave

Cross Slope Shape:

Particle Size Control Section: 3 to 23 cm.

Description origin: Converted from SSL-CMS data

Diagnostic Features: ochric epipedon 0 to 3 cm.
argillic horizon 3 to 43 cm.

Country:

State: California

County: Inyo

MLRA: 29 -- Southern Nevada Basin and Range

Soil Survey Area:

Map Unit:

Quad Name:

Location Description: About 1.25 miles S of intersection with main access road into Wild Horse Mesa, just off N side of dirt rd.

Legal Description: About 500 feet N and 1800 feet E of SW corner of Section 31, Township 22S , Range 41E

Latitude: 35 degrees 58 minutes 26.00 seconds north

Longitude: 117 degrees 35 minutes 42.00 seconds west

Datum:

UTM Zone:

UTM Easting:

UTM Northing:

Primary Earth Cover:

Secondary Earth Cover:

Existing Vegetation: fourwing saltbush, Nevada jointfir, spiny hopsage

Parent Material:

Bedrock Kind:

Bedrock Depth:

Bedrock Hardness:

Bedrock Fracture Interval:

Surface Fragments:

Description database: NSSL

Slope (%)	Elevation (meters)	Aspect (deg)	MAAT (C)	MSAT (C)	MWAT (C)	MAP (mm)	Frost-Free Days	Drainage Class	Slope Length (meters)	Upslope Length (meters)
2.0	1,564.0	270	13.0			18		well		

A--0 to 8 centimeters: brown (10YR 5/3) extremely cobbly very fine sandy loam, dark brown (10YR 3/3), moist; moderate medium platy structure; very friable, slightly hard, slightly sticky, nonplastic; many very fine and fine roots; common fine vesicular and common very fine and fine tubular pores; slight effervescence: slightly alkaline, pH 7.4, Hellige-Truog; clear smooth boundary. Lab sample # 89P1799. very close to a silt loam with high % silts but enough fine sand and medium sand to go with very fine sandy loam and loam; many very fine and fine roots

A bt--8 to 15 centimeters: light brownish gray (10YR 6/2) loam, dark brown (10YR 3/3), moist; moderate medium subangular blocky, and strong coarse platy structure: very friable, slightly hard, moderately sticky, slightly plastic; common very fine and fine roots; many fine vesicular and common fine tubular pores; slight effervescence: slightly alkaline, pH 7.7, Hellige-Truog; abrupt wavy boundary. Lab sample # 89P1800. very close to a silt loam with high % silts but enough fine sand and medium sand to go with very fine sandy loam and loam; common very fine and fine roots

2B t--15 to 46 centimeters: dark yellowish brown (10YR 4/4) clay, brown (10YR 4/3), moist; strong medium and coarse angular blocky, and strong fine and medium prismatic structure; firm, very hard, very sticky, very plastic; few fine and medium roots; common fine tubular pores; slight effervescence: moderately alkaline, pH 8.0, Hellige-Truog; clear wavy boundary. Lab sample # 89P1801. sand and silt sized particles overplacing pressure faces on some ped faces; few fine and medium roots

2B tk--46 to 109 centimeters: dark yellowish brown (10YR 4/4) clay, brown (10YR 4/3), moist; strong medium and coarse angular blocky, and strong fine and medium subangular blocky structure: firm, very hard, very sticky, very plastic; few fine and medium roots; common fine tubular pores; slight effervescent with common to many (25-50% of surface) fine (1-3 mm in length) pred. rod-like and spherical segregated CaCO3 being strongly effervescent; sand and silt sized particles overplacing pressure faces on some ped f; few fine and medium roots 7.9, Hellige-Truog; abrupt wavy boundary. Lab sample # 89P1802. matrix is slightly effervescent with common to many (25-50% of surface) fine (1-3 mm in length) pred.

3B qkm--109 to 152 centimeters: white (10YR 8/1), light brownish gray (10YR 6/2), moist; strong medium platy structure: extremely firm*, extremely hard: violent effervescence. Lab sample # 89P1804. a continuous highly fractured broken indurated duripan with a high % of cemented cobbles and stones; soil material and roots have penetrated fracture zones; a thick (1/2-3/4") laminar cap is continuous throughout pedon but broken and fractur

MATERIALS TESTING REPORT:
SOIL CLASSIFICATION

Project and state __China Lake, California__   Sample location __Inyo County__   Laboratory No. _____

Field sample No. _____   Depth _____   Geologic origin _____

Type of sample _____   Tested at _____   Approved by _____   Date _____

Symbol _____   Description __Wildmesa, Pedon 89325, Bt Horizon, 15-46 cm__

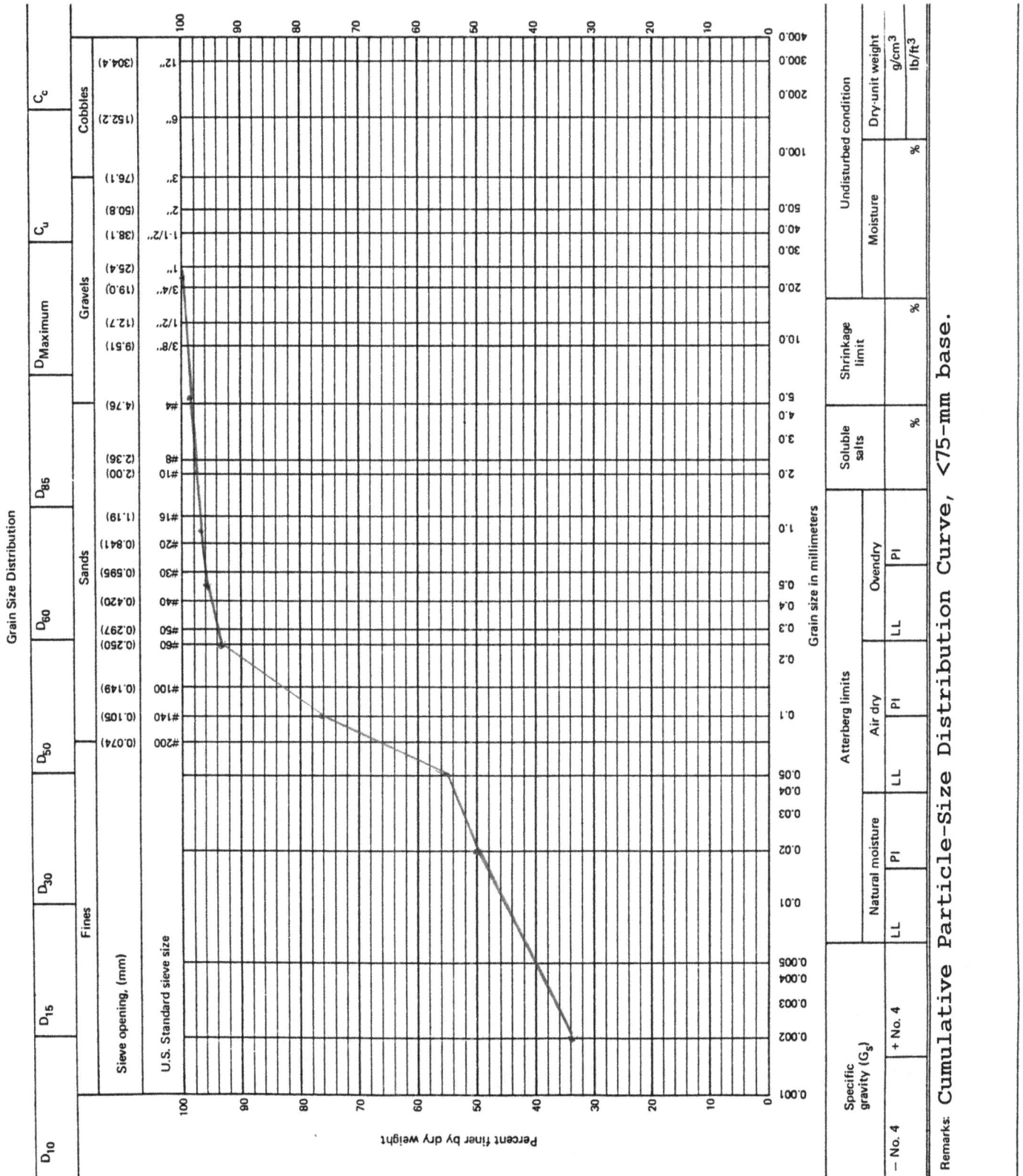

Remarks: Cumulative Particle-Size Distribution Curve, <75-mm base.

ESTIMATED SOIL WATER RETENTION CURVE

Pedon ID: 89CA027004
Layer Natural Key: 89P01799
Horizon: A   0-8.0 cm

<2 mm Fraction

van Genuchten Retention Parameters
from Rosetta*

θ_r	θ_s	$\log_{10}(\alpha)$	$\log_{10}(n)$
0.0272	0.3758	-2.0721	0.1597

Measured Soil Properties

Rock Fragmer	Organic Carbon	Sand	Silt	Clay	Db_{33}	θ_{33}	θ_{1500}
% vol	% wt	% wt	% wt	% wt	g/cc	cm^3/cm^3	cm^3/cm^3
16	0.45	68.7	23.5	7.8	1.47	0.24	0.07

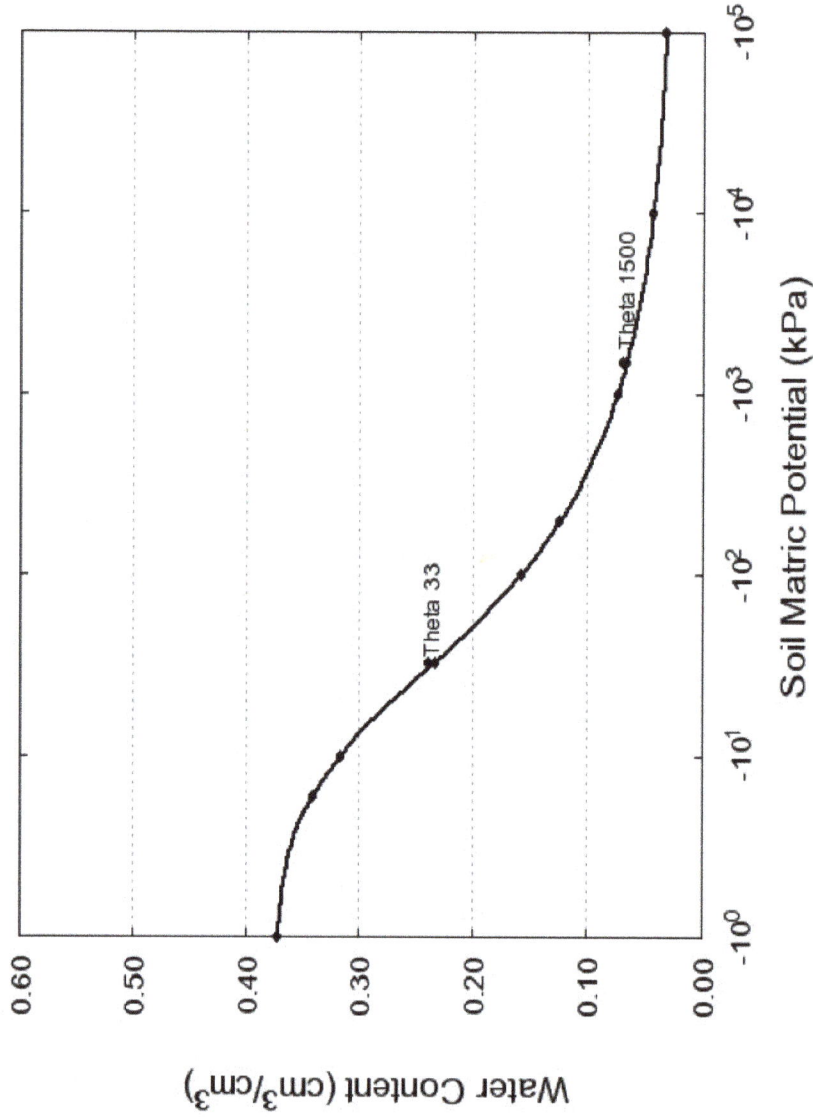

Water Content (cm^3/cm^3) vs Soil Matric Potential (kPa). Theta 33 and Theta 1500 points labeled.

*Rosetta Reference

ESTIMATED SOIL WATER RETENTION CURVE

Pedon ID: 89CA027004
Layer Natural Key: 89P01800
Horizon: AB   8.0-15.0 cm

<2 mm Fraction

van Genuchten Retention Parameters
from Rosetta*

θ_r	θ_s	$\log_{10}(\alpha)$	$\log_{10}(n)$
0.0289	0.3506	-2.0488	0.1535

Measured Soil Properties

Rock Fragmer	Organic Carbon	Sand	Silt	Clay	Db$_{33}$	θ_{33}	θ_{1500}
% vol	% wt	% wt	% wt	% wt	g/cc	cm^3/cm^3	cm^3/cm^3
6	0.15	65.1	24.3	10.6	1.60	0.23	0.08

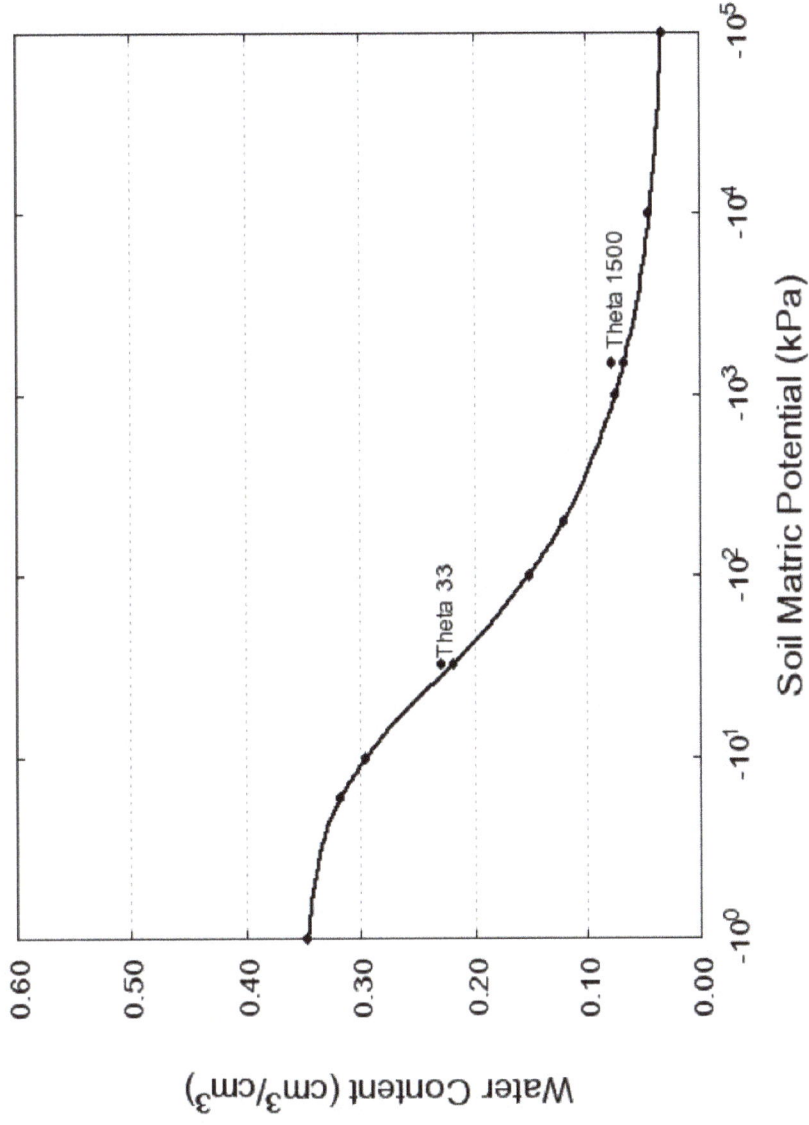

*Rosetta Reference

ESTIMATED SOIL WATER RETENTION CURVE

Pedon ID: 89CA027004
Layer Natural Key: 89P01801
Horizon: 2Bt 15.0-46.0 cm

van Genuchten Retention Parameters
<2 mm Fraction
from Rosetta*

θ_r	θ_s	$\log_{10}(\alpha)$	$\log_{10}(n)$
0.0833	0.4436	-1.5529	0.0802

Measured Soil Properties

Rock Fragmer	Organic Carbon	Sand	Silt	Clay	Db_{33}	θ_{33}	θ_{1500}
% vol	% wt	% wt	% wt	% wt	g/cc	cm^3/cm^3	cm^3/cm^3
1	0.19	43.3	21.8	34.9	1.45	0.35	0.26

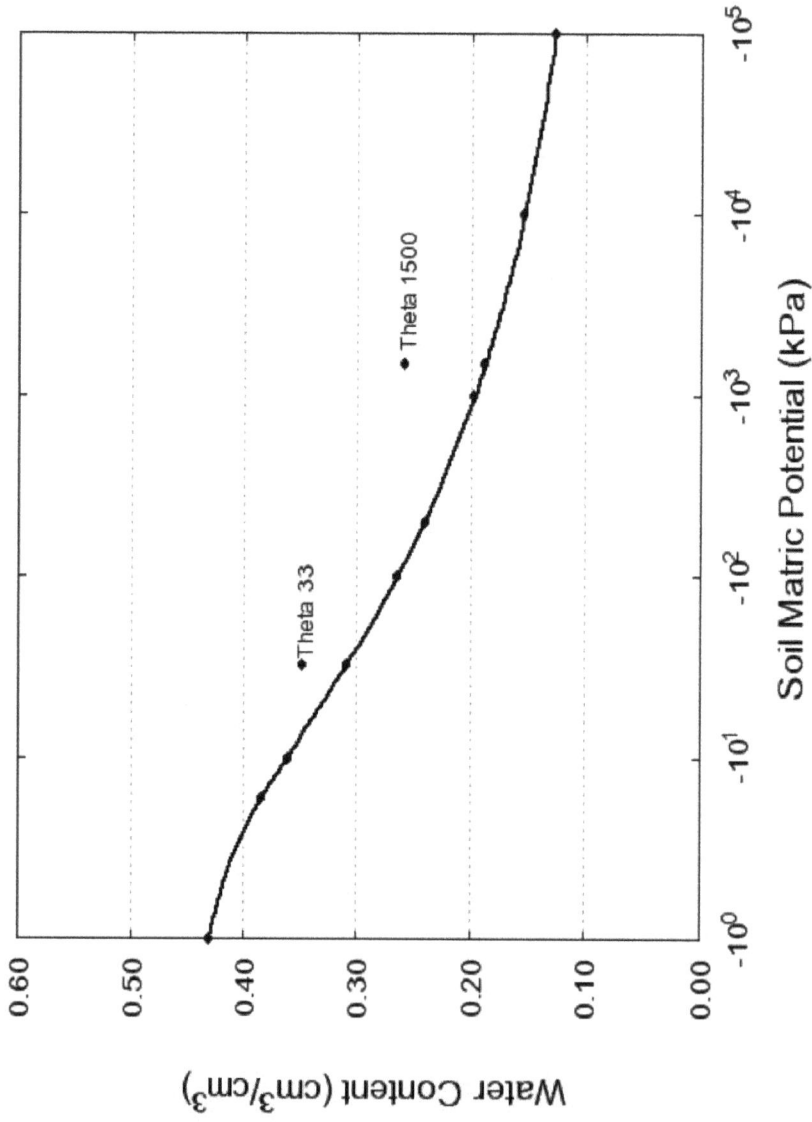

*Rosetta Reference

ESTIMATED SOIL WATER RETENTION CURVE

Pedon ID: 89CA027004
Layer Natural Key: 89P01802
Horizon: 2Btk1   46.0-74.0 cm

<2 mm Fraction

van Genuchten Retention Parameters from Rosetta[*]

θ_r	θ_s	$\log_{10}(\alpha)$	$\log_{10}(n)$
0.0727	0.4703	-2.4042	0.1507

Measured Soil Properties

Rock Fragmer	Organic Carbon	Sand	Silt	Clay	Db$_{33}$	θ_{33}	θ_{1500}
% vol	% wt	% wt	% wt	% wt	g/cc	cm³/cm³	cm³/cm³
0	0.15	35.2	26.7	38.1	1.38	0.41	0.22

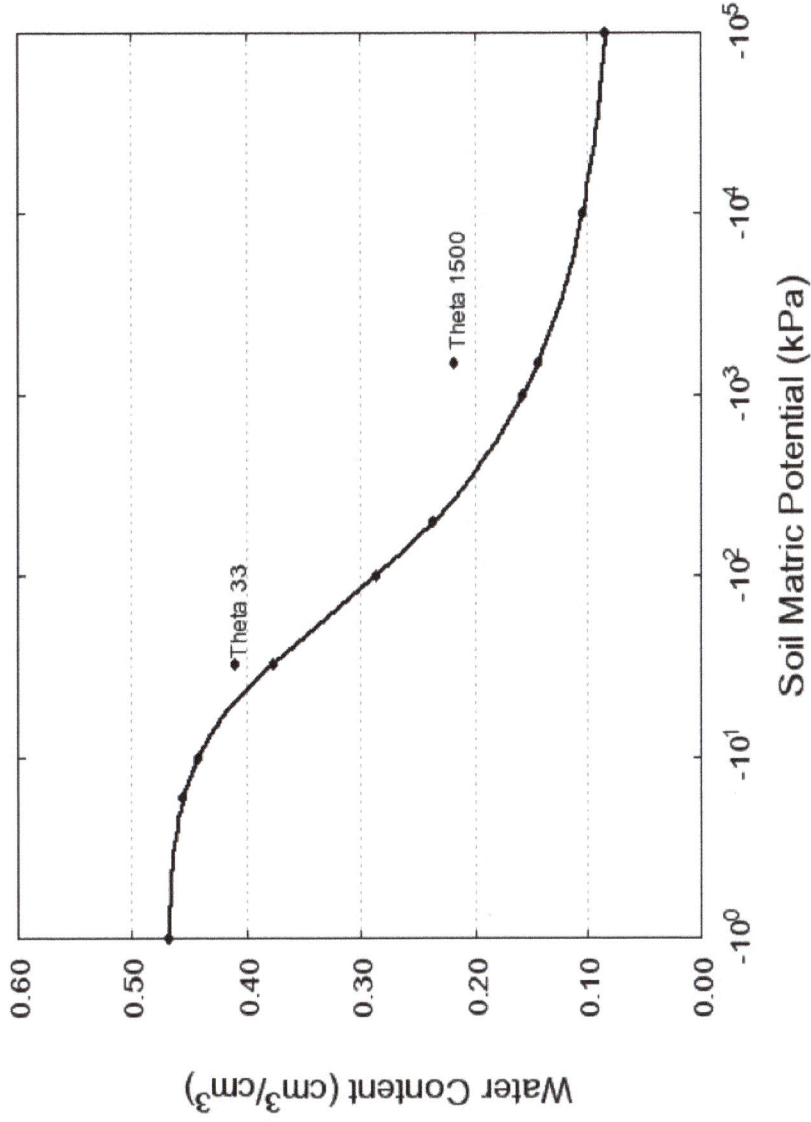

[*]Rosetta Reference

ESTIMATED SOIL WATER RETENTION CURVE

<2 mm Fraction

Pedon ID: 89CA027004
Layer Natural Key: 89P01803
Horizon: 2Btk2 74.0-109.0 cm

Measured Soil Properties

Rock Fragmer	Organic Carbon	Sand	Silt	Clay	Db_{33}	θ_{33}	θ_{1500}
% vol	% wt	% wt	% wt	% wt	g/cc	cm^3/cm^3	cm^3/cm^3
1	0.12	29.5	26.5	44.0	1.26	0.48	0.25

van Genuchten Retention Parameters from Rosetta[*]

θ_r	θ_s	$\log_{10}(\alpha)$	$\log_{10}(n)$
0.0886	0.5209	-2.6858	0.1942

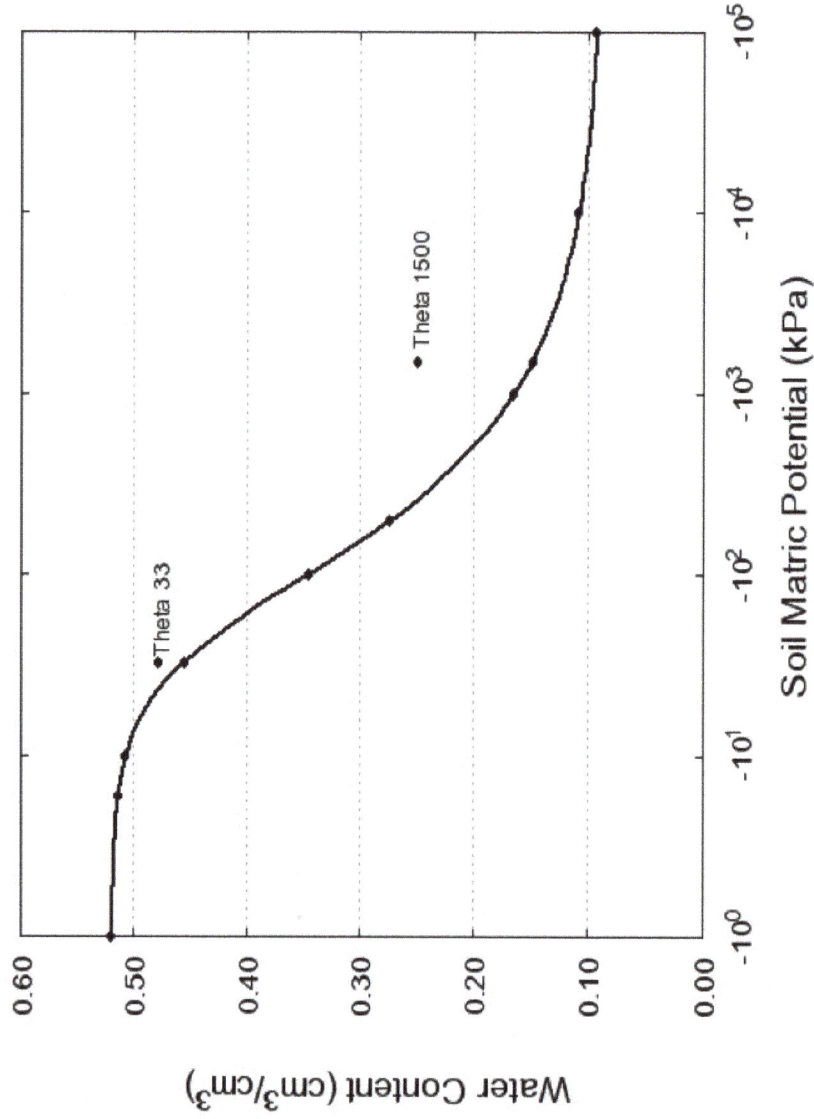

[*]Rosetta Reference

3 Nuvalde

Pedon ID: S09TX307003

*** Primary Characterization Data ***
(McCulloch, Texas)

Print Date: Aug 10 2010 8:51AM

Sampled as :   Nuvalde ; Fine-silty, mixed, superactive, thermic Typic Calciustoll
Revised to :

SSL   - Project   C2009USTX089 KSSPO Sampling Project
      - Site ID   S09TX307003 Lat: 31° 11' 44.75" north Long: 99° 13' 27.70" west NAD83 MLRA: 81B
      - Pedon No.   09N0839
      - General Methods 1B1A, 2A1, 2B

United States Department of Agriculture
Natural Resources Conservation Service
National Soil Survey Center
Soil Survey Laboratory
Lincoln, Nebraska 68508-3866

Layer	Horizon	Orig Hzn	Depth (cm)	Field Label 1	Field Label 2	Field Label 3	Field Texture	Lab Texture
09N02910	Ap1		0-15	S09TX307003-1			SIC	CL
09N02911	Ap2		15-34	S09TX307003-2			SIC	C
09N02912	Bw		34-59	S09TX307003-3			SIC	C
09N02913	Bk1		59-90	S09TX307003-4			SICL	SICL
09N02914	Bk2		90-120	S09TX307003-5			CL	SICL
09N02915	Bk3		120-144	S09TX307003-6			CL	SICL
09N02916	Bk4		144-203	S09TX307003-7			C	SICL

Calculation Name	Result	Units of Measure
Weighted Particles, 0.1-75mm, 75 mm Base	30	% wt
Volume, >2mm, Weighted Average	11	% vol
LE, Whole Soil, Summed to 1m	5	cm/m
Clay, total, Weighted Average	39	% wt
Clay, carbonate free, Weighted Average	28	% wt
CEC Activity, CEC7/Clay, Weighted Average, CECd, Set 1	0.53	(NA)

Pedon Calculations

Weighted averages based on control section: 25-100 cm

PSDA & Rock Fragments

				-1-	-2-	-3-	-4-	-5-	-6-	-7-	-8-	-9-	-10-	-11-	-12-	-13-	-14-	-15-	-16-	-17-
				(----- Total -----)			(-- Clay ---)		(----- Silt ----)		(------------------- Sand -------------------)					(Rock Fragments (mm))				>2 mm
				Clay	Silt	Sand	Fine	CO₃	Fine	Coarse	VF	F	M	C	VC	(------ Weight ------)				wt %
				<	.002	.05	<	<	.002	.02	.05	.10	.25	.5	1	2	5	20	.1-	whole
Layer	Depth (cm)	Horz	Prep	.002	-.05	-2	.0002	.002	-.02	-.05	-.10	-.25	-.50	-1	-2	-5	-20	-75	75	soil
																(------ % of <2mm Mineral Soil ------)			(------ % of <75mm ------)	
09N02910	0-15	Ap1	S	39.8	33.3	26.9		6.0	17.1	16.2	8.2	9.9	6.9	1.2	0.7	1	2	--	21	3
09N02911	15-34	Ap2	S	43.4	30.5	26.1		8.8	17.7	12.8	7.7	9.7	6.0	1.7	1.0	1	1	--	20	2
09N02912	34-59	Bw	S	46.3	31.2	22.5		9.0	18.8	12.4	6.3	7.3	4.2	3.0	1.7	3	2	--	20	5
09N02913	59-90	Bk1	S	34.4	52.0	13.6		14.3	42.7	9.3	3.9	3.8	2.9	2.0	1.0	25	9	--	40	34
09N02914	90-120	Bk2	S	34.2	53.1	12.7		14.6	42.6	10.5	4.4	3.9	2.5	0.8	1.1	16	7	1	30	24
09N02915	120-144	Bk3	S	36.0	52.0	12.0		16.2	42.7	9.3	3.8	4.3	1.7	1.2	1.0	3	1	--	12	4
09N02916	144-203	Bk4	S	36.6	48.3	15.1		12.6	35.9	12.4	5.0	5.4	2.6	1.4	0.7	8	3	--	20	11

Method codes (prep): 3A1a1a

Pedon ID: S09TX307003
Sampled As : Nuvalde
Fine-silty, mixed, superactive, hermic Typic Calciustoll

*** Primary Characterization Data ***
(McCulloch, Texas)

USDA-NRCS-NSSC-Soil Survey Laboratory

; Pedon No. 09N0839

Bulk Density & Moisture

Layer	Depth (cm)	Horz	Prep	-1- (Bulk Density) 33 kPa DbWR1	-2- Oven Dry DbWR1	-3- Cole Whole Soil	-4- 6 kPa	-5- 10 kPa	-6- 33 kPa	-7- 1500 kPa	-8- 1500 kPa Moist	-9- 1500 kPa Ratio AD/OD	-10- WRD Whole Soil	-11- Aggst Stabl 2-0.5mm	-12- CEC7	-13- 1500 kPa
				(--- g cm⁻³ ---) DbWR1 DbWR1			(----------- Water Content -----------) (----- % of <2mm -----) DbWR1 3C2a1a					3D1	cm³ cm⁻³ %		(-- Ratio/Clay --)	
09N02910	0-15	Ap1	S	1.35	1.68	0.075			26.1	16.8		1.049	0.12		0.68	0.42
09N02911	15-34	Ap2	S	1.44	1.74	0.064			23.7	17.2		1.049	0.09		0.60	0.40
09N02912	34-59	Bw	S	1.47	1.77	0.062			24.4	16.7		1.054	0.11		0.61	0.36
09N02913	59-90	Bk1	S	1.27	1.44	0.034			24.7	10.5		1.032	0.14		0.47	0.31
09N02914	90-120	Bk2	S	1.47	1.65	0.033			20.4	10.5		1.031	0.12		0.45	0.31
09N02915	120-144	Bk3	S	1.66	1.85	0.036			17.1	9.5		1.027	0.12		0.40	0.26
09N02916	144-203	Bk4	S	1.60	1.90	0.054			20.2	12.0		1.036	0.12		0.48	0.33

Carbon & Extractions

Layer	Depth (cm)	Horz	Prep	-1- C	-2- N	-3- S	-4- Org C	-5- C/N Ratio	-6- Fe	-7- Al	-8- Mn	-9- Al+½Fe ODOE	-10- Fe	-11- Al	-12- Si	-13- Mn	-14- C	-15-	-16- Fe	-17- Al	-18- Mn
				(----- Total -----) % of <2mm 4H2a 4H2a 4H2a			4H2a		(-- Dith-Cit Ext --) % of <2mm 4G1 4G1 4G1			(----- Ammonium Oxalate Extraction -----) % of <2mm					(--- Na Pyro-Phosphate ---) % of <2mm mg kg⁻¹			
09N02910	0-15	Ap1	S	3.30	0.17	0.02		11	0.3	tr	tr										
09N02911	15-34	Ap2	S	3.17	0.10	tr		12	0.4	tr	tr										
09N02912	34-59	Bw	S	3.24	0.08	0.01		11	0.4	0.1	tr										
09N02913	59-90	Bk1	S	7.09	0.03	tr		12	0.2	--	--										
09N02914	90-120	Bk2	S	6.57	0.03	0.01		8	0.1	--	--										
09N02915	120-144	Bk3	S	6.71	0.04	0.02		1	0.1	--	--										
09N02916	144-203	Bk4	S	5.03	--	0.01		1	0.3	--	--										

*** Primary Characterization Data ***

(McCulloch, Texas)

Fine-silty, mixed, superactive, hermic Typic Calciustoll

Pedon No. 09N0839

CEC & Bases

				-1-	-2-	-3-	-4-	-5-	-6-	-7-	-8-	-9-	-10-	-11-	-12-	-13-	-14-
												CEC8	CEC7	ECEC	Al	(--- Base ---)	
				(----- NH$_4$OAC Extractable Bases -----)				Sum	Acid-	Extr	KCl	Sum	NH$_4$	Bases	Sat	(- Saturation -)	
				Ca	Mg	Na	K	Bases	ity	Al	Mn	Cats	OAC	+Al	Sat	Sum	NH$_4$OAC
Layer	Depth (cm)	Horz	Prep	(------------------------------- cmol(+) kg^{-1} -------------------------------)							mg kg^{-1}	(---- cmol(+) kg^{-1} ----)			(-------- % --------)		
				4B1a1a	4B1a1a	4B1a1a	4B1a1a						4B1a1a				
09N02910	0-15	Ap1	S	63.6*	1.4	tr	1.1	66.1					26.9				100
09N02911	15-34	Ap2	S	65.0*	1.2	tr	0.6	66.8					26.2				100
09N02912	34-59	Bw	S	65.7*	1.4	0.3	0.5	67.9					28.1				100
09N02913	59-90	Bk1	S	57.4*	1.2	0.7	0.3	59.6					16.0				100
09N02914	90-120	Bk2	S	55.1*	1.6	1.4	0.2	58.3					15.3				100
09N02915	120-144	Bk3	S	53.6*	1.9	1.5	0.2	57.2					14.4				100
09N02916	144-203	Bk4	S	55.4*	2.4	1.6	0.3	59.7					17.4				100

* Extractable Ca may contain Ca from calcium carbonate or gypsum., CEC7 base saturation set to 100.

Salt

				-1-	-2-	-3-	-4-	-5-	-6-	-7-	-8-	-9-	-10-	-11-	-12-	-13-	-14-	-15-	-16-	-17-	-18-	-19-	-20-
				(------------------------------------- Water Extracted From Saturated Paste -------------------------------------)																Total	Elec	Pred Elec	Exch	
																				Salts	Cond	Cond	Na	SAR
				Ca	Mg	Na	K	CO$_3$	HCO$_3$	F	Cl	PO$_4$	Br	OAC	SO$_4$	NO$_2$	NO$_3$	H$_2$O				%	
				(---- mmol(+) L^{-1} ----)				(---------------------- mmol(-) L^{-1} ----------------------)										(---- % ----)	(---- % ----)	(-- dS m^{-1} --)		
Layer	Depth (cm)	Horz	Prep	4F2	4F2	4F2	4F2	4F2	4F2	4F2	4F2	4F2	4F2	4F2	4F2	4F2	4F2	4F2		4F2	4F1a1	4F1a1	
09N02910	0-15	Ap1	S	6.4	0.3	-	0.3	-	5.5	0.1	0.5	-	tr	-	0.8	tr	tr	64.3	tr	0.69	0.32		-
09N02911	15-34	Ap2	S																		0.21	tr	
09N02912	34-59	Bw	S																		0.21	1	
09N02913	59-90	Bk1	S																		0.23	5	
09N02914	90-120	Bk2	S	8.0	0.8	8.4	0.1	-	0.6	-	9.6	-	-	-	6.8	-	0.1	49.3	0.1	1.92	0.61		4
09N02915	120-144	Bk3	S	12.1	1.5	10.1	0.1	-	0.3	-	11.1	-	-	-	12.2	-	0.3	47.7	0.1	2.41	0.79		4
09N02916	144-203	Bk4	S	10.6	1.5	8.7	0.2	-	0.4	-	10.7	-	-	-	9.7	-	0.3	52.1	0.1	2.20	0.81		4

*** Primary Characterization Data ***

(McCulloch, Texas)

Pedon ID: S09TX307003

Sampled As : Nuvalde   Fine-silty, mixed, superactive, hermic Typic Calciustoll

USDA-NRCS-NSSC-Soil Survey Laboratory   ;   Pedon No. 09N0839

pH & Carbonates

			Depth	-1-	-2-	-3-	-4-	-5-	-6-	-7-	-8-	-9-	-10-	-11-
					(- - - - - - - - - - - - pH - - - - - - - - - - - -)					(- - Carbonate - -) As $CaCO_3$		(- - Gypsum - - -) As $CaSO_4*2H_2O$		Resist
					$CaCl_2$	H_2O	Sat							
				KCl	0.01M 1:2	1:1	Paste	Sulf	NaF	<2mm	<20mm	<2mm	<20mm	ohms
Layer	Horz	Prep	(cm)		4C1a2a	4C1a2a	4F2			4E1a1a1a1	4E1a1a1a1	4E2a1a1a1		cm^{-1}
										(- - - - - - - - - - - - - % - - - - - - - - - - - - -)				
09N02910	Ap1	S	0-15		7.3	7.6	7.4			12				
09N02911	Ap2	S	15-34		7.4	7.9				16				
09N02912	Bw	S	34-59		7.5	8.0				19				
09N02913	Bk1	S	59-90		7.5	8.2				56				
09N02914	Bk2	S	90-120		7.5	8.1	7.8			53				
09N02915	Bk3	S	120-144		7.6	7.8	7.8			56		--		
09N02916	Bk4	S	144-203		7.6	7.8	7.7			39		--		

Pedon ID: S09TX307003
Sampled As   :   Nuvalde
USDA-NRCS-NSSC-Soil Survey Laboratory

*** Primary Characterization Data ***
(McCulloch, Texas)
Fine-silty, mixed, superactive, hermic Typic Calciustoll
;   Pedon No. 09N0839

Print Date: Aug 10 2010 8:51AM

Clay Mineralogy (<.002 mm)

		Fract ion				X-Ray 7A1a1				Thermal 7A4a		Elemental						EGME Retn	Inter preta tion
			-1-	-2-	-3-	-4-	-5-	-6-	-7-	-8-	-9-	-10-	-11-	-12-	-13-	-14-	-15-	-16-	-17-	-18-
	Depth		(----- peak size -----)							(--- % ---)		SiO₂	Al₂O₃	Fe₂O₃	MgO	CaO	K₂O	Na₂O	mg g⁻¹	
Layer	(cm)	Horz										(------------------------- % -------------------------)							
09N02910	0-15	Ap1	tcly	MT 3	CA 3	KK 2	MI 2														SMEC
09N02912	34-59	Bw	tcly	KK 4	GI 3	VR 2	HE 2	GE 1													CMIX
09N02914	90-120	Bk2	tcly	MT 3	CA 3	KK 2	MI 2				KK 47	GI 15									SMEC
09N02916	144-203	Bk4	tcly	MT 3	CA 2	KK 2	MI 2														SMEC

FRACTION INTERPRETATION:
tcly - Total Clay, <0.002 mm

MINERAL INTERPRETATION:

CA - Calcite	GE - Goethite	GI - Gibbsite	HE - Hematite	KK - Kaolinite
MI - Mica	MT - Montmorillonite	VR - Vermiculite		

RELATIVE PEAK SIZE:

5 Very Large	4 Large	3 Medium	2 Small	1 Very Small	6 No Peaks

Pedon ID: S09TX307003

*** Supplementary Characterization Data ***
(McCulloch, Texas)

Sampled as : Nuvalde ; Fine-silty, mixed, superactive, thermic Typic Calciustoll
Revised to :

SSL - Project: C2009USTX089 KSSPO Sampling Project
- Site ID: S09TX307003 Lat: 31° 11' 44.75" north Long: 99° 13' 27.70" west NAD83 MLRA: 81B
- Pedon No.: 09N0839
- General Methods 1B1A, 2A1, 2B

United States Department of Agriculture
Natural Resources Conservation Service
National Soil Survey Center
Soil Survey Laboratory
Lincoln, Nebraska 68508-3866

Tier 1

Columns -1- through -25-

Engineering PSDA – Percentage Passing Sieve (Inches: 3, 2, 3/2, 1, 3/4, 3/8 | Number: 4, 10, 40, 200 | Microns: 20, 5, 2) | Cumulative Curve Fractions (<75mm) USDA Less Than Diameters (mm): 1., .5, .25, .10, .05 | Percentile at: 60, 50, 10 | Atterberg: LL, PI (%) | Gradation: Uniformity CU, Curvature CC

Layer	Depth (cm)	Horz	Prep	3	2	3/2	1	3/4	3/8	4	10	40	200	20	5	2	1.	.5	.25	.10	.05	60	50	10	LL	PI	CU	CC
09N02910	0-15	Ap1	S	100	100	100	100	100	99	98	97	93	75	55	45	39	96	95	88	79	71	0.03	0.010	tr			72.9	0.1
09N02911	15-34	Ap2	S	100	100	100	100	100	100	99	98	94	77	60	49	43	97	95	89	80	72	0.02	0.005	tr			58.7	0.1
09N02912	34-59	Bw	S	100	100	100	100	100	99	98	95	90	77	62	51	44	93	91	87	80	74	0.02	0.004	tr			46.7	0.2
09N02913	59-90	Bk1	S	100	100	100	100	100	96	91	66	60	58	51	34	26	75	64	62	60	57	0.12	0.019	0.001			>100	0.2
09N02914	90-120	Bk2	S	100	100	100	99	99	96	92	76	74	68	58	39	26	75	75	73	70	66	0.02	0.011	tr			49.7	0.6
09N02915	120-144	Bk3	S	100	100	100	100	100	100	99	96	93	87	76	51	35	95	94	92	88	84	0.01	0.005	tr			21.4	0.7
09N02916	144-203	Bk4	S	100	100	100	100	100	99	97	89	87	78	65	45	33	88	87	85	80	76	0.01	0.007	tr			35.6	0.5

Tier 2

Columns -26- through -50-

Weight Fractions — Whole Soil (mm) % of Whole Soil: >2, 250-UP, 250-75, 75-20, 20-5, 5-2, <2 | <75 mm Fraction % of <75 mm: >2, 75-20, 20-5, 5-2, <2 | Weight Per Unit Volume (g cm⁻³): Whole Soil (Soil Sur 33 kPa, Oven-dry, Engineering Moist, Saturated) and <2 mm Fraction (Soil Survey 33 kPa DbWR1, 1500 kPa, Oven-dry DbWR1, Engineering Moist/Saturated) | Void Ratios At 33 kPa: Whole Soil, <2 mm

Layer	Depth (cm)	Horz	Prep	>2	250-UP	250-75	75-20	20-5	5-2	<2	>2	75-20	20-5	5-2	<2	WS 33 kPa	WS Oven-dry	WS Moist	WS Satur	<2 33 kPa	<2 1500 kPa	<2 Oven-dry	<2 Moist/Satur	Void Whole	Void <2
09N02910	0-15	Ap1	S	3	--	--	--	2	1	97	3	--	2	1	97	1.36	1.69	1.70	1.85	1.35	1.49	1.68	1.84	0.95	0.96
09N02911	15-34	Ap2	S	2	--	--	--	1	1	98	2	--	1	1	98	1.45	1.75	1.78	1.90	1.44	1.54	1.74	1.90	0.83	0.84
09N02912	34-59	Bw	S	5	--	--	--	2	3	95	5	--	2	3	95	1.50	1.79	1.85	1.93	1.47	1.59	1.77	1.92	0.77	0.80
09N02913	59-90	Bk1	S	34	--	--	--	9	25	66	34	--	9	25	66	1.54	1.70	1.79	1.96	1.27	1.38	1.44	1.79	0.72	1.09
09N02914	90-120	Bk2	S	24	--	--	1	7	16	76	24	1	7	16	76	1.65	1.81	1.91	2.03	1.47	1.57	1.65	1.92	0.61	0.80
09N02915	120-144	Bk3	S	4	--	--	--	1	3	96	4	--	1	3	96	1.69	1.87	1.96	2.05	1.66	1.76	1.85	2.03	0.57	0.60
09N02916	144-203	Bk4	S	11	--	--	--	3	8	89	11	--	3	8	89	1.67	1.96	1.97	2.04	1.60	1.75	1.90	2.00	0.59	0.66

*** Supplementary Characterization Data ***

Sampled As : Nuvalde

(McCulloch, Texas)

Fine-silty, mixed, superactive, hermic Typic Calciustoll

USDA-NRCS-NSSC-Soil Survey Laboratory   ;   Pedon No. 09N0839

Tier 3

Column guides: -51- through -64-

Volume Fractions — Whole Soil (mm) At 33 kPa — (% of Whole Soil)

Layer	Depth (cm)	Horz	Prep	>2	250 -UP	250 -75	75 -20	20 -5	5 -2	<2	2- .05	.05- .002	LT .002	Pores D	Pores F	C/N Rat-io
09N02910	0-15	Ap1	S	1	–	–	–	–	1	99	13	17	20	15	34	11
09N02911	15-34	Ap2	S	1	–	–	–	–	1	99	14	16	23	12	33	12
09N02912	34-59	Bw	S	2	–	–	–	1	1	98	12	17	25	9	34	11
09N02913	59-90	Bk1	S	20	–	–	–	5	15	80	5	20	13	17	25	12
09N02914	90-120	Bk2	S	15	–	–	–	4	11	85	6	25	16	12	26	8
09N02915	120-144	Bk3	S	3	–	–	–	1	2	97	7	31	22	9	27	1
09N02916	144-203	Bk4	S	7	–	–	–	2	5	93	8	27	20	7	30	1

Column guides: -65- through -75-

Ratios To Clay (<2 mm Fraction), Linear Extensibility (Whole Soil 33 kPa) and WRD (in^3/in^3)

Layer	Fine Clay	CEC Sum Cats	NH4- OAC	1500 H2O	LEP 33 kPa	1500 -dry (33 kPa)	Oven -dry	1500 kPa	Oven -dry	WRD Whole Soil <2mm to %	WRD <2mm Whole Soil Oven
09N02910				0.42	0.191	3.5	7.5	3.3	7.6	0.12	0.13
09N02911				0.40	0.150	2.5	6.5	2.3	6.5	0.09	0.09
09N02912				0.36	0.138	2.6	6.1	2.7	6.4	0.11	0.11
09N02913				0.31	0.125	2.3	3.3	2.8	4.3	0.14	0.18
09N02914				0.31	0.114	1.8	3.1	2.2	3.9	0.12	0.15
09N02915				0.26	0.103	1.9	3.4	2.0	3.7	0.12	0.13
09N02916				0.33	0.161	2.7	5.5	3.0	5.9	0.12	0.13

Tier 4

Column guides: -76- through -89-

Weight Fractions - Clay Free — Whole Soil (- % of >2 mm Sand and Silt -) / <2 mm Fraction (% of Sand and Silt)

Layer	Depth (cm)	Horz	Prep	>2	75 -20	20 -2	2- .05	.05- .002	.002 <	VC	C	M	F	VF	C	F	ay
09N02910	0-15	Ap1	S	5	5	42	53	63	–	1	2	11	16	14	27	28	66
09N02911	15-34	Ap2	S	3	3	44	52	74	–	2	3	11	17	14	23	31	77
09N02912	34-59	Bw	S	9	3	38	53	79	–	3	6	8	14	12	23	35	86
09N02913	59-90	Bk1	S	44	44	12	44	29	–	3	3	4	6	6	14	65	52
09N02914	90-120	Bk2	S	32	32	13	55	35	–	2	1	4	6	7	16	65	52
09N02915	120-144	Bk3	S	6	6	18	76	53	–	2	2	3	7	6	15	67	56
09N02916	144-203	Bk4	S	16	16	20	64	48	–	2	1	4	9	8	20	57	58

Column guides: -90- through -98-

PSDA (mm), pH, Electrical Resistance/Conductivity, Particle Density

PSDA method: 3A1a1a4C1a2a; Particle Density method: 4F2

Layer	Texture	Sand (2- .05)	Silt (.05- .002)	Clay (.002 <)	pH CaCl2 .01M	Elect. Resist. (ohms)	Conduct. dS m^{-1} (<2 mm)	Particle Density g cm^{-3}
09N02910	cl	26.9	33.3	39.8	7.3		0.69	
09N02911	c	26.1	30.5	43.4	7.4			
09N02912	c	22.5	31.2	46.3	7.5			
09N02913	sicl	13.6	52.0	34.4	7.5			
09N02914	sicl	12.7	53.1	34.2	7.5			1.92
09N02915	sicl	12.0	52.0	36.0	7.6			2.41
09N02916	sicl	15.1	48.3	36.6	7.6			2.20

Print Date: Aug 10 2010 8:51AM

*** Taxonomy Characterization Data ***
(McCulloch, Texas)

Sampled as :   Nuvalde ; Fine-silty, mixed, superactive, thermic Typic Calciustoll
Revised to :

United States Department of Agriculture
Natural Resources Conservation Service
National Soil Survey Center
Soil Survey Laboratory
Lincoln, Nebraska 68508-3866

SSL   - Project   C2009USTX089   KSSPO Sampling Project
    - Site ID   S09TX307003   Lat: 31° 11' 44.75" north   Long: 99° 13' 27.70" west   NAD83   MLRA: 81B
    - Pedon No.   09N0839
    - General Methods 1B1A, 2A1, 2B

Taxonomy Tier 1

				-1-	-2-	-3-	-4-	-5-	-6-	-7-	-8-	-9-	-10-
					Fine	CaCO$_3$	1500		.1-75	Bulk	Cole	Vol	Resist
		Depth		Clay <002	Clay <0002	Clay <002	kPa /Clay	Clay Est	mm Frac	Den 33 kPa	Whole Soil	% of Whole	Min
Layer	Horz	(cm)	Prep	(— % of <2 mm —)			(— /Clay —)	(— % —)	%	g cm^{-3} DbWR1	cm cm^{-1}		%
				3A1a1a	3A1a1a	3A1a1a							
09N02910	Ap1	0-15	S	39.8		6.0	0.42		21	1.35	0.075	1	
09N02911	Ap2	15-34	S	43.4		8.8	0.40		20	1.44	0.064	1	
09N02912	Bw	34-59	S	46.3		9.0	0.36		20	1.47	0.062	2	
09N02913	Bk1	59-90	S	34.4		14.3	0.31		40	1.27	0.034	20	
09N02914	Bk2	90-120	S	34.2		14.6	0.31		30	1.47	0.033	15	
09N02915	Bk3	120-144	S	36.0		16.2	0.26		12	1.66	0.036	3	
09N02916	Bk4	144-203	S	36.6		12.6	0.33		20	1.60	0.054	7	

Taxonomy Tier 2

				-1-	-2-	-3-	-4-	-5-	-6-	-7-	-8-	-9-	-10-	-11-	-12-	-13-	-14-	-15-	-15-
				pH	pH	Org	Tot	Al+½ Fe		CO$_3$ as	(— Base Sat —)		NZ	ECEC	CEC7	ECEC	Al	E C	ESP
		Depth		H$_2$O	NaF	C	C	Oxal	ODOE	CaCO$_3$	NH$_4$	Bases	P Ret	cmol(+)	/Clay	/Clay	Sat		
Layer	Horz	(cm)	Prep			(— % —)			%		%			kg^{-1}			%	dS m^{-1} 4F2	%
				4C1a2a			4H2a			4E1a1a1									
09N02910	Ap1	0-15	S	7.6			3.30			12	100*			0.68				0.69	
09N02911	Ap2	15-34	S	7.9			3.17			16	100*			0.60					tr
09N02912	Bw	34-59	S	8.0			3.24			19	100*			0.61					1
09N02913	Bk1	59-90	S	8.2			7.09			56	100*			0.47					5
09N02914	Bk2	90-120	S	8.1			6.57			53	100*			0.45				1.92	
09N02915	Bk3	120-144	S	7.8			6.71			56	100*			0.40				2.41	
09N02916	Bk4	144-203	S	7.8			5.03			39	100*			0.48				2.20	

*Extractable Ca may contain Ca from calcium carbonate or gypsum.

Pedon Calculations

Calculation Name	Result	Units of Measure
Weighted Particles, 0.1-75mm, 75 mm Base	30	% wt
Volume, >2mm, Weighted Average	11	% vol
LE, Whole Soil, Summed to 1m	5	cm/m
Clay, total, Weighted Average	39	% wt
Clay, carbonate free, Weighted Average	28	% wt
CEC Activity, CEC7/Clay, Weighted Average, CECd, Set 1	0.53	(NA)

PEDON DESCRIPTION

Print Date: 08/10/2010

Description Date: 4/22/2009

Describer: rd,alb,jam,wg,cs

Site ID: S09TX307003

Site Note:

Pedon ID: S09TX307003

Pedon Note: All textures listed are texture by feel estimates.

Lab Source ID: SSL

Lab Pedon #: 09N0839

Soil Name as Described/Sampled: Nuvalde

Soil Name as Correlated: NuB

Classification: Fine-silty, mixed, superactive, thermic Typic Calciustolls

Pedon Type: modal pedon for series

Pedon Purpose: full pedon description

Taxon Kind: series

Associated Soils:

Physiographic Division:

Physiographic Province:

Physiographic Section:

State Physiographic Area:

Local Physiographic Area: Hensell Sand

Geomorphic Setting: on shoulder of side slope of alluvial plain remnant on dissected plateau

Upslope Shape: linear

Cross Slope Shape: convex

Particle Size Control Section: 25 to 100 cm.

Description origin:

Diagnostic Features: mollic epipedon 0 to 34 cm.
cambic horizon 34 to 59 cm.
calcic horizon 59 to 203 cm.

Country:

State: Texas

County: McCulloch

MLRA: 81B -- Edwards Plateau, Central Part

Soil Survey Area: TX307 -- McCulloch County, Texas

Map Unit: NuB -- Nuvalde clay loam, 1 to 3 percent slopes

Quad Name: Rochelle, Texas

Location Description:

Legal Description:

Latitude: 31 degrees 11 minutes 44.75 seconds north

Longitude: 99 degrees 13 minutes 27.70 seconds west

Datum: NAD83

UTM Zone: 14

UTM Easting: 478625 meters

UTM Northing: 3451320 meters

Primary Earth Cover: Grass/herbaceous cover

Secondary Earth Cover: Hayland

Existing Vegetation: Bermudagrass

Parent Material: ancient alluvium derived from limestone

Bedrock Kind:

Bedrock Depth:

Bedrock Hardness:

Bedrock Fracture Interval:

Surface Fragments:

Description database: MLRA09_Office

Cont. **Site ID:** S09TX307003

Pedon ID: S09TX307003

Slope (%)	Elevation (meters)	Aspect (deg)	MAAT (C)	MSAT (C)	MWAT (C)	MAP (mm)	Frost-Free Days	Drainage Class	Slope Length (meters)	Upslope Length (meters)
1.0	525.0	220						well		

Ap1--0 to 15 centimeters; dark grayish brown (10YR 4/2) silty clay, very dark brown (10YR 2/2), moist; 10 percent sand; 45 percent silt; 45 percent clay; weak fine subangular blocky structure; friable, slightly hard, slightly sticky, slightly plastic; many fine roots throughout and many very fine roots throughout; common fine low-continuity irregular pores; violent effervescence, by HCl, 3 normal; clear smooth boundary.

Ap2--15 to 34 centimeters; grayish brown (10YR 5/2) silty clay, very dark grayish brown (10YR 3/2), moist; 10 percent sand; 45 percent silt; 45 percent clay; moderate medium subangular blocky structure; firm, slightly hard, slightly sticky, slightly plastic; common fine roots throughout and common very fine roots throughout; common fine low-continuity irregular pores; violent effervescence, by HCl, 3 normal; gradual smooth boundary.

Bw--34 to 59 centimeters; light brownish gray (10YR 6/2) silty clay, dark grayish brown (10YR 4/2), moist; 10 percent sand; 45 percent silt; 45 percent clay; moderate medium subangular blocky structure; firm, slightly hard, moderately sticky, moderately plastic; common fine roots throughout and common very fine roots throughout; common fine pores; 3 percent nonflat subrounded strongly cemented 2- to 5-millimeter limestone fragments; violent effervescence, by HCl, 3 normal; clear wavy boundary.

Bk1--59 to 90 centimeters; pink (7.5YR 8/3) silty clay loam, light brown (7.5YR 6/3), moist; 8 percent sand; 60 percent silt; 32 percent clay; weak medium subangular blocky structure; friable, slightly hard, moderately sticky, moderately plastic; common fine roots throughout; common fine irregular pores; 10 percent fine prominent irregular 10YR 5/6, moist, masses of oxidized iron with clear boundaries in matrix; 30 percent medium distinct irregular 10YR 8/2, moist, carbonate masses with diffuse boundaries in matrix; 5 percent nonflat subrounded strongly cemented 2- to 20-millimeter limestone fragments; violent effervescence, by HCl, 3 normal; clear wavy boundary.

Bk2--90 to 120 centimeters; pink (7.5YR 8/3) clay loam, light brown (7.5YR 6/3), moist; 33 percent sand; 35 percent silt; 32 percent clay; weak fine subangular blocky structure; friable, slightly hard, slightly sticky, slightly plastic; common fine roots throughout and common very fine roots throughout; common fine irregular pores; 2 percent patchy prominent 10YR2), moist, organic stains on vertical faces of peds; 10 percent fine prominent irregular 7.5YR 5/6, moist, masses of oxidized iron with clear boundaries in matrix; 40 percent medium distinct irregular 10YR 8/2, moist, carbonate masses with diffuse boundaries in matrix; 5 percent nonflat subrounded strongly cemented 2- to 20-millimeter limestone fragments; violent effervescence, by HCl, 3 normal; clear wavy boundary.

Bk3--120 to 144 centimeters; very pale brown (10YR 8/4) clay loam, light yellowish brown (10YR 6/4), moist; 33 percent sand; 35 percent silt; 32 percent clay; weak fine subangular blocky structure; friable, slightly hard, slightly sticky, slightly plastic; common fine roots throughout and common very fine roots throughout; common fine irregular pores; 2 percent patchy prominent 10YR2), moist, organic stains on vertical faces of peds; 10 percent fine prominent irregular 10YR 5/6, moist, masses of oxidized iron with clear boundaries in matrix; 50 percent medium distinct irregular 10YR 8/2, moist, carbonate masses with diffuse boundaries in matrix; 5 percent nonflat subrounded strongly cemented 2- to 20-millimeter limestone fragments; violent effervescence, by HCl, 3 normal; clear wavy boundary.

Bk4--144 to 203 centimeters; pink (7.5YR 7/4) clay, brown (7.5YR 5/4), moist; 20 percent sand; 30 percent silt; 50 percent clay; structureless massive; very firm, very hard, slightly sticky, slightly plastic; 2 percent patchy prominent organic stains on vertical faces of peds; 2 percent fine prominent irregular moderately cemented iron-manganese concretions with sharp boundaries in matrix and 10 percent fine prominent irregular 7.5YR 4/6, moist, masses of oxidized iron with clear boundaries in matrix; 20 percent medium distinct irregular 10YR 8/2, moist, carbonate masses with diffuse boundaries in matrix; 5 percent nonflat subrounded strongly cemented 2- to 20-millimeter limestone fragments; violent effervescence, by HCl, 3 normal.

PEDON DESCRIPTION

Country:
State: Texas

Print Date: 08/10/2010
Description Date: 4/22/2009

Describer: rd,alb,jam,wg,cs

Site ID: S09TX307003

Site Note:

Pedon ID: S09TX307003

Pedon Note: All textures listed are texture by feel estimates.

Lab Source ID: SSL

Lab Pedon #: 09N0839

Soil Name as Described/Sampled: Nuvalde

Soil Name as Correlated: NuB

Classification: Fine-silty, mixed, superactive, thermic Typic Calciustolls

Pedon Type: modal pedon for series

Pedon Purpose: full pedon description

Taxon Kind: series

Associated Soils:

Physiographic Division:

Physiographic Province:

Physiographic Section:

State Physiographic Area:

Local Physiographic Area: Hensell Sand

Geomorphic Setting: on shoulder of side slope of alluvial plain remnant on dissected plateau

Upslope Shape: linear

Cross Slope Shape: convex

Particle Size Control Section: 25 to 100 cm.

Description origin:

Diagnostic Features: mollic epipedon 0 to 34 cm.
calcic horizon 34 to 59 cm.
cambic horizon 59 to 203 cm.

County: McCulloch

MLRA: 81B -- Edwards Plateau, Central Part

Soil Survey Area:

Map Unit:

Quad Name: Rochelle, Texas

Location Description:

Legal Description:

Latitude: 31 degrees 11 minutes 44.75 seconds north

Longitude: 99 degrees 13 minutes 27.70 seconds west

Datum: NAD83

UTM Zone: 14

UTM Easting: 478625 meters

UTM Northing: 3451320 meters

Primary Earth Cover: Grass/herbaceous cover

Secondary Earth Cover: Hayland

Existing Vegetation: Bermudagrass

Parent Material: ancient alluvium derived from limestone

Bedrock Kind:

Bedrock Depth:

Bedrock Hardness:

Bedrock Fracture Interval:

Surface Fragments:

Description database: NSSL

Slope (%)	Elevation (meters)	Aspect (deg)	MAAT (C)	MSAT (C)	MWAT (C)	MAP (mm)	Frost-Free Days	Drainage Class	Slope Length (meters)	Upslope Length (meters)
1.0	525.0	220						well		

Ap1--0 to 15 centimeters; dark grayish brown (10YR 4/2) silty clay, very dark brown (10YR 2/2), moist; 10 percent sand; 45 percent silt; 45 percent clay; weak fine subangular blocky structure; friable, slightly hard, slightly sticky, slightly plastic; many fine roots throughout and many very fine roots throughout; common fine low-continuity irregular pores; violent effervescence, by HCl, 3 normal; clear smooth boundary. Lab sample # 09N02910

Ap2--15 to 34 centimeters; grayish brown (10YR 5/2) silty clay, very dark grayish brown (10YR 3/2), moist; 10 percent sand; 45 percent silt; 45 percent clay; moderate medium subangular blocky structure; firm, slightly hard, slightly sticky, slightly plastic; common fine roots throughout and common very fine roots throughout; common fine low-continuity irregular pores; violent effervescence, by HCl, 3 normal; gradual smooth boundary. Lab sample # 09N02911

Bw--34 to 59 centimeters; light brownish gray (10YR 6/2) silty clay, dark grayish brown (10YR 4/2), moist; 10 percent sand; 45 percent silt; 45 percent clay; moderate medium subangular blocky structure; firm, slightly hard, moderately sticky, moderately plastic; common fine roots throughout and common very fine roots throughout; common fine pores; 3 percent nonflat subrounded strongly cemented 2- to 5-millimeter limestone fragments; violent effervescence, by HCl, 3 normal; clear wavy boundary. Lab sample # 09N02912

Bk1--59 to 90 centimeters; pink (7.5YR 8/3) silty clay loam, light brown (7.5YR 6/3), moist; 18 percent sand; 60 percent silt; 22 percent clay; weak medium subangular blocky structure; friable, slightly hard, moderately sticky, moderately plastic; common fine roots throughout and common very fine roots throughout; common fine irregular pores; 10 percent fine prominent irregular 10YR 5/6, moist, masses of oxidized iron with clear boundaries in matrix; 30 percent medium distinct irregular 10YR 8/2, moist, carbonate masses with diffuse boundaries in matrix; 5 percent nonflat subrounded strongly cemented 2- to 20-millimeter limestone fragments; violent effervescence, by HCl, 3 normal; clear wavy boundary. Lab sample # 09N02913

Bk2--90 to 120 centimeters; pink (7.5YR 8/3) clay loam, light brown (7.5YR 6/3), moist; 33 percent sand; 35 percent silt; 32 percent clay; weak fine subangular blocky structure; friable, slightly hard, slightly sticky, slightly plastic; common fine roots throughout and common very fine roots throughout; common fine irregular pores; 2 percent patchy prominent 10YR2), moist, organic stains on vertical faces of peds; 10 percent fine prominent irregular 7.5YR 5/6, moist, masses of oxidized iron with clear boundaries in matrix; 40 percent medium distinct irregular 10YR 8/2, moist, carbonate masses with diffuse boundaries in matrix; 5 percent nonflat subrounded strongly cemented 2- to 20-millimeter limestone fragments; violent effervescence, by HCl, 3 normal; clear wavy boundary. Lab sample # 09N02914

Bk3--120 to 144 centimeters; very pale brown (10YR 8/4) clay loam, light yellowish brown (10YR 6/4), moist; 33 percent sand; 35 percent silt; 32 percent clay; weak fine subangular blocky structure; friable, slightly hard, slightly sticky, slightly plastic; common fine roots throughout and common very fine roots throughout; common fine irregular pores; 2 percent patchy prominent 10YR2), moist, organic stains on vertical faces of peds; 10 percent fine prominent irregular 10YR 5/6, moist, masses of oxidized iron with clear boundaries in matrix; 50 percent medium distinct irregular 10YR 8/2, moist, carbonate masses with diffuse boundaries in matrix; 5 percent nonflat subrounded strongly cemented 2- to 20-millimeter limestone fragments; violent effervescence, by HCl, 3 normal; clear wavy boundary. Lab sample # 09N02915

Bk4--144 to 203 centimeters; pink (7.5YR 7/4) clay, brown (7.5YR 5/4), moist; 20 percent sand; 30 percent silt; 50 percent clay; structureless massive; very firm, very hard, slightly sticky, slightly plastic; 2 percent patchy prominent organic stains on vertical faces of peds; 2 percent fine prominent irregular moderately cemented iron-manganese concretions with sharp boundaries in matrix and 10 percent fine prominent irregular 7.5YR 4/6, moist, masses of oxidized iron with clear boundaries in matrix; 20 percent medium distinct irregular 10YR 8/2, moist, carbonate masses with diffuse boundaries in matrix; 5 percent nonflat subrounded strongly cemented 2- to 20-millimeter limestone fragments; violent effervescence, by HCl, 3 normal. Lab sample # 09N02916

4 Sverdrup

*** Primary Characterization Data ***
(Grant, Minnesota)

Sampled as :   Sverdrup ; Sandy, mixed Udic Haploboroll
Revised to :

United States Department of Agriculture
Natural Resources Conservation Service
National Soil Survey Center
Soil Survey Laboratory
Lincoln, Nebraska 68508-3866

SSL   - Project   CP87MN190   WEPP MINNESOTA
    - Site ID   87MN051001   Lat: 45° 59' 47.00" north   Long: 95° 52' 17.00" west   MLRA: 102A
    - Pedon No.   87P0576
    - General Methods 1B1A, 2A1, 2B

Layer	Horizon	Orig Hzn	Depth (cm)	Field Label 1	Field Label 2	Field Label 3	Field Texture	Lab Texture
87P03026	Ap	AP	0-23				SL	SL
87P03027	A	A	23-36				SL	SL
87P03028	Bw1	BW1	36-46				SL	FSL
87P03029	Bw2	BW2	46-64				LS	FS
87P03030	Bw3	BW3	64-94				S	S
87P03031	BC	BC	94-102				COS	COS
87P03032	C	C	102-127				COS	COS

Pedon Calculations

Calculation Name	Result	Units of Measure
CEC Activity, CEC7/Clay, Weighted Average	0.91	(NA)
Clay, carbonate free, Weighted Average	5	% wt
Weighted Particles, 0.1-75mm, 75 mm Base	86	% wt
Volume, >2mm, Weighted Average	6	% vol
Clay, total, Weighted Average	5	% wt

Weighted averages based on control section: 25-100 cm

PSDA & Rock Fragments

			-1-	-2-	-3-	-4-	-5-	-6-	-7-	-8-	-9-	-10-	-11-	-12-					
			(- - - - Total - - - - -)			(- - Clay - - -)		(- - - Silt - - - -)		(- - - - - - - - - - Sand - - - - - - - - - -)									
			Clay	Silt	Sand	Fine	CO₃	Fine	Coarse	VF	F	M	C	VC					
			< .002	.002 -.05	.05 -2	< .0002	< .002	.002 -.02	.02 -.05	.05 -.10	.10 -.25	.25 -.50	.5 -1	1 -2					
Layer	Depth (cm)	Horz	Prep																
			3A1	3A1	3A1	3A1	.002	3A1	3A1	3A1	3A1	3A1	3A1	3A1					
							% of <2mm Mineral Soil												
87P03026	0-23	Ap	S	7.9	16.8	75.3	5.5		9.4	7.4	3.7	29.0	27.3	13.4	1.9					
87P03027	23-36	A	S	8.6	15.0	76.4	6.3		8.8	6.2	2.8	29.3	28.3	13.6	2.4					
87P03028	36-46	Bw1	S	8.4	13.7	77.9	5.8		7.5	6.2	5.4	44.1	19.6	6.4	2.4					
87P03029	46-64	Bw2	S	5.4	3.2	91.4	3.9		1.9	1.3	4.5	58.6	23.4	3.8	1.1					
87P03030	64-94	Bw3	S	3.6	2.9	93.5	1.7		1.4	1.5	2.5	40.2	28.3	15.6	6.9					
87P03031	94-102	BC	S	3.9	8.8	87.3	1.1		4.4	4.4	2.7	11.4	20.3	40.2	12.7					
87P03032	102-127	C	S	7.6	1.3	91.1	1.4		0.5	0.8	1.3	4.2	14.3	56.6	14.7					

			-13-	-14-	-15-	-16-	-17-
			(Rock Fragments (mm))				
			(- - - - - - - -Weight- - - - - - - -)				>2 mm
			2 -5	5 -20	20 -75	.1- 75	wt %
			(- - - - % of <75mm - - - - -)				whole
			3B1	3B1	3B1	3B1	soil
87P03026	0-23	Ap	tr	tr	--	72	tr
87P03027	23-36	A	tr	tr	--	74	tr
87P03028	36-46	Bw1	1	tr	--	73	1
87P03029	46-64	Bw2	1	tr	--	87	1
87P03030	64-94	Bw3	5	4	--	92	9
87P03031	94-102	BC	8	20	31	94	59
87P03032	102-127	C	10	7	16	93	33

*** Primary Characterization Data ***

(Grant, Minnesota)
Sandy, mixed Udic Haploboroll
Pedon No. 87P0576

Water Dispersible PSDA

Layer	Depth (cm)	Horz	Prep	-1-	-2-	-3-	-4-	-5-	-6-	-7-	-8-	-9-	-10-	-11-	-12-
				(---- Total ----)			(- - Clay - - -)		(- - - - Silt - - - -)		(- - - - - - - - Sand - - - - - - - -)				
				Clay < .002	Silt .002 -.05	Sand .05 -2	F < .002 .0002	CO3 < .0002 .002	F .002 -.02	C .02 -.05	VF .05 -.10	F .10 -.25	M .25 -.50	C .5 -1	VC 1 -2
							(---- % of <2mm ----)								
				3A1c	3A1c	3A1c	3A1c		3A1c	3A1c	3A1c	3A1c	3A1c	3A1c	3A1c
87P03026	0-23	Ap	S	2.8	21.7	75.5			12.2	8.8	3.7	29.0	27.7	13.2	2.2

Bulk Density & Moisture

Layer	Depth (cm)	Horz	Prep	-1-	-2-	-3-	-4-	-5-	-6-	-7-	-8-	-9-	-10-	-11-	-12-	-13-
				(Bulk Density)		Cole Whole Soil	(- - - - - - - - Water Content - - - - - - - -)					1500 kPa AD/OD Ratio	WRD Whole Soil	Aggst Stabl 2-0.5mm %	(- - Ratio/Clay - -)	
				33 kPa	Oven Dry		6 kPa	10 kPa	33 kPa	1500 kPa	1500 kPa Moist				CEC7	1500 kPa
				(- - - g cm⁻³ - - -)			(- - - - - - - - % of < 2mm - - - - - - - -)						cm³ cm⁻³			
				4A1d	4A1h		4B1c			4B2a		4B5	4C1	4G1	8D1	8D1
87P03026	0-23	Ap	S	1.60	1.67	0.014			13.4	5.4		1.009	0.13		1.39	0.68
87P03027	23-36	A	S	1.53	1.62	0.019			12.5	5.2		1.009	0.11		1.24	0.60
87P03028	36-46	Bw1	S	1.22	1.22	--			15.3	4.2		1.007	0.13	19	0.87	0.50
87P03029	46-64	Bw2	S	1.62	1.62	--			6.1	2.7		1.004	0 05		0.83	0.50
87P03030	64-94	Bw3	S							1.9		1.002			0.81	0.53
87P03031	94-102	BC	S							2.8		1.006			1.10	0.72
87P03032	102-127	C	S							3.8		1.004			0.46	0.50

*** Primary Characterization Data ***

Sampled As   :   Sverdrup
(Grant, Minnesota)
Sandy, mixed Udic Haploboroll

USDA-NRCS-NSSC-Soil Survey Laboratory
;   Pedon No. 87P0576

Water Content

Layer	Depth (cm)	Horz	Prep	Atterberg Limits LL -1-	PI -2-	Bulk Density Field -3-	Recon 33 kPa -4-	Recon Oven Dry -5-	Water Content Field -6-	Recon 33 kPa -7-	Sieved Samples 6 kPa -8-	10 kPa -9-	33 kPa -10-	100 kPa -11-	200 kPa -12-	500 kPa -13-
				pct <0.4mm 4F1	4F	(- - - - - - - g cm⁻³ - - - - - - -)			(- - - - - - - - - - - - - - - % of <2mm - - - - - - - - - - -)						4B1a	
87P03026	0-23	Ap	S	24	10										7.4	
87P03027	23-36	A	S												7.6	
87P03028	36-46	Bw1	S		NP										6.6	
87P03029	46-64	Bw2	S												3.5	
87P03030	64-94	Bw3	S												2.4	
87P03031	94-102	BC	S												4.1	
87P03032	102-127	C	S												3.5	

Carbon & Extractions

Layer	Depth (cm)	Horz	Prep	Total C -1-	N -2-	S -3-	Org C -4-	C/N Ratio -5-	Dith-Cit Ext Fe -6-	Al -7-	Mn -8-	Al+½Fe -9-	Ammonium Oxalate Extraction ODOE -10-	Fe -11-	Al -12-	Si -13-	Mn -14-	Na Pyro-Phosphate C -15-	Fe -16-	Al -17-	Mn -18-
				6A2d	6B3a	(- % of <2mm -)	6A1c		6C2b	6G7a		(- - - - - - % of <2mm - - - - - -)	8J	6C9a	6G12	6V2	6D5b mg kg⁻¹	(- - % of <2mm - -)			
87P03026	0-23	Ap	S	1.14	0.108		1.28	12	0.5	0.1		0.12	0.04	0.10	0.07	0.02	300.0				
87P03027	23-36	A	S	1.11	0.110		1.21	11	0.5	0.1		0.14	0.05	0.14	0.07	0.02	300.0				
87P03028	36-46	Bw1	S	0.42	0.043		0.45	10	0.6	0.1		0.17	0.04	0.17	0.08	0.02	200.0				
87P03029	46-64	Bw2	S	0.24	0.023		0.21	9	0.5	tr		0.12	0.03	0.15	0.04	0.02	200.0				
87P03030	64-94	Bw3	S	0.33	0.014		0.15	11	0.6	tr		0.09	0.02	0.14	0.02	0.02	400.0				
87P03031	94-102	BC	S	1.64	0.028		0.36	13	1.1	tr		0.22	0.04	0.39	0.02	0.06	300.0				
87P03032	102-127	C	S	2.49	0.011		0.13	11	0.3	tr		0.06	0.02	0.07	0.02	0.02	600.0				

*** Primary Characterization Data ***

(Grant, Minnesota)
Sandy, mixed Udic Haploboroll
; Pedon No. 87P0576

CEC & Bases

				-1-	-2-	-3-	-4-	-5-	-6-	-7-	-8-	-9-	-10-	-11-	-12-	-13-	-14-
				(----- NH₄OAC Extractable Bases -----)								CEC8 Sum Cats	CEC7 NH₄ OAC	ECEC Bases +Al	Al Sat	(--- Base ---) (- Saturation -)	
				Ca	Mg	Na	K	Sum Bases	Acid-ity	Extr Al	KCl Mn					Sum	NH₄OAC
Layer	Depth (cm)	Horz	Prep	$(----- cmol(+)\,kg^{-1}$						$----)$	$mg\,kg^{-1}$	$(--- cmol(+)\,kg^{-1} ---)$			$(----- \% -----)$		
				6N2e	6O2d	6P2b	6Q2b	6H5a				5A3a	5A8b			5C3	5C1
87P03026	0-23	Ap	S	9.5*	2.1	--	0.2	11.8	2.8			14.6	11.0			81	100
87P03027	23-36	A	S	8.7*	2.1	0.2	0.2	11.2	3.6			14.8	10.7			76	100
87P03028	36-46	Bw1	S	7.1*	1.7	--	0.1	8.9	2.1			11.0	7.3			81	100
87P03029	46-64	Bw2	S	3.7*	1.1	--	0.1	4.9	0.6			5.5	4.5			89	100
87P03030	64-94	Bw3	S	7.3*	3.7	--	tr		0.3				2.9			97	100
87P03031	94-102	BC	S	19.8*	3.6	tr	0.1						4.3			100	100
87P03032	102-127	C	S	23.4*	2.3	--	0.1						3.5			100	100

* Extractable Ca may contain Ca from calcium carbonate or gypsum., CEC7 base saturation set to 100.

Salt

				-1-	-2-	-3-	-4-	-5-	-6-	-7-	-8-	-9-	-10-	-11-	-12-	-13-	-14-	-15-	-16-	-17-	-18-	-19-	-20-
				(-- Water Extracted From Saturated Paste --)														H₂O	Total Salts	Elec Cond	Pred Elec Cond	Exch Na	SAR
				Ca	Mg	Na	K	CO₃	HCO₃	F	Cl	PO₄	Br	OAC	SO₄	NO₂	NO₃						
Layer	Depth (cm)	Horz	Prep	$(--- mmol(+)\,L^{-1} ---)$				$(------------------------ mmol(-)\,L^{-1} ------------------------)$									$(--- \% ---)$	$(--- \% ---)$	$(-- dS\,m^{-1} --)$		$\%$	
				6N1b	6O1b	6P1b	6Q1b	6I1b	6J1b	6U1a	6K1c			6L1c	6W1a	6M1c	8A	8D5	8A3a	8I	5D2	5E
87P03026	0-23	Ap	S	2.4	1.0	0.2	0.1	--	2.3	0.1	0.4			--	0.6		tr	27.1	tr	0.35	0.16	--	tr
87P03027	23-36	A	S																		0.15	2	
87P03028	36-46	Bw1	S																		0.07	--	
87P03029	46-64	Bw2	S																		0.04	--	
87P03030	64-94	Bw3	S																		0.09	--	
87P03031	94-102	BC	S																		0.12	--	
87P03032	102-127	C	S																		0.09	--	

Pedon ID: 87MN051001

Sampled As : Sverdrup

USDA-NRCS-NSSC-Soil Survey Laboratory

Print Date: Aug 10 2010 8:41AM

*** Primary Characterization Data ***
(Grant, Minnesota)
Sandy, mixed Udic Haploboroll

Pedon No. 87P0576

pH & Carbonates

			-1-	-2-	-3-	-4-	-5-	-6-	-7-	-8-	-9-	-10-	-11-
			(----------------------- pH -----------------------)						(-- Carbonate --)		(-- Gypsum ---)		
				$CaCl_2$					As $CaCO_3$		As $CaSO_4 {\cdot} 2H_2O$		Resist
				0.01M	H_2O	Sat			<2mm	<20mm	<2mm	<20mm	ohms
	Depth		KCl	1:2	1:1	Paste	Sulf	NaF	(--------------------- % ---------------------)				cm^{-1}
Layer	(cm)	Prep	Horz	4C1a2a	4C1a2a	8C1b			6E1g				
87P03026	0-23	S	Ap	6.0	6.5	6.6							
87P03027	23-36	S	A	5.8	6.3								
87P03028	36-46	S	Bw1	6.3	6.9								
87P03029	46-64	S	Bw2	6.3	7.0								
87P03030	64-94	S	Bw3	7.3	8.0				2				
87P03031	94-102	S	BC	7.7	8.2				10				
87P03032	102-127	S	C	7.8	8.5				16				

*** Primary Characterization Data ***

(Grant, Minnesota)
Sandy, mixed Udic Haploboroll
;   Pedon No. 87P0576

Clay Mineralogy (<.002 mm)

				-1-	-2-	-3-	-4-	-5-	-6-	-7-	-8-	-9-	-10-	-11-	-12-	-13-	-14-	-15-	-16-	-17-	-18-
				X-Ray						Thermal			Elemental							EGME	Inter
	Depth		Fract			7A2i							SiO$_2$	Al$_2$O$_3$	Fe$_2$O$_3$	MgO 7C3	CaO	K$_2$O	Na$_2$O	Retn	preta tion
Layer	(cm)	Horz	ion	(- - - - - - - peak size - - - - - - - -)						(- - - - - - - - % - - - - - - - -)			(- - - - - - - - - - - - % - - - - - - - - - - - -)							mg g^{-1}	
87P03026	0-23	Ap	tcly	MT 2	MI 2	KK 1									5.6			1.1			
87P03028	36-46	Bw1	tcly	MT 3	MI 2	KK 1									7.3			1.1			
87P03030	64-94	Bw3	tcly	MT 3	MI 2	KK 1									7.0			1.0			

FRACTION INTERPRETATION:
tcly - Total Clay, <0.002 mm

MINERAL INTERPRETATION:
KK - Kaolinite   MI - Mica   MT - Montmorillonite

RELATIVE PEAK SIZE:
5 Very Large   4 Large   3 Medium   2 Small   1 Very Small   6 No Peaks

Sampled As : Sverdrup

Sandy, mixed Udic Haploboroll

USDA-NRCS-NSSC-Soil Survey Laboratory

; Pedon No. 87P0576

*** Primary Characterization Data ***

(Grant, Minnesota)

Sand - Silt Mineralogy (2 0-0.002 mm)

Layer	Depth (cm)	Horz	Fraction	-1-	-2-	-3-	-4-	-5-	-6-	-7-	-8-	-9-	-10-	-11-	-12-	-13-	-14-	-15-	-16-	-17-	-18-
				X-Ray					Thermal				Tot Re	Optical Grain Count 7B1a						EGME Retn	Interpretation
				(------ peak size ------)					(------ % ------)					(------------------- % -------------------)						mg g^{-1}	
87P03028	36-46	Bw1	vfs										80	QZ 75	FK 13	OP 4	OT 3	BT 1	MS 1		
														AM 1	GN 1	TM tr	HN tr	FP tr	PR tr		
														ZR tr							
87P03030	64-94	Bw3	vfs										66	QZ 59	FK 10	CA 7	BT 7	OP 6	OT 6		
														GN 1	MS 1	PR 1	HN 1	CL tr	AM tr		
														RU tr	CB tr	FP tr	GS tr	ZR tr	PO tr		
														TM tr							

FRACTION INTERPRETATION:

vfs - Very Fine Sand, 0.05-0.1 mm

MINERAL INTERPRETATION:

AM - Amphibole

FK - Potassium Feldspar

MS - Muscovite

QZ - Quartz

BT - Biotite

FP - Plagioclase Feldspar

OP - Opaques

RU - Rutile

CA - Calcite

GN - Garnet

OT - Other

TM - Tourmaline

CB - Carbonate Aggregates

GS - Glass

PO - Plant Opal

ZR - Zircon

CL - Chlorite

HN - Hornblende

PR - Pyroxene

*** Primary Characterization Data ***

(Grant, Minnesota)

Fine Earth Mineralogy (<2.0 mm)

	-1-	-2-	-3-	-4-	-5-	-6-	-7-	-8-	-9-	-10-	-11-	-12-	-13-	-14-	-15-	-16-	-17-	-18-
			X-Ray					Thermal					Elemental				EGME	Inter
										SiO_2	Al_2O_3	Fe_2O_3	MgO	CaO	K_2O	Na_2O	Retn	preta
																	7D2	tion
	Fract																$mg\ g^{-1}$	
Depth	ion	(- - - - - - - - peak size - - - - - - - -)(- - - - - - - % - - - - - - -)					(- - - - - - - - - - - - - - - - % - - - - - - - - - - - - - - - - -)										
Layer (cm) Horz	feth																	
87P03026 0-23 Ap																	8.0	

FRACTION INTERPRETATION:

feth - Fine Earth, <2 mm

MINERAL INTERPRETATION:

eg - Surface Area

Pedon ID: 87MN051001

Print Date: Aug 10 2010 8:41AM

*** Supplementary Characterization Data ***
(Grant, Minnesota)

Sampled as : Sverdrup ; Sandy, mixed Udic Haploboroll
Revised to :

SSL - Project   CP87MN190   WEPP MINNESOTA
     - Site ID   87MN051001 Lat: 45° 59' 47.00" north Long: 95° 52' 17.00" west MLRA: 102A
     - Pedon No.   87P0576
     - General Methods 1B1A, 2A1, 2B

United States Department of Agriculture
Natural Resources Conservation Service
National Soil Survey Center
Soil Survey Laboratory
Lincoln, Nebraska 68508-3866

Tier 1

Layer	Depth (cm)	Horz	Prep	-1- 3	-2- 2	-3- 3/2	-4- 1	-5- 3/4	-6- 3/8	-7- 4	-8- 10	-9- 40	-10- 200	-11- 20	-12- 5	-13- 2	-14- 1.	-15- .5	-16- .25	-17- .10	-18- .05	-19- 60	-20- 50	-21- 10	-22- LL 4F1	-23- PI 4F	-24- CU	-25- CC
				(— Engineering PSDA, Percentage Passing Sieve —) Inches / Number / Microns													(— Cumulative Curve Fractions (<75mm) USDA Less Than Diameters (mm) at —) Millimeter / Percentile								(Atterberg) %		(Gradation)	
87P03026	0-23	Ap	S	100	100	100	100	100	100	100	100	78	27	17	12	8	98	85	57	28	25	0.27	0.198	0.003	24	10	79.8	12.4
87P03027	23-36	A	S	100	100	100	100	100	100	100	100	77	25	17	12	9	98	84	56	26	24	0.28	0.209	0.003			96.3	15.6
87P03028	36-46	Bw1	S	100	100	100	100	100	100	100	99	85	25	16	11	9	97	90	71	27	22	0.20	0.161	0.003		NP	59.0	16.8
87P03029	46-64	Bw2	S	100	100	100	100	100	100	100	99	88	11	7	6	5	98	94	71	13	9	0.21	0.179	0.063			3.3	1.3
87P03030	64-94	Bw3	S	100	100	100	100	100	98	96	91	64	7	5	4	3	85	71	45	8	6	0.38	0.288	0.105			3.6	0.8
87P03031	94-102	BC	S	100	91	85	75	69	59	49	41	17	6	3	2	2	36	19	11	6	5	10.13	5.040	0.206			49.2	0.3
87P03032	102-127	C	S	100	95	92	87	84	81	77	67	17	6	5	5	5	57	19	10	7	6	1.22	0.877	0.256			4.8	1.2

Tier 2

Layer	Depth (cm)	Horz	Prep	-26- >2	-27- 250/-UP	-28- 250/-75	-29- 75/-20	-30- 20/-5	-31- 5/-2	-32- <2	-33- 75/-2 (3B1)	-34- 20/-5 (3B1)	-35- 5/-2 (3B1)	-36- <2	-37-	-38- <2
				(— Weight Fractions, Whole Soil (mm), % of Whole Soil —)							(— <75 mm Fraction, % of <75 mm —)					
87P03026	0-23	Ap	S	tr	—	—	—	—	tr	100	—	tr	tr	tr		100
87P03027	23-36	A	S	tr	—	—	—	—	tr	100	—	tr	tr	tr		100
87P03028	36-46	Bw1	S	1	—	—	1	—	1	99	—	—	1	1		99
87P03029	46-64	Bw2	S	1	—	—	1	tr	1	99	—	tr	tr	1		99
87P03030	64-94	Bw3	S	9	—	—	9	4	5	91	—	4	5	4		91
87P03031	94-102	BC	S	59	—	—	59	31	20	41	31	31	20	59		41
87P03032	102-127	C	S	33	—	—	33	16	7	67	16	16	7	33		67

Layer	-39- Whole Soil 33 kPa (4A1d)	-40- Whole Soil Oven-dry	-41- Whole Soil Moist	-42- Whole Soil Saturated	-43- <2mm 33 kPa (4A1d)	-44- <2mm 1500 kPa	-45- <2mm Oven-dry (4A1h)	-46- <2mm Moist	-47- <2mm Saturated	-48- Ratios Whole Soil	-49- Ratios <2 mm	-50- Void
	(— Weight Per Unit Volume (g cm⁻³), Whole Soil, Soil Survey / Engineering —)				(— <2 mm Fraction, Soil Survey / Engineering —)					(Ratios At 33 kPa)		(Void)
87P03026	1.61	1.68	1.82	2.00	1.60	1.64	1.67	1.81	2.00	0.65	0.66	
87P03027	1.54	1.63	1.74	1.96	1.53	1.59	1.62	1.72	1.95	0.72	0.73	
87P03028	1.23	1.23	1.41	1.77	1.22	1.22	1.22	1.41	1.76	1.15	1.17	
87P03029	1.63	1.63	1.73	2.01	1.62	1.62	1.62	1.72	2.01	0.63	0.64	
87P03030	1.50				1.62							
87P03031	1.97											
87P03032	1.70											

*** Supplementary Characterization Data ***

(Grant, Minnesota)
Sandy, mixed Udic Haploboroll

Pedon No. 87P0576

Tier 3

Volume Fractions – Whole Soil (mm) At 33 kPa (% of Whole Soil); Pores; C/N Ratio

Layer	Depth (cm)	Horz	Prep	>2	250-UP	250-75	75-20	20-5	5-2	<2	2-.05	.05-.002	<.002	Pores D	Pores F	C/N Ratio
87P03026	0-23	Ap	S	tr	–	–	–	–	tr	100	46	10	5	18	21	12
87P03027	23-36	A	S	tr	–	–	–	tr	tr	100	44	9	5	22	20	11
87P03028	36-46	Bw1	S	1	–	–	1	1	1	100	36	6	4	36	18	10
87P03029	46-64	Bw2	S	1	–	–	1	tr	1	99	56	2	3	28	10	9
87P03030	64-94	Bw3	S	5	–	5	5	2	3	95	49	2	2	43		11
87P03031	94-102	BC	S	44	–	44	44	23	15	56	27	3	1	26		13
87P03032	102-127	C	S	21	–	21	21	10	5	79	40	1	3	36		11

Ratios To Clay (<2 mm Fraction); Linear Extensibility; WRD

Layer	Fine Clay	CEC (Cats Sum)	NH4 H_2O (8D1)	LEP 33 kPa	Linear Ext. Whole Soil 33 kPa -dry	Oven -dry	WRD <2 mm to % (4C1)	WRD Whole Soil <2 mm (4C1)
87P03026	0.70	1.85	0.68	0.180	1.0	0.8	0.13	0.13
87P03027	0.73	1.72	0.60	0.220	1.1	1.3	0.11	0.11
87P03028	0.69	1.31	0.50				0.13	0.14
87P03029	0.72	1.02	0.50				0.05	0.06
87P03030	0.47		0.53					
87P03031	0.28		0.72					
87P03032	0.18		0.50					

Tier 4

Weight Fractions – Clay Free

Layer	Depth (cm)	Horz	Prep	>2	75-20	20-2	2-.05 (Sand)	.05-.002 (Silt)	VC	C	M	F	VF	C (silt)	F (silt)	Clay
87P03026	0-23	Ap	S	–			82	18	2	15	30	31	4	8	10	9
87P03027	23-36	A	S	–			84	16	3	15	31	32	3	7	10	9
87P03028	36-46	Bw1	S	1			84	15	3	7	21	48	6	7	8	9
87P03029	46-64	Bw2	S	1			96	3	1	4	25	62	3	1	2	6
87P03030	64-94	Bw3	S	9			88	3	7	16	29	42	3	2	1	4
87P03031	94-102	BC	S	60			36	4	13	42	21	12	3	5	5	4
87P03032	102-127	C	S	35			64	1	16	61	15	5	1	1	1	8

PSDA (% of <2 mm), pH, Electrical Conductivity, Particle Density

Layer	Texture	Sand 2-.05 (3A1)	Silt .05-.002 (3A1)	Clay <.002 (3A1)	pH CaCl2 .01M (4C1a2a)	Elect. Conduct. dS m⁻¹	Particle Density g cm⁻³ (8A3a)
87P03026	sl	75.3	16.8	7.9	6.0	0.35	
87P03027	sl	76.4	15.0	8.6	5.8		
87P03028	fsl	77.9	13.7	8.4	6.3		
87P03029	fs	91.4	3.2	5.4	6.3		
87P03030	s	93.5	2.9	3.6	7.3		
87P03031	cos	87.3	8.8	3.9	7.7		
87P03032	cos	91.1	1.3	7.6	7.8		

Pedon ID: 87MN051001

*** Taxonomy Characterization Data ***
(Grant, Minnesota)

Sampled as :
Revised to : Sverdrup ; Sandy, mixed Udic Haploboroll

SSL - Project   CP87MN190 WEPP MINNESOTA
- Site ID   87MN051001 Lat: 45° 59' 47.00" north Long: 95° 52' 17.00" west MLRA: 102A
- Pedon No.   87P0576
- General Methods 1B1A, 2A1, 2B

United States Department of Agriculture
Natural Resources Conservation Service
National Soil Survey Center
Soil Survey Laboratory
Lincoln, Nebraska 68508-3866

Taxonomy Tier 1

				-1-	-2-	-3-	-4-	-5-	-6-	-7-	-8-	-9-	-10-
				Clay <002	Fine Clay <0002	CaCO3 Clay <002	1500 kPa /Clay	Clay Est	.1-75 mm Frac	Bulk Den 33 kPa	Cole Whole Soil	Vol % of Whole	Resist Min %
				(——— % of <2 mm———)					%	g cm⁻³	cm cm⁻¹		
Layer	Depth (cm)	Horz	Prep	3A1	3A1		8D1			4A1d			7B1a
87P03026	0-23	Ap	S	7.9	5.5		0.68		72	1.60	0.014	tr	
87P03027	23-36	A	S	8.6	6.3		0.60		74	1.53	0.019	tr	
87P03028	36-46	Bw1	S	8.4	5.8		0.50		73	1.22	--	1	80
87P03029	46-64	Bw2	S	5.4	3.9		0.50		87	1.62	--	1	
87P03030	64-94	Bw3	S	3.6	1.7		0.53		92			5	66
87P03031	94-102	BC	S	3.9	1.1		0.72		94			44	
87P03032	102-127	C	S	7.6	1.4		0.50		93			21	

Taxonomy Tier 2

				-1-	-2-	-3-	-4-	-5-	-6-	-7-	-8-	-9-	-10-	-11-	-12-	-13-	-14-	-15-	-15-
				pH	pH	Org	Tot	Al+½ Fe		CO3 as	(—— Base Sat ——)		NZ	ECEC	CEC7	ECEC	Al	E C	ESP
				H2O	NaF	C	C	Oxal	ODOE	CaCO3	NH4	Bases	P Ret	cmol(+)	/Clay	/Clay	Sat		
						%	%		%					kg⁻¹			%	dS m⁻¹	%
Layer	Depth (cm)	Horz	Prep	4C1a2a		6A1c	6A2d		8J	6E1g	5C1	5C3			8D1			8A3a	5D2
87P03026	0-23	Ap	S	6.5		1.28	1.14	0.12	0.04		100*	81			1.39			0.35	--
87P03027	23-36	A	S	6.3		1.21	1.11	0.14	0.05		100*	76			1.24				2
87P03028	36-46	Bw1	S	6.9		0.45	0.42	0.17	0.04		100*	81			0.87				--
87P03029	46-64	Bw2	S	7.0		0.21	0.24	0.12	0.03		100*	89			0.83				--
87P03030	64-94	Bw3	S	8.0		0.15	0.33	0.09	0.02	2	100*	97			0.81				--
87P03031	94-102	BC	S	8.2		0.36	1.64	0.22	0.04	10	100*	100			1.10				--
87P03032	102-127	C	S	8.5		0.13	2.49	0.06	0.02	16	100*	100			0.46				--

*Extractable Ca may contain Ca from calcium carbonate or gypsum.

Pedon Calculations

Calculation Name	Result	Units of Measure
CEC Activity, CEC7/Clay, Weighted Average	0.91	(NA)
Clay, carbonate free, Weighted Average	5	% wt
Weighted Particles, 0.1-75mm, 75 mm Base	86	% wt
Volume, >2mm, Weighted Average	6	% vol
Clay, total, Weighted Average	5	% wt

PEDON DESCRIPTION

Print Date: 08/10/2010

Description Date: 6/1/1987

Describer: Gorton, Yeck, DeMartalaere

Site ID: 87MN051001

Site Note:

Pedon ID: 87MN051001

Pedon Note: Primary Sverdrup pedon, desiganted number 33 on the plot map.

Lab Source ID: SSL

Lab Pedon #: 87P0576

Soil Name as Described/Sampled: Sverdrup

Soil Name as Correlated:

Classification:

Pedon Type:

Pedon Purpose: full pedon description

Taxon Kind:

Associated Soils:

Physiographic Division:

Physiographic Province:

Physiographic Section:

State Physiographic Area:

Local Physiographic Area:

Geomorphic Setting: None Assigned

Upslope Shape: convex

Cross Slope Shape:

Particle Size Control Section:

Description origin: Converted from SSL-CMS data

Diagnostic Features: to cm.

Country:

State: Minnesota

County: Grant

MLRA: 102A -- Rolling Till Prairie

Soil Survey Area:

Map Unit:

Quad Name:

Location Description:

Legal Description: NW 1/4 of the SW 1/4 of Section 8, Township 129N , Range 41W

Latitude: 45 degrees 59 minutes 47.00 seconds north

Longitude: 95 degrees 52 minutes 17.00 seconds west

Datum:

UTM Zone:

UTM Easting:

UTM Northing:

Primary Earth Cover:

Secondary Earth Cover:

Existing Vegetation:

Parent Material:

Bedrock Kind:

Bedrock Depth:

Bedrock Hardness:

Bedrock Fracture Interval:

Surface Fragments:

Description database: NSSL

Cont. **Site ID:** 87MN051001

Pedon ID: 87MN051001

Slope (%)	Elevation (meters)	Aspect (deg)	MAAT (C)	MSAT (C)	MWAT (C)	MAP (mm)	Frost-Free Days	Drainage Class	Slope Length (meters)	Upslope Length (meters)
4.0		180						well		

1A p--0 to 23 centimeters; very dark gray (10YR 3/1) interior sandy loam, dark gray (10YR 4/1) interior, dry; weak very fine and fine granular, and weak fine subangular blocky structure; friable; common fine roots; friable; common fine roots; abrupt smooth boundary. Lab sample # 87P3026. Less than 1% coarse fragments less than 76 mm.; common fine roots

1A--23 to 36 centimeters; very dark gray (10YR 3/1) interior sandy loam; weak very fine and fine granular, and weak fine subangular blocky structure; friable; few fine roots; abrupt wavy boundary. Lab sample # 87P3027. Less than 1% coarse fragments less than 76 mm.; few fine roots

1B w1--36 to 46 centimeters; dark brown (10YR 3/3) interior sandy loam; weak fine subangular blocky structure; friable; few very fine roots; clear wavy boundary. Lab sample # 87P3028. Less than 1% coarse fragments less than 76 mm.; few very fine roots

1B w2--46 to 64 centimeters; dark yellowish brown (10YR 4/4) interior loamy sand; single grain; loose; few very fine roots; clear wavy boundary. Lab sample # 87P3029. Less than 1% coarse fragments less than 76 mm.; few very fine roots

1B w3--64 to 239 centimeters; yellowish brown (10YR 5/4) interior sand; single grain; loose; abrupt wavy boundary. Lab sample # 87P3030. Less than 1% coarse fragments less than 76 mm.

1BC--94 to 102 centimeters; dark reddish brown (2.5YR 3/4) interior very gravelly loamy coarse sand; single grain; loose. Lab sample # 87P3031. 20-35% coarse fragments less than 76 mm.

1C--102 to 127 centimeters; yellowish brown (10YR 5/4) interior coarse sand; single grain; loose; slight effervescence. by HCl, 1 normal. Lab sample # 87P3032. 10-15% coarse fragments less than 76 mm.

5 Caribou

*** Primary Characterization Data ***
(Aroostook, Maine)

Sampled as on Jun 01, 1988 : Caribou : Fine-loamy, mixed, frigid Typic Haplorthod
Revised to correlated on Sep 01, 1991 : Caribou ; Loamy-skeletal, mixed, ac ive, frigid Typic Dystrochrept

SSL - Project CP88ME188 WEPP MAINE
- Site ID 88ME003001 Lat: 46° 0' 55.00" north Long: 68° 1' 11.00" west MLRA: 146
- Pedon No. 88P0722
- General Methods 1B1A, 2A1, 2B

United States Department of Agriculture
Natural Resources Conservation Service
National Soil Survey Center
Soil Survey Laboratory
Lincoln, Nebraska 68508-3866

Layer	Horizon	Orig Hzn	Depth (cm)	Field Label 1	Field Label 2	Field Label 3	Field Texture	Lab Texture
88P03855	Ap1	AP1	0-11					L
88P03856	Ap2	AP2	11-24					L
88P03857	Bs	BS	24-39					L
88P03858	C1	C1	39-66					COSL
88P03859	C2	C2	66-104					L
88P03860	C3	C3	104-143					

Calculation Name		Result	Units of Measure
Pedon Calculations			
CEC Activity, CEC7/Clay, Weighted Average		0.78	(NA)
Clay, carbonate free, Weighted Average		12	% wt
Weighted Particles, 0.1-75mm, 75 mm Base		67	% wt
Volume, >2mm, Weighted Average		43	% vol
Clay, total, Weighted Average		12	% wt

Weighted averages based on control section: 25-100 cm

PSDA & Rock Fragments

			-1-	-2-	-3-	-4-	-5-	-6-	-7-	-8-	-9-	-10-	-11-	-12-	-13-	-14-	-15-	-16-	-17-
			(- - - Total - - - -)			(- - Clay - - -)		(- - - Silt - - - -)		(- - - - - - - - - - - Sand - - - - - - - - - -)					(Rock Fragments (mm))				>2 mm
			Clay	Silt	Sand	Fine	CO$_3$	Fine	Coarse	VF	F	M	C	VC	(- - - - - - - - - Weight - - - - - - - - -)				wt %
			<	.002	.05	<	<	.002	.02	.05	.10	.25	.5	1	2	5	20	.1-	whole
			.002	-.05	-2	.0002	.002	-.02	-.05	-.10	-.25	-.50	-1	-2	-5	-20	-75	75	soil
Layer	Depth (cm)	Horz	(- - - - - - - - - - - - - - - - - - - % of <2mm Mineral Soil -)												(- - - % of <75mm - - - -)				
		Prep	3A1	3A1	3A1	3A1		3A1	3A1	3A1	3A1	3A1	3A1	3A1	3B1	3B1	3B1	3B1	
88P03855	0-11	Ap1	S	12.2	40.8	47.0	1.5		25.6	15.2	11.5	11.0	7.8	7.0	9.7	9	18	18	65	45
88P03856	11-24	Ap2	S	12.1	42.8	45.1	1.0		26.3	16.5	11.1	12.0	8.6	6.7	6.7	10	19	9	59	38
88P03857	24-39	Bs	S	12.3	44.5	43.2	1.4		28.9	15.6	12.1	11.8	7.9	6.5	4.9	5	7	41	68	53
88P03858	39-66	C1	S	13.3	44.9	41.8	3.3		26.2	18.7	13.1	12.1	8.0	5.1	3.5	4	1	4	35	13
88P03859	66-104	C2	S	11.3	17.6	71.1	3.4		12.4	5.2	2.8	3.7	9.3	19.6	35.7	15	29	34	93	83
88P03860	104-143	C3	S	23.2	45.2	31.6	7.3		26.9	18.3	6.5	7.1	6.3	6.2	5.5	8	10	20	54	46

Pedon ID: 88ME003001
Sampled As : Caribou
USDA-NRCS-NSSC-Soil Survey Laboratory

Print Date: Aug 10 2010 8:45AM

*** Primary Characterization Data ***

(Aroostook, Maine)
Fine-loamy, mixed, frigid Typic Haplorthod

Pedon No. 88P0722

Water Dispersible PSDA

				-1-	-2-	-3-	-4-	-5-	-6-	-7-	-8-	-9-	-10-	-11-	-12-
				(-- Total --)			(- Clay -)		(- Silt -)		(- Sand -)				
				Clay	Silt	Sand	F	CO$_3$	F	C	VF	F	M	C	VC
				< .002	.002-.05	.05-2	< .0002	< .002	.002-.02	.02-.05	.05-.10	.10-.25	.25-.50	.5-1	1-2
							.0002-.002								
							-- % of <2mm --									
Layer	Depth (cm)	Horz	Prep												
				3A1c	3A1c	3A1c									
88P03855	0-11	Ap1	S	7.7	41.9	50.4									

Bulk Density & Moisture

				-1-	-2-	-3-	-4-	-5-	-6-	-7-	-8-	-9-	-10-	-11-	-12-	-13-
				(Bulk Density) g cm^{-3}		Cole Whole Soil	(--- Water Content % of < 2mm ---)				1500 kPa Moist	1500 kPa Ratio AD/OD	WRD Whole Soil cm^3 cm^{-3}	Aggst Stabl 2-0.5mm %	(-- Ratio/Clay --) CEC7	1500 kPa
				33 kPa	Oven Dry		6 kPa	10 kPa	33 kPa	1500 kPa							
Layer	Depth (cm)	Horz	Prep	4A1d	4A1h				4B1c	4B2a	4B2b	4B5	4C1		8D1	8D1
88P03855	0-11	Ap1	S													
88P03855	0-11	Ap1	M							8.9	9.9	1.010			1.05	0.73
88P03856	11-24	Ap2	S	1.24	1.29	0.010			26.9				0.17			
88P03856	11-24	Ap2	M							8.6	9.9	1.010			0.97	0.71
88P03857	24-39	Bs	S	1.47	1.50	0.004			20.7				0.08			
88P03857	24-39	Bs	M							7.8	11.6	1.009			0.80	0.63
88P03858	39-66	C1	S	1.81	1.84	0.005			13.8				0.14			
88P03858	39-66	C1	M							10.7	5.1	1.004			0.44	0.80
88P03859	66-104	C2	S													
88P03859	66-104	C2	M							6.6	6.9	1.008			1.04	0.58
88P03860	104-143	C3	S	1.63	1.69	0.008			19.3				0.03			
88P03860	104-143	C3	M							12.0	16.3	1.010			0.44	0.52

*** Primary Characterization Data ***

Pedon ID: 88ME003001
Sampled As   :   Caribou
USDA-NRCS-NSSC-Soil Survey Laboratory

(Aroostook, Maine)
Fine-loamy, mixed, frigid Typic Haplorthod
;   Pedon No. 88P0722

Print Date: Aug 10 2010 8:45AM

Water Content

				-1-	-2-	-3-	-4-	-5-	-6-	-7-	-8-	-9-	-10-	-11-	-12-	-13-
				(-- Atterberg --)		(- - - - Bulk Density - - - - -)			(------------------- Water Content -------------------)								
				(--- Limits ---)		Field	Recon 33 kPa	Recon Oven Dry	Field	Recon 33 kPa	6 kPa	10 kPa	33 kPa	100 kPa	200 kPa	500 kPa
				LL	PI											
	Depth (cm)	Horz	Prep	pct <0.4mm		(-- - - - - - - - - - g cm⁻³ - - - - - - - - -)			(- - - - - - - - - - - - - % of < 2mm - - - - - - - - - - - - -)							
				4F1	4F											
Layer																
88P03855	0-11	Ap1	S	32	28											
88P03858	39-66	C1	S	3	2											

Carbon & Extractions

			-1-	-2-	-2-	-3-	-4-	-5-	-6-	-7-	-8-	-9-	-10-	-11-	-12-	-13-	-14-	-15-	-16-	-17-	-18-
			(- - - - Total - - - - -)			Org C	C/N Ratio	(-- Dith-Cit Ext ---)			Al+½Fe	(- - - - Ammonium Oxalate Extraction - - - - -)					(- - - Na Pyro-Phosphate - - -)			
			C	N	S			Fe	Al	Mn		ODOE	Fe	Al	Si	Mn	C	Fe	Al	Mn
	Depth (cm)	Horz	Prep	(- - - - - % of <2 mm - - - - -)					(--- % of < 2mm ---)							mg kg⁻¹	(- - - % of < 2mm - - -)		
				6A2d	6B3a		6A1c		6C2b	6G7a			8J	6C9a	6G12	6V2	6D5b	6C8a		6G10
Layer																				
88P03855	0-11	Ap1	S	1.98	0.193		2.28	12	1.3	0.4		1.11	0.16	1.07	0.58	0.09	424.8	0.4		0.4
88P03856	11-24	Ap2	S		0.170		1.80	11	1.1	0.3		1.08	0.15	1.07	0.55	0.09	531.1	0.5		0.5
88P03857	24-39	Bs	S				0.93		1.4	0.3		1.21	0.13	1.39	0.52	0.08	487.5	0.5		0.3
88P03858	39-66	C1	S				0.16		0.9	0.1		0.47	0.06	0.49	0.20	0.06	450.8	0.1		0.1
88P03859	66-104	C2	S				0.16		1.4	0.1		0.57	0.05	0.74	0.21	0.09	1077.7	0.1		0.1
88P03860	104-143	C3	S				0.24		2.0	0.1		0.59		0.77		0.12	642.5	tr		tr

CEC & Bases

				-1-	-2-	-3-	-4-	-5-	-6-	-7-	-8-	-9-	-10-	-11-	-12-	-13-	-14-
				(- - - - - - NH₄OAC Extractable Bases - - - - - -)								CEC8	CEC7	ECEC		(- - - Base - - - -)	
								Sum Bases	Acid-ity	Extr Al	KCl Mn	Sum Cats	NH₄ OAC	Bases +Al	Al Sat	(- Saturation -)	
				Ca	Mg	Na	K									Sum	NH₄OAC
	Depth (cm)	Horz	Prep	(- - - - - - - - cmol(+) kg⁻¹ - - - - - - - -)							mg kg⁻¹	(- - - cmol(+) kg⁻¹ - - -)			(- - - - % - - - -)		
				6N2e	6O2d	6P2b	6Q2b		6H5a	6G9b		5A3a	5A8b	5A3b	5G1	5C3	5C1
Layer																	
88P03855	0-11	Ap1	S	5.5	1.5	0.1	0.6	7.7	14.0			21.7	12.8			35	60
88P03856	11-24	Ap2	S	3.4	0.9	0.1	0.3	4.7	15.0	0.9		19.7	11.7	5.6	16	24	40
88P03857	24-39	Bs	S	1.2	0.3	0.1	0.2	1.8	12.7	1.5		14.5	9.8	3.3	45	12	18
88P03858	39-66	C1	S	1.5	0.4	0.1	0.2	2.2	6.6	0.5		8.8	5.9	2.7	19	25	37
88P03859	66-104	C2	S	4.0	0.6	0.1	0.2	4.9	4.5	0.1		9.4	11.7	5.0	2	52	42
88P03860	104-143	C3	S	9.5*	0.6	0.1	0.1	10.3	2.5			12.8	10.2			80	100

* Extractable Ca may contain Ca from calcium carbonate or gypsum., CEC7 base saturation set to 100.

Pedon ID: 88ME003001

Sampled As : Caribou

USDA-NRCS-NSSC-Soil Survey Laboratory

*** Primary Characterization Data ***

(Aroostook, Maine)

Fine-loamy, mixed, frigid Typic Haplorthod

; Pedon No. 88P0722

Salt

Water Extracted From Saturated Paste

Layer	Depth (cm)	Horz	Prep	-1- Ca	-2- Mg	-3- Na	-4- K	-5- CO3	-6- HCO3	-7- F	-8- Cl	-9- PO4	-10- Br	-11- OAC	-12- SO4	-13- NO2	-14- NO3	-15- H2O	-16- Total Salts	-17- Elec Cond	-18- Pred Elec Cond	-19- Exch Na	-20- SAR
				($---$ mmol(+) L^{-1} $---$)				($-------------$ mmol(-) L^{-1} $-----------$)										($---$ % $---$)	($--$ dS m^{-1} $--$)	% 5D2	
88P03855	0-11	Ap1	S																			1	
88P03856	11-24	Ap2	S																			1	
88P03857	24-39	Bs	S																			1	
88P03858	39-66	C1	S																			2	
88P03859	66-104	C2	S																			1	
88P03860	104-143	C3	S																			1	

pH & Carbonates

Layer	Depth (cm)	Horz	Prep	-1- KCl	-2- CaCl2 0.01M 1:2 4C1a2a	-3- H2O 1:1 4C1a2a	-4- Sat Paste	-5- Sulf	-6- NaF	-7-	-8- As CaCO3 <2mm	-9- (Carbonate) <20mm	-10- As CaCO3*2H2O <2mm	-11- (Gypsum) <20mm Resist ohms cm^{-1}
88P03855	0-11	Ap1	S		5.3	5.6								
88P03856	11-24	Ap2	S		4.8	5.2								
88P03857	24-39	Bs	S		4.6	4.9								
88P03858	39-66	C1	S		4.5	4.9								
88P03859	66-104	C2	S		4.9	5.4								
88P03860	104-143	C3	S		6.9	7.4								

*** Primary Characterization Data ***

(Aroostook, Maine)
Fine-loamy, mixed, frigid Typic Haplorthod
;   Pedon No. 88P0722

Pedon ID: 88ME003001
Sampled As   :   Caribou
USDA-NRCS-NSSC-Soil Survey Laboratory

Print Date: Aug 10 2010 8:45AM

Clay Mineralogy (<.002 mm)

Layer	Depth (cm)	Horz	Fract ion	-1-	-2-	-3-	-4-	-5-	-6-	-7-	-8-	-9-	-10-	-11-	-12-	-13-	-14-	-15-	-16-	-17-	-18-
						X-Ray 7A2i					Thermal					Elemental 7C3					EGME Retn	Inter preta tion
													SiO_2	Al_2O_3	Fe_2O_3	MgO	CaO	K_2O	Na_2O	$mg\ g^{-1}$	
				(- - - - - - - peak size - - - - - - -)						(- - - - - - - % - - - - - - -)			(- - - - - - - - - - - - - - - - % - - - - - - - - - - - - - - - -)								
88P03855	0-11	Ap1	tcly	KK 2	VR 2	MI 1									5.7			1.4			
88P03857	24-39	Bs	tcly	VR 3	KK 2	MI 2									10.4			1.7			

FRACTION INTERPRETATION:
tcly - Total Clay, <0.002 mm

MINERAL INTERPRETATION:
KK - Kaolinite   MI - Mica   VR - Vermiculite

RELATIVE PEAK SIZE: 5 Very Large   4 Large   3 Medium   2 Small   1 Very Small   6 No Peaks

*** Primary Characterization Data ***

Pedon ID: 88ME003001
Sampled As   :   Caribou
USDA-NRCS-NSSC-Soil Survey Laboratory

(Aroostook, Maine)
Fine-loamy, mixed, frigid Typic Haplorthod
;   Pedon No. 88P0722

Sand - Silt Mineralogy (2 0-0.002 mm)

		-1-	-2-	-3-	-4-	-5-	-6-	-7-	-8-	-9-	-10-	-11-	-12-	-13-	-14-	-15-	-16-	-17-	-18-
				X-Ray				Thermal			Tot Re			Optical Grain Count 7B1a				EGME Retn	Inter preta tion
Depth	Fract																		
Layer (cm) Horz	ion	(- - - - - - - - peak size - - - - - - - -)						(- - - - - - - % - - - - - - -)			(- - - - - - - - - - - - - - % - - - - - - - - - - - - - - -)							mg g^{-1}	
88P03857 24-39 Bs	vfs										93	QZ 90	OT 5	OP 3	BT 1	FK 1	FP 1		
												CL 1	ZR tr	PR tr	GN tr	HN tr			

FRACTION INTERPRETATION:
vfs - Very Fine Sand, 0.05-0.1 mm

MINERAL INTERPRETATION:

BT - Bio ite	FK - Potassium Feldspar	FP - Plagioclase Feldspar	GN - Garnet
HN - Hornblende	OT - Other	PR - Pyroxene	QZ - Quartz
ZR - Zircon			
CL - Chlorite			
OP - Opaques			

*** Primary Characterization Data ***

Pedon ID: 88ME003001

Sampled As   :   Caribou

USDA-NRCS-NSSC-Soil Survey Laboratory

(Aroostook, Maine)

Fine-loamy, mixed, frigid Typic Haplorthod

;   Pedon No. 88P0722

Fine Earth Mineralogy (<2.0 mm)

Layer	Depth (cm)	Horz	Fraction	-1-	-2-	-3-	-4- X-Ray	-5-	-6-	-7-	-8- Thermal	-9-	-10-	-11-	-12-	-13- Elemental	-14-	-15-	-16-	-17- EGME	-18- Interpretation
													SiO$_2$	Al$_2$O$_3$	Fe$_2$O$_3$	MgO	CaO	K$_2$O	Na$_2$O	Retn 7D2	
				(-------- peak size --------)(------ % ------)									(------------------ % ------------------)							mg g^{-1}	
88P03855	0-11	Ap1	feth																		15.0	

FRACTION INTERPRETATION:

feth - Fine Earth, <2 mm

MINERAL INTERPRETATION:

eg - Surface Area

*** Supplementary Characterization Data ***
(Aroostook, Maine)

Sampled as on Jun 01, 1988 :
Revised to correlated on Sep 01, 1991 :

Caribou ; Fine-loamy, mixed, frigid Typic Haplorthod
Caribou ; Loamy-skeletal, mixed, ac ive, frigid Typic Dystrochrept

SSL
- Project   CP88ME188   WEPP MAINE
- Site ID   88ME003001   Lat 46° 0' 55.00" north   Long: 68° 1' 11.00" west   MLRA: 146
- Pedon No.   88P0722
- General Methods 1B1A, 2A1, 2B

United States Department of Agriculture
Natural Resources Conservation Service
National Soil Survey Center
Soil Survey Laboratory
Lincoln, Nebraska 68508-3866

Tier 1

Columns: -1- through -18- Engineering PSDA Percentage Passing Sieve (Inches: 3, 2, 3/2, 1, 3/4, 3/8; Number: 4, 10, 40, 200; Microns: 20, 5, 2; Millimeter/Percentile: 1., .5, .25, .10, .05). -19- to -21- Cumulative Curve Fractions (<75mm) USDA Less Than Diameters (mm) at Percentile (60, 50, 10). -22- -23- Atterberg (LL, PI %). -24- Uni-fmty CU. -25- Cur-vtur CC.

Layer	Depth (cm)	Horz	Prep	-1-	-2-	-3-	-4-	-5-	-6-	-7-	-8-	-9-	-10-	-11-	-12-	-13-	-14-	-15-	-16-	-17-	-18-	-19-	-20-	-21-	-22-	-23-	-24-	-25-
88P03855	0-11	Ap1	S	100	95	91	86	82	73	64	55	45	33	21	12	7	50	46	42	35	29	3.21	1.044	0.003	32	3	>100	0.3
88P03856	11-24	Ap2	S	100	97	96	93	91	82	72	62	52	38	24	14	8	58	54	48	41	34	1.43	0.309	0.003			>100	0.3
88P03857	24-39	Bs	S	100	88	80	68	59	56	52	47	41	30	19	11	6	45	42	38	32	27	19.65	3.339	0.004			>100	0.1
88P03858	39-66	C1	S	100	99	98	97	96	96	95	91	81	60	36	22	12	88	83	76	65	53	0.08	0.043	0.001	28	2	56.2	1.3
88P03859	66-104	C2	S	100	90	83	73	66	52	37	22	9	7	5	4	2	14	10	8	7	6	14.23	8.791	0.514			27.7	1.4
88P03860	104-143	C3	S	100	94	90	84	80	75	70	62	54	45	31	21	14	59	55	51	46	42	1.33	0.210	0.001			>100	0.2

Tier 2

Columns -26- to -38- Weight Fractions. Whole Soil (mm) % of Whole Soil: -26- >2, -27- 250-UP, -28- 250-75, -29- 75-2, -30- 75-20, -31- 20-5, -32- 5-2, -33- <2. <75 mm Fraction % of <75 mm (3B1): -34- 75-2, -35- 75-20, -36- 20-5, -37- 5-2, -38- <2. Columns -40- to -50- Weight Per Unit Volume (g cm⁻³): Whole Soil (Soil Survey 33 kPa; Engineering Oven-dry, Moist, Saturated — 4A1d), <2 mm Fraction (Soil Survey 33 kPa-dry 4A1d, 1500 kPa 4A1h; Engineering Oven-dry, Moist, Saturated), Ratios At 33 kPa (Whole Soil, <2 Soil), Void (<2 mm).

Layer	Depth (cm)	Horz	Prep	-26-	-27-	-28-	-29-	-30-	-31-	-32-	-33-	-34-	-35-	-36-	-37-	-38-	-40-	-41-	-42-	-43-	-44-	-45-	-46-	-47-	-48-	-49-	-50-
88P03855	0-11	Ap1	S	45	--	--	45	18	18	9	55	45	18	18	9	55	1.82	1.59	1.80	1.96	1.24	1.27	1.29	1.57	1.77	0.72	1.14
88P03856	11-24	Ap2	S	38	--	--	38	9	19	10	62	38	9	19	10	62	1.54	1.94	2.10	2.19	1.47	1.48	1.50	1.77	1.92	0.39	0.80
88P03857	24-39	Bs	S	53	--	--	53	41	7	5	47	53	41	7	5	47	1.91	1.91	2.11	2.17	1.81	1.83	1.84	2.06	2.13	0.41	0.46
88P03858	39-66	C1	S	13	--	4	9	4	1	4	87	9	4	1	4	91	1.88										
88P03859	66-104	C2	S	83	--	23	60	26	22	12	17	78	34	29	15	22	2.32	2.18	2.23		1.63	1.64	1.69	1.94	2.01	0.34	0.63
88P03860	104-143	C3	S	46	--	13	33	17	9	7	54	38	20	10	8	62	1.98										

*** Supplementary Characterization Data ***

Pedon ID: 88ME003001
Sampled As : Caribou
USDA-NRCS-NSSC-Soil Survey Laboratory

(Aroostook, Maine)
Fine-loamy, mixed, frigid Typic Haplorthod

; Pedon No. 88P0722

Tier 3

Columns -51- through -75-

Volume Fractions — Whole Soil (mm) At 33 kPa (% of Whole Soil); Pores; C/N; Ratios To Clay; Linear Extensibility; WRD

Layer	Depth (cm)	Horz	Prep	>2	250-UP	250-75	75-20	20-5	5-2	<2	2-.05	.05-.002	.002 (LT)	Pores D	Pores F	C/N Rat	Fine Clay	Ratio CEC Sum Cats	1500 (8D1)	LEP 33 kPa	LE Whole Soil 33 kPa	LE <2 mm	WRD Whole Soil	WRD <2 mm
88P03855	0-11	Ap1	S	31	--	--	12	12	6	69	18	15	5	31	26	12	0.12	1.78	0.73	0.110	0.6	0.8	0.17	0.21
88P03856	11-24	Ap2	S	22	--	--	5	11	6	78	17	16	4	16	19	11	0.08	1.63	0.71	0.060	0.2	0.2	0.08	0.13
88P03857	24-39	Bs	S	38	--	--	29	5	4	62	15	15	4	9	23		0.11	1.18	0.63	0.050	0.4	0.4	0.14	0.16
88P03858	39-66	C1	S	9	--	3	3	1	1	91	26	27	8	6			0.25	0.66	0.80					
88P03859	66-104	C2	S	73	--	20	23	19	11	27	11	3	2	12	20		0.30	0.83	0.58	0.050	0.2	0.2	0.03	0.05
88P03860	104-143	C3	S	35	--	10	13	7	5	65	13	18	9	5			0.31	0.55	0.52					

Method codes: WRD — 4C1; 1500 — 8D1

Tier 4

Columns -76- through -98-

Weight Fractions — Clay Free (% of Whole Soil / % of >2 mm Sand and Silt); PSDA; pH; Electrical Resistance; Conductivity; Particle Density

Layer	Depth (cm)	Horz	Prep	>2	75-20	20-2	2-.05	.05-.002	.002<	VC	C	M	F	VF	Silt C	Silt F	Clay	Texture by PSDA	PSDA Sand 2-.05	PSDA Silt .05-.002	PSDA Clay .002<	pH CaCl₂ .01M
88P03855	0-11	Ap1	S	48	29	28	24	7	11	8	9	13	13	17	29	14	14	—	47.0	40.8	12.2	5.3
88P03856	11-24	Ap2	S	41	31	30	29	8	8	8	10	13	13	19	30	14	14	—	45.1	42.8	12.1	4.8
88P03857	24-39	Bs	S	56	13	22	22	6	7	7	8	13	13	18	33	14	14	—	43.2	44.5	12.3	4.6
88P03858	39-66	C1	S	15	6	12	44	13	4	6	9	14	15	22	30	15	15		41.8	44.9	13.3	4.5
88P03859	66-104	C2	S	85	35	20	3	2	40	22	10	4	3	6	14	13	13	cosl	71.1	17.6	11.3	4.9
88P03860	104-143	C3	S	53	18	20	28	14	7	8	8	9	8	24	35	30	30	—	31.6	45.2	23.2	6.9

Method codes: PSDA — 3A1; pH — 4C1a2a

Print Date: Aug 10 2010 8:45AM

*** Taxonomy Characterization Data ***
(Aroostook, Maine)

Sampled as on Jun 01, 1988 : Caribou ; Fine-loamy, mixed, frigid Typic Haplorthod
Revised to correlated on Sep 01, 1991 : Caribou ; Loamy-skeletal, mixed, active, frigid Typic Dystrochrept

SSL - Project CP88ME188 WEPP MAINE
- Site ID 88ME003001 Lat: 46° 0' 55.00" north Long: 68° 1' 11.00" west MLRA: 146
- Pedon No. 88P0722
- General Methods 1B1A, 2A1, 2B

United States Department of Agriculture
Natural Resources Conservation Service
National Soil Survey Center
Soil Survey Laboratory
Lincoln, Nebraska 68508-3866

Taxonomy Tier 1

Layer	Depth (cm)	Horz	Prep	-1- Clay <.002	-2- Fine Clay <.0002	-3- CaCO3 Clay <.002	-4- 1500 kPa /Clay	-5- Clay Est	-6- .1-75 mm Frac	-7- Bulk Den 33 kPa	-8- Cole Whole Soil	-9- Vol % of Whole	-10- Resist Min %
				3A1	3A1	6A1c	8D1			4A1d			7B1a
				(------ % of <2 mm ------)	(------ % of <2mm ------)			(---- % ----)	---- % ----	g cm⁻³	cm cm⁻¹		
88P03855	0-11	Ap1	S	12.2	1.5	2.28	0.73	1.11	65			31	
88P03856	11-24	Ap2	S	12.1	1.0	1.80	0.71	1.08	59	1.24	0.010	22	
88P03857	24-39	Bs	S	12.3	1.4	0.93	0.63	1.21	68	1.47	0.004	38	93
88P03858	39-66	C1	S	13.3	3.3	0.16	0.80	0.47	35	1.81	0.005	9	
88P03859	66-104	C2	S	11.3	3.4	0.16	0.58	0.57	93			73	
88P03860	104-143	C3	S	23.2	7.3	0.24	0.52	0.59	54	1.63	0.008	35	

Taxonomy Tier 2

Layer	Depth (cm)	Horz	Prep	-1- pH H2O	-2- pH NaF	-3- Org C	-4- Tot C	-5- Al+½ Fe Oxal	-6- ODOE	-7- CO3 as CaCO3	-8- NH4	-9- Bases	-10- NZ P Ret	-11- ECEC	-12- CEC7 /Clay	-13- ECEC /Clay	-14- Al Sat %	-15- E C dS m⁻¹	-15- ESP %
				4C1a2a		6A1c	6A2d		8J		5C1	5C3		5A3b	8D1		5G1		5D2
						(---- % ----)		%		(--- Base Sat ---)			kg⁻¹	cmol(+) /Clay	/Clay	%	dS m⁻¹	%
88P03855	0-11	Ap1	S	5.6		2.28	1.98	1.11	0.16		60	35		5.6	1.05	0.46	16		1
88P03856	11-24	Ap2	S	5.2		1.80		1.08	0.15		40	24		3.3	0.97	0.27	45		1
88P03857	24-39	Bs	S	4.9		0.93		1.21	0.13		18	12		2.7	0.80	0.20	19		1
88P03858	39-66	C1	S	4.9		0.16		0.47	0.08		37	25		5.0	0.44	0.44	2		2
88P03859	66-104	C2	S	5.4		0.16		0.57	0.06		42	52			1.04				1
88P03860	104-143	C3	S	7.4		0.24		0.59	0.05		100*	80			0.44				

*Extractable Ca may contain Ca from calcium carbonate or gypsum.

Pedon Calculations

Calculation Name	Result	Units of Measure
CEC Activity, CEC7/Clay, Weighted Average	0.78	(NA)
Clay, carbonate free, Weighted Average	12	% wt
Weighted Particles, 0.1-75mm, 75 mm Base	67	% wt
Volume, >2mm, Weighted Average	43	% vol
Clay, total, Weighted Average	12	% wt

Weighted averages based on control section: 25-100 cm

PEDON DESCRIPTION

Print Date: 08/10/2010

Description Date: 6/1/1988

Describer: Lytle, Olson and Grossman

Site ID: 88ME003001

Site Note:

Pedon ID: 88ME003001

Pedon Note: Complete characterization profile. Water Erosion Prediction Project.

Lab Source ID: SSL

Lab Pedon #: 88P0722

Soil Name as Described/Sampled: Caribou

Soil Name as Correlated: Caribou

Classification: Loamy-skeletal, mixed, active, frigid Typic Dystrochrepts

Pedon Type:

Pedon Purpose: full pedon description

Taxon Kind: taxadjunct

Associated Soils:

Physiographic Division:

Physiographic Province:

Physiographic Section:

State Physiographic Area:

Local Physiographic Area:

Geomorphic Setting: on backslope of side slope of upland

Upslope Shape: convex

Cross Slope Shape: convex

Particle Size Control Section: 25 to 100 cm.

Description origin: Converted from SSL-CMS data

Diagnostic Features: ochric epipedon 0 to 24 cm.
spodic horizon 24 to 39 cm.

Country:

State: Maine

County: Aroostook

MLRA: 146 -- Aroostook Area

Soil Survey Area:

Map Unit:

Quad Name:

Location Description: 2 mi N of Presque Isle on Rt. 1, 1.1 mi E on Rt. 210, 2300 feet N on field Rd, 100 feet E.

Legal Description:

Latitude: 46 degrees 0 minutes 55.00 seconds north

Longitude: 68 degrees 1 minutes 11.00 seconds west

Datum:

UTM Zone:

UTM Easting:

UTM Northing:

Primary Earth Cover: Crop cover

Secondary Earth Cover:

Existing Vegetation:

Parent Material:

Bedrock Kind:

Bedrock Depth:

Bedrock Hardness:

Bedrock Fracture Interval:

Surface Fragments:

Description database: NSSL

Slope (%)	Elevation (meters)	Aspect (deg)	MAAT (C)	MSAT (C)	MWAT (C)	MAP (mm)	Frost-Free Days	Drainage Class	Slope Length (meters)	Upslope Length (meters)
6.0	190.0	0	6.0			102		well		

1A p1--0 to 11 centimeters; dark yellowish brown (10YR 4/4) interior gravelly loam, pale brown (10YR 6/3) interior, dry; moderate medium granular, and moderate medium subangular blocky structure; very friable, slightly hard, slightly sticky, slightly plastic; common very fine and fine roots; common very fine and fine interstitial pores; 30 percent 2- to 75-millimeter mixed rock fragments; clear wavy boundary. Lab sample # 88P3855. common very fine and fine roots

1A p2--11 to 24 centimeters; dark yellowish brown (10YR 4/4) interior gravelly loam; moderate medium subangular blocky structure; friable, slightly sticky, slightly plastic; common very fine and fine roots; very fine and fine interstitial pores; 30 percent 2- to 75-millimeter mixed rock fragments; abrupt smooth boundary. Lab sample # 88P3856. common very fine and fine roots

1B s--24 to 39 centimeters; dark yellowish brown (10YR 4/6) interior gravelly fine sandy loam; moderate medium and coarse subangular blocky structure; firm, slightly sticky, nonplastic; few very fine and fine roots; few very fine and fine interstitial and few very fine and fine tubular pores; 20 percent 2- to 75-millimeter mixed rock fragments; abrupt wavy boundary. Lab sample # 88P3857. Pockets of (10YR 6/2) fine sandy loam E horzion in a few areas.; few very fine and fine roots

1C--39 to 66 centimeters; dark grayish brown (10YR 4/2) interior gravelly fine sandy loam; massive; friable, nonsticky, nonplastic; very few very fine roots; few very fine and fine tubular pores; 3 percent 75- to 250-millimeter mixed rock fragments and 30 percent 2- to 75-millimeter mixed rock fragments; abrupt wavy boundary. Lab sample # 88P3858. Soil temperature 60 degrees F at 50 cm.; very few very fine roots

2C 2--66 to 104 centimeters; brown (10YR 4/3) interior extremely gravelly sand; single grain; loose, nonsticky, nonplastic; 20 percent 75- to 250-millimeter mixed rock fragments and 50 percent 2- to 75-millimeter mixed rock fragments; gradual wavy boundary. Lab sample # 88P3859

3C 3--104 to 143 centimeters; dark yellowish brown (10YR 4/6) interior and brown (10YR 4/3) interior very gravelly silt loam; massive; firm, moderately sticky, slightly plastic; 10 percent 75- to 250-millimeter mixed rock fragments and 30 percent 2- to 75-millimeter mixed rock fragments. Lab sample # 88P3860

MATERIALS TESTING REPORT:
SOIL CLASSIFICATION

Project and state **WEPP Maine**　　Sample location **Aroostook County**　　Laboratory No. ____

Field sample No. ____　Depth ____　Geologic origin ____

Type of sample ____　Tested at ____　Approved by ____　Date ____

Symbol ____　Description **Caribou, Pedon 88722, Ap2 Horizon, 11-24 cm**

Grain Size Distribution

Percent finer by dry weight

Grain size in millimeters

Remarks: Cumulative Particle-Size Distribution Curve, <75-mm base.

ESTIMATED SOIL WATER RETENTION CURVE

<2 mm Fraction

Pedon ID: 88ME003001
Layer Natural Key: 88P03856
Horizon: Ap2   11.0–24.0 cm

van Genuchten Retention Parameters
from Rosetta*

θ_r	θ_s	$\log_{10}(\alpha)$	$\log_{10}(n)$
0.0449	0.4301	-2.488	0.2233

Measured Soil Properties

	Rock Fragmer	Organic Carbon	Sand	Silt	Clay	Db_{33}	θ_{33}	θ_{1500}
	% vol	% wt	% wt	% wt	% wt	g/cc	cm^3/cm^3	cm^3/cm^3
	22	1.80	45.1	42.8	12.1	1.24	0.33	0.11

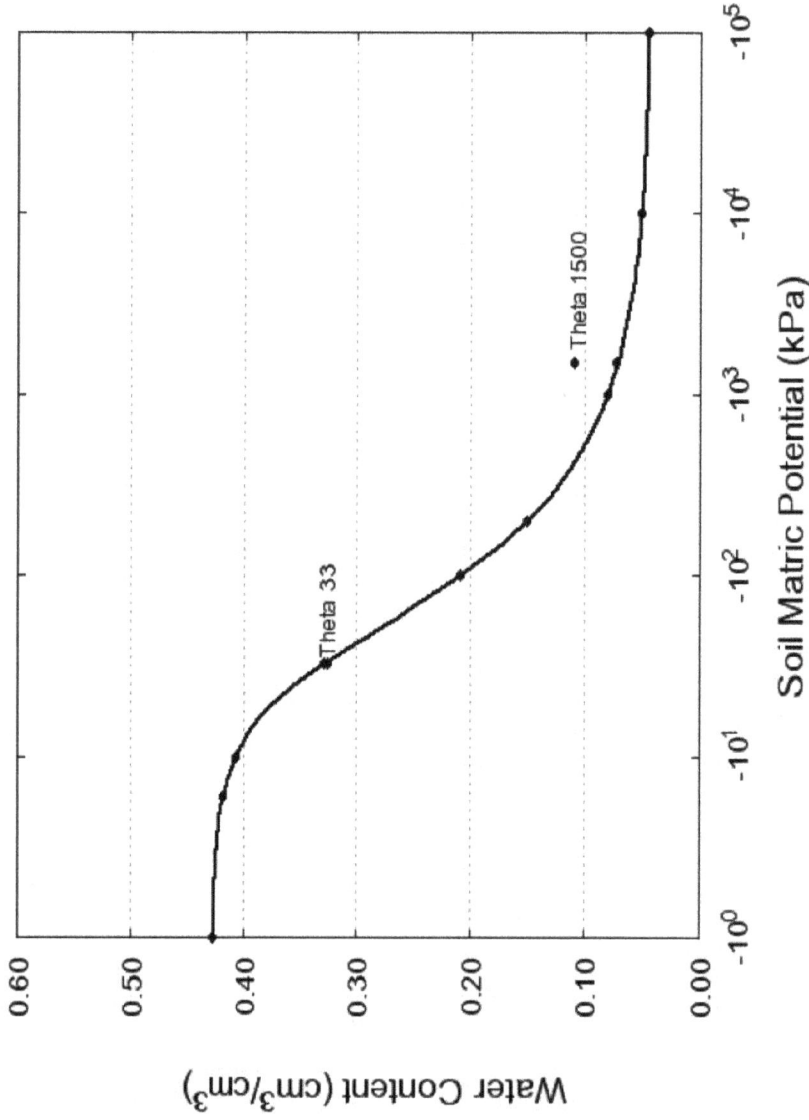

Soil Matric Potential (kPa)

Water Content (cm^3/cm^3)

Theta 33

Theta 1500

*Rosetta Reference

ESTIMATED SOIL WATER RETENTION CURVE

Pedon ID: 88ME003001
Layer Natural Key: 88P03857
Horizon: Bs   24.0-39.0 cm

<2 mm Fraction

van Genuchten Retention Parameters
from Rosetta[*]

θ_r	θ_s	$log_{10}(\alpha)$	$log_{10}(n)$
0.0434	0.3855	-2.3867	0.1948

Measured Soil Properties

Rock Fragmen	Organic Carbon	Sand	Silt	Clay	Db_{33}	θ_{33}	θ_{1500}
% vol	% wt	% wt	% wt	% wt	g/cc	cm^3/cm^3	cm^3/cm^3
38	0.93	43.2	44.5	12.3	1.47	0.3	0.11

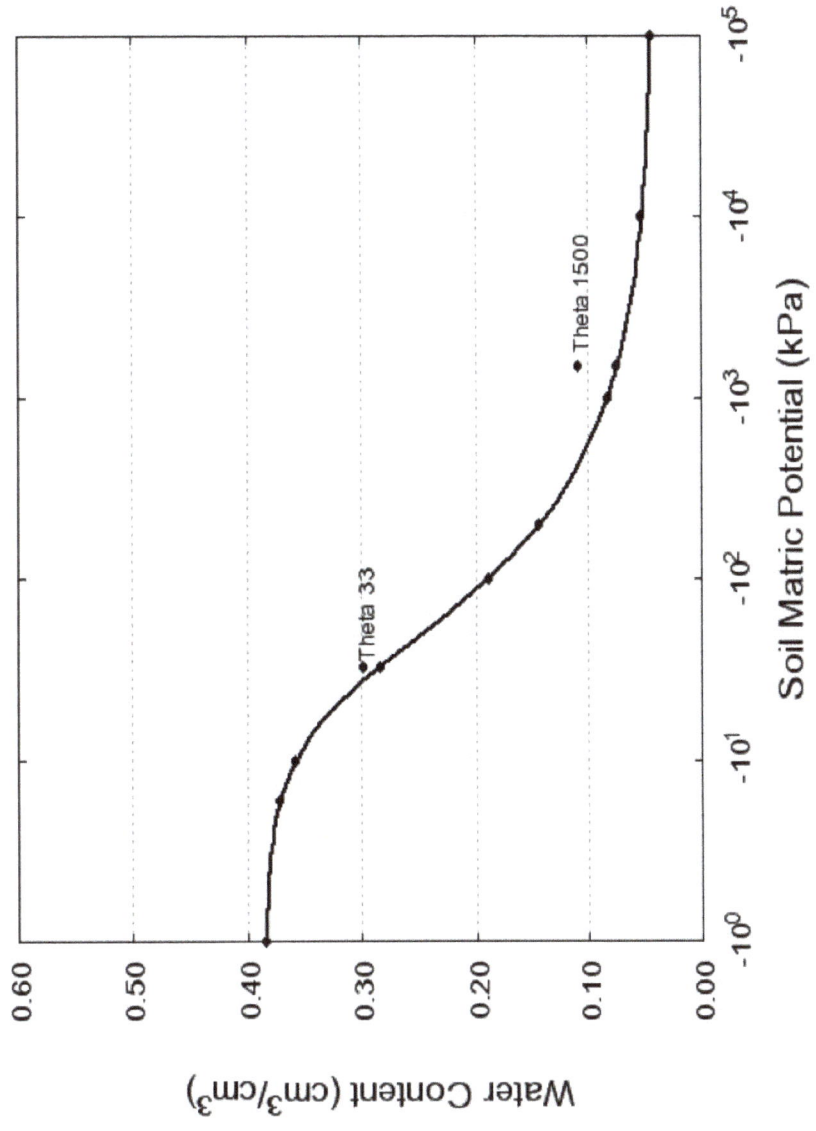

[*]Rosetta Reference

ESTIMATED SOIL WATER RETENTION CURVE

<2 mm Fraction

Pedon ID: 88ME003001
Layer Natural Key: 88P03858
Horizon: C1   39.0-66.0 cm

van Genuchten Retention Parameters
from Rosetta[*]

θ_r	θ_s	$\log_{10}(\alpha)$	$\log_{10}(n)$
0.0705	0.3146	-1.3157	0.0948

Measured Soil Properties

Rock Fragmer	Organic Carbon	Sand	Silt	Clay	Db_{33}	θ_{33}	θ_{1500}
% vol	% wt	% wt	% wt	% wt	g/cc	cm^3/cm^3	cm^3/cm^3
9	0.16	41.8	44.9	13.3	1.81	0.25	0.19

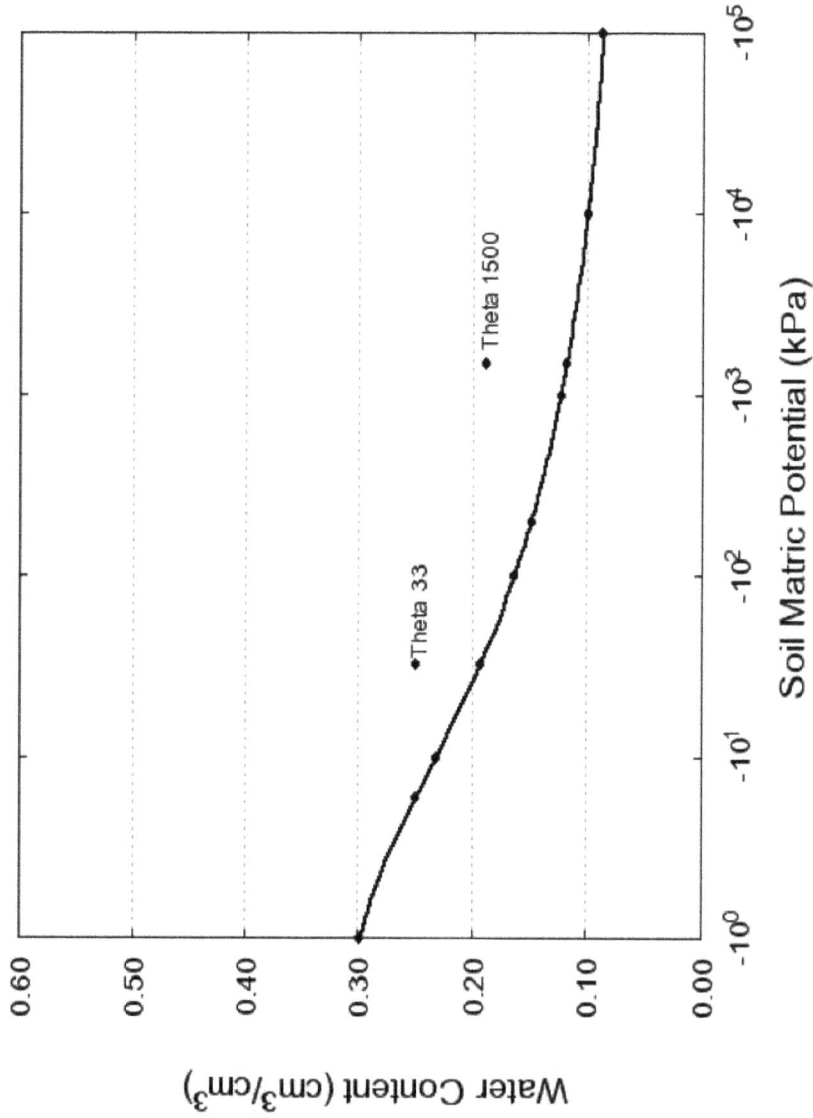

Soil Matric Potential (kPa)

Water Content (cm^3/cm^3)

Theta 33
Theta 1500

[*]Rosetta Reference

ESTIMATED SOIL WATER RETENTION CURVE

<2 mm Fraction

Pedon ID: 88ME003001
Layer Natural Key: 88P03860
Horizon: C3   104.0-143.0 cm

Measured Soil Properties

	Rock Fragments	Organic Carbon	Sand	Silt	Clay	Db_{33}	θ_{33}	θ_{1500}
	% vol	% wt	% wt	% wt	% wt	g/cc	cm^3/cm^3	cm^3/cm^3
	35	0.24	31.6	45.2	23.2	1.63	0.31	0.2

van Genuchten Retention Parameters
from Rosetta[*]

θ_r	θ_s	$\log_{10}(\alpha)$	$\log_{10}(n)$
0.0589	0.3798	-1.8472	0.0952

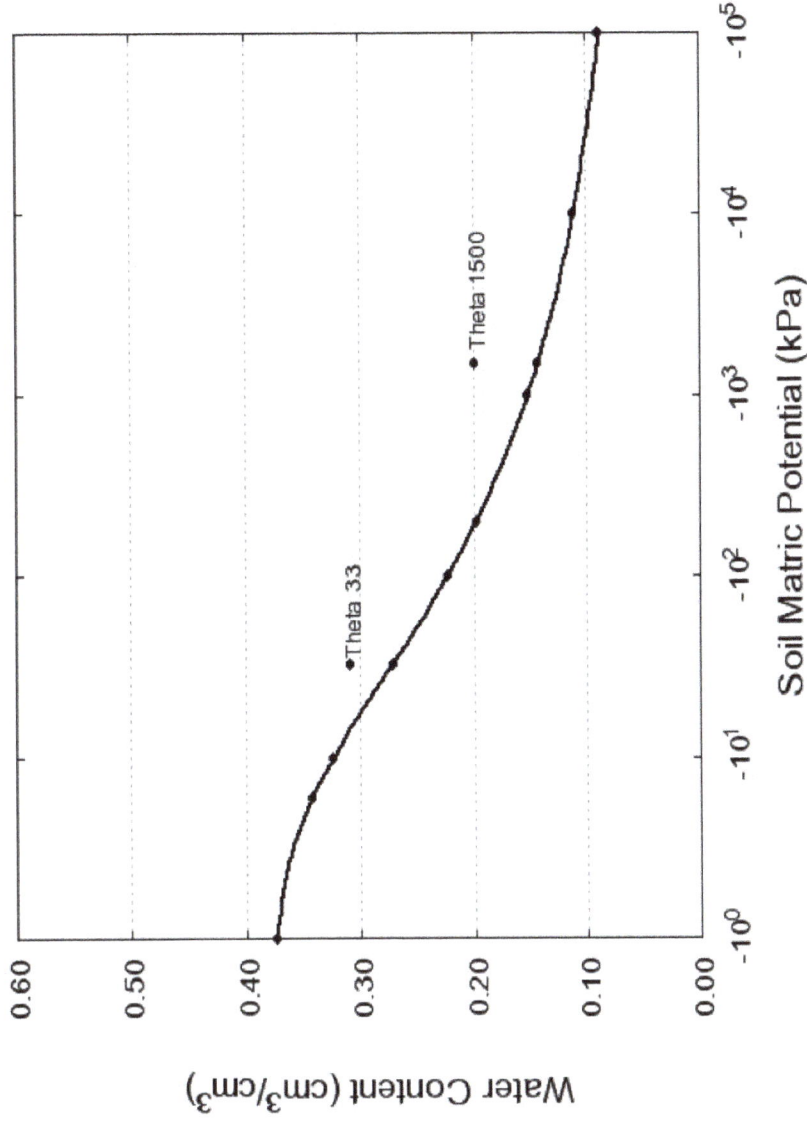

[*]Rosetta Reference

6 Guide for Textural Classification

Guide for Textural Classification in Soil Families

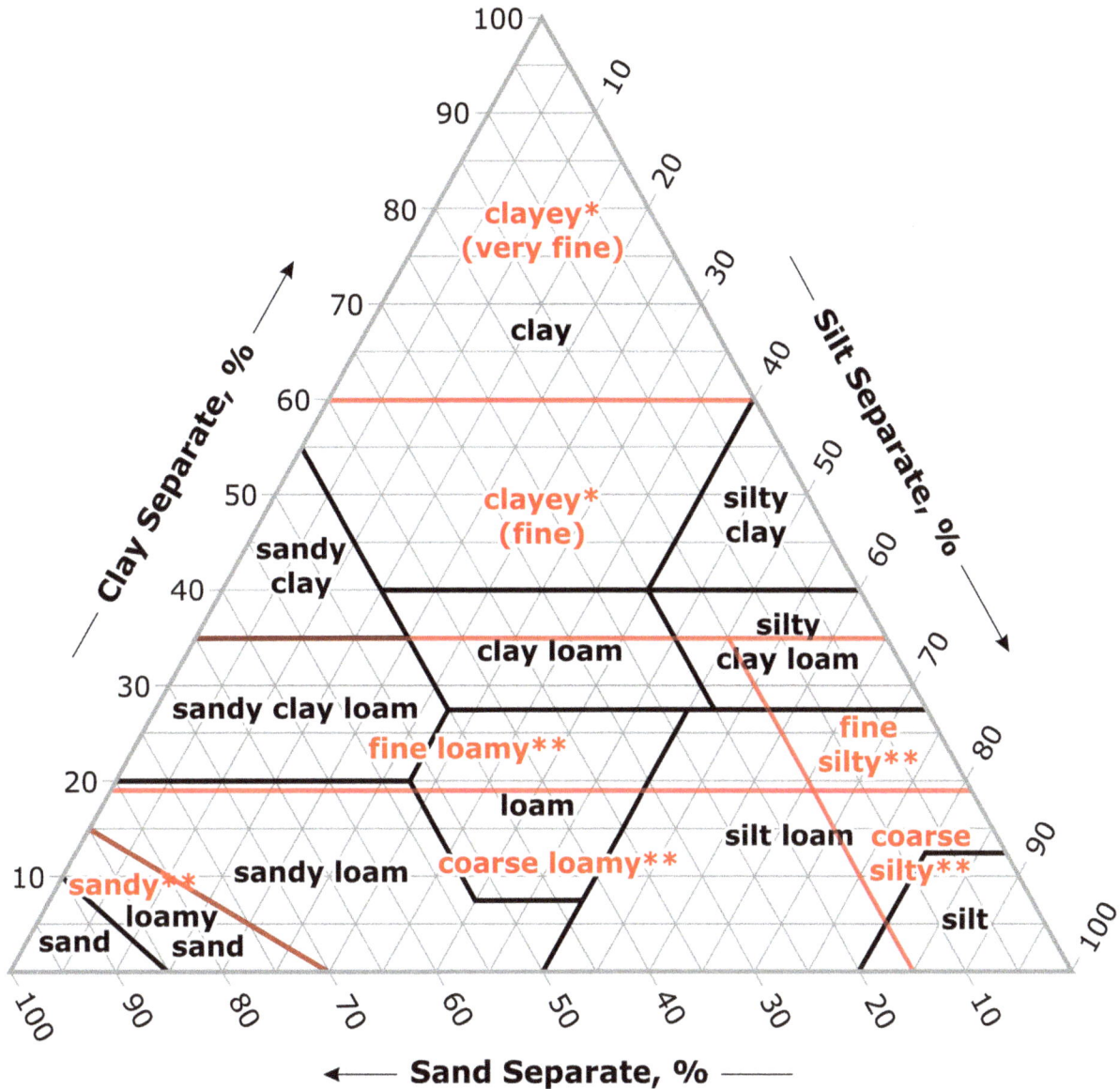

* Clay-size carbonate is treated as silt.
** Very fine sand (0.05 - 0.1) is treated as silt for family groupings; coarse fragments are considered the equivalent of coarse sand in the boundary between the silty and loamy classes.

Comparison of Particle Size Scales

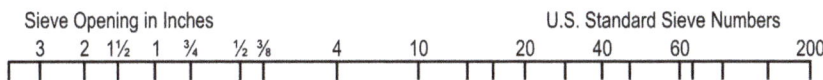

Sieve Opening in Inches

U.S. Standard Sieve Numbers

USDA	GRAVEL			SAND					SILT		CLAY
	Coarse	Medium	Fine	Very Coarse	Coarse	Medium	Fine	Very Fine	Coarse	Fine	

UNIFIED	GRAVEL		SAND			SILT OR CLAY	
	Coarse	Fine	Coarse	Medium	Fine		

AASTHO	GRAVEL OR STONE			SAND		SILT - CLAY	
	Coarse	Medium	Fine	Coarse	Fine	Silt	Clay

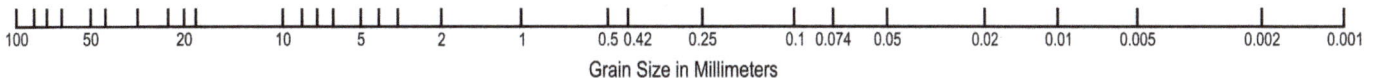

Grain Size in Millimeters

7   Newport

Pedon ID: S08RI005001

Print Date: Sep 22 2010 1:55PM

*** Primary Characterization Data ***
(Newport, Rhode Island)

Sampled as : Newport ; Coarse-loamy, mixed, active, mesic Typic Dystrudept
Revised to :

United States Department of Agriculture
Natural Resources Conservation Service
National Soil Survey Center
Soil Survey Laboratory
Lincoln, Nebraska 68508-3866

SSL - Project   C2008USRI121 Newport & Washington Co.s
     - Site ID   S08RI005-001 Lat: 41° 31' 37.69" north Long: 71° 22' 3.76" west NAD83 MLRA: 144A
     - Pedon No.   08N0528
     - General Methods 1B1A, 2A1, 2B

Layer	Horizon	Orig Hzn	Depth (cm)	Field Label 1	Field Label 2	Field Label 3	Field Texture	Lab Texture
08N03442	Ap		0-20	S08RI005-001-1			FSL	FSL
08N03443	Bw		20-53	S08RI005-001-2			CN-FSL	FSL
08N03444	2Cd1		53-87	S08RI005-001-3			SIL	FSL
08N03445	2Cd2		87-121	S08RI005-001-4			SIL	L
08N03446	2Cd3		121-200	S08RI005-001-5			SIL	L

Calculation Name	Result	Units of Measure
Pedon Calculations		
Weighted Particles, 0.1-75mm, 75 mm Base	50	% wt
Volume, >2mm, Weighted Average	7	% vol
Clay, total, Weighted Average	5	% wt
Clay, carbonate free, Weighted Average	5	% wt
CEC Activity, CEC7/Clay, Weighted Average, CECd, Set 1	0.71	(NA)
LE, Whole Soil, Summed to 1m	1	cm/m

Weighted averages based on control section: 0-53 cm

PSDA & Rock Fragments

			-1-	-2-	-3-	-4-	-5-	-6-	-7-	-8-	-9-	-10-	-11-	-12-	-13-	-14-	-15-	-16-	-17-
			(- - - Total - - - -)			(- - Clay - -)		(- - - Silt - - - -)		(- - - - - - - - - - - Sand - - - - - - - - - - -)					(Rock Fragments (mm))				>2 mm
			Clay	Silt	Sand	Fine	CO₃	Fine	Coarse	VF	F	M	C	VC		(- - - - - - - - Weight - - - - - - - -)			wt %
	Depth (cm)		< .002	.002 -.05	.05 -2	< .0002	< .002	.002 -.02	.02 -.05	.05 -.10	.10 -.25	.25 -.50	.5 -1	1 -2	2 -5	5 -20	20 -75	.1- 75	whole soil
Layer		Horz	Prep																	
			(- % of <2mm Mineral Soil - - - - - - - - - - - - - - - -)												(- - - % of <75mm - - - -)			
			3A1a1a					3A1a1a		3A1a1a	3A1a1a	3A1a1a	3A1a1a	3A1a1a					
08N03442	0-20	Ap	S	7.9	39.7	52.4			17.7	22.0	16.5	19.2	10.4	4.6	1.7	3	7	2	2	44	12
08N03443	20-53	Bw	S	4.0	32.0	64.0			12.5	19.5	17.5	27.6	12.6	4.3	2.0	4	7	2	3	53	13
08N03444	53-87	2Cd1	S	7.5	38.4	54.1			20.6	17.8	16.1	20.3	10.2	4.7	2.8	5	5	1	4	45	11
08N03445	87-121	2Cd2	S	7.0	41.4	51.6			22.3	19.1	16.1	17.9	10.1	4.7	2.8	6	7	2	5	45	15
08N03446	121-200	2Cd3	S	9.0	46.1	44.9			28.0	18.1	12.1	13.3	8.7	6.6	4.2	7	12	7	7	50	26

*** Primary Characterization Data ***

(Newport, Rhode Island)

Coarse-loamy, mixed, active, mesic Typic Dystrudept

; Pedon No. 08N0528

Bulk Density & Moisture

Layer	Depth (cm)	Horz	Prep	-1- (Bulk Density) 33 kPa DbWR1	-2- Oven Dry DbWR1	-3- Cole Whole Soil	-4- 6 kPa	-5- 10 kPa	-6- 33 kPa	-7- 1500 kPa Moist	-8- 1500 kPa	-9- 1500 kPa Ratio AD/OD 3D1	-10- WRD Whole Soil	-11- Aggst Stabl 2-0.5mm	-12- CEC7	-13- 1500 kPa
				(- - g cm⁻³ - -) DbWR1		Whole Soil	(- - - - - - % of <2mm - - - - - -) DbWR1 3C2a1a						cm³ cm⁻³	%	(- - Ratio/Clay - -)	
08N03442	0-20	Ap	S	1.36	1.43	0.016			19.4	6.1		1.013	0.17		0.76	0.77
08N03443	20-53	Bw	S	1.50	1.53	0.006			14.0	3.6		1.007	0.14		0.68	0.90
08N03444	53-87	2Cd1	S	1.77	1.82	0.009			13.4	3.4		1.006	0.16		0.35	0.45
08N03445	87-121	2Cd2	S	1.84	1.86	0.003			12.5	3.6		1.006	0.15		0.50	0.51
08N03446	121-200	2Cd3	S	1.76	1.79	0.005			15.8	4.5		1.007	0.16		0.41	0.50

Carbon & Extractions

Layer	Depth (cm)	Horz	Prep	-1- C	-2- N	-3- S	-4- Org C	-5- C/N Ratio	-6- Fe	-7- Al	-8- Mn	-9- Al+½Fe	-10- ODOE	-11- Fe	-12- Al	-13- Si	-14- Mn	-15- C	-16- Fe	-17- Al	-18- Mn
				(- - Total - - -) % of <2mm 4H2a					(- - Dith-Cit Ext - -) 4G1			(- - - - Ammonium Oxalate Extraction - - - -) % of < 2mm 4G2a					mg kg⁻¹ 4G2a	(- - - Na Pyro-Phosphate - - -) % of < 2mm			
08N03442	0-20	Ap	S	1.92	0.13	0.01		15	1.1	0.4	tr	0.79	0.09	0.34	0.62	tr	151.6				
08N03443	20-53	Bw	S	0.64	0.07	—		9	0.8	0.3	tr	0.64	0.04	0.26	0.51	0.01	48.7				
08N03444	53-87	2Cd1	S	0.88	0.01	0.01		68	0.9	0.2	tr	0.43	0.02	0.16	0.35	0.01	43.5				
08N03445	87-121	2Cd2	S	1.05	0.06	tr		17	1.0	0.2	tr	0.48	0.02	0.16	0.40	0.01	68.0				
08N03446	121-200	2Cd3	S	1.23	0.05	tr		26	1.4	0.2	tr	0.39	0.01	0.21	0.29	tr	29.7				

CEC & Bases

Layer	Depth (cm)	Horz	Prep	-1- Ca	-2- Mg	-3- Na	-4- K	-5- Sum Bases	-6- Acidity	-7- Extr Al	-8- KCl Mn	-9- CEC8 Sum Cats	-10- CEC7 NH₄ OAC	-11- ECEC Bases +Al	-12- Al Sat	-13- Sum	-14- NH₄OAC
				(- - - - - NH₄OAC Extractable Bases - - - - -) cmol(+) kg⁻¹ 4B1a1a					(- - cmol(+) kg⁻¹ - -) 4B2b1a1 4B3a1a		mg kg⁻¹ 4B3a1a	(- - - cmol(+) kg⁻¹ - - -) 4B1a1a			(- - - Base Saturation - - -) %		
08N03442	0-20	Ap	S	0.8	0.3	--	tr	1.1	12.2	1.1	0.4	13.3	6.0	2.2	50	8	18
08N03443	20-53	Bw	S	0.2	tr	--	—	0.2	5.7	0.7	0.1	5.9	2.7	0.9	78	3	7
08N03444	53-87	2Cd1	S	0.4	tr	--	—	0.4	4.0	0.8	tr	4.4	2.6	1.2	67	9	15
08N03445	87-121	2Cd2	S	0.4	tr	--	—	0.4	4.3	0.7	tr	4.7	3.5	1.1	64	9	11
08N03446	121-200	2Cd3	S	1.0	0.3	--	—	1.3	4.5	0.8	tr	5.8	3.7	2.1	38	22	35

Pedon ID: S08RI005001

Sampled As   :   Newport

USDA-NRCS-NSSC-Soil Survey Laboratory

Print Date: Sep 22 2010 1:55PM

*** Primary Characterization Data ***
(Newport, Rhode Island)
Coarse-loamy, mixed, active, mesic Typic Dystrudept
;   Pedon No. 08N0528

Salt

Water Extracted From Saturated Paste

			-1-	-2-	-3-	-4-	-5-	-6-	-7-	-8-	-9-	-10-	-11-	-12-	-13-	-14-	-15-	-16-	-17-	-18-	-19-	-20-
			Ca	Mg	Na	K	CO$_3$	HCO$_3$	F	Cl	PO$_4$	Br	OAC	SO$_4$	NO$_2$	NO$_3$	H$_2$O	Total Salts	Elec Cond	Pred Elec Cond	Exch Na	SAR
			(--- mmol(+) L^{-1} ---)				(--- mmol(-) L^{-1} ---)											(--- % ---)	(-- dS m^{-1} --)		%	
Layer	Depth (cm)	Horz	Prep																			
08N03442	0-20	Ap	S																		--	
08N03443	20-53	Bw	S																		--	
08N03444	53-87	2Cd1	S																		--	
08N03445	87-121	2Cd2	S																		--	
08N03446	121-200	2Cd3	S																		--	

pH & Carbonates

			-1-	-2-	-3-	-4-	-5-	-6-	-7-	-8-	-9-	-10-	-11-
				CaCl$_2$		Sat			(-- Carbonate --) As CaCO$_3$		(-- Gypsum --) As CaSO$_4$*2H$_2$O		Resist
			KCl	0.01M	H$_2$O	Paste	Sulf	NaF	<2mm	<20mm	<2mm	<20mm	ohms cm^{-1}
				1:2	1:1				(------ % ------)				
Layer	Depth (cm)	Horz	Prep	4C1a2a	4C1a2a								
08N03442	0-20	Ap	S		4.6	5.0								
08N03444	53-87	2Cd1	S		4.7	5.7								
08N03445	87-121	2Cd2	S		4.9	5.7								
08N03446	121-200	2Cd3	S		4.7	5.7								

Phosphorous

			-1-	-2-	-3-	-4-	-5-	-6-	-7-	-8-	-9-	-10-
			Melanic Index	NZ	Acid Oxal	Bray 1	Bray 2	Olsen	H$_2$O	Citric Acid	Mehlich III	Extr NO$_3$
			%		(--------------- mg kg^{-1} --------------- mg kg^{-1})							
Layer	Depth (cm)	Horz	Prep			4G2a	4D3						
08N03442	0-20	Ap	S		--		56.5						
08N03443	20-53	Bw	S		--		84.3						
08N03444	53-87	2Cd1	S		--		75.0						
08N03445	87-121	2Cd2	S		--		134.1						
08N03446	121-200	2Cd3	S		--		62.7						

*** Primary Characterization Data ***

(Newport, Rhode Island)

Coarse-loamy, mixed, active, mesic Typic Dystrudept

; Pedon No. 08N0528

Trace Elements Tier 1

Layer	Depth (cm)	Horz	Prep	-1- Ag mg/kg 4H1a	-2- As mg/kg 4H1a	-3- Ba mg/kg 4H1a	-4- Be mg/kg 4H1a	-5- Cd mg/kg 4H1a	-6- Co mg/kg 4H1a	-7- Cr mg/kg 4H1a	-8- Cu mg/kg 4H1a	-9- Mn mg/kg 4H1a	-10- Mo mg/kg 4H1a	-11-	-12- Hg ug/kg 4H1a
08N03442	0-20	Ap	HM	0.11	3.67	71.12	0.79	0.10	5.62	18.87	9.39	471.51	0.64		42
08N03443	20-53	Bw	HM	0.05	3.28	43.54	0.51	0.05	5.65	17.27	8.94	282.61	0.46		27
08N03444	53-87	2Cd1	HM	--	2.87	70.00	0.68	0.04	8.41	21.38	16.82	306.19	0.63		8
08N03445	87-121	2Cd2	HM	--	2.90	65.37	0.71	0.03	9.27	21.92	22.31	289.21	0.74		5
08N03446	121-200	2Cd3	HM	--	2.62	39.66	0.63	0.05	11.59	24.83	22.08	300.94	0.86		7

Trace Elements Tier 2

Layer	Depth (cm)	Horz	Prep	-1- Ni mg/kg 4H1a	-2- P mg/kg 4H1a	-3- Pb mg/kg 4H1a	-4- Sb mg/kg 4H1a	-5- Se ug/kg 4H1a	-6- Sn mg/kg 4H1a	-7- Sr mg/kg 4H1a	-8- Tl mg/kg	-9- V mg/kg 4H1a	-10- W mg/kg 4H1a	-11- Zn mg/kg 4H1a
08N03442	0-20	Ap	HM	21.33	698.29	16.22	0.14	896.96	1.14	20.07		32.10	0.12	46.52
08N03443	20-53	Bw	HM	20.69	444.76	7.05	0.07	624.03	0.65	14.61		26.39	0.11	32.74
08N03444	53-87	2Cd1	HM	34.80	405.15	8.96	0.09	445.51	0.53	19.72		30.04	0.14	43.07
08N03445	87-121	2Cd2	HM	26.62	548.85	10.09	0.08	447.21	0.47	17.77		29.38	0.13	46.84
08N03446	121-200	2Cd3	HM	48.35	568.10	11.26	0.06	416.65	0.28	12.61		29.13	0.13	59.31

Major Elements

Layer	Depth (cm)	Horz	Prep	-1- Al mg/kg 4H1b	-2- Ca mg/kg 4H1b	-3- Fe mg/kg 4H1b	-4- K mg/kg 4H1b	-5- Mg mg/kg 4H1b	-6- Mn mg/kg 4H1b	-7- Na mg/kg 4H1b	-8- P mg/kg 4H1b	-9- Si mg/kg 4H1b	-10- Sr mg/kg 4H1b	-11- Ti mg/kg 4H1b	-12- Zr mg/kg 4H1b
08N03442	0-20	Ap	HM	52242	5370	26293	14766	3468	711	9200	1045	324625	93	5531	111
08N03443	20-53	Bw	HM	47936	6042	26160	15220	3489	606	10517	660	330610	98	5354	121
08N03444	53-87	2Cd1	HM	46854	3842	34379	16934	4799	532	7982	369	275141	79	6670	116
08N03445	87-121	2Cd2	HM	58459	3910	35013	17250	5544	539	6999	634	316182	89	6371	88
08N03446	121-200	2Cd3	HM	56757	2525	43454	17620	5601	555	5210	583	298033	60	7801	136

Pedon ID: S08RI005001

Sampled As : Newport

USDA-NRCS-NSSC-Soil Survey Laboratory

*** Primary Characterization Data ***

(Newport, Rhode Island)

Coarse-loamy, mixed, active, mesic Typic Dystrudept

; Pedon No. 08N0528

Clay Mineralogy (<.002 mm)

			-1-	-2-	-3-	-4-	-5-	-6-	-7-	-8-	-9-	-10-	-11-	-12-	-13-	-14-	-15-	-16-	-17-	-18-
					X-Ray						Thermal				Elemental				EGME Retn	Interpretation
													SiO$_2$	Al$_2$O$_3$	Fe$_2$O$_3$	MgO	CaO	K$_2$O	Na$_2$O	mg g^{-1}	
	Depth				7A1a1															
Layer	(cm)	Horz	Fract ion		(- - - - - - - peak size - - - - - - - - - - -)						(- - - - - - - % - - - - - - -)		(- - - - - - - - - - - - - - % - - - - - - - - - - - - -)								
08N03442	0-20	Ap	tcly	CL 2	KK 1	GI 1															CMIX
08N03444	53-87	2Cd1	tcly	KK 4	MI 3	CL 3															CMIX
08N03446	121-200	2Cd3	tcly	KK 4	MI 3	CL 3															CMIX

FRACTION INTERPRETATION:

tcly - Total Clay, <0.002 mm

MINERAL INTERPRETATION:

CL - Chlorite   GI - Gibbsite   KK - Kaolinite   MI - Mica

RELATIVE PEAK SIZE:

5 Very Large   4 Large   3 Medium   2 Small   1 Very Small   6 No Peaks

*** Primary Characterization Data ***

(Newport, Rhode Island)
Coarse-loamy, mixed, active, mesic Typic Dystrudept

Pedon ID: S08RI005001
Sampled As   :   Newport
USDA-NRCS-NSSC-Soil Survey Laboratory

;   Pedon No. 08N0528

Sand - Silt Mineralogy (2.0-0.002 mm)

Layer	Depth (cm)	Horz	Fraction	-1- X-Ray	-2-	-3-	-4-	-5-	-6- Thermal	-7-	-8-	-9-	-10- Tot Re	-11-	-12-	-13- Optical Grain Count 7B1a2	-14-	-15-	-16-	-17- EGME Retn mg g⁻¹	-18- Interpretation
				(- - - - - - - - peak size - - - - - - - -)					(- - - - - % - - - - -)					(- - - - - - - - - - - - - - - - - - % - - - - - - - - - - - - - - - - - -)						
08N03442	0-20	Ap	csi										68	QZ 65 AR 1 GN tr	FK 12 HN 1 FE tr	MS 11 OP 1 BY tr	BT 4 PO tr	PR 3 RU tr	CD 2 ZR tr		SMIX
08N03443	20-53	Bw	fs										75	QZ 69 OP 2	FK 17 PR 1	AR 3 MS 1	BT 3 FP 1	CD 2 GN tr	FE 2 HN tr		SMIX

FRACTION INTERPRETATION:

csi - Coarse Silt, 0.02-0.05 mm   fs - Fine Sand, 0.1-0.25 mm

MINERAL INTERPRETATION:

AR - Weatherable Aggregates	BT - Biotite	BY - Beryl	CD - Chert (Chalcedony	FE - Iron Oxides (Goethite
FK - Potassium Feldspar	FP - Plagioclase Feldspar	GN - Garnet	HN - Hornblende	MS - Muscovite
OP - Opaques	PO - Plant Opal	PR - Pyroxene	QZ - Quartz	RU - Rutile
ZR - Zircon				

INTERPRETATION (BY HORIZON):

SMIX - Mixed Sand

*** Supplementary Characterization Data ***
(Newport, Rhode Island)

Sampled as : Newport ; Coarse-loamy, mixed, active, mesic Typic Dystrudept
Revised to :

SSL — Project C2008USRI121 Newport & Washington Co.s
— Site ID S08RI005-001 Lat: 41° 31' 37.69" north Long: 71° 22' 3.76" west NAD83 MLRA: 144A
— Pedon No. 08N0528
— General Methods 1B1A, 2A1, 2B

United States Department of Agriculture
Natural Resources Conservation Service
National Soil Survey Center
Soil Survey Laboratory
Lincoln, Nebraska 68508-3866

Tier 1

Engineering PSDA — Percentage Passing Sieve (columns -1- to -13-); Cumulative Curve Fractions (<75mm) USDA Less Than Diameters (mm) at percentile (-14- to -21-); Atterberg (-22-, -23-); Gradation (-24-, -25-)

Layer	Depth (cm)	Horz	Prep	-1- 3	-2- 2	-3- 3/2	-4- 1	-5- 3/4	-6- 3/8	-7- 4	-8- 10	-9- 40	-10- 200	-11- 20	-12- 5	-13- 2	-14- 1.	-15- .5	-16- .25	-17- .10	-18- .05	-19- 60	-20- 50	-21- 10	-22- LL	-23- PI	-24- CU	-25- CC
08N03442	0-20	Ap	S	100	99	99	98	98	95	91	88	80	50	23	13	7	87	82	73	56	42	0.12	0.074	0.003			38.7	2.1
08N03443	20-53	Bw	S	100	99	99	98	98	95	91	87	79	40	14	8	3	85	82	71	47	31	0.17	0.114	0.008			21.0	1.6
08N03444	53-87	2Cd1	S	100	100	100	99	99	97	94	89	80	49	25	14	7	87	82	73	55	41	0.13	0.078	0.003			42.1	1.8
08N03445	87-121	2Cd2	S	100	100	99	99	98	95	91	85	76	49	25	13	6	83	79	70	55	41	0.14	0.078	0.003			41.7	1.6
08N03446	121-200	2Cd3	S	100	98	97	94	93	87	81	74	64	46	27	15	7	71	66	60	50	41	0.26	0.103	0.003			90.3	0.8

Tier 2

Weight Fractions — Whole Soil (mm) % of Whole Soil (-26- to -32-); <75 mm Fraction % of <75 mm (-33- to -38-)

Layer	Depth (cm)	Horz	Prep	-26- >2	-27- 250-UP	-28- 250-75	-29- 75-20	-30- 20-5	-31- 5-2	-32- <2	-33- 75-2	-34- 75-20	-35- 20-5	-36- 5-2	-37- <2	-38-
08N03442	0-20	Ap	S	12			2	7	3	88	12	2	7	3	88	
08N03443	20-53	Bw	S	13			2	7	4	87	13	2	7	4	87	
08N03444	53-87	2Cd1	S	11			1	5	5	89	11	1	5	5	89	
08N03445	87-121	2Cd2	S	15			2	7	6	85	15	2	7	6	85	
08N03446	121-200	2Cd3	S	26			7	12	7	74	26	7	12	7	74	

Weight Per Unit Volume (g cm^{-3}) — Whole Soil (-40- to -43-); <2 mm Fraction (-44- to -48-); Void Ratios At 33 kPa (-49-, -50-)

Layer	Depth (cm)	Horz	Prep	-40- Soil Sur 33 kPa	-41- Oven-dry	-42- Eng Moist	-43- Eng Satur-ated	-44- 33 kPa DbWR1	-45- 1500 kPa	-46- Oven-dry DbWR1	-47- Eng Moist	-48- Eng Satur-ated	-49- Whole Soil	-50- <2 Soil
08N03442	0-20	Ap	S	1.44	1.51	1.68	1.90	1.36	1.41	1.43	1.62	1.85	0.84	0.95
08N03443	20-53	Bw	S	1.58	1.61	1.77	1.98	1.50	1.52	1.53	1.71	1.93	0.68	0.77
08N03444	53-87	2Cd1	S	1.83	1.88	2.05	2.14	1.77	1.81	1.82	2.01	2.10	0.45	0.50
08N03445	87-121	2Cd2	S	1.93	1.95	2.14	2.20	1.84	1.85	1.86	2.07	2.15	0.37	0.44
08N03446	121-200	2Cd3	S	1.92	1.95	2.15	2.20	1.76	1.78	1.79	2.04	2.10	0.38	0.51

*** Supplementary Characterization Data ***

(Newport, Rhode Island)

Sampled As　　　:　　Newport

Coarse-loamy, mixed, active, mesic Typic Dystrudept

USDA-NRCS-NSSC-Soil Survey Laboratory　　　　;　　Pedon No. 08N0528

Tier 3

Layer	Horz	Depth (cm)	Prep	-51- >2	-52- 250-UP	-53- 250-75	-54- 75-2	-55- 75-20	-56- 20-5	-57- 5-2	-58- <2	-59- 2-.05	-60- .05-.002	-61- LT.002	-62- Pores D	-63- Pores F	-64- C/N Ratio	-66- CEC Sum Cats	-68- 1500 H₂O	-69- LEP 33 kPa	-70- 33 kPa-1500	-71- Oven-dry	-72- 1500 kPa	-73- Oven-dry	-74- Whole Soil WRD	-75- <2 mm WRD
08N03442	Ap	0-20	S	6	--	--	6	1	4	2	94	25	19	4	22	24	15	1.68	0.77	0.215	1.1	1.6	1.2	1.7	0.17	0.18
08N03443	Bw	20-53	S	7	--	--	8	1	4	2	93	33	17	2	21	19	9	1.48	0.90	0.175	0.4	0.6	0.4	0.7	0.14	0.16
08N03444	2Cd1	53-87	S	7	--	--	8	1	3	3	93	33	24	5	9	22	68	0.59	0.45	0.120	0.7	0.9	0.7	0.9	0.16	0.18
08N03445	2Cd2	87-121	S	11	--	--	11	1	5	4	89	32	26	4	6	21	17	0.67	0.51	0.057	0.2	0.3	0.2	0.4	0.15	0.16
08N03446	2Cd3	121-200	S	18	--	--	19	5	9	5	82	24	25	5	5	23	26	0.64	0.50	0.067	0.3	0.5	0.4	0.6	0.16	0.20

Column groups: Volume Fractions — Whole Soil (mm) At 33 kPa (% of Whole Soil) [-51- to -63-]; C/N Ratio [-64-]; Ratios To Clay, <2 mm Fraction [-65- to -69-]; Linear Extensibility, Whole Soil / <2 mm to % [-70- to -73-]; WRD Whole Soil / <2 mm [-74-, -75-].

Tier 4

Layer	Horz	Depth (cm)	Prep	-76- >2	-77- 75-20	-78- 20-2	-79- 2-.05	-80- .05-.002	-81- .002 <	-82- VC	-83- C	-84- M	-85- F	-86- VF	-87- C	-88- F	-89- Clay	-90- Texture	-91- Sand	-92- Silt	-93- Clay	-95- pH Ca Cl₂ .01M
08N03442	Ap	0-20	S	13	13	11	50	38	7	2	5	11	21	18	24	19	9	fsl	52.4	39.7	7.9	4.6
08N03443	Bw	20-53	S	13	13	11	58	29	4	2	4	13	29	18	20	13	4	fsl	64.0	32.0	4.0	
08N03444	2Cd1	53-87	S	12	12	11	52	37	7	3	5	11	22	17	19	22	8	fsl	54.1	38.4	7.5	4.7
08N03445	2Cd2	87-121	S	16	16	14	47	37	6	3	5	11	19	17	21	24	8	l	51.6	41.4	7.0	4.9
08N03446	2Cd3	121-200	S	28	28	20	36	37	7	5	7	10	15	13	20	31	10	l	44.9	46.1	9.0	4.7

Column groups: Weight Fractions - Clay Free, <2 mm Fraction — Whole Soil [-76- to -81-]; Sands VC,C,M,F,VF (% of Sand and Silt) [-82- to -86-]; Silts C,F [-87-,-88-]; Clay [-89-]; Texture by PSDA, <2 mm (-- % of 2 mm --) Sand .05 / Silt .002 / Clay .002 [-90- to -94-], method 3A1a1a4C1a2a; pH Ca Cl₂ .01M [-95-]; Elect. Resist. ohms [-96-]; Conductivity dS m⁻¹ [-97-]; Particle Density g cm⁻³ [-98-].

Pedon ID: S08RI005001

Print Date: Sep 22 2010 2:05PM

*** Taxonomy Characterization Data ***
(Newport, Rhode Island)

Sampled as :   Newport ; Coarse-loamy, mixed, active, mesic Typic Dystrudept
Revised to :

SSL
- Project   C2008USRI121   Newport & Washington Co.s
- Site ID   S08RI005-001   Lat: 41° 31' 37.69" north   Long: 71° 22' 3.76" west   NAD83   MLRA: 144A
- Pedon No.   08N0528
- General Methods 1B1A, 2A1, 2B

United States Department of Agriculture
Natural Resources Conservation Service
National Soil Survey Center
Soil Survey Laboratory
Lincoln, Nebraska 68508-3866

Taxonomy Tier 1

Layer	Depth (cm)	Horz	Prep	-1- Clay <.002	-2- Fine Clay <.0002	-3- CaCO₃ Clay <.002	-4- 1500 kPa /Clay	-5- Clay Est	-6- .1-75 mm Frac	-7- Bulk Den 33 kPa DbWR1	-8- Cole Whole Soil	-9- Vol % of Whole	-10- Resist Min %
				(------% of <2 mm------)			(----% ----)	(----% ----)		g cm⁻³	cm cm⁻¹		
			3A1a1a										
08N03442	0-20	Ap	S	7.9			0.77	10.5	44	1.36	0.016	6	68
08N03443	20-53	Bw	S	4.0			0.90	7.4	53	1.50	0.006	7	75
08N03444	53-87	2Cd1	S	7.5			0.45		45	1.77	0.009	7	
08N03445	87-121	2Cd2	S	7.0			0.51		45	1.84	0.003	11	
08N03446	121-200	2Cd3	S	9.0			0.50		50	1.76	0.005	18	

Taxonomy Tier 2

Layer	Depth (cm)	Horz	Prep	-1- pH H₂O	-2- pH NaF	-3- Org C	-4- Tot C	-5- Al+½Fe Oxal	-6- ODOE	-7- CO₃ as CaCO₃	-8- NH₄ (Base Sat)	-9- Bases	-10- NZ P Ret	-11- ECEC cmol(+) kg⁻¹	-12- CEC7 /Clay	-13- ECEC /Clay	-14- Al Sat %	-15- E C dS m⁻¹	ESP %
				(--------------- % ---------------)															
			4C1a2a			4H2a		4G2a											
08N03442	0-20	Ap	S	5.0			1.92	0.79	0.09		18	8			0.76				
08N03443	20-53	Bw	S				0.64	0.64	0.04		7	3			0.68				
08N03444	53-87	2Cd1	S	5.7			0.88	0.43	0.02		15	9			0.35				
08N03445	87-121	2Cd2	S	5.7			1.05	0.48	0.02		11	9			0.50				
08N03446	121-200	2Cd3	S	5.7			1.23	0.39	0.01		35	22			0.41				

Pedon Calculations

Calculation Name	Result	Units of Measure
Weighted Particles, 0.1-75mm, 75 mm Base	50	% wt
Volume, >2mm, Weighted Average	7	% vol
Clay, total, Weighted Average	5	% wt
Clay, carbonate free, Weighted Average	5	% wt
CEC Activity, CEC7/Clay, Weighted Average, CECd, Set 1	0.71	(NA)
LE, Whole Soil, Summed to 1m	1	cm/m

Weighted averages based on control section: 0-53 cm

PEDON DESCRIPTION

Print Date: 09/22/2010

Description Date: 6/10/2008

Describer: Jim Turenne, Donald Parizek, Debbie Surabian

Site ID: S08-RI005-001-Newport

Site Note: Backhoe pit used for National Soil Judging Contest. Pit dug to 2 meters, west face used for sample. Backhoe had difficulty digging into lodgement till.; Soil was at or near saturated conditions below 1.5 meters but no water was ever found in pit.

Pedon ID: S08-RI005-001-Newport

Pedon Note: Benchmark Series, need to check lab data to see if it fits the ROC for Newport.

Lab Source ID: SSL

Lab Pedon #: 08N0528

Soil Name as Described/Sampled: Newport

Soil Name as Correlated:

Classification: Coarse-loamy, mixed, active, mesic Typic Dystrudepts

Pedon Type: within range of series

Pedon Purpose: full pedon description

Taxon Kind: series

Associated Soils:

Physiographic Division: Appalachian Highlands

Physiographic Province: New England Province

Physiographic Section: Seaboard lowland section

State Physiographic Area:

Local Physiographic Area: Narragansett Basin

Geomorphic Setting: on summit of drumlin on summit of upland

Upslope Shape: convex

Cross Slope Shape: convex

Particle Size Control Section: 0 to 53 cm.

Description origin:

Country:

State: Rhode Island

County: Newport

MLRA: 144A -- New England and Eastern New York Upland, Southern Part

Soil Survey Area: RI600 -- State of Rhode Island: Bristol, Kent, Newport, Providence, and Washington Counties

Map Unit: NeA -- Newport silt loam, 0 to 3 percent slopes

Quad Name: Prudence Island, Rhode Island

Location Description: Town of Jamestown, RI, 250 ft north of Eldridge Road, 630 feet west of East Shore Road, Community Gardens (Ceppie Farm).

Legal Description:

Latitude: 41 degrees 31 minutes 37.69 seconds north

Longitude: 71 degrees 22 minutes 3.76 seconds west

Datum: NAD83

UTM Zone: 19

UTM Easting: 302460 meters

UTM Northing: 4599984 meters

Primary Earth Cover: Grass/herbaceous cover

Secondary Earth Cover: Hayland

Existing Vegetation:

Parent Material: slightly weathered, loamy eolian deposits derived from sandstone and siltstone over slightly weathered, loamy lodgment till derived from sandstone and siltstone

Bedrock Kind:

Bedrock Depth:

Bedrock Hardness:

Bedrock Fracture Interval:

Surface Fragments: 0.0 percent

Description database: NSSL

Diagnostic Features: densic contact 53 to 200 cm.

Top Depth (cm)	Bottom Depth (cm)	Restriction Kind	Restriction Hardness
53	200	densic material	noncemented

Pedon ID: S08-RI005-001-Newport

Cont. Site ID: S08-RI005-001-Newport

Slope (%)	Elevation (meters)	Aspect (deg)	MAAT (C)	MSAT (C)	MWAT (C)	MAP (mm)	Frost-Free Days	Drainage Class	Slope Length (meters)	Upslope Length (meters)
2.0	28.0	240						well		

Ap--0 to 20 centimeters; very dark grayish brown (10YR 3/2) fine sandy loam; structureless coarse subangular blocky, and moderate medium granular structure; friable; many fine roots throughout and many very fine roots throughout; 2 percent nonflat subrounded mixed rock fragments and 5 percent flat subangular mixed rock fragments; clear smooth boundary. Lab sample # 08N03442

Bw--20 to 53 centimeters; very dark grayish brown (2.5Y 3/2) channery fine sandy loam; weak coarse subangular blocky structure; friable; common fine roots throughout and common very fine roots throughout; 3 percent nonflat subrounded mixed rock fragments and 5 percent nonflat subrounded less than 430-millimeter mixed rock fragments and 8 percent flat subangular mixed rock fragments; gradual wavy boundary. Lab sample # 08N03443

2Cd1--53 to 87 centimeters; black (N 2.5/), silt loam; structureless massive; firm; few fine roots throughout and few medium roots throughout and few very fine roots throughout; 2 percent nonflat subrounded mixed rock fragments and 4 percent flat subangular mixed rock fragments; clear irregular boundary. Lab sample # 08N03444

2Cd2--87 to 121 centimeters; 60 percent very dark gray (N 3/), and 40 percent dark olive gray (5Y 3/2) silt loam; structureless single grain, and structureless massive; firm; 1 percent nonflat subrounded mixed rock fragments and 6 percent flat subangular mixed rock fragments; Horizon had a thin lens of loose loamy sand material found throughout the pit.; clear irregular boundary. Lab sample # 08N03445

2Cd3--121 to 200 centimeters; black (N 2.5/), silt loam; structureless massive; very firm; 2 percent nonflat subrounded mixed rock fragments and 10 percent flat subangular mixed rock fragments. Lab sample # 08N03446

Subject Index

504